# Rethinking Agri-Food and Marine Waste and Byproducts for Circular and Sustainable Bio-Based Food Packaging

# Rethinking Agri-Food and Marine Waste and Byproducts for Circular and Sustainable Bio-Based Food Packaging

Editors

**Sílvia Petronilho**
**Mario M. Martinez**

MDPI

Basel • Beijing • Wuhan • Barcelona • Belgrade • Novi Sad • Cluj • Manchester

*Editors*

Sílvia Petronilho
University of Aveiro
Aveiro
Portugal

Mario M. Martinez
University of Valladolid
Valladolid
Spain

*Editorial Office*
MDPI AG
Grosspeteranlage 5
4052 Basel, Switzerland

This is a reprint of articles from the Special Issue published online in the open access journal *Foods* (ISSN 2304-8158) (available at: https://www.mdpi.com/journal/foods/special_issues/food_waste_and_byproduct_packaging).

For citation purposes, cite each article independently as indicated on the article page online and as indicated below:

| Lastname, A.A.; Lastname, B.B. Article Title. *Journal Name* **Year**, *Volume Number*, Page Range. |
|---|

**ISBN 978-3-7258-2639-1 (Hbk)**
**ISBN 978-3-7258-2640-7 (PDF)**
**doi.org/10.3390/books978-3-7258-2640-7**

# Contents

# About the Editors

**Sílvia Petronilho**

Sílvia Petronilho, PhD in Chemistry, is a Researcher (Level 1) of LAQV&CICECO—Department of Materials and Ceramic Engineering at the University of Aveiro (Portugal). Previously, she was a post-doctoral researcher at the University of Trás-os-Montes and Alto Douro (Portugal). Since 2017, her research work has focused on the valorization of non-edible agrifood byproducts through the development of advanced biomaterials for food and biomedical applications. Sílvia is also engaged in the biological valorization of lipid-based industrial waste for polyhydroxyalkanoates' production by mixed microbial cultures. She has published several papers in peer-reviewed journals. She has received several international/national awards in recognition of the scientific quality of her research. She belongs to the organizing committees of scientific congresses and workshops and has participated in different I&D projects. She belongs to the Editorial Board of *Foods* as a Topical Advisory Member and is a Guest Editor of two Special Issues. She has international exposure proven through collaborations and internships with different universities and research institutes, and she has been integrated as a member of three international COST Actions.

**Mario M. Martinez**

Mario Martínez is an Associate Professor and Consolidated Researcher (Research Consolidation 2023, AEI) in the Food Technology Division at the University of Valladolid. He began his independent research career at the University of Guelph (ON, Canada), where he received the College of Engineering and Physical Science Assistant Professor Research Excellence Award from the Faculty of Physics and Engineering at the University of Guelph, the Ontario Early Researcher Award from the Government of Ontario, and the Young Scientist Award from the Cereals & Grains Association in the United States. Mario Martínez joined Aarhus University in April 2020, where his research was recognized with prestigious awards and elite grants, including the Nils Foss Talent Prize 2021 (Foss Analytics, Denmark) and the DFF-Sapere Aude excellence project (Denmark). In June 2024, he joined the University of Valladolid as an Associate Professor through two talent-attraction programs. In addition to his main affiliation at the University of Valladolid, Mario holds secondary positions as an Associate Professor in the Department of Food Science at Aarhus University and as an Adjunct Professor at the Whistler Center for Carbohydrate Research (Purdue University, USA). Mario has been an editor of the International Journal of Biological Macromolecules since September 2023.

*foods*

MDPI

*Article*

# Liquid and Pressure-Sensitive Adhesives Based on Cassava Starch and Gelatin Capsule Residue: Green Alternatives for the Packaging Industry

**Yuliana Monroy [1], Sandra Rivero [1,2] and María Alejandra García [1,2,*]**

1 Centro de Investigación y Desarrollo en Criotecnología de Alimentos (CIDCA-CONICET), 47 y 116 S/N, La Plata B1900AJJ, Buenos Aires, Argentina; monroyuliana92@gmail.com (Y.M.); sandragmrivero@gmail.com (S.R.)

2 Facultad de Ciencias Exactas, Universidad Nacional de La Plata (UNLP), 47 y 115 S/N, La Plata B1900AJJ, Buenos Aires, Argentina

* Correspondence: magarcia@quimica.unlp.edu.ar

**Abstract:** Natural polymer-based adhesives are *green* alternatives, necessary to reduce the problems impacted by synthetic adhesives. Starch and gelatin have extraordinary potential for the synthesis of biobased adhesives. Citric acid (CA), a natural acid, induces the crosslinking and hydrolyzing of both gelatin and starch. In this sense, this work deals with the use of gelatin capsule residues as a promising material to produce biobased adhesives in combination with cassava starch in the presence of different CA concentrations characterizing their mechanical, physicochemical and microstructural properties. Depending on CA concentration, formulations adjusted to different applications can be obtained such as liquid and pressure-sensitive adhesive films. The inclusion of CA allows us not only to improve the applicability of the system since it modifies the flowability of the adhesives as evidenced by the observed changes in the viscosity (from 158.3 to 90.3 for formulations with 20 and 80% CA, respectively). In addition, mechanical profiles showed that the inclusion of CA increased the adhesive bond strength (from 2230.7 to 2638.7 for formulations with 20 and 80% CA, respectively). Structural modifications induced by CA in adhesive formulations were highlighted by ATR-FTIR analysis.

**Keywords:** biobased adhesives; rheological behavior; citric acid; adhesive capacity

**Citation:** Monroy, Y.; Rivero, S.; García, M.A. Liquid and Pressure-Sensitive Adhesives Based on Cassava Starch and Gelatin Capsule Residue: Green Alternatives for the Packaging Industry. *Foods* **2023**, *12*, 3982. https://doi.org/10.3390/foods12213982

Academic Editor: Baki Hazer

Received: 29 September 2023
Revised: 27 October 2023
Accepted: 28 October 2023
Published: 31 October 2023



## 1. Introduction

Greater awareness of the need to preserve non-renewable natural resources has led to the development of more ecological high-performance polymers with new functionalities [1]. In recent years, there have been increasing efforts in the adhesive industry to improve the sustainability of processes and products. This is due, on the one hand, to the interest in developing eco-compatible formulations and, on the other, the impending shortage of petroleum, from which many chemical products are derived. Natural polymer-based adhesives are "green" alternatives, necessary to reduce the problems caused by synthetic adhesives [2,3]. In the adhesive industry, this has manifested itself through the gradual shift from solvent adhesives to water-based or high-solid ones and renewed interest in adhesives based on natural polymers. Therefore, the development of eco-compatible adhesives that allow easy and quick washing is seen as an alternative with potential functionality to minimize environmental impacts, also contributing to the sustainability of the processes.

One of the alternatives for the production of biodegradable systems is the use of polymers obtained from renewable resources or waste and industrial by-products. Uranga et al. [4] reported that the recovery of biomass waste and its subsequent introduction into different production chains allow for the more efficient use of resources and contribute to the sustainability of the processes, in line with the premises of the circular economy. In this sense, biobased adhesives can be obtained from natural resources like vegetable or animal products such as starch and their derivatives, cellulose, proteins, casein, collagen and gelatin among others [5–7].

In this way, starch and gelatin have extraordinary potential for the synthesis of various formulations due to the versatility of functional groups which in turn offer unlimited opportunities to introduce modifications in their structure [6,8].

Currently, one interesting source of protein to develop biodegradable films is the residue generated by the industrialization of gelatin capsules, which are used to deliver bio-active compounds in the nutraceutical sector [5,9,10]. Therefore, the valorization of industrial by-products to produce bioadhesive materials may contribute to scaling down the amount of residue disposal reducing the use of non-biodegradable synthetic polymers [11].

On the other hand, starch has advantages such as being biodegradable, low cost and renewable, and it contains hydroxyl groups with strong intra- and intermolecular hydrogen bonding [7]. Citric acid is a natural acid, and it has been already demonstrated to be suitable for the crosslinking of gelatin and starch [12,13].

Adhesive formulations in the packaging industries are essential to give structure to most paper, board or glass packaging, and therefore, the design of pressure-sensitive adhesives (PSA) represents an innovative development. A pressure-sensitive adhesive is an adhesive system that is permanently tacky and adheres to a variety of surfaces [14]. To the best of our knowledge, studies related to the design of biodegradable pressure-sensitive adhesives produced from biopolymers are scarce [15]. In this sense, Sartori et al. [15] developed biodegradable pressure-sensitive adhesives produced from wheat gluten.

In addition, the incorporation of an economical and non-toxic polycarboxylic acid such as CA is proposed as an overcoming strategy, which allows for improving the system and introducing distinctive adhesive characteristics, compared to commercially available products. In line with the current needs and requirements, the novelty of the present work is focused on the design of versatile liquid and/or pressure-sensitive adhesive formulations for bonding different substrates such as paper and glass, among others, with potential applications in food packaging and/or in the labeling processes.

The aims of this research were to develop biobased adhesives (liquid or solid) from the use of gelatin capsule residues as a promising biopolymer in combination with cassava starch in the presence of different CA concentrations characterizing their mechanical, physicochemical and microstructural properties. An in-depth study of the structural modifications induced by the addition of citric acid and their correlation with techno-functional properties was carried out.

## 2. Materials and Methods

### 2.1. Formulation of Biobased Adhesives

Water-based composite adhesives were developed from mixtures of cassava starch (*Manihot esculenta*, Montecarlo, Misiones) and gelatin from discarded drug capsules (Bagó, Argentina), containing 2.91% $w/w$ of $TiO_2$. The starch–gelatin dispersions were hydrated and treated in a thermostatic bath in the presence of citric acid (CA, purity degree 99,5%) (Biopack, Argentina), for 20 min at 90 °C. The composite adhesives were formulated with 5% $w/w$ starch and 15% $w/w$ gelatin, to obtain a final solids concentration of 20% $w/w$. Citric acid was incorporated into the formulations at 20, 40 or 80% $w/w$ based on the concentration of the polymer blend [16]. Nomenclature used for each system was CA-20, CA-40 and CA-80, respectively.

#### 2.1.1. Rheological Characterization

Rheological analyses were performed in a G2 TA Instrument (New Castle, DE, USA) rheometer using a plate–plate sensor of parallel standard steel plates of 40 mm at controlled temperature (20 °C) with a gap of 1 mm. Tests of the viscoelastic behavior of the different adhesives were carried out in dynamic mode. First, stress sweeps (0–100 Pa) were performed at constant frequency (1 Hz) to determine the range of linear viscoelasticity.

Frequency sweeps (0.01 to 100 Hz) were then performed at a constant stress value (1 Pa). The dynamic rheological parameters recorded were the storage modulus (G′), the loss modulus (G″) and the tangent of the offset angle (tan δ = G″/G′). Mechanical spectra were obtained by plotting G′ and G″ as a function of frequency (ω). Rheological tests were performed at least in duplicate.

#### 2.1.2. Back Extrusion

Formulation consistency, as well as their changes during storage time, was studied by the back extrusion technique in a TA-XT2i texturometer (Stable Micro Systems, Surrey, UK) at 25 °C [17]. The back extrusion rig P/75 probe was used.

### 2.2. Adhesive Capacity of Liquid Formulations

The ASTM D6195–03 (2019) standard procedure was adapted to evaluate the tacking capacities of formulated bioadhesives. A loop of Kraft paper 25 mm wide with a minimum length of 175 mm was formed and held with the A/TG tension clamp probe in a TA-XT2i texturometer. The paper loop was placed in contact (25 × 25 mm contact area) with the other substrate (paper or glass) containing the adhesive layer to be evaluated which was kept fixed on the base of the equipment and peeled off at a speed of 5 mm $min^{-1}$. Tests were carried out on 8 replicates at least.

Likewise, formulations were applied forming a thin and uniform adhesive layer on the paper strips, as described in a previous work [18]. The roughness of the samples (Kraft paper or glass) was determined by using a PCE-RT 1200 roughness tester (Schwyz, Switzerland). The parameters of roughness were expressed as the average roughness (Ra) and the mean peak-to-valley height (Rz), [19].

Peel strength tensile tests (*peel T* assays) were performed in a TA.XT2i—Stable Micro Systems (Surrey, UK) texturometer, using A/TG uniaxial tensile grips at a test speed of 10 mm $s^{-1}$ [18]. Kraft paper specimens (at least 10 per sample) were prepared according to ASTM F904 with a bond area of 50 mm × 25 mm (paper or glass).

### 2.3. Analysis of the Microstructure by ATR-FTIR and SEM

The effect of CA concentration on the adhesive formulation was studied by ATR-FTIR and SEM. Fourier transform infrared spectroscopy (FTIR) spectra were recorded in Nicolet iS10 Thermo Scientific equipment (Madison, WI, USA). Films of each formulation, obtained by drying in an oven at 37 °C, were placed on the diamond crystal accessory to obtain FTIR spectra combined with attenuated total reflectance (ATR). Likewise, the native cassava starch suspension, gelatin capsules and CA were also analyzed. Spectra were obtained (as the average of at least 32 scans) between 4000 and 400 $cm^{-1}$, with a resolution of 4 $cm^{-1}$, which were analyzed with baseline correction using OMNIC™ software (version 8.3, ThermoFisher Scientific, Waltham, MA, USA).

The morphology of the bioadhesive films was studied by SEM using a Quanta200 FEI model electron microscope (Eindhoven, The Netherlands). The accelerating voltage used was 20 kV, under low-vacuum conditions. In order to study the microstructure of adhesive films, the samples were bonded to bronze blocks with double-sided carbon tape, which allowed visualization of the surface and its cross-section.

### 2.4. Contact Angle

Contact angle measurements were made in a model 190 Goniometer (Ramé-Hart, Instrument Co., New Jersey, USA) on substrates with bioadhesives after drying. For this, a Kraft paper specimen was placed on a slide, which was fixed to the support with double-sided tape, to avoid any waviness in the paper. A uniform layer of adhesive was applied to the cellulosic substrate and then dried at room temperature. For the determination, milliQ water was used in the syringe of the goniometer, and at least 10 measurements were performed on each sample [18].

### 2.5. Evaluation of the Adhesive Capacity of Pressure-Sensitive Formulation

The combination starch: gelatin: citric acid allowed us to obtain a film with self-adhesive properties whose nomenclature is CA-80. The adhesive (100 g) film was performed using 24 cm diameter silicone molds and dried at 37 °C for 24 h. The films were unmolded, and 2 × 4 cm specimens were cut; then, the adhesion was tested on different surfaces: ceramic, metal, glass, plastic, paper, cardboard and Durlock™. Adhesion was studied using the cross-cut test following the ASTM D3359 [20] standard, with slight modifications. This is a standardized test that allows qualitative estimation of the adhesion of a coating on a substrate using a test tape. It is a widely used, fast and inexpensive visual comparison method for evaluating the adhesion of film coatings to metal substrates by applying and removing a pressure-sensitive tape over grid cuts in the film.

In accordance with the standard, the adhesive film was cut into small squares (2 × 2 mm), and to evaluate the adhesion according to ISO 2409 and ASTM D 3559-B [20], a standard test method for measuring adhesion by tape test was used. To make the cut, a sharp tool was used to obtain a cross-cut of 36 squares. A commercial adhesive (3M™ 3903 and Double A™, recognized for their adhesive properties on a wide variety of substrates) tape was placed over the traced grid which was then quickly pressed and removed. The adhesion of the adhesive film was determined according to the classification criteria of images and descriptions based on the aforementioned standard.

### 2.6. Statistical Analysis

All experiments were performed at least in duplicates. Analysis of variance (ANOVA) and comparison of means with Fisher's least significant difference (LSD) test were conducted, at a significance level of $\alpha = 0.05$. The InfoStat Software, version 2020 (InfoStat Group, Agricultural Sciences College, National University of Cordoba, Cordoba, Argentina), was used.

## 3. Results and Discussion

### 3.1. Rheological Behavior of Composite Adhesive Formulations

Composite bioadhesives formulated with cassava starch and gelatin chemically modified in the presence of different concentrations of citric acid were obtained, taking advantage of the functionality and capacity of citric acid to modify these polymers.

The gelatinization of cassava starch in the presence of underutilized gelatin and citric acid allowed obtaining polymeric blends with pseudoplastic behavior. All the adhesive formulations had a white appearance due to the $TiO_2$ coloration contained in the underutilized capsules of the pharmaceutical industry which presented acidic characteristics with pH values below 4. The pH decreased with the CA content, with values close to 4 for samples formulated with 20% *w*/*w* CA, in the range of 2–3 for bioadhesives prepared with 40% *w*/*w* CA and 1–2 for the maximum concentration of the CA tested (80% *w*/*w*). Regarding the stability of the adhesives over time, no syneresis (obvious sign of starch retrogradation) was observed even after one week of storage at room temperature, which demonstrates the degree of stability of the developed adhesives. In this sense, it is important to stress that adhesives formulated without CA solidify and are unmanageable, and, in addition, they do not have the ability to stick.

Dynamic-mechanical spectra of the composite adhesive formulations are shown in Figure 1. The bioadhesives exhibited a behavior dependent on the citric acid content. Figure 1 shows a typical spectrum of a gel for the samples containing 20 and 40% *w*/*w* CA, where the elastic modulus (G′) was higher than the viscous one (G″) throughout the frequency range analyzed.

Firoozmand et al. [21] reported that gelatin achieves maximum elasticity by lowering the pH below its isoelectric point (pI = 4–4.5). Likewise, the authors pointed out that in a composite matrix, the pH value influences the degree of compatibility and the nature of the interactions established between the biopolymers, which affects the microstructure and viscoelastic properties of the resulting gel. In accordance with Goudie et al. [22], a dense and compact structure was promoted at a pH close to the isoelectric point, where the interactions within the gelatin network are equilibrated.

On the other hand, with the addition of 80% CA, the rheological behavior corresponds to that of a concentrated solution (Figure 1). This could be explained by the hydrolysis that gelatin undergoes at the working pH (1–2). According to Wang et al. [23], citric acid hydrolyzes gelatin, and its effect increased with the CA concentration leading to a formulation that weakened its gelling capacity.

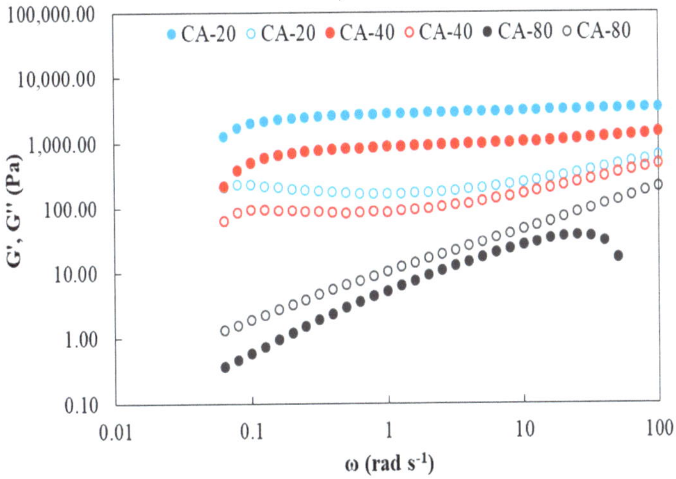


**Figure 1.** Effect of the citric acid concentration on the evolution of the elastic modulus (G′, filled circles) and viscous (G″, empty circles) of bioadhesives based on cassava starch and gelatin. Nomenclature used: CA = citric acid, the following number (20, 40, 80) indicates the concentration of CA incorporated into the adhesive % (g acid/100 g of polymer mixture).

In addition, consistency studies were conducted to complement the rheological analysis. Table 1 shows the values of the parameters obtained for the different adhesive formulations developed. From the analysis of the back extrusion test results obtained for the biobased formulations, the effect of citric acid is evident (Table 1). The firmness and consistency of the adhesive decreased significantly ($p < 0.05$) at high concentrations of CA, indicating its predominantly hydrolyzing action on the polymer mixture. A similar trend was observed for the cohesivity and viscosity index.

**Table 1.** Back extrusion profile parameters of biobased starch–gelatin composite adhesives.

| Samples | Firmness (g) | Consistency ($g\ s^{-1}$) | Cohesiveness ($\lvert g \rvert$) | Viscosity Index ($\lvert g\ s^{-1} \rvert$) |
|---|---|---|---|---|
| CA-20 | $32.1 \pm 0.6$ [b] | $279.4 \pm 2.5$ [b] | $17.2 \pm 0.5$ [b] | $158.3 \pm 2.3$ [c] |
| CA-40 | $30.5 \pm 1.3$ [b] | $262.3 \pm 10.9$ [b] | $16.4 \pm 0.3$ [b] | $137.8 \pm 5.1$ [b] |
| CA-80 | $22.9 \pm 1.0$ [a] | $215.4 \pm 4.7$ [a] | $13.2 \pm 0.4$ [a] | $90.3 \pm 3.6$ [a] |

Mean values ± standard deviation are presented. Different letters within the same column indicate significant differences ($p < 0.05$).

### 3.2. Adhesive Capacity of Composite Formulations

The development of adhesive formulations was proposed for their potential use in gluing cellulosic and glass substrates, thinking about their subsequent application in the bottle packaging and/or labeling area.

Tack is the ability of an adhesive to wet the surface and adhere to the substrate and depends on the mobility of the adhesive's constituents and their ability to wet the substrate.

When evaluating peel strength, external pressure is applied to allow complete wetting of the adhesive on the substrate [24]. Loop adhesion testing is routinely performed on commercial adhesives and is used for the quality control of adhesive tapes and pressure-sensitive adhesives to evaluate the tacking. The adhesive formulations registered a decrease in adhesiveness and so in stickiness when the acid concentration increased (Figure 2), in close relationship with the evolution of the rheological characteristics evaluated through the back extrusion test (Table 1). Kumar et al. [25] reported that CA is an excellent crosslinking agent that favors the interaction between citric acid carboxyl groups and starch hydroxyl groups.

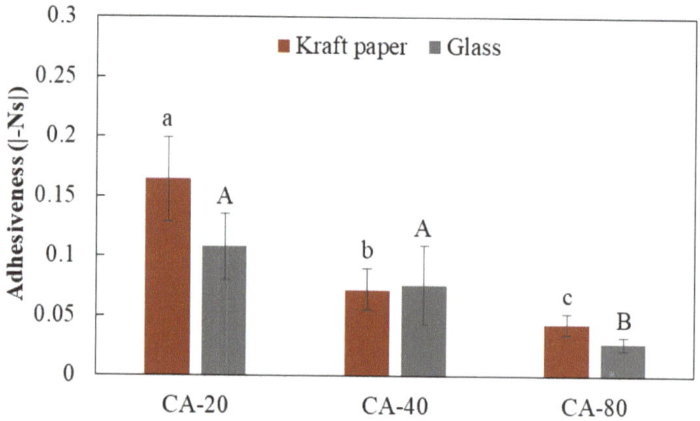


**Figure 2.** Adhesiveness determined from loop adhesion tests of the developed formulations as a function of the CA concentration. Different letters (lowercase for Kraft and uppercase for glass) indicate significant differences ($p < 0.05$) for each substrate studied.

There was evidence of a significant decrease ($p < 0.05$) in adhesiveness with the concentration of CA (Figure 2). The difference in values between the substrates was mainly due to the roughness of each one of them. Dohr and Hirn [26] pointed out that the roughness of the adherent surface strengthens the bond with the adhesive by increasing the bond area and promoting mechanical interlocking. The glass presented a value of Ra = $0.04 \pm 0.01$ µm and Rz = $0.3 \pm 0.04$ µm while the Kraft paper registered Ra and Rz values of $3.1 \pm 0.4$ µm and $30.1 \pm 0.8$ µm, respectively. At higher roughness, the adhesive can penetrate the structure of the substrate and improve the adhesiveness and characteristics of the adhesive–substrate bond. Therefore, characteristics of the substrate such as its greater roughness and porosity are factors that promote greater adhesion. It is necessary to ensure a good wettability of the substrate by the adhesive, since the cavities not reached by it constitute potential points of initiation of rupture of the adhesive bond.

When an adhesive is formulated, it is intended to provide an adhesive bond of adequate strength. This means that the failure mode is not intended to occur at the substrate–adhesive interface but rather that this interface is as strong as the substrate [27]. In this sense, while adhesive and cohesive failure is undesirable, since it indicates a weak adhesive bond, substrate failure indicates that the adhesive bond is strong, reaching the maximum strength of the material [26].

The mechanical adhesive behavior of the different formulations developed for bonding cellulosic Kraft paper substrates is shown in Figure 3. With CA concentration, stress values of Kraft paper samples bonded with the bioadhesive did not change significantly ($p > 0.05$). Meanwhile, tensile *peel* tests on Kraft paper samples bonded to glass showed a variation with the CA concentration, with the CA-40 formulation being the one that showed a significant increase ($p < 0.05$) in the stress values (Figure 3).

Regardless of the bioadhesive formulation, compatibility was observed between the system under study: glass–adhesive–Kraft paper, which could benefit from the interactions with the silanol groups that make up the glass and the carboxylic groups of the bioadhesives, in addition to possible interactions by hydrogen bonding that occurs between the cellulosic material and the adhesive (Figure 3b).

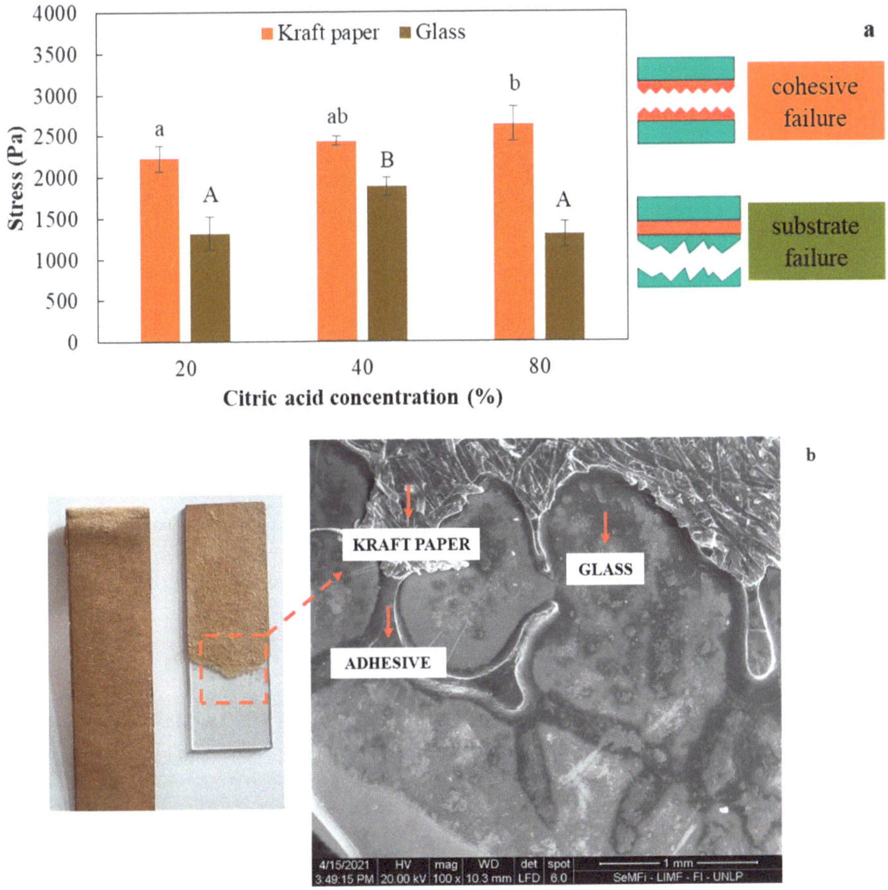


**Figure 3.** (**a**) Effect of citric acid concentration (CA, % *w*/*w*) on the bonding strength of different substrates (Kraft paper–Kraft paper or Kraft paper–glass). Different letters (lowercase for Kraft and uppercase for glass) indicate significant differences ($p < 0.05$) for each substrate studied; (**b**) SEM micrographs of the adhesive–substrate interface after *T-peel* test.

All the formulations exhibited filmogenic capacity (the ability to form a film by drying), and the film obtained only with the formulation with 5% starch-15% gelatin and 80% citric acid/100 g of polymer mixture (P1-80) proved to have self-adhesive capacity.

As is known, both the surface tension of the adhesive and the surface energy of the substrate affect the contact angle. The adhesive's hydrophobicity showed a significant increase ($p < 0.05$) with CA concentration in the formulation (Table 2). These results could be attributed to the crosslinking effect of CA (at 40 and 80%) that favors the crosslinking of the polymeric mixture during heat treatment, which leads to obtaining matrices with hydrophobic characteristics.

**Table 2.** Contact angle of adhesive formulations with 5% *w/w* starch-15% *w/w* gelatin and different CA concentrations on Kraft paper.

| CA-20 | CA-40 | CA-80 |
| --- | --- | --- |
| 76.7 ± 2.6 [a] | 81.3 ± 3.3 [b] | 85.9 ± 3.7 [c] |
|  |  |  |

The reported results correspond to the mean ± standard deviation. Different letters indicate significant differences ($p < 0.05$). CA: 20, 40 and 80 correspond to % CA.

### 3.3. Structural Modifications Induced in Adhesive Formulations: ATR-FTIR Studies

The structural modification of the starch–gelatin system in the presence of CA was studied using the ATR-FTIR technique.

From the gelatin capsule (G) spectra, the characteristic absorption peaks of the amide band I (corresponding to the stretching vibration of the carbonyl C=O) and the amide II band (corresponding to the bending vibration of the C-N-H bond) were detected at 1627 and 1530 $cm^{-1}$, respectively. Similar results were reported by He et al. [28]. Furthermore, the absorption peak at 1240 $cm^{-1}$ was assigned to the amide III band, corresponding to the C-N stretching vibration. All the bioadhesives exhibited a broad band located at 3500–3000 $cm^{-1}$ ascribed to the free and bound -NH and -OH absorption groups capable of forming hydrogen bonds with both gelatin and starch [4].

All bioadhesives exhibited the broad band at 3500–3000 $cm^{-1}$ adsorbed by free and bound -NH and -OH groups capable of forming hydrogen bonds with both gelatin and starch. On the other hand, the band ascribed to the symmetric and asymmetric stretching vibration of methylene (-$CH_2$) was identified at 2850 and 2919 $cm^{-1}$, respectively. The addition of CA was correlated with the appearance of a peak at 1710 $cm^{-1}$, corresponding to the C=O stretching vibration as a result of the interactions that it establishes with the components of the adhesive formulation. Likewise, this peak was magnified with the increase in CA concentration (Figure 4a). Similar results were reported by Kumar et al. [29] who studied biodegradable films based on modified corn starch and gelatin. In addition, Reddy and Yang [30] and Brandelero et al. [31] reported that with the inclusion of citric acid in starch-based matrices, the appearance of a band at 1724 $cm^{-1}$ could be elucidated, corresponding to the association between the carboxyl and hydroxyl groups. In addition, Kumar et al. [29] demonstrated that citric acid esterified the hydroxyl groups of starch, inducing strong interactions between the polymer chains. In addition, this interaction could also occur between citric acid and hydroxyl groups present in gelatin. Similar assignments were made by He et al. [28] and Kumar et al. [29] for composite systems based on gelatin and starch.

Due to its functional groups (-$NH_2$, -OH, -COOH/-COO−), gelatin is able to interact through hydrogen bonding and electrostatic interactions, further improving the compatibility and chemistry of the resulting systems [32]. Bands at 1646 $cm^{-1}$ and 1363 $cm^{-1}$ corresponded to the $\gamma$-dicarbonyl and C-N stretching of the amide groups, respectively [33].

Another difference detected from the spectra against the addition of CA is related to the changes observed in the band at 1638 $cm^{-1}$, which would indicate that a chemical association between the system components and citric acid was established. According to Singh and Maitra [34], these changes are correlated with modifications in the configuration of the glycosidic ring as a consequence of the association of citric acid groups with starch.

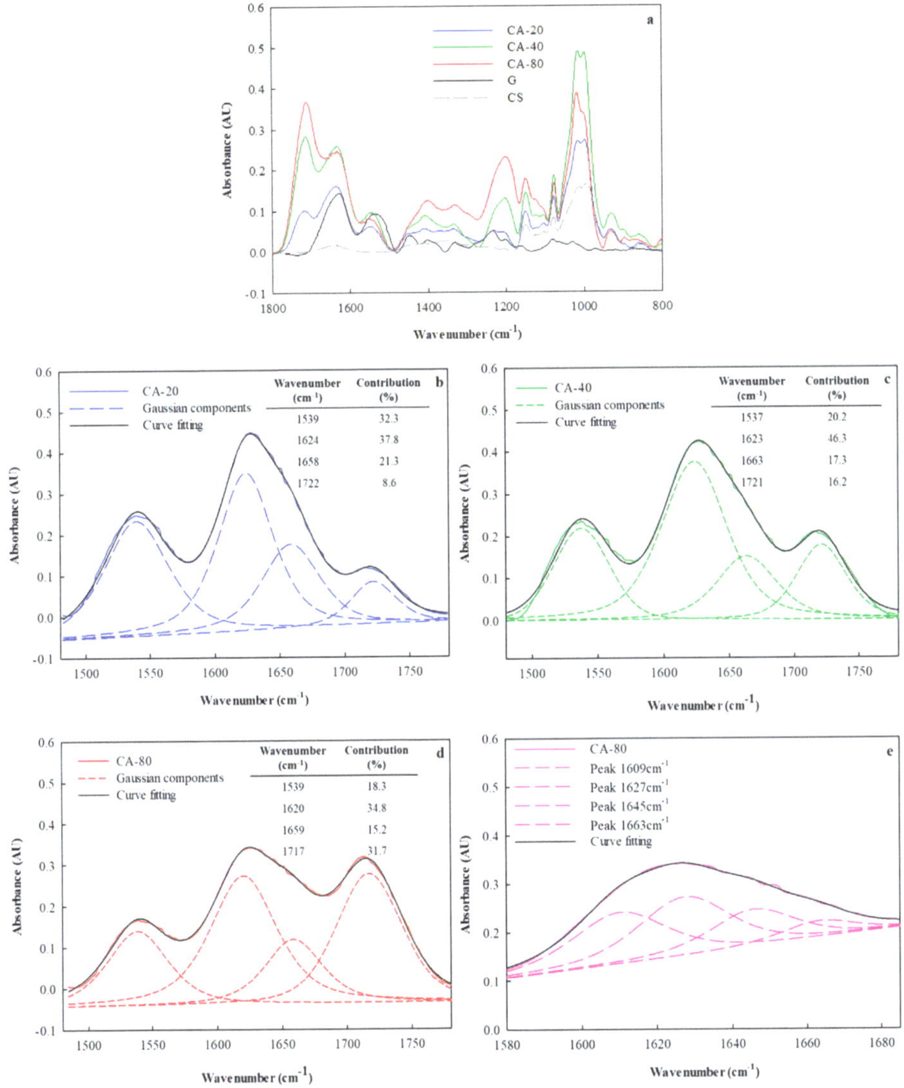


**Figure 4.** (**a**) ATR-FTIR spectra of adhesive films formulated in the presence of CA at 20%, 40% and 80% *w*/*w* (g acid/100 g polymer mixture). Second derivative curves in the spectral region from 1450 to 1850 $cm^{-1}$ for adhesive containing different CA concentrations: (**b**) CA-20, (**c**) CA-40 and (**d**) CA-80; and (**e**) second derivative curves in the spectral region from amide I for the sample CA-80. Experimental data (solid colored lines), individual components of the Gaussian (dotted lines) and curve fitting (solid black lines).

Spectral deconvolution was performed to improve the resolution of the spectra, and in this way, it was possible to identify the exact position of the peaks in the ATR-FTIR spectra (Figure 4). In the adhesive formulations, a more marked change in the relative contribution of the peak located at 1620 $cm^{-1}$ was observed with increasing citric acid concentration, confirming modifications in the secondary structure, in coincidence with that reported by Liguori et al. [12]. In contrast, a higher contribution of the peak at 1720 $cm^{-1}$ was observed for systems with higher CA content (CA-80 vs. CA-40), which would indicate that crosslinking promoted by CA would occur in these formulations (Table inserted in Figure 4a–c).

According to Stanca et al. [35], the secondary structure components of a protein were found at different positions: β-sheet (1610–1635 $cm^{-1}$), random coil (1635–1645 $cm^{-1}$), α-helix (1645–1665 $cm^{-1}$), β-turn (1662–1682 $cm^{-1}$) and β-sheet antiparallel (1682–1689 $cm^{-1}$). As is well known, the second derivative of the ATR-FTIR spectra is a tool that allows magnifying some spectral features that appear as weak. Spectral deconvolution was performed to ensure a more accurate estimation of the amide I component bands. From curve fitting, four Gaussian components of the amide I band in the ATR-FTIR spectrum were detected (Figure 4e). The peak assignment and the percentage contributions are shown in Table 3. CA concentrations changed their relative contributions substantially.

**Table 3.** Secondary structure contribution (%) obtained by deconvolution of the amide I region taking into account the Gaussian components of the ATR-FTIR spectra.

| Wavenumber ($cm^{-1}$) | Amide I—Contribution (%) | | |
|---|---|---|---|
| | CA-20 | CA-40 | CA-80 |
| 1609 | 4.7 | 17.7 | 32.9 |
| 1626 | 46.5 | 45.9 | 36.5 |
| 1646 | 43.3 | 21.4 | 21.7 |
| 1664 | 5.5 | 14.9 | 8.8 |

According to Koochakzaei [36], after exposure to acid addition, the triple helix structure of gelatin loses its conformation. Therefore, the conversion of α-helix to β-sheet due to stability is another exchange due to changes in the secondary structure of the gelatin.

These results revealed that the CA-80 film exhibited self-adhesive properties not only to the touch but also to the aforementioned substrates. A possible explanation for these effects would be related to the conformational changes that are revealed in the gelatin in the presence of CA, evidenced by ATR-FTIR, as a consequence of which functional groups are exposed that would be responsible for the adhesive characteristics to the touch of the gelatin. A similar hypothesis was proposed by Sartori et al. [15] to explain the adhesive capacity of gluten-protein-based adhesives.

### 3.4. Pressure-Sensitive Adhesive: Characteristics and Potential Applications

In order to investigate the possibility of using the adhesive film on different substrates, a test tape adhesion assay was carried out. The CA-80 adhesive film showed a greater adhesion on the metal (Figure 5a) and plastic (Figure 5b) sheets. Both Double A™ and 3M™ test tapes showed no peeling of any film grid, indicating their high adhesion to these substrates. According to the ASTM D3359-02 [20] standard, the adhesion of the CA-80 film is classified as 5B, since the corners of the cuts are intact, that is, no square of the film has come off.

When evaluating the Kraft paper, it was observed that part of the bioadhesive film was dragged by the Double A™ tape (Figure 5c). According to the specifications of the standard, the adhesion to this substrate is classified as 1B; the adhesive film was folded and peeled at all intersections and most of the squares, revealing an affected area of 35 to 65%.

In the ceramic (Figure 5d) and in the glass (Figure 5e), the bioadhesive film was adhered to both substrates, and the tapes could not lift the adhesive. In corrugated cardboard, the Double A™ tape was again able to lift part of the CA-80 film (Figure 5f), in which the adhesive–cellulose-fiber interaction is less than that of the adhesive–test tape. This could be due to the low compatibility with the substrate and the poor penetration of the adhesive into materials. Finally, the Durlock™ test was performed (Figure 5g) to evaluate its potential use as a self-adhesive film, considering possible applications as a wall adhesive, and again, the strong adhesive capacity of the studied bioadhesive film to adhere to this substrate was observed, with different characteristics from those mentioned above.

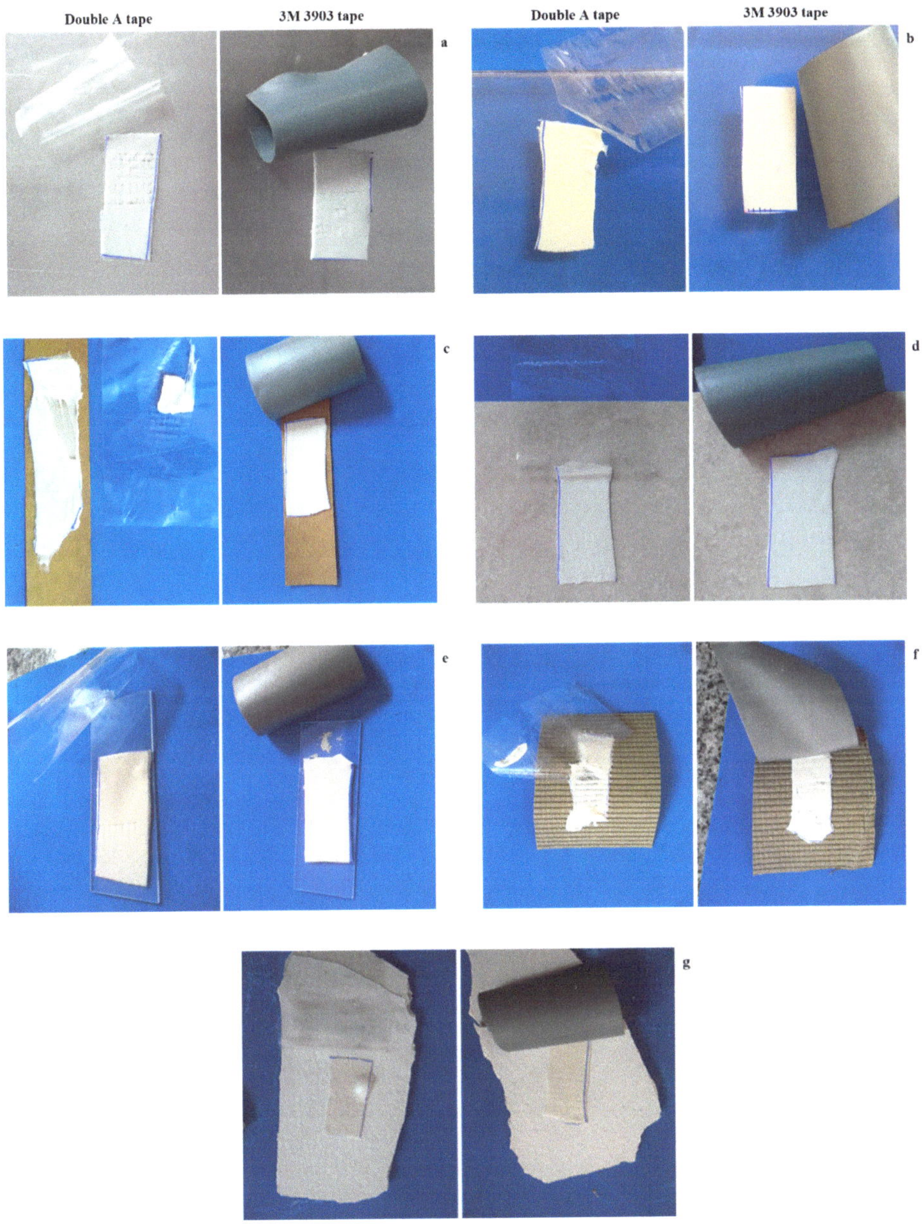


**Figure 5.** Photographs taken during the adhesion test on different substrates: (**a**) metal, (**b**) plastic, (**c**) Kraft paper, (**d**) ceramic, (**e**) glass, (**f**) corrugated cardboard and (**g**) Durlock$^{TM}$.

Thus, the results obtained through the standardized test, although qualitative, indicate the potential and versatility of the bioadhesive film for bonding different types of substrates.

Complementary analysis by SEM could allow us to study the morphology of the pressure-sensitive adhesive. Figure 6 shows the SEM micrograph of the film obtained with the CA-80 formulation. With a compact, dense and smooth structure, the filmogenic capacity developed by the formulation can be observed, which is expected to have a positive impact on the adhesive characteristics and on the resistance to sticking of different substrates.

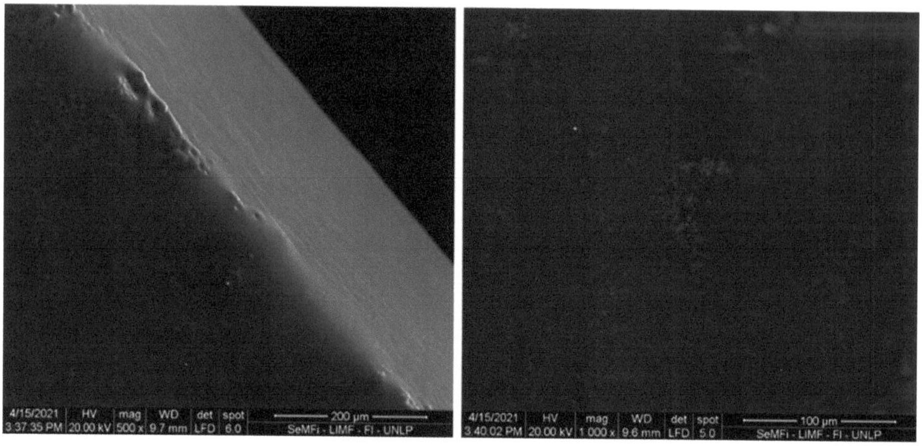


**Figure 6.** SEM micrograph of the film obtained from the formulation based on 5% *w*/*w* starch-15% *w*/*w* gelatin and 80% *w*/*w* CA (CA-80). On the left is the cross-section, and on the right is the surface. The magnification is indicated in the micrographs.

## 4. Conclusions

Composite adhesive formulations were developed using starch, gelatin and citric acid by heat treatment under controlled conditions. Depending on the concentration of citric acid, formulations adapted to different applications can be obtained. The inclusion of citric acid not only improves the applicability of the system by modifying the flowability of the adhesives, as evidenced by the observed changes in consistency/viscosity, but also provides increased adhesive bond strength.

In the case of liquid adhesive formulations, the results revealed that the functional properties of the starch–gelatin blends were significantly improved by the addition of citric acid up to a level of 40%. The structural changes induced by the addition of CA were revealed by ATR-FTIR analysis and allowed explaining the adhesive behavior of the different formulations.

On the other hand, the highly adhesive properties of the material obtained by drying the CA-80 formulation revolutionize the studies and invite us to further deepen the analysis of pressure-sensitive adhesives, opening up possible industrial applications. In this sense, it is important to highlight that although liquid adhesive formulations containing 80%CA have low adhesion and tack, once dried, they allow a self-adhesive tape to be obtained. This bioadhesive film, based on chemically modified natural polymers (cassava starch and gelatin), is able to bond substrates of different nature (cardboard, paper, glass, Durlock™, ceramic, plastic).

Therefore, by tailoring the formulation of the adhesive based on cassava starch and gelatin, it was possible to obtain different types of adhesives, from liquid adhesives for Kraft paper and glass widespread use in the packaging industry to a self-adhesive tape, demonstrating the versatility of the formulations.

## 5. Patent

Monroy Y., Rivero S.M.G, García M.A. (2022). Composiciones adhesivas ecocompatibles y película adhesiva bifaz, y procedimientos de elaboración. Acta N° 20220102305.

**Author Contributions:** The manuscript was written with equal contributions of all authors. All authors have read and agreed to the published version of the manuscript.

**Funding:** Authors acknowledge the financial support of Agencia Nacional de Promoción Científica y Tecnológica (ANPCyT), projects PICT 2019-2827 and PICT 2019-03380, CONICET and the University of La Plata (UNLP).

**Data Availability Statement:** The data presented in this study are available on request from the corresponding author.

**Acknowledgments:** Authors acknowledge the financial support of Agencia Nacional de Promoción Científica y Tecnológica and the University of La Plata (UNLP). Authors wish to thank María Fernanda Hamet (CIDCA) for her valuable help in texture tests.

**Conflicts of Interest:** Authors declare no conflict of interest.

## References

1. Ortega, F.; Versino, F.; López, O.V.; García, M.A. Biobased composites from agro-industrial wastes and by-products. *Emergent Mater.* **2022**, *5*, 873–921. [CrossRef]
2. Dunky, M. Applications and Industrial Implementations of Naturally-Based Adhesives. In *Biobased Adhesives: Sources, Characteristics and Applications*; Scrivener Publishing LLC: Beverly, MA, USA, 2023; pp. 659–704.
3. Rizky, S.; Wintoko, J. Modification of bioadhesive based on crosslinked alginate and gelatin. *Mater. Today Proc.* **2023**. [CrossRef]
4. Uranga, J.; Nguyen, B.T.; Si, T.T.; Guerrero, P.; de la Caba, K. The effect of cross-linking with citric acid on the properties of agar/fish gelatin films. *Polymers* **2020**, *12*, 291. [CrossRef] [PubMed]
5. Vargas, C.G.; Mercali, G.D.; de Oliveira Rios, A.; Rahier, H.; Flôres, S.H. Effect of moderate electric field on the properties of gelatin capsule residue-based films. *Food Hydrocoll.* **2019**, *89*, 29–35.
6. Versino, F.; Ortega, F.; Monroy, Y.; Rivero, S.; López, O.V.; García, M.A. Sustainable and bio-based food packaging: A review on past and current design innovations. *Foods* **2023**, *12*, 1057. [CrossRef]
7. Wangtueai, S.; Chaiyaso, T.; Rachtanapun, P.; Jantrawut, P.; Ruksiriwanich, W.; Seesuriyachan, P.; Leksawasdi, N.; Phimolsiripol, Y.; Techapun, C.; Phongthai, S.; et al. Thermoplastic cassava starch blend with polyethylene-grafted-maleic anhydride and gelatin core-shell structure compatibilizer. *Int. J. Biol. Macromol.* **2022**, *197*, 49–54. [CrossRef]
8. Luque, G.C.; Stürtz, R.; Passeggi Jr, M.C.; Gugliotta, L.M.; Gonzalez, V.D.; Minari, R.J. New hybrid acrylic/collagen nanocomposites and their potential use as bio-adhesives. *Int. J. Adhes. Adhes.* **2020**, *100*, 102624. [CrossRef]
9. Costa TM, H.; de Oliveira Rios, A.; Flôres, S.H. Residues of minimally processed carrot and gelatin capsules: Potential materials for packaging films. *Ind. Crops Prod.* **2015**, *76*, 1071–1078.
10. Iahnke, A.O.E.S.; Stoll, L.; Bellé, A.S.; Hertz, P.F.; Rios, A.D.O.; Rahier, H.; Flôres, S.H. Gelatin capsule residue-based films crosslinked with the natural agent genipin. *Packag. Technol. Sci.* **2020**, *33*, 15–26. [CrossRef]
11. García-Campo, M.J.; Boronat, T.; Quiles-Carrillo, L.; Balart, R.; Montanes, N. Manufacturing and characterization of toughened poly (lactic acid)(PLA) formulations by ternary blends with biopolyesters. *Polymers* **2017**, *10*, 3. [CrossRef]
12. Liguori, A.; Uranga, J.; Panzavolta, S.; Guerrero, P.; de la Caba, K.; Focarete, M.L. Electrospinning of fish gelatin solution containing citric acid: An environmentally friendly approach to prepare crosslinked gelatin fibers. *Materials* **2019**, *12*, 2808. [CrossRef] [PubMed]
13. Monroy, Y. Desarrollo de Bioadhesivos a Base de Almidones Modificados Con Aplicaciones Potenciales en el Área de Envases. Doctoral Dissertation, Universidad Nacional de La Plata, La Plata, Argentina, 2021.
14. Cohen, E.; Binshtok, O.; Dotan, A.; Dodiuk, H. Prospective materials for biodegradable and/or biobased pressure-sensitive adhesives: A review. *J. Adhes. Sci. Technol.* **2013**, *27*, 1998–2013. [CrossRef]
15. Sartori, T.; Feltre, G.; do Amaral Sobral, P.J.; da Cunha, R.L.; Menegalli, F.C. Biodegradable pressure sensitive adhesives produced from vital wheat gluten: Effect of glycerol as plasticizer. *Colloids Surf. A Physicochem. Eng. Asp.* **2019**, *560*, 42–49. [CrossRef]
16. Monroy, Y.; Rivero, S.M.G.; García, M.A. *Composiciones Adhesivas Ecocompatibles y Película Adhesiva Bifaz, y Procedimientos de Elaboración*; Acta No. 20220102305; INPI: Buenos Aires, Argentina, 2022.
17. Monroy, Y.; Cabezas, D.M.; Rivero, S.; García, M.A. Structural analysis and adhesive capacity of cassava starch modified with NaOH: Urea mixtures. *Int. J. Adhes. Adhes.* **2023**, *126*, 103470. [CrossRef]
18. Monroy, Y.; Hamet, M.F.; Rivero, S.; García, M.A. Tailor-made starch-based adhesives chemically modified with NaOH: Urea and their applications on a cellulosic substrate. *Int. J. Biol. Macromol.* **2023**, *247*, 125423. [CrossRef]
19. Monroy, Y.; Rivero, S.; García, M.A. Sustainable panels design based on modified cassava starch bioadhesives and wood processing byproducts. *Ind. Crop. Prod.* **2019**, *137*, 171–179. [CrossRef]
20. ASTM Committee D-1 on Paint and Related Coatings, Materials, and Applications. In *Standard Test Methods for Measuring Adhesion by Tape Test*; ASTM International: West Conshohocken, PA, USA, 2009.
21. Firoozmand, H.; Murray, B.S.; Dickinson, E. Microstructure elastic modulus of mixed gels of gelatin-oxidized starch: Effect of, p.H. *Food Hydrocoll.* **2012**, *26*, 286–292. [CrossRef]
22. Goudie, K.J.; McCreath, S.J.; Parkinson, J.A.; Davidson, C.M.; Liggat, J.J. Investigation of the influence of pH on the properties and morphology of gelatin hydrogels. *J. Polym. Sci.* **2023**, *61*, 2316–2332. [CrossRef]

23. Wang, R.; Hartel, R.W. Citric acid and heating on gelatin hydrolysis and gelation in confectionery gels. *Food Hydrocoll.* **2022**, *129*, 107642. [CrossRef]
24. Fang, C.; Wu, C.; Zhao, X. Preparation and characterization of high solid content acrylate latex pressure sensitive adhesives with difunctional cross-linker EGDMA. *Int. J. Adhes. Adhes.* **2023**, *125*, 103403. [CrossRef]
25. Kumar, R.; Ghoshal, G.; Goyal, M. Synthesis and functional properties of gelatin/CA–starch composite film: Excellent food packaging material. *J. Food Sci. Technol.* **2019**, *56*, 1954–1965. [CrossRef] [PubMed]
26. Dohr, C.A.; Hirn, U. Influence of paper properties on adhesive strength of starch gluing. *Nord. Pulp Pap. Res. J.* **2022**, *37*, 120–129. [CrossRef]
27. Ocaña López, R. Degradación Ambiental y en Condiciones Adversas de Adhesivos Estructurales: Análisis y Consideraciones técnicas Para su Aplicación Industrial. Doctoral Dissertation, Universidad Politécnica de Madrid, Madrid, España, 2017.
28. He, Q.; Zhang, Y.; Cai, X.; Wang, S. Fabrication of gelatin–TiO2 nanocomposite film and its structural, antibacterial and physical properties. *Int. J. Biol. Macromol.* **2016**, *84*, 153–160. [CrossRef] [PubMed]
29. Kumar, R.; Ghoshal, G.; Goyal, M. Biodegradable composite films/coatings of modified corn starch/gelatin for shelf life improvement of cucumber. *J. Food Sci. Technol.* **2020**, *58*, 1227–1237. [CrossRef]
30. Reddy, N.; Yang, Y. Citric acid cross-linking of starch films. *Food Chem.* **2010**, *118*, 702. [CrossRef]
31. Brandelero RP, H.; Alfaro, A.T.; Marques, P.T.; Brandelero, E.M. New Approach of Starch and Chitosan Films as Biodegradable Mulching. *Rev. Virtual De Química* **2019**, *11*, 686–698. [CrossRef]
32. Sethi, S.; Kaith, B.S. A review on chitosan-gelatin nanocomposites: Synthesis, characterization and biomedical applications. *React. Funct. Polym.* **2022**, *179*, 105362. [CrossRef]
33. Shi, C.; Tao, F.; Cui, Y. New starch ester/gelatin based films: Developed and physicochemical characterization. *Int. J. Biol. Macromol.* **2018**, *109*, 863–871. [CrossRef]
34. Singh, N.; Maitra, J. Antibacterial Evaluation of Starch and Chitosan Based Polymeric Blend. *J. Appl. Chem.* **2015**, *8*, 26.
35. Stanca, M.; Gaidau, C.; Zaharescu, T.; Balan, G.A.; Matei, I.; Precupas, A.; Leonties, A.R.; Ionita, G. Physico-Chemical Changes Induced by Gamma Irradiation on Some Structural Protein Extracts. *Biomolecules* **2023**, *13*, 774. [CrossRef]
36. Koochakzaei, A. Determination of Sulfuric Acid Effects on Degradation and Structural Changes of Gelatin Using Fourier-Transform Infrared Spectroscopy and Peak Deconvolution Analysis. *Infrared Spectrosc.* **2023**, *38*, 5–11. [CrossRef]

foods

MDPI

Article

# Subcritical Water Extraction for Valorisation of Almond Skin from Almond Industrial Processing

**Pedro A. V. Freitas *, Laia Martín-Pérez, Irene Gil-Guillén , Chelo González-Martínez and Amparo Chiralt**

Institute of Food Engineering FoodUPV, Universitat Poltècnica de València, 46022 Valencia, Spain
* Correspondence: pedvidef@doctor.upv.es

**Abstract:** Almond skin (AS) is an agro-industrial residue from almond processing that has a high potential for valorisation. In this study, subcritical water extraction (SWE) was applied at two temperatures (160 and 180 °C) to obtain phenolic-rich extracts (water-soluble fraction) and cellulose fibres (insoluble fraction) from AS. The extraction conditions affected the composition and properties of both valorised fractions. The dry extracts obtained at 180 °C were richer in phenolics (161 vs. 101 mg GAE. $g^{-1}$ defatted almond skin (DAS)), with greater antioxidant potential (1.063 vs. 1.490 mg DAS.$mg^{-1}$ DPPH) and showed greater antibacterial effect (lower MIC values) against *L. innocua* (34 vs. 90 mg·$mL^{-1}$) and *E. coli* (48 vs. 90 mg·$mL^{-1}$) than those obtained at 160 °C, despite the lower total solid yield (21 vs. 29%) obtained in the SWE process. The purification of cellulose from the SWE residues, using hydrogen peroxide ($H_2O_2$), revealed that AS is not a good source of cellulose material since the bleached fractions showed low yields (20–21%) and low cellulose purity (40–50%), even after four bleaching cycles (1 h) at pH 12 and 8% $H_2O_2$. Nevertheless, the application of a green, scalable, and toxic solvent-free SWE process was highly useful for obtaining AS bioactive extracts for different food, cosmetic, or pharmaceutical applications.

**Keywords:** phenolic compounds; cellulose fibres; integral fractionation; active compounds; bleaching optimisation; bioactive properties

**Citation:** Freitas, P.A.V.; Martín-Pérez, L.; Gil-Guillén, I.; González-Martínez, C.; Chiralt, A. Subcritical Water Extraction for Valorisation of Almond Skin from Almond Industrial Processing. *Foods* **2023**, *12*, 3759. https://doi.org/10.3390/foods12203759

Academic Editors: Mario M. Martinez and Sílvia Petronilho

Received: 27 September 2023
Revised: 6 October 2023
Accepted: 9 October 2023
Published: 13 October 2023



## 1. Introduction

Almonds are a very important crop throughout the world's temperate regions, with a worldwide almond production of 4.6 Mt in 2021, with the USA being the first producer with about 2.0 Mt per year (*International Nuts and Dried Fruits Statistical Yearbook*). Fresh almond fruit consists of four portions: kernel or meat (~11%), middle shell (~33%), outer green shell cover or almond hull (~52%), and a thin leathery layer known as brown skin (~4%) [1]. The processing by-products, shells, and hulls, account for more than 50% of the dry weight of the almond fruits, which, in the past, were used as animal feed or burned for fuel production and energy use. Likewise, in the industrial process of almond kernel, the brown skin is removed by blanching, generating a residual product that constitutes about 6–8% wt. of the seed in large amounts, which has been also used as animal feed or burned as fuel in processing plants [2].

However, almond skins (AS) are known to have a great number of nutritional and health benefits, mainly based on the presence of polyphenols and the high dietary fibre content [3]. Different functional properties have been reported for these phenolic compounds, such as inhibiting lipid oxidation by scavenging free radicals, chelating metals, activating antioxidant enzymes, reducing tocopherol radicals, or inhibiting enzymes that cause oxidation reactions [4]. The composition of AS with high phenolic content and different functional properties represents a high added value for this industrial by-product. Several authors [1,4,5] have reviewed the phenolic composition of AS reported in different studies using different extraction methods. The major components of AS are proanthocyanidins (dimers or larger molecules) (mean content: 6.98 mg.100 $g^{-1}$), hydrolysable tannins, flavonoids (non-isoflavone) (71.3 mg.100 $g^{-1}$), and phenolic acids and aldehydes (11.6 mg.100 $g^{-1}$) [5]. In the complex phenolic profile of AS, different flavonol glycosides have been identified, namely, isorhamnetin rutinoside, isorhamnetin glucoside, kaempferol rutinoside, kaempferol glucoside, quercetin glycosylated to glucose, galactose and rhamnose, kaempferol, naringenin, catechin, protocatechuic acid, vanillic acid, and a benzoic acid derivative [4]. Flavanols and flavonol glycosides were the most abundant phenolic compounds in AS, representing up to 38–57% and 14–35% of the total quantified phenolics, respectively [6]. Due to their potential antioxidant properties, AS can be considered to be a value-added by-product for potential use as dietary antioxidant ingredients. Esfahlan et al. (2010) [4] reported antioxidant activities of ethanolic extracts from AS using different free radical scavenging assays revealing that the Trolox equivalent antioxidant capacity of the AS extract was 13 times greater than that of the whole seed extract at the same extract concentration.

AS also has antimicrobial properties that may arise from the action of phenolic compounds or their synergistic interactions. Methanolic extracts of defatted AS exhibited antibacterial activity against *L. monocytogenes* and *S. aureus* in the range of 250–500 µg $mL^{-1}$, with non-blanched skins showing antimicrobial potential against the Gram-negative *S. enterica* [7]. The aglycones epicatechin and catechin isolated from AS showed great activity against *S. aureus* ATCC 6538P (MIC values of 0.078–0.15 and 0.15 mg.$mL^{-1}$, respectively) [8]. Extracts from liquid-nitrogen-blanched skin also inhibited *Escherichia coli*, *Serratia marcescens*, and *Streptococcus mutans* [9].

In terms of health benefits, phenolic compounds of AS are bioavailable and extensively biotransformed by the microbiota and host tissue upon consumption [10]. Although the precise function of polyphenol metabolites in preventing chronic disease is emerging, these have been described as protective agents against inflammatory processes [3], cancer [11], herpes [8,12], or cardiovascular disease [13,14], while modulating plasma biomarkers of oxidative stress [15]. AS also exhibits prebiotic activity [16] due to the polyphenol delivery to the colon [17,18] and increased probiotic bacteria in the culture, including bifidobacteria and eubacteria [19]. Thus, almond polyphenols may contribute to gut health through modulation of microbiota.

On the basis of the described properties of AS and their availability as industrial waste from almond processing, obtaining active extracts from AS for pharmaceutical cosmetics or food industry applications represents an interesting strategy for its valorisation. In this sense, sustainable, ecologic, and scalable extraction methods should be proposed and developed. At the laboratory scale, solid–liquid extraction and liquid–liquid extraction steps have been applied using organic solvents, such as methanol, acetone, or ethyl acetate to obtain phenolic-rich extracts [20]. Microwave-assisted extraction of AS has been optimised to increase the phenolic yield, using ethanol as the extraction solvent [21]. Phenolic compounds from gamma-irradiated AS were extracted with 40% ethanol, obtaining an increase in the yield of total phenolic content and enhanced antioxidant activity of the extracts [22]. Nevertheless, a chemical-free extraction process would be more interesting from the point of view of its ecology and sustainability. The overuse of organic solvents causes problems, such as the negative environmental impact through the generation of toxic wastes, the solvents' effect on human health, and the labour-intensiveness and time requirements of the extraction methods [23]. Subcritical water extraction (SWE), also known as pressurised hot-water extraction or superheated water extraction, is a green technique for the isolation of organic compounds and hemicellulose fractions with preserved molecular functionalities and high molecular weight [24]. Under subcritical conditions, the water solvent properties are similar to those of organic solvents, being able to dissolve different components of medium and low polarity [25] due to the changes in the water properties, depending on different factors. Temperature is the most important factor affecting the extraction efficiency and selectivity. When temperature increases, the dielectric constant and the polarity of water decrease, which enhances the solubility of less polar compounds in water. Thus, subcritical water can behave similarly to methanol or ethanol [26]. This makes subcritical water a green extraction fluid used for a variety of organic species. Compared with organic solvents, subcritical water not only has advantages in ecology, economy, and safety, but also its density, ion product, and dielectric constant can be adjusted by temperature [25]. Likewise, it is faster and scalable, does not require pre-treatments, and produces a low degree of sugar degradation [25–27]. It has been applied to extract polyphenols from several natural products such as pistachio hulls [28], chestnut shells [29], comfrey root [30], or pomegranate peel [31]. The optimal extraction temperature and time deduced from different studies were between 130 and 200 °C and 15 and 35 min [26].

When applied to extract natural products, the extraction residue could also be used to obtain insoluble polysaccharides, such as cellulose for different applications, enabling the integral fractionation of material. Different studies have been carried out using SWE for the integral fractionation of plant products, such as rice husk, rice straw, or grapevine by-products, in order to achieve a comprehensive use of the starting material [23,32–34]. Nevertheless, no previous studies were found on the AS by-product fractionation by SWE to obtain value-added components for specific applications.

The aim of this work was to fractionate blanched AS industrial residue using SWE at two temperatures (160 and 180 °C) to obtain phenolic-rich extracts and polysaccharide-rich solid residues for different uses. Phenolic content, antioxidant, and antibacterial properties were characterised in the extracts, while the potential use of the residue as a source of cellulose fibres has been explored, through cellulose purification by a green bleaching process using hydrogen peroxide. Active extracts and cellulose fibres could be used to obtain active, reinforced packaging materials.

## 2. Materials and Methods

### 2.1. Chemicals

Folin–Ciocalteu reagent (2 N), gallic acid, methanol, 2,2-Diphenyl-1-picrylhydrazyl (DPPH), sodium hydroxide (NaOH), glucose, and arabinose were purchased from Sigma-Aldrich (USA). Ethanol (98%), hydrogen peroxide ($H_2O_2$, 30%), sulphuric acid ($H_2SO_4$, 98%), and sodium carbonate ($Na_2CO_3$, 99.5%) were obtained from Panreac Quimica S.L.U (Castellar del Vallés, Barcelona, Spain). Phosphorous pentoxide ($P_2O_5$, 98.2%) was obtained from VWR Chemicals (Leuven, Belgium). D(+)-Xylose was supplied by Merck KGaA (Darmstadt, Germany). For microbiological analyses, tryptone soy broth (TSB), bacteriological agar, and peptone water were purchased from Scharlab (Barcelona, Spain). Strains of *Escherichia coli* (CECT 101) and *Listeria innocua* (CECT910) were obtained from the Spanish Type Collection (CECT, Universitat de València, València, Spain).

### 2.2. Plant Material Preparation

Blanched ASs (*Prunus dulcis, Nonpareil var.* from California) were kindly provided by Importaco S.A (Valencia, Spain) from its almond peeled production line as an agro-industrial by-product from the harvest of 2022. The plant material was dried at 40 ± 2 °C for 3 days in a forced-air oven (S. P. Selecta, s. a., Barcelona, Spain). Afterwards, the dried AS was milled using a mill (Model SM300 stainless, Retsch GmbH, Haan, Germany) and sieved to obtain particles with dimensions of under 0.55 mm and stored until further use.

Given the high oil content, the dried AS was defatted using petroleum ether (40–60 °C bp) with Soxhlet. The residual solvent in the defatted almond skins (DAS) was eliminated at 40 °C for 3 days in a fume extraction cabin. Meanwhile, the almond oil (AO) (soluble fraction) was recovered using a vacuum rotary evaporator (Rotary Evaporators, Heidolph Instruments GmbH & Co. KG, Walpersdorfer, Germany) at 35 °C, and the total oil content of AS was quantified.

### 2.3. Obtaining AS Fractions by Applying SWE

DAS was subjected to subcritical water extraction (SWE) using a pressure reactor (Model 1-TAP-CE, 5 L capacity, Amar Equipment PVT. LTD, Mumbai, India) at a DAS: distilled water ratio of 1:6 (*w/w*). Two extraction conditions were applied according to previous studies into phenolic extraction of plant products [15,34]: 160 °C, 7 bar, 150 rpm, for 30 min and 180 °C, 15 bar, 150 rpm, for 30 min. After each extraction procedure, the DAS dispersions were filtered (Filterlab, Barcelona, Spain, pore size < 0.5 mm) and the solid extracts were obtained by freeze-drying the soluble fractions at −60 °C, 0.8 mbar for 72 h. The obtained extracts, named E-160 and E-180, were stored in desiccators ($P_2O_5$, 0% relative humidity (RH)) at 4 °C until further use. In every SWE condition, the total solid yield of the extracts (TSY, g extract solids/100 g DAS) was determined by sampling two aliquots of the liquid extracts and drying at 105 °C until constant weight to determine the solid:water ratio and the corresponding total solids extracted from DAS, considering the total water mass in the reactor. The insoluble residues (R-160 and R-180) were washed with distilled water, filtered, and dried at 40 °C for 48 h, to determine the yield in the extraction residue, and afterwards stored at 5 °C until further use.

### 2.4. Bleaching Process of the Extraction Residue

The extraction residues (R-160 and R180) were submitted to a bleaching process in order to obtain cellulose fibres, following the methodology described by Li et al. (2011) [35], with some modifications. In order to optimise the bleaching process, hydrogen peroxide solutions at 4% and 8% were used in a bleaching solution, with a solid residue ratio of 30:1. The temperature of the dispersions was kept at 40 °C (using a water bath), while the tested reaction times were 1, 2, and 3 h, fitting the solution pH values at 11 (0.75% wt. NaOH) and 12 (2.15% wt. NaOH). After the bleaching time, the cellulose fractions were filtered and washed with abundant distilled water to remove residual bleaching solution and dried at room temperature for 2 days. To determine the bleaching efficiency of each condition applied, the bleached samples were characterised in terms of yield (Equation (1)) and whiteness index (*WI*). For this, the colour coordinates $L^*$ (lightness), $a^*$ (red-green), and $b^*$ (yellowish-blue) of each bleached fraction were obtained with a CM-3600d spectro-colourimeter (Minolta Co., Tokyo, Japan), and the *WI* was calculated according to Equation (2).

$$Yield = \frac{Weight\ of\ bleached\ cellulose}{Weight\ of\ material\ before\ bleaching} \tag{1}$$

$$WI = 100 - \sqrt{(100 - L^*)^2 + a^{*2} + b^{*2}} \tag{2}$$

Once the optimum bleaching conditions (lowest yield and highest *WI*) were selected, four bleaching cycles were applied to the solid fractions to enhance the cellulose fibre purification process.

### 2.5. Characterisation of Cellulosic Residues

#### 2.5.1. Chemical Composition

The chemical composition of the raw material (DAS), extraction residues (R160 and R180), and bleached fractions were analysed in terms of structural carbohydrates, lignin, and ashes. Before the quantification, the extractives were determined using the standard NREL method (NREL/TP-510-42619—2008) [36]. This procedure was performed using a Soxhlet set-up, which consists of two stages: a first extraction with water for 6 h, followed by a second extraction with ethanol at 60 °C for 6 h. Once the material was free of extractives, structural carbohydrates, and lignin were determined following the procedure described in the standard NREL method (NREL/TP-510-42618—2008) [37], which consists of a two-step acid hydrolysis with $H_2SO_4$. The insoluble fraction was used to gravimetrically determine the Klason lignin content. In addition, the soluble fraction was used to quantify the sugar content (glucose, xylose, and arabinose) using high-performance liquid chromatography (HPLC, Agilent Technologies, model 1120 Compact LC, Waldbronn, Germany). A RezexTM RCM-Monosaccharide $Ca^{2+}$ column (150 × 7.8 mm) and an evaporative light scattering detector (ELSD Agilent Technologies 1200 Series, Waldbronn, Germany) were used. The mobile phase was deionised water, in an isocratic mode, at a flow rate of 0.4 mL.min$^{-1}$. The detector conditions were: 40 °C, 3.0 bar gas pressure ($N_2$), and a gain of 3. The software used was ChemStation (version LTS 01.11, Agilent Technologies, Waldbronn, Germany) and the data were analysed with the Origin program (version OriginPro 2021, OriginLab Corporation, Northampton, MA, USA) by applying the Gaussian model for peak area determination. The cellulose content was determined from the glucose content, while the hemicellulose content was obtained from the sum of xylose and arabinose sugars.

The ash content of the cellulosic fractions was determined by sample incineration at 575 °C for 24 h. The procedure was performed in duplicate.

#### 2.5.2. Thermogravimetric Analysis (TGA)

The thermal stability of different samples was analysed using a thermogravimetric analyser (TGA 1 Stare System analyser, Mettler-Toledo, Greifensee, Switzerland). In each case, 4–5 mg of conditioned samples (in desiccators containing $P_2O_5$, at 25 °C for two weeks) were weighted in alumina pans and heated from 25 °C to 900 °C under a constant flow of nitrogen (10 mL.min$^{-1}$) and a 10 °C.min$^{-1}$. Samples were analysed in duplicate and TGA and DTGA curves were obtained using the STAR$^e$ Evaluation Software (version V12.00a, Mettler-Toledo, Inc., Greifensee, Switzerland).

### 2.6. *Characterisation of SWE Extracts*

Phenolic-rich extracts obtained by SWE (E-160 and E-180) were characterised according to total phenolic content, antioxidant activity, and antibacterial activity.

#### 2.6.1. Determination of Total Phenolic Content (TPC)

Total phenolic content (TPC) of the solid extracts was determined using the Folin–Ciocalteu method, according to the procedure described by Freitas et al. (2020) [38]. Briefly, 0.5 mL of each extract was mixed with 6 mL of distilled water and 0.5 mL of Folin reagent (2 N). After one minute, 1.5 mL of $Na_2CO_3$ at 20% (*w*/*v*) was added and the final volume was adjusted to 10 mL with distilled water and kept in the dark at room temperature for 2 h. Afterwards, the absorbance was measured using a UV-Vis spectrophotometer (model Evolution 201, Thermo Scientific, Waltham, MA, USA) at 747 nm. The total phenolic content was determined using a standard curve ($R^2$ = 0.9991) of gallic acid (2–20 mg·L$^{-1}$) and expressed as mg gallic acid equivalents (GAE).100 g$^{-1}$ extract. The measurements were performed in triplicate.

#### 2.6.2. Radical Scavenging Activity (DPPH)

The anti-radical activity of the extracts was determined using the free radical 2,2-diphenyl-1-pikryl-hydrazyl (DPPH) method according to a modified procedure described by Freitas et al. (2020) [38]. For each extract, methanolic DPPH solution at $6.22 \times 10^{-2}$ mM (Abs at 515 nm = $0.7 \pm 0.2$) was mixed with different extract concentrations to a final volume of 4 mL. The resultant solutions were kept in the dark at room temperature for 12 h, and then the absorbance at 515 nm was measured. The initial and final concentration of DPPH in the reaction medium was calculated from a calibration curve fitted by linear regression ($R^2$ = 0.9992). The anti-radical activity was evaluated by means of $EC_{50}$, defined as the amount of antioxidants required to reduce the initial concentration of DPPH by 50%, when reaction stability is reached, and expressed as mg freeze-dried.mg$^{-1}$ DPPH. The $EC_{50}$ values were determined from graphics of $\%[DPPH]_{remaining}$ vs. mg solid extract/mg DPPH:

$$\%[DPPH]_{remaining} = \frac{[DPPH]_t}{[DPPH]_{t=0}} \times 100 \quad (3)$$

where $[DPPH]_t$ is the DDPH concentration value when the reaction was stable, and $[DPPH]_{t=0}$ is the DPPH initial concentration. Each determination was carried out in duplicate.

#### 2.6.3. Antibacterial Activity

The minimum inhibitory concentration (MIC) of the extracts (E-160 and E-180) was determined for two bacteria: a Gram-positive bacterium *Listeria innocua* and a Gram-negative bacterium *Escherichia coli*. This analysis was carried out following the method described by Requena et al. (2019) [32] and Freitas et al. (2020) [38]. For this analysis, standard 96-well microtiter plates with a total volume of 200 µL were used.

For both bacterial strains, previously stored at −20 °C, a stock solution was prepared by transferring amounts of bacteria using an inoculation loop to a volume of 10 mL TSB and incubated at 37 °C for 24 h. A volume of 10 µL of the stock solution was taken and transferred to a tube containing 10 mL of TSB to prepare the corresponding working solution with $10^5$ CFU.mL$^{-1}$, which was confirmed by serial dilution and counting. For each bacterium, 100 µL of the bacterial solution with an initial concentration of $10^5$ CFU.mL$^{-1}$ was added to each well. Different volumes of each extract solution at 200 mg.mL$^{-1}$ were added, and then the final volume of each well was made up to 200 µL with TSB, and the plates were incubated at 37 °C for 24 h. Afterwards, 100 µL of each well was transferred in TSA plates and incubated at 37 °C for 24 h for final counting. For each extract, the MIC was determined as the lowest extract concentration at which no bacterial growth was observed on the TSA plate. This analysis was performed in duplicate.

#### 2.6.4. Protein Content

The protein content was determined by the Dumas method using a Leco instrument calibrated according to the manufacturers. The analysis was performed in duplicate for DAS, extracts (E-160, E-180), and residues (R-160 and R-180) obtained from SWE. The factor to convert total nitrogen in protein content was 6.25.

### 2.7. Statistical Analysis

The experimental data were subjected to an analysis of variance (ANOVA) using Minitab statistical software (version 17), considering a confidence level of 95%. Differences between treatment responses were determined by the Fisher test, using the least significant difference (α) of 5%.

## 3. Results and Discussion

### 3.1. Yields of the Process Steps

Figure 1 shows the flow chart of the process applied to AS fractionation indicating the obtained yield of each process step. In the first defatting step, 20.8 ± 0.7% of almond oil was obtained from the dried AS, which represents an important amount of oil with high value in the food and cosmetic market. This content was similar to that reported in other studies using AS [39]. The lipid composition of blanched AS has been reported [39], where monounsaturated fatty acids (mainly oleic acid) are the principal component (56%), followed by polyunsaturated (34%, mainly linoleic acid) and saturated fatty acids (10.3%, mainly palmitic acid). The reported total vitamin E was 13 mg.100 g$^{-1}$, of which α-tocopherol represented 99%; only traces of β- and γ-tocopherol were detected. These data were consistent for samples from different sources, almond cultivars, and varieties [39]. The lipid profile of AS is similar to that observed in almond kernel oil and could be extracted for different applications [40].

The yields in terms of solid extracts (E-160 and E-180) and dried residues (R-160 and R-180) of the SWE process performed at 160 and 180 °C are shown in Figure 1. In contrast with that obtained for other plant materials, where extraction yield increased when temperature rose [28,34,41], the extract yield from DAS was lower at 180 °C (21 g dry extract.100 g$^{-1}$ DAS) than at 160 °C (29 g dry extract.100 g$^{-1}$ DAS). Likewise, this trend was also observed for the yield in residues after extraction (56 g dried insoluble fraction.100 g$^{-1}$ DAS and 55 g dried insoluble fraction.100 g$^{-1}$ DAS). It is also remarkable that the sum of both yields at a given temperature does not close the mass balance, in which about 15% and 24% of dry matter was not recovered at 160 °C and 180 °C, respectively. On the other hand, the pressure in the extraction reactor exceeded the water vapour pressure at the set point temperature by 2 and 5 bars, for 160 and 180 °C, respectively. This result suggests the formation of gases associated with a certain degree of mineralisation of the organic matter present in the reactor at the process conditions, forming $CO_2$. In fact, SWE has been applied with high efficiency in the mineralisation of contaminant organic compounds, using hydrogen peroxide at different concentrations to effectively remove many hazardous compounds in wastewater [42,43]. Although no oxidising agent was added in the extraction carried out, and it took place at lower temperatures, the demonstrated prooxidant action of many phenolic compounds [44], depending on the medium conditions, could induce partial mineralisation of DAS samples under the subcritical water conditions applied in the study; the higher the extraction temperature, the greater the substrate degradation.

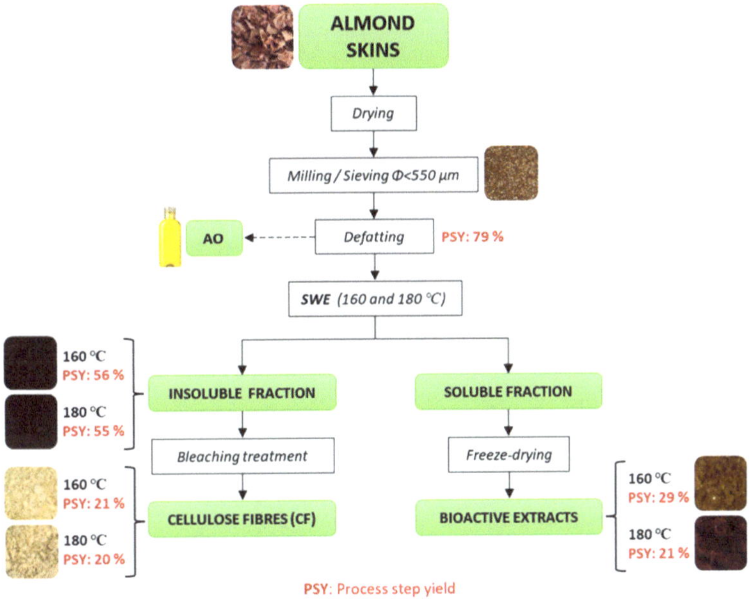


**Figure 1.** Flow chart of the process used for almond skin (AS) fractionation, showing the process step yields (PSY: g outgoing solids. 100 $g^{-1}$ of incoming dried material) for the defatting step, SWE at 160 and 180 °C (soluble extracts: E-160 and E-180; insoluble residues: R-160 and R-180), and blanching treatment of insoluble residues (BR-160 and BR-180).

Other authors also observed thermo-degradation of different compounds when temperature increased in the SWE. For instance, procyanidin dimers were identified in pistachio extracts, indicating the thermal degradation of compounds of larger molecular weights [28]. The application of high temperature (190 °C) and long extraction time led to the modification or degradation of stilbenes from grapevine by-products [23]. Likewise, 5-hydroxymethylfurfural (HMF) formation was observed above 150 °C due to thermo-oxidation or Maillard reactions occurring at high temperatures [45,46]. The formation of HMF at high temperatures might be attributed to the decomposition of monomeric sugars formed through the hydrolyses of polysaccharides, such as hemicelluloses, which is promoted by SWE at high temperatures [27,32,47].

The thermogravimetric analysis of the samples (DAS, extracts, and residues), shown in Figure 2, was coherent with the partition of components between the liquid and solid phases during extraction. The raw DAS presents three thermal degradation steps: the first one between 40 and 180 °C corresponds to the loss of adsorbed water and reaction products due to the caramelisation of free sugars present in the plant matrix [33]; the second one (190–450 °C) is associated to the degradation of polymeric materials, which are constitutional polysaccharides such as hemicellulose (150–350 °C) and cellulose (275–350 °C), and a fraction of the total lignin (160–900 °C) [48,49]; and the third one (460–670 °C) relates to the degradation of the residual lignin fraction and the products from the fragmentation of the organic structures thermo-degraded at lower temperatures [50]. Malayil et al. (2022) [51] observed similar thermal degradation behaviour for the untreated AS. The extracts (Figure 2a) were enriched in low molecular weight compounds with lower degradation temperatures, while the insoluble residues (Figure 2b) showed an increase in polymeric material, as deduced from their degradation at higher temperatures. In both cases, an increase in mass loss was observed for the temperature range of lignin degradation, which is consistent with the formation of degraded/condensed compounds during the high-temperature extraction process. An increment in the final mass residue assignable to ash is also observed for both extracts and insoluble residues. However, under nitrogen flow, this final residue could, in part, correspond to degraded/condensed organic matter more temperature-resistant to oxidative processes. Therefore, thermal analysis was coherent with the selective fractionation of DAS by applying SWE, giving rise to aqueous extracts richer in different compounds of lower molecular (sugars, phenolic compounds, and minerals) and polysaccharide-rich insoluble residues.

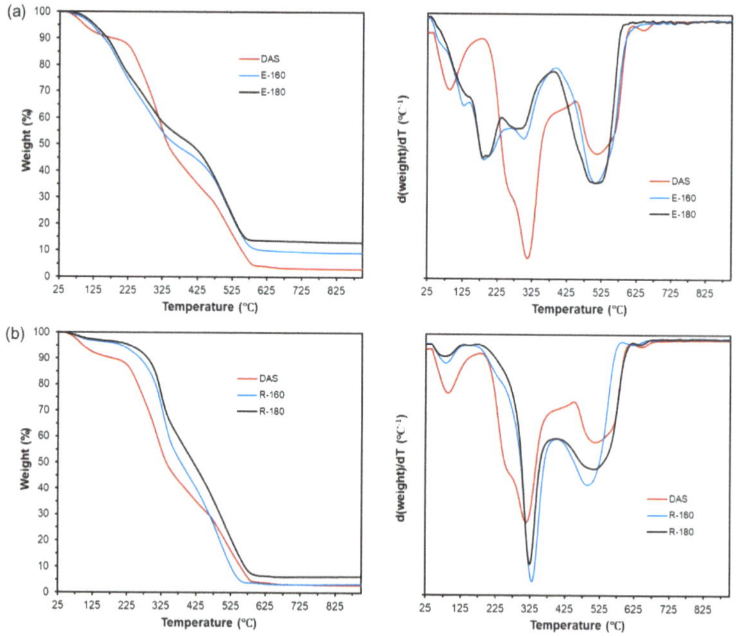


**Figure 2.** TGA and DTGA curves of the DAS, the active extracts (E-160 and E-180) (**a**), and the insoluble fractions (R-160 and R-180) (**b**) obtained from SWE at 160 and 180 °C.

The ash content of the DAS, extracts, and residues (Tables 1 and 2), determined by sample incineration, indicated that minerals from DAS were mainly present in the extracts (13.1 and 15.5%, respectively, for E-160 and E-180) whereas small amounts remained in the residues (2.3 and 2.8%, respectively, for R-160 and R-180). Considering these values and the respective yields in extracts and residues, the predicted ash content was similar to that obtained for the DAS sample (5.09%). This value was in the range of the ash content reported by other authors for almond skin [39], which is rich in manganese, calcium, zinc, iron, magnesium, copper, phosphorus, and potassium [52,53].

**Table 1.** Total phenolic (TPC) and ash contents, antioxidant activity (EC50), and minimal inhibitory concentration (MIC) of the DAS extracts (E-160 and E-180) (values ± standard deviation).

| | E-160 | E-180 |
|---|---|---|
| $TPC_1$ (mg GAE.g$^{-1}$ dried extract) | 100.9 ± 0.5 [b] | 161 ± 4 [a] |
| $TPC_2$ (mg GAE.g$^{-1}$ DAS) | 29.6 ± 0.15 [b] | 33.8 ± 0.9 [a] |
| $EC_{50}$ (mg extract.mg$^{-1}$ DPPH) | 1.490 ± 0.003 [a] | 1.063 ± 0.012 [b] |
| $EC_{50}$ (mg DAS.mg$^{-1}$ DPPH) | 5.13 ± 0.01 [a] | 5.06 ± 0.06 [b] |
| MIC *L. innocua* (mg.mL$^{-1}$) | 90 | 34 |
| MIC *E. coli* (mg.mL$^{-1}$) | 90 | 48 |
| Ash (g.100 g$^{-1}$ solid extract) | 13.1 ± 0.1 [b] | 15.5 ± 0.7 [a] |

Different letters in the same row indicate significant differences between films by Fisher test (α = 0.05).

**Table 2.** Chemical composition in terms of cellulose, hemicellulose, Klason lignin, and ash content of the defatted almond skin DAS, the insoluble residues (R-160 and R-180) obtained from SWE, and the bleached residues (BR-160 and BR-180) submitted to different bleaching cycles (1–4) (mean values ± standard deviation).

| Sample | Cellulose (%wt.) | Hemicellulose (%wt.) | Klason Lignin (%wt.) | Ash (%wt.) |
|---|---|---|---|---|
| DAS | 9.8 ± 0.3 | 11.6 ± 0.5 | 17.4 ± 0.2 | 5.1 ± 0.1 |
| R-160 | 18.6 ± 0.9 [b] | 6.4 ± 0.3 [a] | 45.2 ± 0.5 [b] | 2.3 ± 0.1 [a] |
| R-180 | 20.9 ± 0.2 [a] | 5.1 ± 0.1 [b] | 51.2 ± 0.3 [a] | 2.7 ± 0.6 [a] |
| BR-160-1 | 35.0 ± 3.0 [a,1] | 13.4 ± 0.9 [b,1] | 22.0 ± 3.0 [a,2] | 15.0 ± 1.0 [c,1] |
| BR-160-2 | 40.0 ± 4.0 [a,2] | 17.5 ± 1.9 [a,1] | 18.9 ± 1.9 [b,2] | 17.0 ± 4.0 [bc,1] |
| BR-160-3 | 40.0 ± 4.0 [a,2] | 15.0 ± 3.0 [ab,1] | 15.0 ± 2.0 [c,2] | 18.0 ± 4.0 [b,2] |
| BR-160-4 | 40.0 ± 3.0 [a,2] | 13.6 ± 0.7 [ab,1] | 13.6 ± 0.1 [c,2] | 25.0 ± 3.0 [a,1] |
| BR-180-1 | 35.6 ± 1.8 [b,1] | 9.3 ± 1.2 [b,2] | 33.0 ± 2.0 [a,1] | 10.4 ± 0.1 [c,2] |
| BR-180-2 | 52.0 ± 2.0 [a,1] | 13.5 ± 0.5 [a,2] | 22.5 ± 0.7 [b,1] | 17.0 ± 5.0 [b,1] |
| BR-180-3 | 49.9 ± 0.8 [a,1] | 10.0 ± 2.0 [b,1] | 22.0 ± 4.0 [bc,1] | 26.3 ± 0.5 [a,1] |
| BR-180-4 | 49.6 ± 0.6 [a,1] | 13.9 ± 1.2 [a,1] | 16.7 ± 0.7 [c,1] | 22.0 ± 7.0 [ab,1] |

Different subscript letters indicate significant differences between samples of the same group (insoluble residues R; bleached residues BR-160 or BR-180). Different numbers indicated significant differences between BR-160 and BR-180 samples at the same bleaching cycle (Fisher test, $p < 0.05$).

Subcritical extraction also promoted partition of the protein content of DAS (13.2 ± 0.1 g.100 g$^{-1}$ DAS), which was higher in the extracts (13.6 ± 0.4 and 24.1 ± 0.3 g.100 g$^{-1}$ dried extract, for E-160 and E-180, respectively) whereas the extraction residues became poorer in protein at 180 °C (13.4 ± 0.3 and 12.7 ± 0.4 g.100 g$^{-1}$ dried residue, respectively, for R-160 and R-180). The protein content of DAS was in the range of previously reported values for almond skin [39]. The enrichment of the extraction residues in protein indicates the low solubility of almond skin protein under the used water subcritical conditions.

### 3.2. Antioxidant and Antibacterial Properties of the SWE Extracts

The obtained extracts at 160 and 180 °C were characterised according to their antioxidant and antibacterial properties in order to know their potential applications as functional ingredients in the food, pharmaceutical, or cosmetic industry. Table 1 summarises the total phenolic content (TPC) determined by the Folin–Ciocalteu method and the DPPH radical scavenging capacity (by means of $EC_{50}$), as well as the minimal inhibitory concentration (MIC) for *L. innocua* and *E. coli*.

The TPC determined in the solid extracts ($TPC_1$) was also referred to per mass unit of DAS ($TPC_2$). Higher values were obtained at 180 °C than at 160 °C, which is in line with that observed by other authors for the phenolic extraction of different plant residues submitted to SWE [25,28,34]. Ersan et al. (2018) [28] observed a very efficient selective extraction of different phenols in pistachio hulls by SWE at 150–170 °C; total gallotannin yielded up to 33 mg.g$^{-1}$, where gallic acid (22.2 mg.g$^{-1}$) and penta-*O*-galloyl-$\beta$-d-glucose

(9.77 mg.g$^{-1}$) levels were 13- and 11-fold higher than those in the aqueous methanol extracts. However, neo-formed antioxidant compounds during extraction at the highest temperature [45,46] could also contribute to the quantified TPC by the unspecific Folin–Ciocalteu method. Vladic et al. (2020) [30] reported high levels of HMF when SWE was applied to pomegranate peel at temperatures above 130 °C for 20 min. The authors recommended lower temperatures of extraction because they provide a high content of phenols and minimise the presence of HMF. Considering the reported maximum non-toxic HMF intake levels (80–100 mg.kg$^{-1}$ per day) [54], the formed levels of this compound should be considered carefully when the extracts are intended to be used as a food additive.

In terms of the TPC extracted from DAS, 2.96 and 3.38 g GAE.100 g$^{-1}$ DAS were obtained at 160 and 180 °C, respectively. These values were higher than those reported by other authors for blanched DAS using different extraction methods based on organic solvents. Mandalari et al. (2010) [39] reported 0.279 g GAE.100 g$^{-1}$ blanched skins, this value is lower than that obtained for non-blanched skins (3.474 g GAE.100 g$^{-1}$), which is mainly due to the removal of water-soluble flavonoids and phenolic acids during the industrial blanching and is similar to that reported by Garrido et al. (2008) [55]. Other authors [56] also reported significant losses (89%) in the TPC of peanut skins during the blanching process. Phenolic compounds from AS also showed oxidative and other degradative phenomena during drying thermal treatment. The total amounts of phenolic compounds extracted from thermally dried skins were always lower than those extracted from non-dried skins, with the most evident decrease affecting (+)-catechin. Therefore, antioxidant capacity may be affected by the degradation of bioactive compounds at high temperatures, generating other products with antioxidant properties, as observed for dried blanched skins and roasted samples [55]. It is worth mentioning that the Folin–Ciocalteu method is based on a redox reaction, where compounds other than phenolics, such as ascorbic acid, reducing sugars, or some amino acids, can reduce the Folin–Ciocalteu reagent, thus overestimating the TPC content [57,58]. Nonetheless, this method is useful for indicating the redox power of extracts obtained under different extraction conditions, which may be related to their global antioxidant capacity. In addition, despite the possible partial modification of the phenolic fraction during SWE, especially at the highest temperature, the process allows for obtaining highly rich extracts in antioxidant compounds, which is probably promoted by removing bonded phenolics in the skin cell structure under subcritical water hydrolysis.

In agreement with the TPC values, the DPPH radical scavenging capacity was also very high. It increased when the extraction temperature rose, exhibiting values near the most known antioxidants, such as ascorbic acid or α-tocopherol ($EC_{50}$: 0.12 and 0.26 mg compound.mg$^{-1}$ DPPH, respectively) [59]. Therefore, the extracts obtained under sub-critical conditions, especially the E-180, exhibited a remarkable DPPH radical scavenging capacity, which was close to that of the strong antioxidants. Nevertheless, as previously commented for the TPC values, the increased antioxidant capacities at higher temperatures could be attributed not only to the higher extraction efficiency for antioxidant compounds but also to the neo-formed antioxidants via Maillard, caramelisation, and thermo-oxidation reactions [45,46]. Other antioxidant tests, such as the ABTS radical scavenging capacity or FRAP assay, could be carried out to verify the actual antioxidant potential of the extracts. Nevertheless, for their use as active components of food packaging materials, their effectiveness at controlling food oxidation reactions, such as oxidation of unsaturated lipids, should be proven when incorporated into these materials and brought into contact with the target food.

DAS extracts (E-160 and E-180) also exhibited antibacterial effects against Gram-positive and Gram-negative bacteria. *E. coli* is a Gram-negative bacterium found in the intestines of humans and animals as well as in untreated water and foods, which cause serious food poisoning. Likewise, *Listeria* innocua is a Gram-positive bacterial strain that has previously been used to model the food-borne pathogen *L. monocytogenes* in inactivation experiments [60]. Table 1 shows the MIC values of both extracts for *L. innocua*

(90 and 34 mg.mL$^{-1}$ for E-160 and E-180, respectively) and *E. coli* (90 and 48 mg.mL$^{-1}$ for E-160 and E-180, respectively). E-180, with higher TPC values, was more effective against both bacteria since it had lower MIC values, whereas both bacteria exhibited similar sensitivity for each extract. Other authors also reported the antibacterial activity of AS extracts obtained by different procedures, which have been attributed to the action of the constitutive phenolic compounds. Phenolic fractions of almond skin exhibited antibacterial activity against *L. monocytogenes* and *S. aureus* in the range of 250–500 μg.mL$^{-1}$ [7]. The aglycones epicatechin and catechin isolated from AS showed great activity against *S. aureus* ATCC 6538P (MIC values of 0.078–0.15 and 0.15 mg·mL$^{-1}$, respectively) [61]. Smeriglio et al. (2016) [9] report very low MIC values of phenolic extracts from AS against several Gram-positive and Gram-negative strains. Of the Gram-positive bacteria, *S. aureus* showed the highest sensitivity, followed by *S. epidermidis* ATCC 35984 and *S. mutans*. The obtained SWE extracts from DAS could be considered to be effective antibacterial products useful for controlling bacterial growth in different systems, with MIC values lower than that observed for other SWE plant extracts, such as rice straw [34] or rice husk [32].

Therefore, DAS extracts from the almond industrial process, obtained by SWE, can be considered functional ingredients to control oxidative or microbial spoilage of different products, such as foods or pharmaceuticals. Nevertheless, the formation of potentially toxic compounds makes the toxicological study necessary. Likewise, different health-promoting properties of the extracts should be analysed to obtain useful information about their potential uses.

### 3.3. Bleaching of the Extraction Residues

The extraction residues (R-160 and R180) were treated in order to recover the cellulose fraction, which can be used to obtain fibres or nanomaterials useful for different applications [62]. Thus, a green bleaching process using hydrogen peroxide was optimised to purify cellulose, varying the pH, concentration of the oxidant agent, and treatment time. The efficiency of the different bleaching treatments was evaluated by monitoring the yield (BY: mass of bleached residue/initial mass, %) and the whiteness index (*WI*) of the cellulosic material. The yield informs about the total amount of compounds removed in the oxidative process, while the whiteness index accounts for the removal of coloured compounds.

The first experimental bleaching series was carried out with 4% hydrogen peroxide at different pH values (11 and 12) to analyse its effect on the efficiency of the process for 1, 2, or 3 h of treatment. The *WI* and BY values obtained after each procedure are shown in Figure 3a. A multifactorial analysis of variance (MANOVA), with the factors pH, extraction temperature, and bleaching time, revealed a significant effect of pH on BY (F-value: 19) and a significant effect of pH (F-value: 19) and SWE temperature (F-value: 25) on *WI*, with no significant effect of bleaching time in any case. The higher the pH value, the lower the yield (higher removal of oxidisable compounds) and the higher the *WI* (higher removal of coloured compounds). The WI was lower for samples obtained at 180 °C that also showed a lower initial *WI* value.

Based on these results, treatments at pH 12, with 2, 4, and 8% hydrogen peroxide, for 1, 2, and 3 h were carried out. The BY and *WI* values obtained for this experimental series are shown in Figure 3b. The MANOVA showed significantly lower values of BY when the concentration of hydrogen peroxide (F-value: 24) and bleaching time (F-value: 10) rose, while the BY values were significantly lower for the R-180 sample, compared with R-160. Likewise, *WI* showed a significant increase when the oxidant concentration rose, with no significant effect of the bleaching time, exhibiting significantly higher values for the residue obtained at 160 °C. Therefore, the residue obtained at 180 °C was more resistant to bleaching than that obtained at 160 °C, and, in both cases, greater efficiency (lower BY and higher *WI*) was observed with the higher concentration of hydrogen peroxide, but without significant effects of the bleaching time. On the basis of these results, four bleaching cycles were applied at the most effective bleaching conditions (8% hydrogen peroxide, pH 12 for 1 h) to promote cellulose purification, evaluating the BY and *WI* (Figure 4), as well as the composition of the residue in each cycle, in terms of cellulose, hemicellulose, lignin, and ash contents (Table 2). The contents in these components of the raw material (DAS) and the extraction residues (R-160 and R180) were also quantified and are also shown in Table 2.

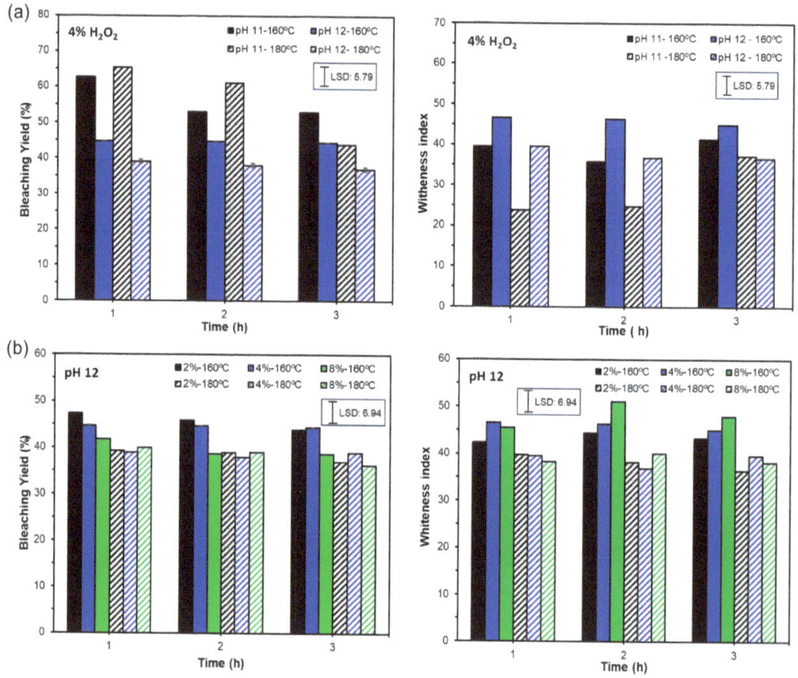


**Figure 3.** Values of bleaching yield and whiteness index reached in the different bleaching steps of the extraction residues obtained at 160 and 180 °C (R-160 and R-180 samples). (**a**) Treatments with 4% hydrogen peroxide, varying pH (11 and 12), and bleaching time (1, 2, and 3 h). (**b**) Treatments at pH 12, varying hydrogen peroxide concentration (2, 4, and 8%), and bleaching time (1, 2, and 3 h).

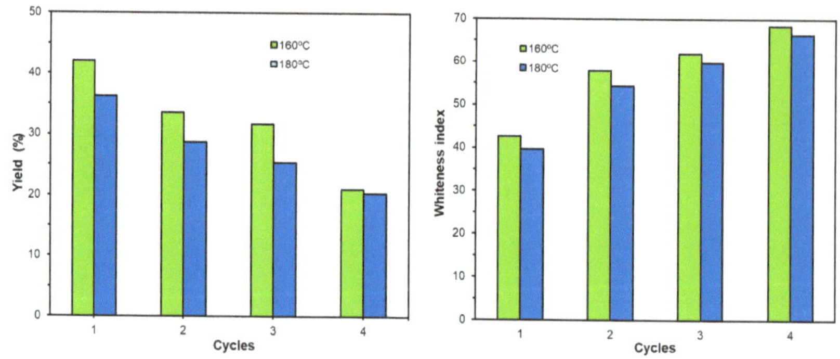


**Figure 4.** Values of bleaching yield and whiteness index reached in the different bleaching cycles of the extraction residues obtained at 160 and 180 °C (R-160 and R-180 samples).

The application of four successive bleaching cycles significantly decreased BY while *WI* increased, reflecting the progressive purification of cellulose in each cycle (Figure 4). The cellulose purification progress was monitored through the analysis of lignin and sugars through the NREL method. After removal of the water (which includes soluble sugars) and ethanol extractives in the DAS, R-160, and R-180 samples (23.5, 18.5, and 17.3% wt., in total, respectively), glucose, xylose, and arabinose were quantified in the different materials. Glucose was the major component, and arabinose was only present in the DAS sample. Mandalari et al. (2010) [39] reported that carbohydrates comprised 45% of almond skin in which the amounts of different monomeric sugars (as a percentage of total sugar content) were galacturonic acid (36%), glucose (29%), arabinose (18%), and xylose (7.6%). Considering the defatted basis in the DAS sample and the possible free sugars elimination in the previous water extraction, similar contents of glucose (7.7 g.100 $g^{-1}$ DAS), xylose (4.1 g.100 $g^{-1}$ DAS), and arabinose (5.0 g.100 $g^{-1}$ DAS) were obtained in the present study. In the extraction residues (R-160 and R-180), no arabinose was detected, which indicates that it was dissolved under the subcritical water conditions. Hemicellulose content was considered as total xylose content (or xylose and arabinose in the DAS sample), whereas total glucose was attributed to the cellulose content [37].

The percentage of cellulose increased in both extraction residues (R-160 and R-180) (Table 2) with respect to the untreated DAS, and significant differences in cellulose richness were observed for R-160 and R-180, with the former being poorer in cellulose (Table 2). Hemicellulose was more extensively removed from the DAS matrix treated at 180 °C, but lignin and ash contents increased to a greater extent in this case (Table 2). Cellulose content significantly increased in the first and second bleaching cycles for both R-160 and R-180 samples, but no subsequent increase was observed in the successive cycles. This suggests that a part of the quantified glucose was not cellulosic but hemicellulosic and was also progressively removed during blanching.

The Klason lignin content reported for blanched AS was 20.6% wt., but it may be overestimated as the residue representing the Klason lignin is a mixture of lignin, residual protein, suberin, and ash [39]. The quantified lignin content of the DAS sample was in the range of reported values for AS. The high content of Klason lignin determined in both residues (R-160 and R-180) is remarkable, the magnitude of which is unexpected from a mass balance, considering the content of the initial DAS sample. This indicates that the SWE process generated acid-insoluble compounds that were quantified as lignin fractions. Thus, during the partial degradation reactions that occurred in SWE, neo-formed compounds seem to have similar acid-solubilisation behaviour to the initial lignin content, significantly increasing the gravimetrically quantified lignin. This content decreased in the bleached samples, but these still showed very high values, even after the four bleaching cycles, especially in the BR-180 samples obtained a higher extraction temperature, where greater matter degradation was observed in the SWE treatment. The lignin content progressively decreased with bleaching cycles, but the highest initial content of R-180 samples makes its effective removal with four bleaching cycles more difficult.

The enrichment of the samples in the hemicellulose fraction during the first and second bleaching cycles, with no subsequent decrease in the following cycles, is also remarkable. This indicates an inherent difficulty of the material for cellulose purification and removal of other polysaccharides in the lignocellulosic matrix. The maximum percentage of cellulose obtained in the bleached residues was about 50% for BR-180 and 40% for BR-160 by applying only two cycles with no significant subsequent purification in the successive cycles. In fact, only the lignin content progressively decreased throughout the bleaching cycles.

The TGA and their DTGA curves of the insoluble and bleached residues are shown in Figure 5. Compared to the non-bleached residues, the TGA curves of the bleached fractions showed significant differences in the thermal stability associated with the compositional changes that occurred in the bleaching step. The increment of degradable compounds at low temperatures (between 45 and 160 °C) after bleaching, which may correspond to mono or oligosaccharides resulting from the alkaline hydrolysis of the present polysaccharides that are not removed in the subsequent washing step, is remarkable. The degradation of bleached polymeric material is also reflected in the decrease of the degradation temperature of structural polysaccharides. The bleached residues showed a displacement of the main degradation peak from 330 (R-160) and 323 °C (R-180) to a range of 244–267 °C, suggesting the depolymerisation of constitutive polysaccharides during the bleaching process. In fact, the decomposition of the active species ($HOO^-$) during the bleaching reactions into its intermediates such as hydroxyl radical (HO•) and superoxide anion ($O_2^{-}$•) can lead to the attack of the cellulose and hemicellulose chains, thus worsening the thermal stability of the bleached fractions [35,63]. Thermal analysis also showed, the formation of compounds with high thermal stability (600–800 °C) in the bleached residues while decreasing the typical peak of lignin present in the initial residue (460–670 °C). This was observed in all bleached fractions, where the intensity of the characteristic lignin peak decreased as the bleaching cycles progressed, whereas other degradation peaks at higher temperatures (about 725 °C) appeared due to the newly formed condensed compounds. The bleached residues also had a high final residual mass, which increased as the bleaching cycles progressed. This behaviour was similar for both R-160 and R-180 samples when submitted to bleaching. Therefore, the oxidative process applied with hydrogen peroxide in an alkaline medium seems to degrade some constituents that remain bonded to the matrix and do not leach during the bleaching process or subsequent washing, making cellulose purification difficult.

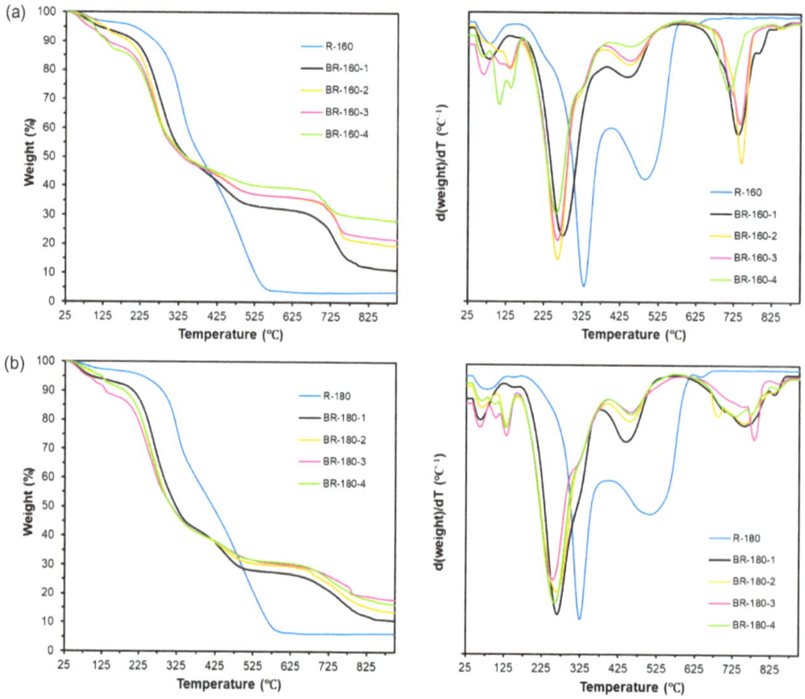


**Figure 5.** TGA and DTGA curves of the insoluble residues (R) and the bleached residues (BR) obtained from SWE at 160 (**a**) and 180 °C (**b**) submitted to different bleaching cycles (1–4).

The ash contents of the residues, determined by sample incineration, progressively increased after the successive bleaching cycles, as also observed for the TGA residual mass, reaching values of 21–24% after four bleaching cycles. This behaviour suggests that sodium ions of an alkaline bleaching medium were entrapped in bleached residues, increasing the final ash content of the samples. Other authors [64,65] also observed that when some ligno-cellulosic matrices were submitted to alkali treatments with NaOH solutions, to promote chemical degradation and removal of amorphous components, Na ions penetrate into the cellulose crystalline network, forming antiparallel crystalline soda–cellulose complexes with modified crystallisation patterns. This allows for the swelling of interfibrillar regions and the rearrangement of the cellulose microfibrils.

Therefore, AS was not a good source of cellulose due to its relatively low content and the difficulties in its purification, in comparison with other agro-industrial residues such as rice straw [33]. In fact, considering the yields of the different steps applied to obtain cellulose (Figure 1), only 11–12% wt. of DAS could be transformed in partially purified cellulose (40–50% purity) by applying SWE and bleaching processes. The highest temperature of SWE allows for the highest purity, but the very low total yield makes the process unprofitable.

## 4. Conclusions

The current increasing trend of almond production and its industrial processing has been accompanied by an increment in bleached skin waste generation, which requires seeking solutions to its reutilisation. The high content of polyphenols in this waste and their bioactivity offer an interesting opportunity for obtaining phenol-rich extracts, using a green and scalable process without using toxic solvents. Subcritical water extraction at 160 or 180 °C allowed for obtaining bioactive extracts (21–29% of defatted almond skin) with high phenolic content and antioxidant and antibacterial properties, which could be used for food, pharmaceutical, or cosmetic applications. The highest temperature provided the extracts with higher bioactive properties. Nevertheless, the thermo-formation of new antioxidant compounds during the partial degradation of extracts could be responsible for this enhanced activity. Therefore, a toxicological study of the extracts is required to ensure their safety for different applications, such as the development of active packaging materials.

Cellulose recovery from the extraction residues (55–56% of defatted almond skin) was not effective enough due to the low yield obtained in the blanching process (20–21%) and the low purity of cellulose (40–50%) reached in the blanched fractions after four bleaching cycles (1 h) at pH 12 and 8% hydrogen peroxide. Therefore, almond skin is not a good source of cellulosic materials.

**Author Contributions:** P.A.V.F.: conceptualization, methodology, formal analysis, investigation, writing—original draft preparation, and writing—review and editing; L.M.-P.: conceptualization, methodology, investigation, writing—original draft preparation, and writing—review and editing; I.G.-G.: conceptualization, methodology, investigation, writing—original draft preparation, and writing—review and editing; C.G.-M.: conceptualization, methodology, investigation, writing—original draft preparation, and writing—review and editing; A.C.: conceptualization, methodology, investigation, writing—original draft preparation, and writing—review and editing. All authors have read and agreed to the published version of the manuscript.

**Funding:** This study forms part of the AGROALNEXT programme (AGROALNEXT/2022/026) and was supported by MCIN with funding from the European Union NextGenerationEU (PRTR-C17.I1) and the project CIPROM/2021/071 (Generalitat Valenciana).

**Data Availability Statement:** The data used to support the findings of this study can be made available by the corresponding author upon request.

**Acknowledgments:** Authors L.M.P and I.G.G. are grateful to Generalitat Valenciana, NextGenerationEU, and Plan de recuperación transformación y resiliencia (Gobierno de España) for the INVESTIGO 2022 grant (INVEST/2022/4).

**Conflicts of Interest:** The authors declare no conflict of interest.

## References

1. Barral-Martinez, M.; Fraga-Corral, M.; Garcia-Perez, P.; Simal-Gandara, J.; Prieto, M.A. Almond By-Products: Valorization for Sustainability and Competitiveness of the Industry. *Foods* **2021**, *10*, 1793. [CrossRef]
2. González, J.F.; Gañán, J.; Ramiro, A.; González-García, C.M.; Encinar, J.M.; Sabio, E.; Román, S. Almond Residues Gasification Plant for Generation of Electric Power. Preliminary Study. *Fuel Proces. Technol.* **2006**, *87*, 149–155. [CrossRef]
3. Mandalari, G.; Bisignano, C.; Genovese, T.; Mazzon, E.; Wickham, M.S.J.; Paterniti, I.; Cuzzocrea, S. Natural Almond Skin Reduced Oxidative Stress and Inflammation in an Experimental Model of Inflammatory Bowel Disease. *Int. Immunopharmacol.* **2011**, *11*, 915–924. [CrossRef]
4. Esfahlan, A.J.; Jamei, R.; Esfahlan, R.J. The Importance of Almond (*Prunus amygdalus* L.) and Its by-Products. *Food Chem.* **2010**, *120*, 349–360. [CrossRef]

5. Bolling, B.W. Almond Polyphenols: Methods of Analysis, Contribution to Food Quality, and Health Promotion: Almond Polyphenol. *Compr. Rev. Food Sci. Food Saf.* **2017**, *16*, 346–368. [CrossRef]
6. Monagas, M.; Garrido, I.; Lebrón-Aguilar, R.; Bartolome, B.; Gómez-Cordovés, C. Almond (*Prunus dulcis* (Mill.) D.A. Webb) Skins as a Potential Source of Bioactive Polyphenols. *J. Agric. Food Chem.* **2007**, *55*, 8498–8507. [CrossRef]
7. Mandalari, G.; Bisignano, C.; D'Arrigo, M.; Ginestra, G.; Arena, A.; Tomaino, A.; Wickham, M.S.J. Antimicrobial Potential of Polyphenols Extracted from Almond Skins: Antimicrobials from Almond Skins. *Lett. Appl. Microbiol.* **2010**, *51*, 83–89. [CrossRef]
8. Musarra-Pizzo, M.; Ginestra, G.; Smeriglio, A.; Pennisi, R.; Sciortino, M.T.; Mandalari, G. The Antimicrobial and Antiviral Activity of Polyphenols from Almond (*Prunus dulcis* L.) Skin. *Nutrients* **2019**, *11*, 2355. [CrossRef]
9. Smeriglio, A.; Mandalari, G.; Bisignano, C.; Filocamo, A.; Barreca, D.; Bellocco, E.; Trombetta, D. Polyphenolic Content and Biological Properties of Avola Almond (*Prunus dulcis* Mill. D.A. Webb) Skin and Its Industrial Byproducts. *Ind. Crops Prod.* **2016**, *83*, 283–293. [CrossRef]
10. Garrido, I.; Urpi-Sarda, M.; Monagas, M.; Gómez-Cordovés, C.; Martín-álvarez, P.J.; Llorach, R.; Bartolomé, B.; Andrés-Lacueva, C. Targeted Analysis of Conjugated and Microbial-Derived Phenolic Metabolites in Human Urine After Consumption of an Almond Skin Phenolic Extract. *J. Nutr.* **2010**, *140*, 1799–1807. [CrossRef]
11. García-Pérez, P.; Barreal, M.E.; Rojo-De Dios, L.; Cameselle-Teijeiro, J.F.; Gallego, P.P. Bioactive Natural Products From the Genus Kalanchoe as Cancer Chemopreventive Agents: A Review. In *Studies in Natural Products Chemistry*; Elsevier: Amsterdam, The Netherlands, 2019; Volume 61, pp. 49–84. ISBN 978-0-444-64183-0.
12. Bisignano, C.; Mandalari, G.; Smeriglio, A.; Trombetta, D.; Pizzo, M.; Pennisi, R.; Sciortino, M. Almond Skin Extracts Abrogate HSV-1 Replication by Blocking Virus Binding to the Cell. *Viruses* **2017**, *9*, 178. [CrossRef] [PubMed]
13. Berryman, C.E.; Preston, A.G.; Karmally, W.; Deckelbaum, R.J.; Kris-Etherton, P.M. Effects of Almond Consumption on the Reduction of LDL-Cholesterol: A Discussion of Potential Mechanisms and Future Research Directions. *Nutr. Rev.* **2011**, *69*, 171–185. [CrossRef] [PubMed]
14. Kalita, S.; Khandelwal, S.; Madan, J.; Pandya, H.; Sesikeran, B.; Krishnaswamy, K. Almonds and Cardiovascular Health: A Review. *Nutrients* **2018**, *10*, 468. [CrossRef]
15. Chen, C.-Y.O.; Milbury, P.E.; Blumberg, J.B. Polyphenols in Almond Skins after Blanching Modulate Plasma Biomarkers of Oxidative Stress in Healthy Humans. *Antioxidants* **2019**, *8*, 95. [CrossRef] [PubMed]
16. Liu, Z.; Lin, X.; Huang, G.; Zhang, W.; Rao, P.; Ni, L. Prebiotic Effects of Almonds and Almond Skins on Intestinal Microbiota in Healthy Adult Humans. *Anaerobe* **2014**, *26*, 1–6. [CrossRef]
17. Mandalari, G.; Tomaino, A.; Rich, G.T.; Curto, R.L.; Arcoraci, T.; Martorana, M.; Bisignano, C.; Sajia, A.; Parker, M.L.; Waldron, K.W.; et al. Polyphenol and nutrient release from skin of almonds during simulated human digestion. *Food Chem.* **2010**, *122*, 1083–1088. [CrossRef]
18. Mandalari, G.; Vardakou, M.; Faulks, R.; Bisignano, C.; Martorana, M.; Smeriglio, A.; Trombetta, D. Food Matrix Effects of Polyphenol Bioaccessibility from Almond Skin during Simulated Human Digestion. *Nutrients* **2016**, *8*, 568. [CrossRef]
19. Mandalari, G.; Faulks, R.M.; Bisignano, C.; Waldron, K.W.; Narbad, A.; Wickham, M.S.J. In Vitro Evaluation of the Prebiotic Properties of Almond Skins (*Amygdalus communis* L.). *FEMS Microbiol. Lett.* **2010**, *304*, 116–122. [CrossRef]
20. Pasqualone, A.; Laddomada, B.; Spina, A.; Todaro, A.; Guzmàn, C.; Summo, C.; Mita, G.; Giannone, V. Almond By-Products: Extraction and Characterization of Phenolic Compounds and Evaluation of Their Potential Use in Composite Dough with Wheat Flour. *LWT* **2018**, *89*, 299–306. [CrossRef]
21. Valdés, A.; Vidal, L.; Beltrán, A.; Canals, A.; Garrigós, M.C. Microwave-Assisted Extraction of Phenolic Compounds from Almond Skin Byproducts (*Prunus amygdalus*): A Multivariate Analysis Approach. *J. Agric. Food Chem.* **2015**, *63*, 5395–5402. [CrossRef]
22. Harrison, K.; Were, L. Effect of Gamma Irradiation on Total Phenolic Content Yield and Antioxidant Capacity of Almond Skin Extracts. *Food Chem.* **2007**, *102*, 932–937. [CrossRef]
23. Gabaston, J.; Leborgne, C.; Valls, J.; Renouf, E.; Richard, T.; Waffo-Teguo, P.; Mérillon, J.-M. Subcritical Water Extraction of Stilbenes from Grapevine By-Products: A New Green Chemistry Approach. *Ind. Crops Prod.* **2018**, *126*, 272–279. [CrossRef]
24. Ruthes, A.C.; Martínez-Abad, A.; Tan, H.-T.; Bulone, V.; Vilaplana, F. Sequential Fractionation of Feruloylated Hemicelluloses and Oligosaccharides from Wheat Bran Using Subcritical Water and Xylanolytic Enzymes. *Green Chem.* **2017**, *19*, 1919–1931. [CrossRef]
25. Zhang, J.; Wen, C.; Zhang, H.; Duan, Y.; Ma, H. Recent Advances in the Extraction of Bioactive Compounds with Subcritical Water: A Review. *Trends Food Sci. Technol.* **2020**, *95*, 183–195. [CrossRef]
26. Cheng, Y.; Xue, F.; Yu, S.; Du, S.; Yang, Y. Subcritical Water Extraction of Natural Products. *Molecules* **2021**, *26*, 4004. [CrossRef]
27. Hanim, S.S.; Norsyabilah, R.; Suhaila, M.H.N.; Noraishah, A.; Kartina, A.K.S. Effects of Temperature, Time and Pressure on the Hemicelluloses Yield Extracted Using Subcritical Water Extraction. *Procedia Eng.* **2012**, *42*, 562–565. [CrossRef]
28. Erşan, S.; Güçlü Üstündağ, Ö.; Carle, R.; Schweiggert, R.M. Subcritical Water Extraction of Phenolic and Antioxidant Constituents from Pistachio (*Pistacia vera* L.) Hulls. *Food Chem.* **2018**, *253*, 46–54. [CrossRef]
29. Pinto, D.; Vieira, E.F.; Peixoto, A.F.; Freire, C.; Freitas, V.; Costa, P.; Delerue-Matos, C.; Rodrigues, F. Optimizing the Extraction of Phenolic Antioxidants from Chestnut Shells by Subcritical Water Extraction Using Response Surface Methodology. *Food Chem.* **2021**, *334*, 127521. [CrossRef]
30. Vladic, J.; Nastic, N.; Stanojkovic, T.; Zizak, Z.; Cakarevic, J.; Popovic, L.; Vidovic, S. Subcritical Water for Recovery of Polyphenols from Comfrey Root and Biological Activities of Extracts. *ACSi* **2019**, *66*, 473–783. [CrossRef]

31. Vladić, J.; Janković, T.; Živković, J.; Tomić, M.; Zdunić, G.; Šavikin, K.; Vidović, S. Comparative Study of Subcritical Water and Microwave-Assisted Extraction Techniques Impact on the Phenolic Compounds and 5-Hydroxymethylfurfural Content in Pomegranate Peel. *Plant Foods Hum. Nutr.* **2020**, *75*, 553–560. [CrossRef]
32. Requena, R.; Jiménez-Quero, A.; Vargas, M.; Moriana, R.; Chiralt, A.; Vilaplana, F. Integral Fractionation of Rice Husks into Bioactive Arabinoxylans, Cellulose Nanocrystals, and Silica Particles. *ACS Sustain. Chem. Eng.* **2019**, *7*, 6275–6286. [CrossRef]
33. Freitas, P.A.V.; González-Martínez, C.; Chiralt, A. Influence of the Cellulose Purification Process on the Properties of Aerogels Obtained from Rice Straw. *Carbohydr. Polym.* **2023**, *312*, 120805. [CrossRef] [PubMed]
34. Freitas, P.A.V.; González-Martínez, C.; Chiralt, A. Active Poly (Lactic Acid) Films with Rice Straw Aqueous Extracts for Meat Preservation Purposes. *Food Bioprocess. Technol.* **2023**, *16*, 2635–2650. [CrossRef]
35. Li, L.; Lee, S.; Lee, H.L.; Youn, H.J. Hydrogen Peroxide Bleaching of Hardwood Kraft Pulp with Adsorbed Birch Xylan and Its Effect on Paper Properties. *BioRes* **2011**, *6*, 721–736. [CrossRef]
36. Sluiter, A. *Determination of Extractives in Biomass: Laboratory Analytical Procedure (LAP)*; Technical Report; National Renewable Energy Laboratory: Golden, CO, USA, 7 July 2005; p. 12.
37. Sluiter, A. *Determination of Structural Carbohydrates and Lignin in Biomass: Laboratory Analytical Procedure (LAP)*; Technical Report, Version 07-08-2011; National Renewable Energy Laboratory: Golden, CO, USA, 2008; p. 18.
38. Freitas, P.A.V.; González-Martínez, C.; Chiralt, A. Application of Ultrasound Pre-Treatment for Enhancing Extraction of Bioactive Compounds from Rice Straw. *Foods* **2020**, *9*, 1657. [CrossRef] [PubMed]
39. Mandalari, G.; Tomaino, A.; Arcoraci, T.; Martorana, M.; Turco, V.L.; Cacciola, F.; Rich, G.T.; Bisignano, C.; Saija, A.; Dugo, P.; et al. Characterization of Polyphenols, Lipids and Dietary Fibre from Almond Skins (*Amygdalus communis* L.). *J. Food Compos. Anal.* **2010**, *23*, 166–174. [CrossRef]
40. Ouzir, M.; Bernoussi, S.E.; Tabyaoui, M.; Taghzouti, K. Almond Oil: A Comprehensive Review of Chemical Composition, Extraction Methods, Preservation Conditions, Potential Health Benefits, and Safety. *Comp. Rev. Food Sci. Food Saf.* **2021**, *20*, 3344–3387. [CrossRef]
41. Çalhan, S.D.; Meryemoğlu, B.; Eroğlu, P.; Saçlı, B.; Kalderis, D. Subcritical Water Extraction of Onosma Mutabilis: Process Optimization and Chemical Profile of the Extracts. *Molecules* **2023**, *28*, 2314. [CrossRef]
42. Yabalak, E.; Gizir, A.M. Treatment of Agrochemical Wastewater by Subcritical Water Oxidation Method: Chemical Composition and Ion Analysis of Treated and Untreated Samples. *J. Environ. Sci. Health Part A* **2020**, *55*, 1424–1435. [CrossRef]
43. Yabalak, E. Water Recovery from Textile Bath Wastewater Using Combined Subcritical Water Oxidation and Nanofiltration. *J. Clean. Prod.* **2021**, *290*, 125207. [CrossRef]
44. Spiegel, M.; Cel, K.; Sroka, Z. The Mechanistic Insights into the Role of pH and Solvent on Antiradical and Prooxidant Properties of Polyphenols—Nine Compounds Case Study. *Food Chem.* **2023**, *407*, 134677. [CrossRef]
45. Plaza, M.; Amigo-Benavent, M.; Del Castillo, M.D.; Ibáñez, E.; Herrero, M. Neoformation of Antioxidants in Glycation Model Systems Treated under Subcritical Water Extraction Conditions. *Food Res. Int.* **2010**, *43*, 1123–1129. [CrossRef]
46. Plaza, M.; Amigo-Benavent, M.; Del Castillo, M.D.; Ibáñez, E.; Herrero, M. Facts about the Formation of New Antioxidants in Natural Samples after Subcritical Water Extraction. *Food Res. Int.* **2010**, *43*, 2341–2348. [CrossRef]
47. Singh, R.D.; Nadar, C.G.; Muir, J.; Arora, A. Green and Clean Process to Obtain Low Degree of Polymerisation Xylooligosaccharides from Almond Shell. *J. Clean. Prod.* **2019**, *241*, 118237. [CrossRef]
48. Collazo-Bigliardi, S. Isolation and Characterisation of Microcrystalline Cellulose and Cellulose Nanocrystals from Coffee Husk and Comparative Study with Rice Husk. *Carbohydr. Polym.* **2018**, *11*, 205–215. [CrossRef]
49. Mansaray, G.; Ghaly, A.E. Thermal degradation of rice husks in nitrogen atmosphere. *Bioresour. Technol.* **1998**, *65*, 13–20. [CrossRef]
50. Theng, D.; Arbat, G.; Delgado-Aguilar, M.; Ngo, B.; Labonne, L.; Evon, P.; Mutjé, P. Comparison between Two Different Pretreatment Technologies of Rice Straw Fibers Prior to Fiberboard Manufacturing: Twin-Screw Extrusion and Digestion plus Defibration. *Ind. Crops Prod.* **2017**, *107*, 184–197. [CrossRef]
51. Malayil, S.; Surendran, A.N.; Kate, K.; Satyavolu, J. Impact of Acid Hydrolysis on Composition, Morphology and Xylose Recovery from Almond Biomass (Skin and Shell). *Bioresour. Technol. Rep.* **2022**, *19*, 101150. [CrossRef]
52. Mohammed, D.; Freije, A.; Abdulhussain, H.; Khonji, A.; Hasan, M.; Ferraris, C.; Gasparri, C.; Aziz Aljar, M.A.; Ali Redha, A.; Giacosa, A.; et al. Analysis of the Antioxidant Activity, Lipid Profile, and Minerals of the Skin and Seed of Hazelnuts (*Corylus avellana* L.), Pistachios (*Pistacia vera*) and Almonds (*Prunus dulcis*)—A Comparative Analysis. *Appl. Chem* **2023**, *3*, 110–118. [CrossRef]
53. Prgomet, I.; Gonçalves, B.; Domínguez-Perles, R.; Pascual-Seva, N.; Barros, A. Valorization Challenges to Almond Residues: Phytochemical Composition and Functional Application. *Molecules* **2017**, *22*, 1774. [CrossRef]
54. Abraham, K.; Gürtler, R.; Berg, K.; Heinemeyer, G.; Lampen, A.; Appel, K.E. Toxicology and Risk Assessment of 5-Hydroxymethylfurfural in Food. *Mol. Nutr. Food Res.* **2011**, *55*, 667–678. [CrossRef] [PubMed]
55. Garrido, I.; Monagas, M.; Gómez-Cordovés, C.; Bartolomé, B. Polyphenols and Antioxidant Properties of Almond Skins: Influence of Industrial Processing. *J. Food Sci.* **2008**, *73*, C106–C115. [CrossRef] [PubMed]
56. Yu, J.; Ahmedna, M.; Goktepe, I.; Dai, J. Peanut Skin Procyanidins: Composition and Antioxidant Activities as Affected by Processing. *J. Food Compos. Anal.* **2006**, *19*, 364–371. [CrossRef]
57. Lawag, I.L.; Nolden, E.S.; Schaper, A.A.M.; Lim, L.Y.; Locher, C. A Modified Folin-Ciocalteu Assay for the Determination of Total Phenolics Content in Honey. *Appl. Sci.* **2023**, *13*, 2135. [CrossRef]

58. Singleton, V.L.; Orthofer, R.; Lamuela-Raventós, R.M. Analysis of Total Phenols and Other Oxidation Substrates and Antioxidants by Means of Folin-Ciocalteu Reagent. In *Methods in Enzymology*; Elsevier: Amsterdam, The Netherlands, 1999; Volume 299, pp. 152–178; ISBN 978-0-12-182200-2.
59. Brand-Williams, W.; Cuvelier, M.E.; Berset, C. Use of a Free Radical Method to Evaluate Antioxidant Activity. *LWT—Food Sci. Technol.* **1995**, *28*, 25–30. [CrossRef]
60. Mohan, V.; Wibisono, R.; De Hoop, L.; Summers, G.; Fletcher, G.C. Identifying Suitable Listeria Innocua Strains as Surrogates for Listeria Monocytogenes for Horticultural Products. *Front. Microbiol.* **2019**, *10*, 2281. [CrossRef]
61. Moon, R.J.; Schueneman, G.T.; Simonsen, J. Overview of Cellulose Nanomaterials, Their Capabilities and Applications. *JOM* **2016**, *68*, 2383–2394. [CrossRef]
62. Potůček, F.; Milichovský, M. Kraft Pulp Bleaching with Hydrogen Peroxide and Peracetic Acid. *Chem. Pap.* **2000**, *54*, 406–411.
63. Liu, Y.; Hu, H. X-ray Diffraction Study of Bamboo Fibers Treated with NaOH. *Fibers Polym.* **2008**, *9*, 735–739. [CrossRef]
64. Yue, Y.; Zhou, C.; French, A.D.; Xia, G.; Han, G.; Wang, Q.; Wu, Q. Comparative Properties of Cellulose Nano-Crystals from Native and Mercerized Cotton Fibers. *Cellulose* **2012**, *19*, 1173–1187. [CrossRef]
65. Budtova, T.; Navard, P. Cellulose in NaOH-Water Based Solvents: A Review. *Cellulose* **2016**, *23*, 5–55. [CrossRef]

*foods*

MDPI

*Article*

# Bay Leaves Extracts as Active Additive for Food Protective Coatings

**Manuel Peña-Ortiz [1,2], Luis Serrano [2,*], Antonio A. Romero [1] and Araceli García [1,*]**

1 FQM-383 NANOVAL Group, Organic Chemistry Department, University of Córdoba, Campus de Rabanales, Marie Curie Building, Ctra. Nnal. IV-A, Km 396, 14014 Córdoba, Spain; b52penom@uco.es (M.P.-O.); qo1rorea@uco.es (A.A.R.)

2 BioPrEn Group (RNM 940), Chemical Engineering Department, Instituto Químico para la Energía y el Medioambiente (IQUEMA), Faculty of Science, University of Córdoba, 14014 Córdoba, Spain

* Correspondence: iq3secal@uco.es (L.S.); qo2ganua@uco.es (A.G.)

**Abstract:** Ethanolic extracts of bay leaves were obtained using the Soxhlet method (extraction yield of 22.3 ± 1.2%) and further analyzed through different methods, thus determining the chemical composition with gas chromatography, phenolic content with the Folin–Ciocalteu technique (11.8 ± 0.4% wt.) and antioxidant power with the radical 2,2′-Azino-bis(3-ethylbenzothiazoline-6-sulfonic acid) method (75.06%). Furthermore, its effect on the growth of two bacteria, *Escherichia coli* and *Bacillus cereus*, and on two yeasts, *Candida glabrata* and *Saccharomyces cerevisiae*, was determined, showing a minimum inhibitory concentration of 0.65 mg/mL on the growth of *B. cereus*. Finally, edible films were prepared using different polymers (carboxymethyl cellulose, gum Arabic, polyvinyl pyrrolidone, and polyvinyl alcohol) containing 0, 5, 10, or 15% wt. of bay leaf extract as troubleshooting for perishable fruits, specifically for cultivated strawberry. The prepared composites presented reduced water vapor permeabilities (up to $4.3 \times 10^{-7}$ g·Pa$^{-1}$·m$^{-1}$·h$^{-1}$), high specific transparencies (≈30%/mm), as well as the effective blocking of ultraviolet radiation (>99.9%). In vivo tests showed that the most suitable treatment for strawberry protection was the impregnation with a composite comprising polyvinyl alcohol and a 15% wt. bay leaf extract, resulting in a noteworthy reduction in mass loss (22% after 6 days). It can be asserted that food packaging with the designed composites would be an effective alternative for the reduction in postharvest losses.

**Keywords:** *Laurus nobilis*; strawberry; edible coating; antioxidant; antimicrobial

**Citation:** Peña-Ortiz, M.; Serrano, L.; Romero, A.A.; García, A. Bay Leaves Extracts as Active Additive for Food Protective Coatings. *Foods* **2023**, *12*, 3741. https://doi.org/10.3390/foods12203741

Academic Editors: Mario M. Martinez and Sílvia Petronilho

Received: 20 September 2023
Revised: 9 October 2023
Accepted: 10 October 2023
Published: 11 October 2023



## 1. Introduction

According to United Nations projections, the world's population could rise to 9.15 billion by 2050, meaning that the supply of food products will only be possible by increasing their production by more than 70% [1]. The development of molecular biology and the intensive use of fertilizers are currently seeking to meet this objective, although new approaches to minimizing postharvest losses through affordable packaging procedures are also being studied.

Fruits, one of the most perishable food groups, require proper handling since harvesting and subsequent refrigeration, which is directly associated with the problem of energy expenditure during transport and storage [2]. One of the perishable fruits par excellence, *Fragaria* × *ananassa* (commonly known as cultivated strawberry), has a shelf life of 5–7 days to ensure its quality and safety. Being the most consumed berry globally for its multiple properties and benefits, as a non-climacteric fruit, the increase in the concentration of the abscisic acid phytohormone after its harvest, in addition to its high respiration rate, determines the generation of water loss, mechanical damage, and fungal growth, meaning this short postharvest life [3].

Against these drawbacks, research has explored multiple materials with the ability to significantly extend the shelf life of strawberries while being less harmful to the environment, including their packaging and coating by using bioplastics such as polyvinyl alcohol (PVA) and chitosan [4]. Edible coatings are wrappers that form a thin, semi-permeable barrier on the surface of foods. They serve multiple purposes, enhancing barrier properties (reducing transpiration, loss of water...), microbiological safety (minimizing the risk of contamination), and structural integrity (mitigating mechanical damage from punctures and handling during storage) [5,6]. Notably, transparent coatings that preserve the original color of food are often preferred, resulting in a glossy surface that enhances the visual appeal of the foodstuff [7].

To date, edible coatings have been based on different polymers (carboxymethyl cellulose, gum Arabic...), together with other additives. As for the latter, the growing interest in the valorization of agro-industrial byproducts (whey proteins, collagen, pomegranate peel...) and other natural substances, such as aloe vera leaves, organic acids, and vegetable essential oils (EOs), has prompted research into their utilization for this purpose, seeking to use their inherent bioactive properties to enhance the multifunctionality of the designed products [8,9].

EOs are a group of generally aromatic liquids extracted from different parts of plant species that have been widely used to improve the shelf life of food products due to their rich content of bioactive molecules. These extracts have generally been obtained through their affinity for some organic solvents. In the case of the Soxhlet method, the reflux of the liquid phase ensures the contact of fresh solvent with the solid sample, which has been used routinely for the extraction of EOs from plant biomass in multiple investigations [10].

Laurel (*Laurus nobilis*), also called bay leaf, is a small evergreen tree belonging to the Lauraceae family, native to the Mediterranean area. Its EO is mainly composed of tannins, flavonoids, alkaloids, anthocyanins, traces of fats, and many minerals, among others, which provide a strong and pleasant smell to the oil, as well as biological activities such as antioxidant, antimicrobial, insect repellent, healing, immunostimulant, and anti-inflammatory [11]. For all these reasons, laurel has been cultivated extensively.

Considering all the above, we set out to use the bay leaf EO as a natural additive in different polymeric matrices (i.e., carboxymethyl cellulose, gum Arabic, polyvinylpyrrolidone, and polyvinyl alcohol) as a method to extend the shelf life of cultivated strawberry by edible coating, trying to demonstrate the potential suitability of the designed treatment for strawberries as well as other traded perishable foods.

In terms of economic feasibility, the substantial byproducts produced by *L. nobilis* during pruning (over 90% of the annual laurel production is discarded) underscore the potential for their utilization within the framework of a biorefinery process that aligns with the principles of a circular economy [12]. Furthermore, multifaceted edible coatings help to maintain the physiological and physicochemical properties of food over an extended period of time, leading to a reduction in post-harvest losses. However, the choice of the materials comprising them is crucial, since it is estimated that raw material costs contribute to 70% of the total edible coating expenditure. Thus, the availability, extraction method, and cost of additives play a key role in this food protection approach [13].

## 2. Materials and Methods

### 2.1. Materials

Bay (*L. nobilis*) leaves were collected as pruning residue and kindly provided by an independent farmer from Espejo (Córdoba, Spain). Before use, they were air-dried under controlled conditions (21 ± 2 °C, 40 ± 5% relative humidity) until constant weight (approx. 10 days) and milled in a home conventional grinder. Sample fraction under 20 mesh size was rejected. Strawberries were purchased from a local supermarket.

The chemical characterization of the samples involved the use of ethanol absolute pure (99.5%, M.W. = 46.07, Panreac, Barcelona, Spain), toluene purissimum (99.5%, M.W. = 92.14, Panreac, Spain), sulfuric acid (95–98%, USP-NF, BP, Ph. Eur., PRS-CODEX, M.W. = 98.08, Panreac, Spain), acetic acid glacial (99.7%, Reag. USP, Ph. Eur., for analysis, ACS, ISO, M.W. = 60.05, Panreac, Spain), sodium chlorite (solution 25% *w*/*w*. for synthesis, M.W. = 90.44,

Panreac, Spain), sodium hydroxide (≥97%, pellets EMPLURA®, M.W. = 40.00, Merck, Madrid, Spain), 2,2′-Azino-bis(3-ethylbenzothiazoline-6-sulfonic acid) diammonium salt (ABTS, ≥98%, M.W. = 548.68, Sigma-Aldrich, Shanghai, China), potassium persulfate (≥99%, ACS reagent, M.W. = 270.32, Sigma-Aldrich, Darmstadt, Germany), Folin and Ciocalteu's phenol reagent (2 M with respect to acid, Sigma-Aldrich, Buchs, Switzerland), gallic acid (≥98%, anhydrous for synthesis, M.W. = 170.12, Merck, Darmstadt, Germany), and sodium carbonate (≥99.5%, ACS reagent, M.W. = 105.99, Sigma-Aldrich, Germany).

Culture media were prepared using bacteriological peptone (LP0037B, Oxoid, Hampshire, UK), yeast extract (LP0021B, Oxoid, Dardilly, France), bacteriological agar (Agar No. 1, LP0011B, Oxoid, Madrid, Spain), D(+)-Glucose (Reag. Ph Eur, anhydrous for biochemistry, M.W. = 180.16, Merck, Martillac, France), and sodium chloride (pure, pharma grade, USP, BP, Ph. Eur., JP, M.W. = 58.44, Panreac, Barcelona, Spain).

For composites preparation, carboxymethyl cellulose (sodium salt, average M.W. = 250,000, DS = 0.9, Acros Organics, Illkirch-Graffenstaden, France), gum Arabic (general purpose grade, powder, Fisher Scientific, Loughborough, UK), polyvinylpyrrolidone (M.W. = 360,000, powder, Alfa Aesar, Karlsruhe, Germany), and polyvinyl alcohol (98–99% hydrolyzed, low molecular weight, Alfa Aesar, Germany) were procured.

### 2.2. Bay Leaves Extraction Process

Bay leaf extract (BLE) was obtained following the method of Rincón et al., 2019 [14]. Thus, a conventional Soxhlet apparatus was used to perform an ethanolic extraction of bay leaves, introducing 5 g of sample in cellulose thimbles and adding 200 mL of absolute ethanol, keeping the process up to 5 h. The liquid fraction was then concentrated under vacuum (rotary evaporator Laborota 4000, Heidolph Instruments GmbH&Co, Shwabach, Germany) to around 10 mL, and both the extraction yield and the final concentration were determined. The resulting oily extract was stored for further analysis. Regarding the solid fraction obtained, it was air-dried and subsequently characterized to determine the inorganic matter, extractives, acid-soluble lignin, α-cellulose, and hemicelluloses contents [15]. For this procedure, 6 replicates were made.

### 2.3. Bay Leaves Extract Characterization

The main components in BLE were analyzed on an Agilent Technologies GC–MS Instruments 5977B GC/MSD, Inc., Santa Clara, CA, USA, US equipped with an HP-5 ms Ultra Inert column (30 m × 250 μm × 0.25 μm). GC-MS spectra were obtained using the following conditions: carrier gas, He; flow rate, 1.8 mL/min; split, 1:20; injection volume, 1 μL; injection temperature, 300 °C; oven temperature, 200 °C; the ionization mode used was electronic impact at 70 eV. The identification of the components was achieved by comparing mass spectral fragmentation patterns with those stored in the data bank NIST Standard Reference Database 1Av17 (Data Version NIST v17, Software Version 2.3). The relative content (referred to as the % of the total area) of the main compounds was determined by the integration of peaks according to the method followed by Koudehi et al., 2020 [16].

Total Phenolic Content (TPC) in BLE was determined according to the Folin–Ciocalteu method previously described by García et al., 2017 [15] with slight modifications, thus using gallic acid as the standard compound (calibration curve) and ethanol as solvent. Briefly, 0.5 mL of sample was mixed with 2.5 mL of Folin–Ciocalteu reagent, 5 mL of 0.2 M $Na_2CO_3$, and distilled water up to 50 mL. After 30 min in a thermostatic bath at 40 ± 2 °C, the sample was measured in a Perkin Elmer UV/VIS Lambda 365 spectrophotometer, Inc Shelton, US at 750 nm against a reference (containing 0.5 mL of ethanol). TPC was determined by triplicate and expressed as mg of gallic acid equivalents by g of sample (mg GAE/g).

Antioxidant power (AOP) of BLE was evaluated according to the ABTS (2,2′-Azino-bis(3-ethylbenzothiazoline-6-sulphonic acid)) spectrophotometric assay at 734 nm in a Perkin Elmer UV/VIS Lambda 365 spectrophotometer [15]. Briefly, 7 mM ABTS in 2.45 mM of potassium persulfate was prepared and left overnight in darkness. For the measurement, 0.02 mL of BLE was added to 2 mL of the radical solution in 50% vol. ethanol-water (with an absorbance of 0.7 ± 0.02 at 734 nm). A blank sample (0.02 mL of ethanol) was also measured, both against the radical solution in 50% vol. ethanol-water. *AOP* calculation was done as follows:

$$AOP(\%) = \frac{A^{734}_{blank6} - A^{734}_{BLE6}}{A^{734}_{ABTS0}} \times 100$$

where ${A^{734}}_{ABTS0}$ is the absorbance of the radical solution at 0 min, and ${A^{734}}_{blank6}$, ${A^{734}}_{BLE6}$ are the absorbances after 6 min room incubation of the blank and BLE samples, respectively.

The antimicrobial activity of the BLE was evaluated against four different microbial strains provided by the Department of Microbiology of the University of Córdoba (i.e., *Escherichia coli*, *Bacillus cereus*, *Candida glabrata*, and *Saccharomyces cerevisiae*). Growth inhibition was determined following three different approaches. In the first one, the method followed by Diba et al., 2011 [17] was adapted. Thus, serial volumetric dilutions of BLE (1/20 to 1/160) were prepared on nutrient broth for bacteria (0.3% yeast extract, 0.5% peptone, 0.5% NaCl, 2% agar) and YPD broth for yeasts (1% yeast extract, 2% peptone, 2% glucose, 2% agar). Then, inoculation was carried out, incubating them at 28 ± 2 °C for *B. cereus*, *C. glabrata*, *S. cerevisiae*, and at 37 ± 2 °C for *E. coli* for a duration of 24 h. After incubation, it was qualitatively determined if there was microbial inhibition through light microscopy (Gram staining for bacteria and simple staining for yeasts). In the second approach, the disk-diffusion method originally introduced by Kirby-Bauer was followed, with minor adjustments made to the procedure described by Șachir et al., 2022 [18]. Briefly, after the inoculation of each culture medium, disks were impregnated with 50 μL of each previously prepared BLE dilution and subsequently incubated under the same conditions as in the first approach. Finally, the third test consisted of the agar-well method. To accomplish this, the procedure outlined by Chau et al., 2022 [19] was followed to assess whether the nature of the disks interfered with the diffusion capacity of BLE in the culture medium. All assays were conducted in triplicate, using culture medium lacking BLE as negative controls. The final aim was to determine the minimum inhibitory concentration (MIC), defined as the lowest concentration of the substance that can inhibit the visible growth of the microbial strain after being incubated for 24 h at its optimal temperature growth. BLE was subsequently categorized based on its MIC by using the classification method described by Aligiannis et al., 2001 [20]. Thus, BLE can be considered a strong inhibitor (MIC up to 0.5 mg/mL), a moderate inhibitor (MIC ranging from 0.6 mg/mL to 1.5 mg/mL), or a weak inhibitor (MIC above 1.6 mg/mL).

### 2.4. Edible Films Preparation and Characterization

Edible films were prepared using the solvent casting method, dissolving 0.5 g of different polymeric matrices (i.e., carboxymethyl cellulose (CMC), polyvinyl pyrrolidone (PVP), polyvinyl alcohol (PVA), and gum Arabic (GA)) in 20 mL of solvent containing 0, 5, 10 and 15% wt. of BLE respect to polymer. Water was used as the solvent for CMC, PVA, and GA films, whereas ethanol was chosen for dissolving PVP [21,22]. To reduce GA composite fragility, glycerol (10% wt. respect to polymer) was used as a plasticizer. Subsequently, casting was conducted for 72 h at 21 ± 1 °C, performing a final vacuum drying (Vacuo-Temp JPSELECTA) at 50 ± 2 °C for total solvent removal. All the films were prepared in triplicate and individually stored until their characterization, placing them in a conditioned room (25 ± 1 °C, 50% RH) for 72 h before characterization to standardize the starting conditions.

The water vapor permeability (WVP) of films was determined according to the ASTM E96/E96M-10 standard [21]. Shortly, a film square of 4 cm$^2$ was cut from each composite and attached to an aluminium adhesive tape. Then, the system was arranged in plastic jars, whose lids were previously perforated with ≈10 mm diameter holes, and finally, a desiccant material ($CaCl_2$) was added. Subsequently, jars were placed in a controlled chamber (25 ± 1 °C, 50% RH), and their weight was recorded at 0, 1, 2, 3, 4, 5, 6, 7, and

24 h. Three replicates were measured for each sample. Thus, $WVP$ ($g{\cdot}Pa^{-1}{\cdot}m^{-1}{\cdot}h^{-1}$) was calculated as:

$$WVP = \frac{G{\cdot}e}{t{\cdot}A{\cdot}P{\cdot}H}$$

where $G$ (g) is the weight gain slope in a time $t$ (s) through an area $A$ ($m^2$) of a film with a thickness $e$ (m) tested at a pressure $P$ (Pa) and specific humidity $H$ (expressed as per one). $WVP$ measurements were determined by triplicate on each composite.

Furthermore, the light transmission was investigated in a Perkin Elmer UV/VIS Lambda 365 spectrophotometer in the 200–700 nm range to evaluate the opacity and UV-light protective capacity of the prepared films, according to the method of Rincón et al., 2019 [14]. Thus, specific transparency ($ST$, %/mm) and UV-light blocking ($UV_{block}$, %) were calculated as follows:

$$ST = \frac{\log T_{600}}{e}$$

$$UV_{block} = 100{\cdot}\left(1 - \frac{T_{280}}{T_{660}}\right)$$

where $T_{600}$, $T_{660}$, and $T_{280}$ are the transmittances (%) recorded at 600, 660, and 280 nm, respectively, and $e$ (mm) represents the thickness of the tested films. These measurements were performed in triplicate for each sample.

### 2.5. Fruit Protection Tests

The application of edible coatings was performed on similar strawberries (shape, weight, and ripening stage). Each specimen was immersed in its respective emulsion (a concentration of 2.5% wt. of dissolved pure matrix or composite, i.e., of dissolved polymer and with either 0% wt. or 15% wt. of BLE). In the case of GA-based samples, a 10% wt. of glycerol was also considered in the final 2.5% wt. concentration of the prepared emulsion. The immersion process lasted for 10 s, followed by drying on Petri dishes and storage under controlled conditions (21 ± 2 °C and 40 ± 5% RH). Uncoated fruits (UF) were used as controls for comparison purposes. Then, mass loss (%) was determined through gravimetry on days 1, 2, 5, and 6, while the appearance of the samples was evaluated through photographs taken for 12 days, carrying out in both cases three replicates for each specimen.

### 2.6. Statistical Analysis

Data are shown as means ± standard deviation. To carry out the statistical analysis, IBM SPSS software version 25 for Windows was used, performing one-way ANOVA tests to determine the differences between the groups for multiple comparisons, while Tukey post-hoc tests were used to identify which groups differed. The level of significance was fixed at 5%.

## 3. Results and Discussion

### 3.1. BLE Extraction and Characterization

The ethanolic extraction of the raw material resulted in the removal of 22.3 ± 1.2% wt. of the starting components, thus producing an oily extract with a concentration of 0.104 g/mL of total dissolved solids. The obtained extraction yield was observed to be lower than previously reported for similar techniques, solvents, and raw materials [14]. This discrepancy can be attributed to various factors that can influence this value, including the climatic conditions of the collection area and the age of the plant, among others [23]. However, it is noteworthy that our extraction yield was higher when compared to other conventional extraction techniques used on BLE extraction, such as the heat-reflux method (10.34 ± 1.58%) or the microwave-assisted extraction method (10.50 ± 0.57%) [24]. This comparison highlights the substantial variability in extraction yields depending on the chosen extraction method, underscoring its crucial role in achieving optimal extraction efficiency. As shown in Figure 1, most of the extractable and inorganic matter was removed after the ethanolic extraction (around 70 and 44% of the initial content, respectively), in addition to the partial dissolution of hemicelluloses (approximately 14% of the initial content).

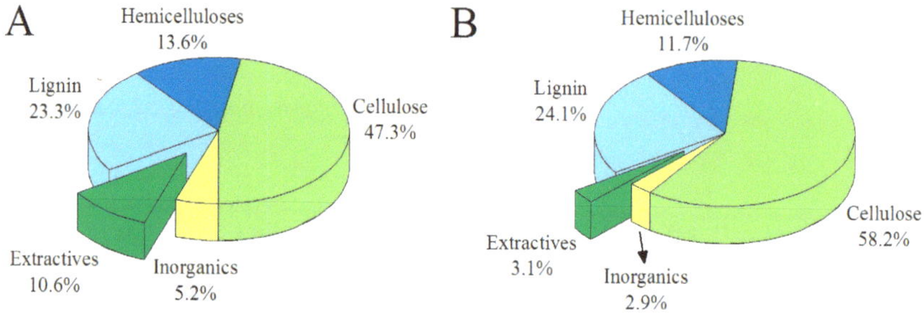


**Figure 1.** Composition (% wt. basis) of bay leaves fibers before (**A**) and after (**B**) ethanolic extraction.

The identification and relative quantification (% relative area, RA%) of the different bioactive compounds in BLE are shown in Table 1, according to the results of the performed GC/MS analysis. In general, the amount and type of bioactive compounds present in EOs depend on many factors such as the extraction method used, the geographical origin of the raw material, whether it is wild or cultivated plants, etc. In the present work, BLE had a significant presence of monoterpenoid compounds (methyl-4,6-O-benzylidene-hexopyranose), also appearing monoterpenoids (1,8-cineole, isopulegol, and α-terpineol), phenylpropanoids (methyl eugenol and elemicin), and sesquiterpenoid compounds (β-acorenol, β-caryophyllene, and cedrandiol), thus coinciding with the composition reported by other authors [25]. Moreover, it has been reported that the presence of methyl-4,6-O-benzylidene-hexopyranose, isopulegol, α-terpineol, methyl eugenol, elemicin, β-acorenol, β-caryophyllene, and cedrandiol can induce synergistic microbiocidal action against most common pathogens that threaten food quality and safety such as *Pseudomonas* spp., *Leuconostoc* spp., *Saccharomyces* spp., *Penicillium* spp., and *Fusarium* spp., among others [26].

**Table 1.** Identification and quantification of bioactive compounds present in BLE sample (RT, retention time; CD, carbohydrate derivative; ME, monoterpenoid ether; MA, monoterpenoid alcohol; Ph, phenylpropanoid; S, sesquiterpenoid; SA, sesquiterpenoid alcohol; relative area, RA %).

| RT (min) | Compound | Type | Structure | RA * (%) | Bioactivity | Refs. |
|---|---|---|---|---|---|---|
| 1.07 | Methyl-4,6-O-benzylidene-hexopyranose | CD | 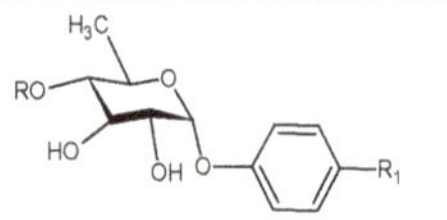 | 8.08 | Antimicrobial | [27] |
| 1.17 | 1,8-cineole | ME | 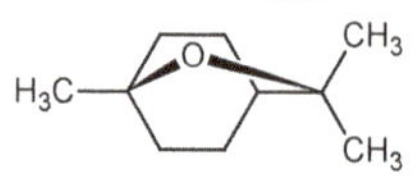 | 17.65 | Insecticidal | [28] |
| 1.28 | Isopulegol | MA | 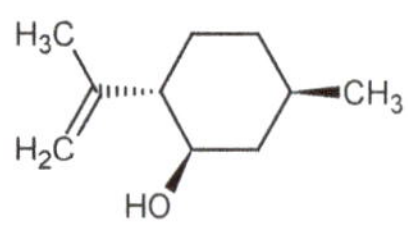 | 5.17 | Antibacterial | [29] |
| 1.45 | α-terpineol | MA | 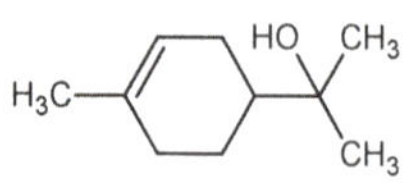 | 4.98 | Antibacterial | [30] |
| 1.53 | Methyleugenol | Ph | 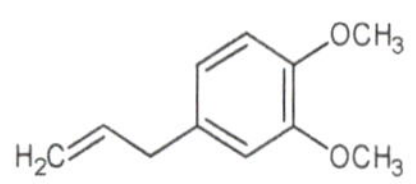 | 5.44 | Antibacterial | [30] |

**Table 1.** *Cont.*

| RT (min) | Compound | Type | Structure | RA * (%) | Bioactivity | Refs. |
|---|---|---|---|---|---|---|
| 1.88 | Elemicin | Ph | 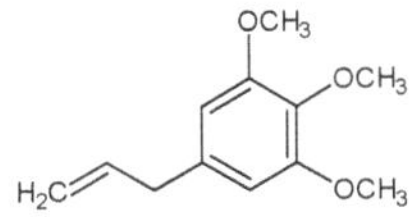 | 2.87 | Antibacterial | [31] |
| 1.65 | β-acorenol | SA | 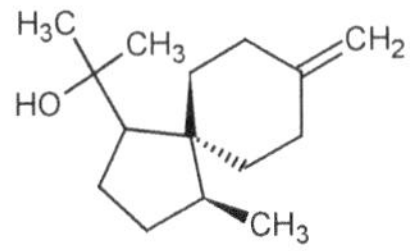 | 3.38 | Antimicrobial | [32] |
| 2.17 | β-caryophyllene | S | 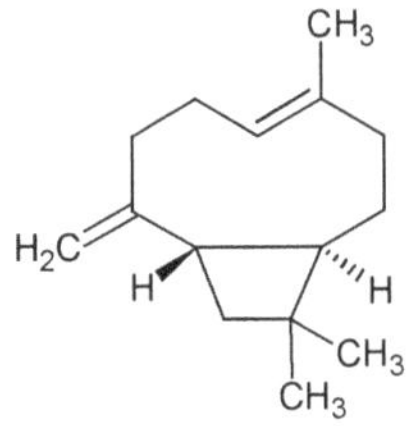 | 1.32 | Antimicrobial | [30] |
| 2.45 | Cedrandiol | SA | 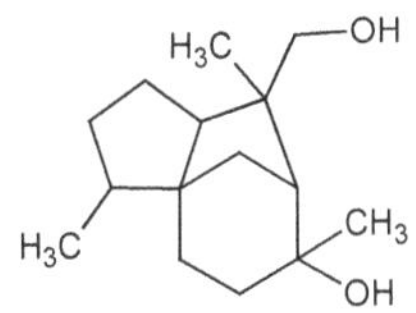 | 0.89 | Antimicrobial | [33] |

* The relative area (RA) on chromatograms obtained from the composition analysis was determined in triplicate, consistently resulting in a standard deviation of less than 5% for its value.

In terms of the TPC of BLE, the value obtained (118 ± 4 mg GAE/g) aligns well with the findings of previous research conducted on *L. nobilis* leaves. For instance, Dhifi et al., 2018 [34] reported a TPC of 174.1 ± 11.6 mg GAE/g for methanolic extracts obtained by stirring. In fact, this value is deeply influenced by both the polarity of the solvent used and the extraction method employed, playing a key role in determining the potential bioactivities associated with BLE and its applicability. The significant presence of phenolic compounds in our BLE extract suggests its potential as a sustainable additive in food coating, given their well-known properties in promoting the preservation and protection of food products.

Indeed, phenolic compounds are often correlated to the AOP due to their capability to act as electron donors in a free radical reaction [35]. In the present research, ABTS radical reduction power of 75.06% was obtained. This value was 1.6 times higher than that obtained by Muñiz-Márquez et al., 2014 [36], although they carried out a heat-reflux extraction, which also supports the conclusions reached above. The obtained results indicate a great antioxidant capacity of BLE, promising for its application in food protection against oxygen free radicals.

According to the microbial inhibition tests (Figure 2), the microscopy analysis of serial dilutions of BLE showed the tolerance of *E. coli* against the ethanolic extract, a behavior corroborated after observing the growth of the bacteria in the solid medium, where no inhibition halos were obtained in any of the two proposed tests. These results match the information cited in the literature [37], so we can rely on the nature of the outer membrane to explain them, as it probably limits the passage of some bioactive compounds present in BLE. Referring to *B. cereus*, a biocidal power of BLE was observed in all dilutions made (MIC = 0.65 mg/mL, moderate inhibitor [20]), observing both a decrease in the number of bacilli and a decrease in forming associations, thus agreeing with the results obtained by Btissam et al., 2018 [38]. This behavior was corroborated through the Kirby-Bauer and agar-well assays, observing inhibition halos in all the dilutions of the ethanolic extract tested. In this way, the results obtained agree with authors who state that Gram-positive bacteria are more sensitive than Gram-negative bacteria to aromatic plants EOs [38]. Finally, considering the results obtained with the two yeasts tested, it was found that BLE hardly caused growth inhibition in any of them. However, other studies have shown inhibition in the growth of *S. cerevisiae* with a considerably lower concentration (0.02%) of laurel EO [39], while other tests with *C. glabrata* have shown that *L. nobilis* EOs are able to inhibit its growth [40]. Nevertheless, in both investigations, they extracted aqueous extracts instead of ethanolic extracts from the plant, so in the absence of previous studies testing ethanolic extracts, our results confirm the presence of levuricidal and/or levuristatic compounds in the water-soluble fraction of bay leaves. Ultimately, these results encourage the search for other approaches to enhance the antimicrobial power of the chemical compounds present in *L. nobilis* EO by modifying the technique, time, or solvent used in the extraction to test its effectiveness against other food spoilage microorganisms.

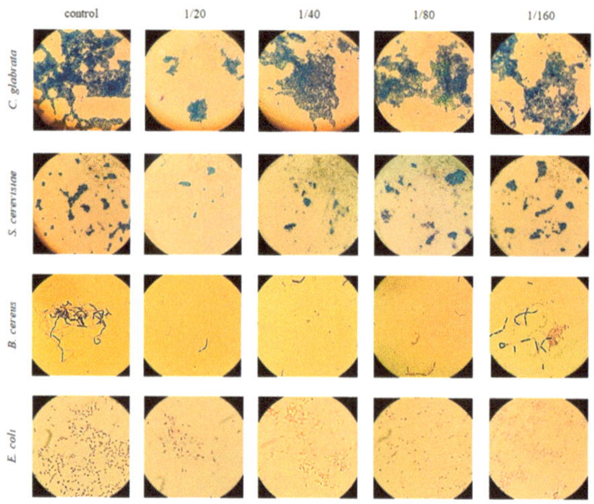


**Figure 2.** Effect of BLE concentration (left to right: dilution increasing) on the growth of the tested microorganisms (top to bottom: *C. glabrata, S. cerevisiae, B. cereus, E. coli*).

### 3.2. Properties of BLE Containing Edible Films

Depending on the food to be protected, a film must allow moisture transfer from the atmosphere to a greater or lesser extent. Considering our results for WVP (Table 2), the incorporation of BLE in CMC-based biocomposites caused a decrease in their permeabilities (between 41% and 48%, referred to the value observed for CMC0), becoming more significant as extract content increased. Regarding GA-based composites, BLE determined a reduction in the WVP of Arabic gum between 40% and 75%. Interestingly, a percolation threshold seemed to appear at GA5, thus determining an increase in its value, although from GA10 onwards the WVP again decreased considerably. The appearance of percolation thresholds in composites is generally due to a heterogeneous distribution of components on the polymer matrix. In this sense, the extremely randomized structure of GA leads to weaker interactions with BLE components. In addition to this, most of its interactions are through the -OH group, much weaker than in the case of CMC, whose hydroxyl group is more polarized by the presence of adjacent carbonyl groups. The abrupt change in WVP observed for GA10 and the great variability in its standard deviation suggest the probable presence of pores or a worse dispersion of the BLE during the preparation of these samples. For PVP-based films, BLE incorporation linearly reduced the water vapor transference through the biocomposite (22%, 43%, and 65% less when 5%, 10%, and 15% of extract was added, respectively). However, these differences were not significant ($p > 0.05$) as indicated by the statistical analysis, probably due to the great variability among the replicates. Lastly, PVA-based biocomposites were those that allowed the transfer of water vapor from the atmosphere to a lesser extent. In them, an apparent reduction in the WVP value (between 47% and 74%) was also observed when BLE was incorporated. Furthermore, a percolation threshold seemed to appear in PVA10, determining a slight increase in the property. In essence, the occupation of free spaces by BLE generated more difficult pathways for water molecules, subsequently reducing WVP [21]. Thus, an increase in the water vapor barrier is related to the content of solids and polyphenols, components that interact with the polymer and make the film matrix more compact.

**Table 2.** Barrier and optical properties of prepared composites using different matrices and %wt. of BLE.

| Composite | WVP ($g{\cdot}Pa^{-1}{\cdot}m^{-1}{\cdot}h^{-1}{\cdot}10^{-6}$) | ST (%/mm) | $UV_{block}$ (%) |
|---|---|---|---|
| CMC0 | 2.89 ± 0.19 | 29.42 ± 0.01 | 37.22 ± 2.63 |
| CMC5 | 1.68 ± 0.39 | 28.05 ± 0.03 (*) | 99.94 ± 0.00 (*) |
| CMC10 | 1.58 ± 0.30 (*) | 30.46 ± 0.00 (*) | 99.93 ± 0.00 (*) |
| CMC15 | 1.52 ± 0.32 (*) | 26.26 ± 0.01 (*) | 99.91 ± 0.00 (*) |
| GA0 | 10.53 ± 1.39 | 23.03 ± 0.01 | 8.25 ± 0.45 |
| GA5 | 2.64 ± 1.47 (*) | 28.21 ± 0.03 (*) | 99.94 ± 0.00 (*) |
| GA10 | 6.28 ± 2.83 | 22.38 ± 0.03 (*) | 99.93 ± 0.00 (*) |
| GA15 | 2.75 ± 1.63 | 31.55 ± 0.02 (*) | 99.92 ± 0.00 (*) |
| PVP0 | 5.55 ± 1.05 | 27.01 ± 0.33 | 17.63 ± 3.07 |
| PVP5 | 4.31 ± 1.16 | 25.28 ± 0.26 | 99.94± 0.00 (*) |
| PVP10 | 3.14 ± 1.21 | 27.26 ± 0.09 | 99.93 ± 0.00 (*) |
| PVP15 | 1.96 ± 0.71 | 28.99 ± 1.03 | 99.92 ± 0.00 (*) |
| PVA0 | 1.64 ± 0.30 | 39.12 ± 0.01 | 47.99 ± 5.95 |
| PVA5 | 0.86 ± 0.41 | 41.38 ± 0.01 (*) | 99.94 ± 0.00 (*) |
| PVA10 | 0.43 ± 0.20 | 37.61 ± 0.01 (*) | 99.92 ± 0.00 (*) |
| PVA15 | 0.58 ± 0.26 | 36.08 ± 0.00 (*) | 99.92 ± 0.00 (*) |

Statistical analyses were performed over each set of composites, comparing the properties of each polymeric matrix with its BLE containing set of samples. (*) denotes significant differences between control and composites using Tukey post-hoc tests ($p < 0.05$).

Given the effect of opacity on the consumer acceptability of a product, it is crucial that the prepared formulations have the highest possible ST. In our case, the addition of BLE determined significant differences in the ST of CMC, GA, and PVA-based films, determining its increase or decrease depending on the concentration and type of polymer matrix, although there was a generalized maintenance of transparency in the presence of all BLE concentrations tested, as shown in Table 2. The clearest example of this fact was found in the PVP-based composites, where the differences found with respect to the control were not statistically significant. Thus, considering the values obtained for ST, the most transparent films were those formed using PVA as the polymer matrix (T660 nm > 90%), followed by CMC, PVP, and GA, respectively. The maintenance of transparency in our composites despite the addition of BLE is in agreement with Oun et al., 2022 [41], who state that the type of polymer used, the concentration of the additive, its dispersibility, and compatibility with the polymer matrix can determine an increase or reduction in the ST, indistinctly, as has been demonstrated in other works to date, although a generalized non-modification of the property is related to a good dispersion of the additive in the polymeric matrix. On the other hand, the reduction in the ST obtained in some of the analyzed biocomposites containing BLE is justified by the presence of a lipid phase capable of scattering light, as reported by other authors [42].

Furthermore, due to long hours of customer service in supermarkets, multiple foods are continuously exposed to fluorescent light, leading to vitamins, pigments, amino acids, and fats photodegradation [43]. Considering our results for $UV_{block}$ improvement (Table 2), the addition of BLE significantly improved the property, achieving total blocking (>99.90%)

in all films, even at 5% wt. The observed increase in GA-based films was particularly remarkable, exhibiting a substantial rise from 8.25% to 99.94%. In contrast, the improvement was less evident in PVA-based composites, as in the absence of BLE they were able to block UV radiation up to 48%. According to previous research, the UV light absorption capacity of ethanolic extracts of *L. nobilis* has been specifically associated with the presence of chromophore groups in them [44]. Similarly, Moustafa et al., 2023 [45] stated that this considerable enhancement of UV blocking ability is generally caused by aromatic rings present in essential oils, making it a broad natural protector against UV radiation across the entire UV-visible spectrum.

### 3.3. Protection/Coating of Fresh Fruit

Analyzing the results obtained for mass loss (Table 3), it can be established that most of the impregnation treatments (both including or not BLE) significantly favored the maintenance of the property over time with respect to the control (46.72 ± 1.96% on day 6), an effect accentuated after 5 and 6 days of testing. Indeed, the only exception was the treatment with GA0, whose application determined a mass loss even slightly higher than the trials in the absence of coating (48.01 ± 1.64% on day 6). This behavior can be explained due to the nature of the polymer, which provides an extra carbon source and determines a greater susceptibility to being attacked by microorganisms that cause fruit spoilage such as *Alicyclobacillus acidoterrestris* or *Byssochlamys nivea* [46]. On the other hand, strawberries impregnated with both CMC and PVP presented a mass loss similar to each other (32.17 ± 2.34% and 29.13 ± 0.36%, respectively, after 6 days), being in any case significantly lower than that obtained for UF. Finally, strawberries impregnated with the PVA15 biocomposite showed the least variation in mass over time (22.13 ± 0.77% on day 6). These results agree with those obtained for WVP, confirming that the most permeable coating material is the one that further determines the greatest mass loss of strawberries, while the least permeable material is the one that has managed to maintain the mass of the strawberries tested to a greater extent. In this regard, a Pearson correlation coefficient of 0.85 was found, indicating a positive and direct relationship between these two parameters. Furthermore, the results obtained in relation to the addition of BLE agree with the conclusions reached after its characterization, since the long-term mass loss value is lower in the biocomposites containing it with respect to those lacking BLE, except in the case of impregnation with CMC, where the values obtained were similar to each other (≈30%).

**Table 3.** Mass loss (%) over time in uncoated (UF), coated lacking BLE (0% wt.) and coated containing BLE (15% wt.) strawberries.

| Coating | Mass Loss (%) | | | |
|---|---|---|---|---|
| | Day 1 | Day 2 | Day 5 | Day 6 |
| UF | 5.22 ± 0.37 | 11.23 ± 0.97 | 34.54 ± 1.78 | 46.72 ± 1.96 |
| CMC0 | 3.64 ± 0.43 (*) | 10.45 ± 0.38 | 23.38 ± 1.75 (*) | 32.17 ± 2.34 (*) |
| CMC15 | 3.32 ± 0.23 | 9.42 ± 0.18 | 22.11 ± 0.29 (*) | 27.82 ± 0.41 (*) |
| GA0 | 5.02 ± 0.40 | 10.04 ± 0.42 | 36.89 ± 1.89 | 48.01 ± 1.64 |
| GA15 | 3.48 ± 0.54 (*) | 8.07 ± 0.31 (*) | 27.23 ± 1.20 (*) | 37.21 ± 0.91 (*) |
| PVP0 | 4.67 ± 0.16 | 10.76 ± 0.51 | 25.30 ± 0.67 (*) | 29.13 ± 0.36 (*) |
| PVP15 | 5.65 ± 0.47 | 11.12 ± 0.66 | 26.60 ± 0.31 (*) | 30.39 ± 0.17 (*) |
| PVA0 | 2.52 ± 0.56 (*) | 6.05 ± 0.88 (*) | 18.92 ± 0.54 (*) | 26.69 ± 0.48 (*) |
| PVA15 | 3.87 ± 0.22 | 8.25 ± 0.05 (*) | 19.60 ± 0.19 (*) | 22.13 ± 0.77 (*) |

Statistical analysis was performed comparing the mass loss of uncoated fruit with those coated with polymer/composite containing BLE. (*) denotes significant differences between the behavior of UF and coated strawberries using Tukey post-hoc tests ($p < 0.05$).

The results obtained for mass loss agree with the appearance observed in each of the specimens (Figure 3). Indeed, the pattern of biodegradation over time showed that strawberries impregnated with any biocomposite (except GA0) presented a much better appearance than UF, thus demonstrating the suitability of their application. In particular, the best-preserved ones both in the short and long term were those coated with PVA (regardless of the presence of BLE), followed by those coated with PVP15. Furthermore, the photographs taken showed that the presence of BLE in composites resulted in better maintenance of the mass of the specimens while preserving, as was also indicated by the mass loss values. Finally, as the values obtained for ST showed, it could be observed that the impregnation treatments did not greatly affect the appearance (color) of the strawberries despite containing BLE, which would not pose a problem in terms of potential consumer acceptability on the market.

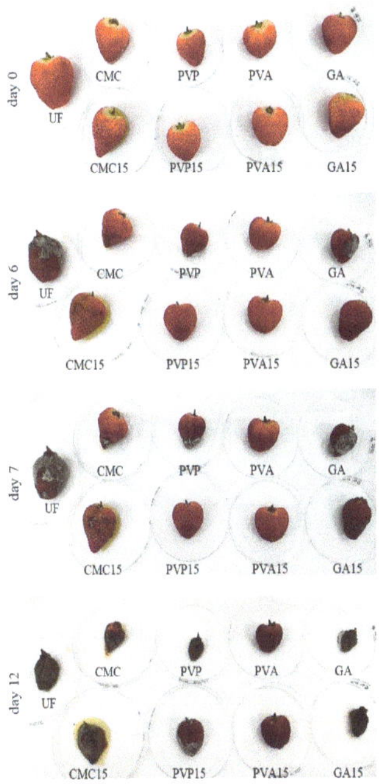


**Figure 3.** Biodegradation process under controlled conditions (21 ± 2 °C, 40 ± 5% RH) of both uncoated and coated fruits containing 15% wt. of BLE.

## 4. Conclusions

In this work, a set of edible strawberry coatings based on the composites formed by different polymeric matrices and the ethanolic fraction of *L. nobilis* leaves (BLE) were designed. The identification of several bioactive chemical compounds in BLE justified its moderate inhibition against *B. cereus* during in vitro tests. Furthermore, the addition of this extract into polymeric matrices allowed to block almost completely the UV radiation (>99.9%) while maintaining the transparency of the composites. Also, it determined an improvement in fruit moisture protection during the coating assays (overall decrease in WVP between 47% and 75% compared to the pure matrix). This maintained a better appearance of coated strawberries over time in CMC15, GA15, PVP15, and PVA15. Overall, it can be concluded that the prepared composites are suitable for the protection of fresh fruit. It is expected to continue research on this subject by evaluating other properties such as organoleptic features (odor, flavor...), additional barrier properties ($O_2$ and $CO_2$ permeabilities...), thermomechanical characteristics (DMA analysis, burst test...), as well as their applicability in other food products.

**Author Contributions:** Conceptualization, A.G. and L.S.; methodology, A.G. and A.A.R.; formal analysis, M.P.-O. and A.G.; investigation, M.P.-O.; data curation, M.P.-O. and A.G.; writing—original draft preparation, M.P.-O. and A.G.; writing—review and editing, M.P.-O., L.S., A.A.R. and A.G.; supervision, A.G. and L.S.; funding acquisition, L.S. and A.A.R. All authors have read and agreed to the published version of the manuscript.

**Funding:** This research was funded by Universidad de Córdoba (Contrato Predoctoral Plan Propio de Investigación UCO 2021).

**Data Availability Statement:** The data used to support the findings of this study can be made available by contacting the corresponding author.

**Acknowledgments:** The authors would like to thank Universidad de Córdoba for the financial support during this work (Contrato Predoctoral Plan Propio de Investigación UCO 2021).

**Conflicts of Interest:** The authors declare no conflict of interest.

## References


1. Hosseini, H.; Jafari, S.M. Introducing nano/microencapsulated bioactive ingredients for extending the shelf-life of food products. *Adv. Colloid Interfac.* **2020**, *282*, 102210. [CrossRef] [PubMed]
2. Trinh, B.M.; Smith, M.; Mekonnen, T.H. A nanomaterial-stabilized starch-beeswax Pickering emulsion coating to extend produce shelf-life. *Chem. Eng. J.* **2022**, *431*, 133905. [CrossRef]
3. Pott, D.M.; de Abreu e Lima, F.; Soria, C.; Willmitzer, L.; Fernie, A.R.; Nikoloski, Z.; Osorio, S.; Vallarino, J.G. Metabolic reconfiguration of strawberry physiology in response to postharvest practices. *Food Chem.* **2020**, *321*, 126747. [CrossRef]
4. Al-Asmar, A.; Giosafatto, C.V.L.; Sabbah, M.; Sánchez, A.; Santana, R.V.; Mariniello, L. Effect of Mesoporous Silica Nanoparticles on The Physicochemical Properties of Pectin Packaging Material for Strawberry Wrapping. *Nanomaterials* **2020**, *10*, 52. [CrossRef]
5. Salehi, F. Edible Coating of Fruits and Vegetables Using Natural Gums: A Review. *Int. J. Fruit Sci.* **2020**, *20*, S570–S589. [CrossRef]
6. Bamdad, F.; Goli, A.H.; Kadivar, M. Preparation and Characterization of Proteinous Film from Lentil (*Lens culinaris*): Edible Film from Lentil (*Lens culinaris*). *Food Res. Int.* **2006**, *39*, 106–111. [CrossRef]
7. Youseftabar-Miri, N.; Sedaghat, N.; Khoshnoudi-Nia, S. Effect of Active Edible Coating on Quality Properties of Green-Raisin and Ranking the Samples Using Fuzzy Approach. *J. Food Meas. Charact.* **2021**, *15*, 46–58. [CrossRef]
8. Yousuf, B.; Wu, S.; Siddiqui, M.W. Incorporating essential oils or compounds derived thereof into edible coatings: Effect on quality and shelf life of fresh/fresh-cut produce. *Trends Food Sci. Tech.* **2021**, *108*, 245–257. [CrossRef]
9. Chiralt, A.; Menzel, C.; Hernandez-García, E.; Collazo, S.; González-Martínez, C. Use of By-Products in Edible Coatings and Biodegradable Packaging Materials for Food Preservation. In *Sustainability of the Food System: Sovereignty, Waste, and Nutrients Bioavailability*; Betoret, N., Betoret, E., Eds.; Academic Press: Cambridge, MA, USA, 2020; pp. 101–127. [CrossRef]
10. Gao, X.; Lv, S.; Wu, Y.; Li, J.; Zhang, W.; Meng, W.; Wang, C.; Meng, Q. Volatile Components of Essential Oils Extracted from Pu-Erh Ripe Tea by Different Extraction Methods. *Int. J. Food Prop.* **2017**, *20*, S240–S253. [CrossRef]
11. Batool, S.; Khera, R.A.; Hanif, M.A.; Ayub, M.A. Bay Leaf. In *Medicinal Plants of South Asia*; Hanif, M.A., Khan, M.M., Nawaz, H., Byrne, H.J., Eds.; Elsevier: Amsterdam, The Netherlands, 2020; pp. 63–74. [CrossRef]
12. Rincón, E.; Zuliani, A.; Jiménez-Quero, A.; Vilaplana, F.; Luque, R.; Serrano, L.; Balu, A.M. Combined Extraction/Purification-Catalytic Microwave-Assisted Conversion of *Laurus nobilis* L. Pruning Waste Polysaccharides into Methyl Levulinate. *ACS Sustain. Chem. Eng.* **2020**, *8*, 11016–11023. [CrossRef]
13. Yadav, A.; Kumar, N.; Upadhyay, A.; Sethi, S.; Singh, A. Edible Coating as Postharvest Management Strategy for Shelf-Life Extension of Fresh Tomato (*Solanum lycopersicum* L.): An Overview. *J. Food Sci.* **2022**, *87*, 2256–2290. [CrossRef] [PubMed]
14. Rincón, E.; Serrano, L.; Balu, A.M.; Aguilar, J.J.; Luque, R.; García, A. Effect of bay leaves essential oil concentration on the properties of biodegradable carboxymethyl cellulose-based edible films. *Materials* **2019**, *12*, 2356. [CrossRef]
15. García, A.; Spigno, G.; Labidi, J. Antioxidant and biocide behaviour of lignin fractions from apple tree pruning residues. *Ind. Crops Prod.* **2017**, *104*, 242–252. [CrossRef]
16. Koudehi, M.F.; Ardalan, A.A.; Zibaseresht, R. Chemical constituents of an Iranian grown Capsicum annuum and their cytotoxic activities evaluation. *Org. Med. Chem. Int. J.* **2020**, *9*, 148–154. [CrossRef]
17. Diba, K.; Shoar, M.G.; Shabatkhori, M.; Khorshivand, Z. Antifungal activity of alcoholic extract of *Peganum harmala* seeds. *J. Med. Plant Res.* **2011**, *5*, 5550–5554.
18. Şachir, E.E.; Puşcaşu, C.G.; Caraiane, A.; Raftu, G.; Badea, F.C.; Mociu, M.; Albu, C.M.; Sachelarie, L.; Liliana Hurjui, L.; Bartok-Nicolae, C. Studies Regarding the Antibacterial Effect of Plant Extracts Obtained from *Epilobium parviflorum* Schreb. *App. Sci.* **2022**, *12*, 2751. [CrossRef]
19. Chau, T.P.; Veeraragavan, G.R.; Narayanan, M.; Chinnathambi, A.; Alharbi, S.A.; Subramani, B.; Brindhadevi, K.; Pimpimon, T.; Pikulkaew, S. Green synthesis of Zirconium nanoparticles using *Punica granatum* (pomegranate) peel extract and their antimicrobial and antioxidant potency. *Environ. Res.* **2022**, *209*, 112771. [CrossRef]
20. Aligiannis, N.; Kalpoutzakis, E.; Mitaku, S.; Chinou, I.B. Composition and Antimicrobial Activity of the Essential Oils of Two Origanum Species. *J. Agric. Food. Chem.* **2001**, *49*, 4168–4170. [CrossRef]
21. Jannatyha, N.; Shojaee-Aliabadi, S.; Moslehishad, M.; Moradi, E. Comparing mechanical, barrier and antimicrobial properties of nanocellulose/CMC and nanochitosan/CMC composite films. *Int. J. Biol. Macromol.* **2020**, *164*, 2323–2328. [CrossRef]

22. Vecchi, C.F.; Cesar, G.B.; Souza, P.R.; de Caetano, W.; Bruschi, M.L. Mucoadhesive polymeric films comprising polyvinyl alcohol, polyvinylpyrrolidone, and poloxamer 407 for pharmaceutical applications. *Pharm. Dev. Technol.* **2020**, *26*, 138–149. [CrossRef]
23. Mehalaine, S.; Chenchouni, H. Quantifying how climatic factors influence essential oil yield in wild-growing plants. *Arab. J. Geosci.* **2021**, *14*, 1257. [CrossRef]
24. Muñiz-Márquez, D.B.; Wong-Paz, J.E.; Contreras-Esquivel, J.C.; Rodríguez-Herrera, R.; Aguilar, C.N. Bioactive compounds from bay leaves (*Laurus nobilis*) extracted by microwave technology. *Z. Naturforschung C J. Biosci.* **2018**, *73*, 401–407. [CrossRef] [PubMed]
25. Caputo, L.; Nazzaro, F.; Souza, L.F.; Aliberti, L.; De Martino, L.; Fratianni, F.; Coppola, R.; De Feo, V. *Laurus nobilis*: Composition of essential oil and its biological activities. *Molecules* **2017**, *22*, 930. [CrossRef] [PubMed]
26. Rawat, S. Food spoilage: Microorganisms and their prevention. *Asian J. Plant Sci. Res.* **2015**, *5*, 47–56.
27. Kawsar, S.M.; Kabir, A.K.; Bhuiyan, M.M.; Manik, M.M. Synthesis and Characterization of Methyl 4, 6-O-Enzylidene-α-D-Glucopyranoside Derivatives. *J. Bangladesh Chem. Soc.* **2012**, *25*, 101–109. [CrossRef]
28. Rozman, V.; Kalinovic, I.; Korunic, Z. Toxicity of naturally occurring compounds of Lamiaceae and Lauraceae to three stored-product insects. *J. Stored Prod. Res.* **2007**, *43*, 349–355. [CrossRef]
29. Liu, T.T.; Yang, T.S. Antimicrobial impact of the components of essential oil of *Litsea cubeba* from Taiwan and antimicrobial activity of the oil in food systems. *Int. J. Food Microbiol.* **2012**, *156*, 68–75. [CrossRef]
30. Oriola, A.O.; Oyedeji, A.O. Essential Oils and Their Compounds as Potential Anti-Influenza Agents. *Molecules* **2022**, *27*, 7797. [CrossRef]
31. Rossi, P.G.; Bao, L.; Luciani, A.; Panighi, J.; Desjobert, J.M.; Costa, J.; Casanova, J.; Bolla, J.M.; Berti, L. (E)-methylisoeugenol and elemicin: Antibacterial components of *Daucus carota L.* essential oil against *Campylobacter jejuni*. *J. Agric. Food Chem.* **2007**, *55*, 7332–7336. [CrossRef]
32. Rajani, P.; Rajasekaran, C.; Vasanthakumari, M.M.; Olsson, S.B.; Ravikanth, G.; Uma Shaanker, R. Inhibition of plant pathogenic fungi by endophytic *Trichoderma* spp. through mycoparasitism and volatile organic compounds. *Microbiol. Res.* **2021**, *242*, 126595. [CrossRef]
33. Valadbeigi, T.; Rashki, S. GC-MS analysis and anticancer effect against MCF-7 and HT-29 cell lines and antioxidant, antimicrobial and wound healing activities of plant-derived. *J. Basic Res. Med. Sci.* **2015**, *2*, 1–11.
34. Dhifi, W.; Bellili, S.; Jazi, S.; Nasr, S.B.; El Beyrouthy, M.; Mnif, W. Phytochemical composition and antioxidant activity of Tunisian *Laurus nobilis*. *Pak. J. Pharm. Sci.* **2018**, *31*, 2397–2402. [PubMed]
35. Jeyakani, M.; Rajalakshmi, M. Antioxidant activity, total phenolic content of essential oils, and extract determined from natural leaves, in-vitro leaves, and callus sources of *Melaleuca alternifolia*-a comparative study. *Asian J. Pharm. Clin. Res.* **2021**, *14*, 141–143. [CrossRef]
36. Muñiz-Márquez, D.B.; Rodríguez, R.; Balagurusamy, N.; Carrillo, M.L.; Belmares, R.; Contreras, J.C.; Nevárez, G.V.; Aguilar, C.N. Phenolic content and antioxidant capacity of extracts of *Laurus nobilis* L., *Coriandrum sativum* L. and *Amaranthus hybridus* L. *CyTA—J. Food* **2014**, *12*, 271–276. [CrossRef]
37. Ramos, C.; Teixeira, B.; Batista, I.; Matos, O.; Serrano, C.; Neng, N.R.; Nogueira, J.M.F.; Nunes, M.L.; Marques, A. Antioxidant and antibacterial activity of essential oil and extracts of bay laurel *Laurus nobilis Linnaeus* (Lauraceae) from Portugal. *Nat. Prod. Res.* **2012**, *26*, 518–529. [CrossRef]
38. Btissam, R.; Fatima, E.; Mohamed, N. In vitro Study of Antibacterial Activity of Hydro-Alcohol Moroccan Plants Extracts. *Pharmacogn. J.* **2018**, *10*, 519–526. [CrossRef]
39. Erkmen, O.; Özcan, M.M. Antimicrobial effects of Turkish propolis, pollen, and laurel on spoilage and pathogenic food-related microorganisms. *J. Med. Food* **2008**, *11*, 587–592. [CrossRef]
40. Peixoto, L.R.; Rosalen, P.L.; Ferreira, G.L.S.; Freires, I.A.; de Carvalho, F.G.; Castellano, L.R.; de Castro, R.D. Antifungal activity, mode of action and anti-biofilm effects of *Laurus nobilis Linnaeus* essential oil against *Candida* spp. *Arch. Oral Biol.* **2017**, *73*, 179–185. [CrossRef] [PubMed]
41. Oun, A.A.; Shin, G.H.; Kim, J.T. Multifunctional poly(vinyl alcohol) films using cellulose nanocrystals/oregano and cellulose nanocrystals/cinnamon Pickering emulsions: Effect of oil type and concentration. *Int. J. Biol. Macromol.* **2022**, *194*, 736–745. [CrossRef] [PubMed]
42. Cano, A.; Cháfer, M.; Chiralt, A.; González-Martínez, C. Physical and Antimicrobial Properties of Starch-PVA Blend Films as Affected by the Incorporation of Natural Antimicrobial Agents. *Foods* **2015**, *5*, 3. [CrossRef]
43. Hashimoto, A.; Sakamoto, K. UV-blocking film for food storage using titanium dioxide. *Food Sci. Technol. Res.* **2011**, *17*, 199–202. [CrossRef]
44. Fernández, N.J.; Damiani, N.; Podaza, E.A.; Martucci, J.F.; Fasce, D.; Quiroz, F.; Meretta, P.E.; Quintana, S.; Eguaras, M.J.; Gende, L.B. *Laurus nobilis L.* extracts against *Paenibacillus larvae*: Antimicrobial activity, antioxidant capacity, hygienic behavior and colony strength. *Saudi J. Biol. Sci.* **2019**, *26*, 906–912. [CrossRef]

45. Moustafa, H.; El-Sayed, S.M.; Youssef, A.M. Synergistic impact of cumin essential oil on enhancing of UV-blocking and antibacterial activity of biodegradable poly(butylene adipate-co-terephthalate)/clay platelets nanocomposites. *J. Thermoplast. Compos.* **2023**, *36*, 96–117. [CrossRef]
46. Hindi, S.S.Z.; Albureikan, M.O.; Al-Ghamdi, A.A.; Alhummiany, H.; Ansari, M.S. Synthesis, characterization and biodegradation of gum Arabic-based bioplastic membranes. *Nanosci. Nanotechnol. Res.* **2017**, *4*, 32–42. [CrossRef]

foods MDPI

*Article*

# Effect of Biobased Cling Films on Cheese Quality: Color and Aroma Analysis for Sustainable Food Packaging

**Chiara Chirilli and Luisa Torri ***

University of Gastronomic Sciences, Piazza Vittorio Emanuele II 9, 12042 Pollenzo, Italy; c.chirilli@unisg.it
* Correspondence: l.torri@unisg.it; Tel.: +39-0172-458509

**Abstract:** Biobased and biodegradable polymeric materials are a sustainable alternative to the conventional plastics used in food packaging. This study investigated the possible effect of biobased cling films derived from renewable and circular and sustainable sources on key cheese sensory parameters (appearance and odor) able to influence consumer acceptance or rejection of a food product over time. For this purpose, a semi-hard cheese was selected as food model and stored for 14 days at 5 °C wrapped with five cling films: two bio-plastic materials from renewable circular and sustainable sources (R-BP1 and R-BP2), one bio-plastic film from a non-renewable source (NR-BP), and two conventional cling films (LDPE and PVC). Three analytical approaches (image analysis, electronic nose, and sensory test) were applied to evaluate the variation and the acceptability in terms of appearance and odor of the cheese. In preserving cheese color, the R-BP1 and RBP2 films were comparable to LDPE film, while NR-BP film was comparable to PVC film. In terms of odor preservation, R-BP2 film was comparable to LDPE and PVC. The consumer test showed that appearance and odor scores were higher for cheeses stored in R-BP1 and R-BP2 films than NR-BP film. Moreover, in terms of odor, R-BP1 film performed better than conventional films. This study shows how biodegradable cling films from renewable circular and sustainable resources could have comparable performance to conventional plastics (LDPE and PVC) used in the food sector.

**Keywords:** sustainable packaging; biobased material; electronic nose analysis; image analysis; consumer perception

**Citation:** Chirilli, C.; Torri, L. Effect of Biobased Cling Films on Cheese Quality: Color and Aroma Analysis for Sustainable Food Packaging. *Foods* **2023**, *12*, 3672. https://doi.org/10.3390/foods12193672

Academic Editors: Mario M. Martinez and Sílvia Petronilho

Received: 19 September 2023
Revised: 27 September 2023
Accepted: 28 September 2023
Published: 6 October 2023



## 1. Introduction

The environmental concern has promoted the urgency of developing sustainable materials for biodegradable and/or compostable food packaging [1,2]. Indeed, there is a growing interest in packaging made from renewable natural resources, and biobased polymers may be a viable substitute for synthetic plastics due to their ability to biodegrade and the availability of raw materials [3,4], as evidenced by the study by Chirilli et al. [5] who investigated how aware and attentive consumers were to concepts related to the environmental sustainability of food packaging.

Biobased plastics made from renewable resources are supposed to be less environmentally impactful and may represent a great opportunity in the food packaging industry. Reducing and repurposing waste materials, such as industrial by-products and secondary flows from industries, agricultural wastes, food wastes, and lignocellulosic materials, and using them as raw materials can help develop a sustainable circular bioeconomy system that would reduce the use of nonrenewable resources by promoting an environmentally friendly resource flow instead [6]. Therefore, the design and development of sustainable packaging is currently focusing not only on improving the performance of biopolymers and synthesizing new biobased and biodegradable materials but also on finding renewable sources, particularly undervalued raw materials, and implementing material reuse and recycling systems [7].

Among the biodegradable biopolymers that have found greater application in food packaging are cellulose, starch, polylactide (PLA) (a synthetic polyester produced from lactic acid) and polyhydroxyalkanoates (PHAs) (polyesters produced by microorganisms from carbohydrates or lipids), chitosan, soy protein isolate, whey protein isolate, and gluten [8]. Moreover, biobased materials are capable, like conventional packaging, of ensuring food safety and sensory quality by protecting food from moisture, oxygen, lipid oxidation, and loss of volatile components [9–11]. Overall, there are multiple benefits of using biobased materials manufactured from renewable and recyclable sources, as they reduce environmental impact by reducing carbon dioxide emissions, increase consumers' environmental awareness, preserve barrier properties, ensure adequate food shelf life, and increase the sustainability of the products for which they are used [12]. Despite the various positive contributions that these materials make, packaging still mainly consists of polymers such PE, PET, PP, PS, PVC, and PA [13], probably because the latter offer higher performance, better processing, and lower overall costs [14]. Indeed, to prevent the occurrence of enzymatic, chemical, physical, and microbiological changes that result in spoilage reactions in food, appropriate packaging materials and methods should be chosen [15]. Biobased packaging materials must be able to maintain their barrier and mechanical properties in the condition of use and fulfill their preservation role until disposal [16].

Among the most commonly used types of food packaging materials, cling films (or cling wraps) play an important role. These films are generally disposable, which results in the production of large volumes of waste with a considerable effect on the environment. Cling films originated as thin, transparent plastic films employed to wrap food and maintain it for a given storage period. Because traditional synthetic polymers used for packaging are not biodegradable and have a negative impact on the environment, there are many studies on the development of cling films based on natural polymers, generally obtained by combining polysaccharides, including cellulose, chitosan, starch, pectin, alginate, and carrageenan, with lipids or proteins [17,18]. They are fully degraded by microorganisms to water, carbon dioxide, and biomass [19], and they are able to protect food from microbial spoilage, allow adequate storage, and maintain suitable barriers to oxygen and water vapor and good mechanical properties [2]. Piscopo et al. [20] studied a multilayer film composed of PLA coated with a silicon oxide barrier as a method of cheese packaging that proved to be a viable alternative to conventional plastic multilayer films, while Bharathi et al. [21] developed a biodegradable zein-based cling film and obtained good results comparing it with films consisting of synthetic polymeric materials.

In this context, it is therefore important to study solutions that can promote the use of biodegradable cling films for food packaging in order to reduce the production and consumption of disposable materials that have a negative impact on the environment. At the same time, it is essential to verify the consumers' sensory acceptance for the food products wrapped with biodegradable cling films. Indeed, several studies [22–25] revealed that sensory appeal is a key factor influencing consumer perception, purchasing behavior, and satisfaction. In particular, appearance and odor are the first components of sensory appeal considered for the acceptability of a product, before tasting it. In fact, food appearance, influenced by surface color, is the first sensation that consumers perceive and take into account to accept or reject a food [23]. After appearance, odor is the second sensory parameter considered, because not only it helps to understand whether changes have occurred in the product, but also because if the orthonasal olfactory perception is unacceptable, very rarely will the product be tasted by consumers [24]. Therefore, the aim of this work was to evaluate how experimental biodegradable and compostable cling films obtained from renewable sources may have influenced the appearance and odor of cheese and its acceptability during storage, in comparison with biodegradable and compostable cling films obtained from non-renewable sources and those made of the most commercially used synthetic polymers (PVC and LDPE). Thus, the variations in appearance and odor of cheese packaged in the different biobased materials and fossil-based plastics were estimated during storage by combining instrumental techniques (image analysis and electronic nose) and sensory methods (consumer acceptability testing).

## 2. Materials and Methods

### 2.1. Cling Films

Five samples of cling films for food applications were analyzed. In particular, two innovative and experimental bio-plastic materials obtained from renewable sources (R-BP1 and R-BP2) were compared to one commercial bio-plastic film obtained from non-renewable source (NR-BP) and two commercial and conventional cling films (LDPE and PVC). The physical-chemical properties of the cling films are reported in Table 1. R-BP1 and R-BP2 were developed within the context of the "PRIME—Innovative Processes and Products of Green Chemistry" research project (https://www.unisg.it/en/ricerca/prime-innovative-processes-and-products-of-green-chemistry/, accessed on 28 August 2023), while NR-BP, LDPE, and PVC cling films are the commercial references selected as a benchmark and provided by the PRIME research project coordinators. The NR-BP, R-BP1, and R-BP2 cling films consist of biodegradable and compostable biopolyesters synthesized via polycondensation of aliphatic and/or aromatic dicarboxylic acids and diols. In particular, under the PRIME project and following the principles of the circular economy, the two innovative and experimental bio-plastic cling films from renewable sources are based on acids obtained from vegetable oils recovered from waste biomass present in Piedmont. Specifically, the biobased formulations R-BP1 and R-BP2 are composed of two polyesters synthesized from the biobased monomers azelaic acid and butanediol. The two cling films differ in the percentage of the monomers. Moreover, R-BP1 has a higher melting point than R-BP2 in order to ensure greater stability during the filming process, while R-BP2 has a lower crystallization temperature in order to optimize clinging capacity.

**Table 1.** Properties of the cling films used.

| Code | From Renewable and Cicrular Source | Biodegradable and Compostable Material | Thickness (mm) | WVTR [a] (ASTM F1249, 23 °C, 50% RH, g $m^{-2}$ 24 $h^{-1}$) | $O_2$TR (ASTM D3985, 23 °C, 50% RH, $cm^3$ $m^{-2}$ 24 $h^{-1}$ $bar^{-1}$) | Tensile Properties (ASTM D882, 23 °C, 50% RH, $V_0$ 50 mm/min) | | | | | Puncture Test (ASTM D5748, 23 °C, 50% RH- $V_0$ 250 mm/min) | | |
|---|---|---|---|---|---|---|---|---|---|---|---|---|---|
| | | | | | | Direction | sb (Mpa) | eb (%) | E (Mpa) | Enb (kJ/$m^2$) | Fmax (N) | Enb (mJ) | Stroke @Fmax (mm) |
| R-BP1 | Yes | Yes | 20 ± 1 | 77 ± 8 | 5050 ± 837 | MD | 53 ± 2 | 343 ± 16 | 123 ± 15 | 5649 ± 349 | 31 ± 2 | 89 ± 3 | 90 ± 3 |
| | | | | | | TD | 36 ± 3 | 649 ± 39 | 125 ± 12 | 5038 ± 298 | | | |
| R-BP2 | Yes | Yes | 20 ± 3 | 90 ± 11 | 6235 ± 947 | MD | 42 ± 2 | 356 ± 17 | 87 ± 8 | 4923 ± 320 | 29 ± 3 | 84 ± 2 | 102 ± 4 |
| | | | | | | TD | 30 ± 2 | 625 ± 35 | 94 ± 7 | 4476 ± 365 | | | |
| NR-BP | No | Yes | 20 ± 3 | 125 ± 13 | 5022 ± 1066 | MD | 66 ± 3 | 322 ± 15 | 119 ± 10 | 5679 ± 371 | 36 ± 3 | 121 ± 3 | 98 ± 3 |
| | | | | | | TD | 53 ± 4 | 593 ± 42 | 158 ± 17 | 5752 ± 496 | | | |
| PVC | No | No | 20 ± 1 | 39 ± 5 | 3253 ± 554 | MD | 26 ± 1 | 163 ± 7 | 76 ± 6 | 1390 ± 71 | 45 ± 3 | 83 ± 4 | 78 ± 2 |
| | | | | | | TD | 32 ± 2 | 246 ± 13 | 99 ± 8 | 2233 ± 107 | | | |
| LDPE | No | No | 20 ± 2 | 1.0 ± 0.2 | 6738 ± 1002 | MD | 40 ±3 | 393 ± 18 | 231 ± 19 | 4129 ± 289 | 20 ± 2 | 91 ± 3 | 95 ± 1 |
| | | | | | | TD | 22 ±2 | 734 ± 51 | 270 ± 23 | 4124 ± 301 | | | |

[a] Relative value compared to the value obtained for LDPE. sb: strain at break; eb: elongation at break; E: elastic modulus; Enb: 5-ethylidene-2-norbornene content; Fmax: maximum force to puncture; MD: machine direction; TD: transverse direction.

### 2.2. Cheese

A semi-hard cheese was chosen for this study because cheeses belonging to this category were previously used in experimental studies aimed to test cling film packaging [26,27]. In particular, the semi-hard Bra PDO cheese was selected as the food product to compare the effects of the selected cling films on the preservation of food quality during storage because several reasons: (i) it is usually bought or stored at home packaged in cling films; (ii) it has a uniform internal paste in terms of color and appearance (suitable for image analysis); (iii) it has a quite intense odor (suitable for electronic nose analysis); (iv) it is easy to cut into uniform shaped and sized slices (suitable for sensory tests); (v) it is not expensive; (vi) it is produced in the town where the present study was conducted, thus being a local product could be considered a socially and environmentally sustainable choice. The Bra PDO cheese was purchased at a local cheese store (Bra, Italy). According to D.P.R. 54/97 specifications, the Bra PDO cheese is a semi-fat, pressed cheese made from cow's milk, possibly with slight additions of sheep's and/or goat's milk. The color is white or ivory white and the texture of the paste is moderately consistent and elastic, with very small, barely visible, and not too diffused holes. The flavor is pleasantly aromatic, moderately piquant and savory. The percentage of fat in dry matter is a minimum of 32 per cent. Aging period is a minimum of forty-five days. The Bra PDO cheese was cut into 21 slices (≈250 g each), 124 slices (≈10 g), and 560 slices (≈15 g) for image analysis, electronic nose analysis, and sensory evaluation, respectively. Cheese slices were individually wrapped in the five different cling films and stored for 14 days at 5 ± 1 °C. The study was stopped after 14 days, when some mold appeared on the surface of the cheese, thus making unsuitable the cheese for the acceptability test.

### 2.3. Image Analysis

The color variation of cheese wrapped in the five cling films after 0, 4, 7, 9, 10, 11, and 14 days of storage time (T0, T4, T7, T9, T10, T11, T14) was assessed by image analysis. For each film type, three cheese slices (3 replicates) were analyzed at each control time, for a total of 21 slices. At each control storage time, the three replicates of slices of cheese were unwrapped and placed on black cardboard, and images were acquired with a digital camera in jpeg format. After the analysis, the three replicates were rewrapped with the cling film and stored at 5 ± 1 °C. The camera was placed perpendicularly to the sample under controlled lighting conditions. The lighting system consisted of one cool daylight lamp also placed perpendicularly above the sample to avoid irregular reflection on the sample surface. Images were elaborated by the ImageJ version 1.51 software to obtain the intensity mean values of R (red), B (blue), and G (green) color indices. These values were converted as L*, a*, and b* coordinates (CIE L*a*b* color space) with EasyRGB (www.easyrgb.com/en/convert.php, accessed on 8 August 2022) and used to calculate the hue (a/b) and saturation $(a^2+b^2)^{1/2}$ indexes [28].

### 2.4. Electronic Nose

The variation in odor of cheese wrapped in the five cling films at different storage times (T4, T7, T9, T10, T11, T14) was evaluated. For each film type, four cheese slices (four replicates) were analyzed at each control time, for a total of 120 slices. These samples were compared with the aroma profile of the cheese at T0 (four replicates). The total number of slices analyzed was thus 124. The emissions of volatile compounds were assessed by the portable PEN 3 e-nose (Win Muster Airsense Analytic Inc., Schwerim, Germany) equipped with 10 metal oxide semiconductors (MOS) and fully described by Torri et al. [29]. Cheese (10.0 ± 1 g) was placed in 45 mL glass air-tight vials and hermetically sealed with a PTFE/silicone septum and a screw cap. The vials were stored at −18 ± 1 °C until analysis. Then, the vials were equilibrated at 25 ± 1 °C for 1 h and analyzed at the same temperature under standardized conditions. The measurement device aspirated the gaseous compounds from the headspace of the sample through the sensor array at 400 mL/min for 30 s. After sample analysis, the system was purged for 70 s at a flowrate of 600 mL/min with filtered air prior to the injection of the next sample to allow the instrument to re-establish a baseline. Four replications were performed for each sample (124 measurements in total), and the average of the sensor responses was used for the subsequent statistical analysis.

### 2.5. Sensory Evaluation

Fifty-six untrained subjects (M = 20, F = 36; mean age = 26, age range = 20–55; Italians = 37, non-Italians = 19) were recruited among staff and students of the University of Gastronomic Sciences (Bra, Italy). Subjects voluntarily joined the sensory test and signed an informed consent before starting the test. The study was approved by the Ethics Committee of the University of Gastronomic Sciences (Ethics Committee Proceeding no. 2022.01) and conducted according to the guidelines of the Declaration of Helsinki.

Consumers' acceptability for the appearance and odor of the cheese wrapped in the cling films was assessed after 7 and 10 days of storage (T7 and T10, respectively). Since the cling films tested were still experimental and migration data capable of ensuring the safety of tasters were not available, it was deemed appropriate not to evaluate the level of acceptability for taste but to evaluate only the acceptability of appearance and smell of the packaged samples. In order to minimize the possible subjective biases associated with the sensory evaluation due to physiological and psychological factors [30], a standardized procedure of sample presentation and evaluation was adopted. In particular, thirty minutes before the sensory test, cheese samples were taken out of the fridge and kept at room temperature (25 ± 1 °C) until evaluation. One slice of cheese for each type of cling film was unwrapped and put in a 90 cc PLA container sealed with parafilm and codified with three-digit codes. Samples were served in a randomized order across subjects. Participants were instructed to observe and odor the sample but to not taste it, then to rate how much they disliked/liked the appearance and the odor on a nine-point hedonic scale (1 = extremely dislike, 9 = extremely like). A 30 s waiting procedure was enforced between samples. Evaluation was conducted in individual booths under white light. The experimenters verbally introduced the consumers to the computerized data collection procedure (FIZZ Acquisition software, version 2.46A, Biosystèmes, Courtenon, France). The consumers completed their evaluation in approximately 15–20 min.

### 2.6. Data Analysis

The values of the color parameters obtained from the cheese sample wrapped in each cling film at each storage time were subtracted from the color values observed for its reference sample (T0), in order to calculate the single delta values for L* (ΔL), a* (Δa), b* (Δb), hue (Δ hue), and saturation (Δ saturation) indexes and for the total color difference (ΔE).

The matrices of the single delta values obtained for each cling film sample were independently submitted to one-way analysis of variance (ANOVA) models applied to determine the main effects of storage time on the color evolution of sample. Moreover, the matrices of the delta values obtained for each cling film sample were independently submitted to one-way analysis of variance (ANOVA) models applied to determine the overall main effects of cling film (fixed factor; five levels: R-BP1, R-BP2, NR-BP, PVC, and LDPE) and storage time (fixed factor; seven levels: T0, T4, T7, T9, T10, T11, T14) on color parameters (ΔE, ΔL, Δa, Δb, Δ hue, and Δ saturation). The output of the ANOVA models provided the overall mean delta values of the color parameters as a function of both each cling film and storage time. Additionally, the combined effect of the material used for wrapping and the storage time in relation to color parameters (ΔE, ΔL, Δa, Δb, Δ hue, and Δ saturation) was analyzed by means of principal component analysis (PCA).

Concerning the e-nose data, responses from the 10 MOS sensors were submitted to principal PCA to evaluate the variation over time of the aromatic profile of cheese packaged in the five different cling films.

Regarding the sensory data, two-way ANOVA mixed models (fixed factors: variables material and time of storage; random factor: subject) were independently applied to investigate the effect of the cling film and storage time on appearance and odor acceptability of cheese. Subjects' preferences regarding the appearance and odor of the products packaged in the different cling films at the two storage times considered were also submitted to PCA to obtain an internal preference map.

All ANOVA models were followed by Tukey's HSD (honestly significant difference) test. Significance criteria were set at alpha equal to 0.05 for all analyses. All analyses were conducted using the XLSTAT statistical software package version 2021.2.2 (Addinsoft, New York, NY, USA).

## 3. Results and Discussion

### 3.1. Effect of the Cling Films on Cheese Color over Storage Time

The one-way ANOVA models applied to the delta values of the color parameters to analyze the effect of the cling film on the variation in appearance over time permitted the delineation of the gradual trend in the variation of the color parameters over the storage period, identifying the times at which the main variations occur (Table 2). Total color difference ΔE, as the main quality indicator, can be used to represent the overall color change [31], while ΔHue is usually used as a ripeness indicator for visualizing the color of food products effectively [32]. In addition, the hue parameter is among the most discriminating parameters in visual comparison because it is able to be perceived well by the human eye [33]. In the present study, it was found that the color parameters examined varied depending on the film material considered, in accordance with previous research [34–36]. The parameters that varied the most over time for all cling films were Δa and Δ hue. In particular, for the cheese packaged with NR-BP, there are significant differences for the parameters delta a ($p < 0.0001$) and delta hue ($p < 0.0001$), whereby after seven days of storage, the values of both parameters start to increase, resulting in the cheese being significantly different from that stored for 4 days. The Δa and Δhue parameter values remain stable until T14, whereby the highest values for both parameters were found to be significantly different from the other control times. On the other hand, for cheese packaged with R-BP1, the parameters that present significant differences as a function of their variation over time were ΔE ($p = 0.007$) and Δhue ($p = 0.021$), whose significant variations occured between T4 and T14. The cheeses packaged with R-BP2 and PVC showed similar variations over time; specifically, significant differences were found for the parameters Δa ($p = 0.020$ for R-BP2 and 0.0004 for PVC) and Δ hue ($p = 0.009$ and $p = 0.002$, for R-BP2 and PVC, respectively), with the times that were significantly different between them being T4 and T14. With regard to cheeses packaged with LDPE, the parameters with the greatest significant differences were ΔE ($p = 0.001$), Δa ($p = 0.022$), Δb ($p = 0.019$), and Δ hue ($p = 0.012$). For all parameters, the main variations and significant differences were found between T4 and T14.

**Table 2.** Color parameters of cheese samples wrapped in the five tested cling films for different storage times.

| Cling Film | Storage Time | ΔE | ΔL | Δa | Δb | Δ Hue | Δ Saturation |
|---|---|---|---|---|---|---|---|
| R-BP1 | T4 | **3.9 [b]** | 1.6 [a] | −0.2 [a] | 0.03 [a] | **−0.01 [b]** | 0.03 [a] |
| | T7 | **4.1 [b]** | 0.6 [a] | 0.7 [a] | −0.5 [a] | **0.03 [ab]** | −0.5 [a] |
| | T9 | **3.9 [b]** | 2.3 [a] | 0.9 [a] | −0.9 [a] | **0.03 [ab]** | −0.9 [a] |
| | T10 | **4.9 [b]** | 1.1 [a] | 1.3 [a] | −3.1 [a] | **0.1 [ab]** | −3.1 [a] |
| | T11 | **3.3 [ab]** | 0.8 [a] | 1.3 [a] | −2.2 [a] | **0.1 [ab]** | −2.2 [a] |
| | T14 | **9.1 [a]** | 5.1 [a] | 2.3 [a] | −6.1 [a] | **0.1 [a]** | −6.1 [a] |
| *p* | | **0.007** | 0.416 | 0.104 | 0.227 | **0.021** | 0.291 |
| R-BP2 | T4 | 6.8 [a] | 1.3 [a] | **0.2 [b]** | −0.7 [a] | **0.01 [b]** | −0.7 [a] |
| | T7 | 7.5 [a] | 0.7 [a] | **0.5 [b]** | −1.6 [a] | **0.02 [b]** | −1.7 [a] |
| | T9 | 6.9 [a] | 1.7 [a] | **1.2 [ab]** | −2.4 [a] | **0.05 [b]** | −2.4 [a] |
| | T10 | 6.1 [a] | 1.0 [a] | **1.1 [ab]** | −2.7 [a] | **0.04 [b]** | −2.8 [a] |
| | T11 | 6.3 [a] | −1.2 [a] | **1.0 [ab]** | −2.6 [a] | **0.04 [b]** | −2.7 [a] |
| | T14 | 12.5 [a] | −5.3 [a] | **2.3 [a]** | −10.3 [a] | **0.1 [a]** | −10.3 [a] |
| *p* | | 0.557 | 0.647 | **0.020** | 0.522 | **0.0004** | 0.528 |
| NR-BP | T4 | 8.6 [a] | −5.5 [a] | **−0.5 [c]** | 2.9 [a] | **0.02 [c]** | 2.9 [a] |
| | T7 | 7.9 [a] | −2.1 [a] | **0.4 [b]** | 2.3 [a] | **0.02 [b]** | 2.3 [a] |
| | T9 | 9.2 [a] | −3.5 [a] | **0.9 [b]** | 1.6 [a] | **0.04 [b]** | 1.6 [a] |
| | T10 | 9.9 [a] | −4.1 [a] | **0.5 [b]** | −0.4 [a] | **0.02 [b]** | −0.4 [a] |
| | T11 | 11.0 [a] | −6.2 [a] | **1.1 [b]** | 0.3 [a] | **0.1 [b]** | 0.3 [a] |
| | T14 | 9.1 [a] | −2.8 [a] | **2.0 [a]** | −5.5 [a] | **0.1 [a]** | −5.4 [a] |
| *p* | | 0.990 | 0.995 | **<0.0001** | 0.393 | **<0.0001** | 0.401 |

**Table 2.** *Cont.*

| Cling Film | Storage Time | ΔE | ΔL | Δa | Δb | Δ Hue | Δ Saturation |
|---|---|---|---|---|---|---|---|
| PVC | T4 | 6.9 $^{a}$ | −6.4 $^{a}$ | **0.2 $^{b}$** | 1.5 $^{a}$ | **0.01 $^{b}$** | 1.5 $^{a}$ |
| | T7 | 4.8 $^{a}$ | −4.3 $^{a}$ | **0.5 $^{b}$** | 1.1 $^{a}$ | **0.02 $^{b}$** | 1.1 $^{a}$ |
| | T9 | 5.2 $^{a}$ | −3.6 $^{a}$ | **1.0 $^{ab}$** | −0.2 $^{a}$ | **0.05 $^{b}$** | −0.2 $^{a}$ |
| | T10 | 4.1 $^{a}$ | −2.4 $^{a}$ | **1.2 $^{ab}$** | −0.9 $^{a}$ | **0.1 $^{b}$** | −1.0 $^{a}$ |
| | T11 | 4.8 $^{a}$ | −4.2 $^{a}$ | **0.8 $^{ab}$** | 0.3 $^{a}$ | **0.03 $^{ab}$** | 0.3 $^{a}$ |
| | T14 | 7.3 $^{a}$ | −3.1 $^{a}$ | **2.1 $^{a}$** | −5.8 $^{a}$ | **0.1 $^{a}$** | −5.7 $^{a}$ |
| $p$ | | 0.724 | 0.722 | **0.009** | 0.103 | **0.002** | 0.108 |
| LDPE | T4 | **3.7 $^{b}$** | 2.7 $^{a}$ | **−0.6 $^{b}$** | **0.7 $^{a}$** | **−0.02 $^{b}$** | **0.8 $^{a}$** |
| | T7 | **3.8 $^{b}$** | 3.1 $^{a}$ | **−0.02 $^{ab}$** | **1.3 $^{a}$** | **0.002 $^{b}$** | **1.3 $^{a}$** |
| | T9 | **4.7 $^{b}$** | 3.5 $^{a}$ | **0.2 $^{ab}$** | **0.4 $^{a}$** | **0.01 $^{ab}$** | **0.4 $^{a}$** |
| | T10 | **6.1 $^{ab}$** | 5.5 $^{a}$ | **0.4 $^{ab}$** | **0.4 $^{a}$** | **0.02 $^{b}$** | **0.3 $^{a}$** |
| | T11 | **4.3 $^{b}$** | 3.7 $^{a}$ | **0.4 $^{ab}$** | **0.7 $^{a}$** | **0.02 $^{ab}$** | **0.6 $^{a}$** |
| | T14 | **9.7 $^{a}$** | 5.3 $^{a}$ | **1.8 $^{a}$** | **−7.5 $^{b}$** | **0.1 $^{a}$** | **−7.5 $^{b}$** |
| $p$ | | **0.001** | 0.036 | **0.022** | **0.019** | **0.012** | **0.019** |

Text in bold and different lowercase letters in each column indicate statistically significant differences between mean values related to cheese wrapped in the same cling film material and stored for different storage times (Tukey's HSD test, $p < 0.05$).

Specifically, to provide an overview of the effects of cling films over the entire storage period and the effect of all times on the variation of color parameters, the one-way ANOVA models applied to color delta mean values showed a significant effect of the different cling film used on ΔE, ΔL, Δa, and Δ hue parameters, while no significant differences were observed for Δb and Δ saturation (Table 3). The observed color variations could be explained taking into account the oxygen and water vapor barrier properties of the cling films. In particular, concerning the ΔE parameter, the highest mean value was observed for the cheese wrapped in NR-BP film that was significantly different ($p = 0.002$) from the cheese wrapped in R-BP1, PVC, and LDPE cling films. The resulted color differences could be partly determined also by the different WVTR values characterizing the three cling films, since the NR-BP film has higher WVTR values than the films made from renewable and circular resources, R-BP1 and R-BP2. In fact, higher water vapor permeability may affect shelf life and may lead to a variation in food quality, thus influencing weight loss and color variation [37]. In the case of cheese, $a_w$ is responsible for its stability. In particular, water activity depends mainly on moisture and salt content, so the rate of water vapor transmission through the packaging material is crucial in controlling $a_w$ [38]. These results are in agreement with the study by Peixoto et al. [39] in which different ΔE values were found over time between cheese packaged in fossil-based plastic and cheese packaged in biodegradable packaging made from renewable resources in which color changes led to darkening of the cheese, explained by the fact that lower water activity due to the removal of water from the cheese surface probably caused Maillard reactions between lactose and lysine residues of the proteins. Significant differences were found for the ΔL parameter ($p < 0.0001$), with cheese wrapped in R-BP1 and LDPE film having higher values than cheese wrapped in NR-BP and PVC films. The slices of cheese packaged with R-PB1 and R-BP2 showed significant differences ($p = 0.010$) to the slices of cheese packaged with LDPE films regarding the Δa parameter. Only the cheese wrapped in PVC and LDPE films was significantly different for the Δ hue parameter ($p = 0.027$). The clear difference between the behavior of LDPE cling films and PVC cling films could be attributable to a different oxygen transmission rate that distinguish the two types of material (6738 ± 1002 and 3253 ± 554 $cm^3\ m^{-2}\ 24\ h^{-1}\ bar^{-1}$, respectively), confirming what reported in several studies [40,41]. Oxygen permeability is the parameter that could have most influenced the cheese evolution. In fact, exposure of cheese to oxygen can trigger enzymatic oxidation reactions resulting in degradation processes such as discoloration, production of off-flavors, loss of nutrients, and formation of toxic substances. Regarding the main effects of storage times on color parameters, significant differences were found for Δa and Δ hue ($p < 0.0001$) where the cheese stored for 14 days had the highest mean value, while the cheese stored for four days turned out to have a lower value. The only significant difference due to Δb was found for the cheese stored for 14 days with the lowest value. Although no particular significant differences were found in Δb values but only in Δa values, the combined parameter hue may lead to some insights. Carotenoids are the compounds that give the yellow color. Therefore, less oxidation of carotenoids due to a higher oxygen barrier may have helped to maintain a more yellow color, resulting in less variation in hue [32]. This study agrees with this statement because cheeses packaged in PVC, which has the highest oxygen barrier, experienced less hue variation than cheeses packaged in LDPE.

**Table 3.** Overall mean delta values of the color parameters as a function of both each cling film and storage time.

| Variables | ΔE | ΔL | Δa | Δb | Δ Hue | Δ Saturation |
|---|---|---|---|---|---|---|
| *Cling film* [a] | | | | | | |
| R-BP1 | **4.9** [b] | **1.9** [a] | **1.0** [a] | −2.1 [a] | **0.04** [ab] | −2.2 [a] |
| R-BP2 | **7.7** [ab] | **−0.3** [ab] | **1.1** [a] | −3.4 [a] | **0.04** [ab] | −3.4 [a] |
| NR-BP | **9.3** [a] | **−4.0** [b] | **0.7** [ab] | 0.2 [a] | **0.04** [ab] | 0.2 [a] |
| PVC | **5.5** [b] | **−4.0** [b] | **1.0** [ab] | −0.7 [a] | **0.05** [a] | −0.7 [a] |
| LDPE | **5.4** [b] | **4.0** [a] | **0.4** [b] | −0.7 [a] | **0.02** [b] | −0.7 [a] |
| *p* | **0.002** | **<0.0001** | **0.010** | 0.110 | **0.027** | 0.107 |
| *Storage time* [b] | | | | | | |
| T4 | **6.0** [a] | −1.3 [a] | **−0.2** [c] | **0.9** [a] | **−0.01** [c] | **0.9** [a] |
| T7 | **5.6** [a] | −0.4 [a] | **0.4** [bc] | **0.5** [a] | **0.02** [bc] | **0.5** [a] |
| T9 | **6.3** [a] | 0.5 [a] | **0.9** [b] | **−0.3** [a] | **0.04** [b] | **−0.3** [a] |
| T10 | **5.9** [a] | −0.2 [a] | **0.8** [b] | **−1.4** [a] | **0.04** [b] | **−1.4** [a] |
| T11 | **5.9** [a] | −1.4 [a] | **0.9** [b] | **−0.7** [a] | **0.04** [b] | **−0.7** [a] |
| T14 | **9.5** [a] | −0.2 [a] | **2.1** [a] | **−7.1** [b] | **0.11** [a] | **−7.0** [b] |
| *p* | **0.045** | 0.932 | **<0.0001** | **<0.0001** | **<0.0001** | **<0.0001** |

Mean values in bold and different lowercase letters in each column indicate statistically significant differences (Tukey's HSD test, $p < 0.05$).

Furthermore, despite the different WVTR values between the LDPE cling film, characterized by the lowest values of water vapor permeability, and the cling films R-BP1 and R-BP2, there do not seem to be any great differences in the color analysis of the samples packaged with these films. Probably, in this case, these results could be attributable to a balancing of both WVTR and $O_2$TR values, such that the LDPE films present higher values than the R-BP1 and R-BP2 films, confirming that, although biological and biodegradable materials have a higher water vapor permeability than fossil-based plastics, they are still characterized by relatively good gas barrier properties [2,42]. On the contrary, the one-way ANOVA did not show significant differences for ΔE, ΔL, and Δ saturation.

The biplot obtained from the PCA applied to the interaction of the material used for wrapping and the storage time in relation to color parameters is shown in Figure 1. The two first principal components (PC1 and PC2) accounted for 88.57% of the total variance at 66.70% and 21.86%, respectively. The score plot, which illustrates the mutual relationships between samples and color parameters, shows the sample separation according to the storage condition. Indeed, samples were distributed along PC1 and PC2 according to the cling film and storage time, respectively. In particular, according to the distribution of the samples along the PC1, there is a clear separation among the cling films, implying a similarity in performance and in the variation of the color parameters between the NR-BP and PVC packaged cheese samples, which are distinguished from the cheese samples packaged in the other three cling films (R-BP1, R-BP2, and LDPE). The similar trend in Δa variation between cheese packaged in PVC and in NR-BP cling film is partially in contrast to the study of Brandelero et al. [43] in which the lettuce stored in PVC presented lower luminosity (L) and a higher value of the b* parameter than that stored in biodegradable films. In addition, the distribution of materials in the biplot could be due to oxygen barrier values similar for the cling films R-BP1 (5050 ± 837 $cm^3\ m^{-2}\ 24\ h^{-1}\ bar^{-1}$), R-BP2 (6235 ± 947 $cm^3\ m^{-2}\ 24\ h^{-1}\ bar^{-1}$), and LDPE (6738 ± 1002 $cm^3\ m^{-2}\ 24\ h^{-1}\ bar^{-1}$), but higher than that of the PVC cling film (3253 ± 554 $cm^3\ m^{-2}\ 24\ h^{-1}\ bar^{-1}$). The change in color is correlated with the carotenoid content variation [44]. Therefore, higher oxygen permeability could have led to degradation of carotenoids and consequently to color fading [33]. This could confirm the results obtained in this study in which samples packaged with the lowest oxygen barrier materials (LDPE, R-BP1, and R-BP2) experienced a greater change in lightness. A difference in lightness may be due to a denaturation of ß-lactoglobulin and its conjugation to j-casein [33]. Moreover, the second component discriminates samples according to their storage time: taking into consideration that color is considered as an explanatory factor for deterioration [31], on the right-hand side of the biplot are the samples that have been stored the longest. This would indicate that higher values of Δb, Δ saturation, and ΔL describe the samples during the first days of storage. In contrast, higher values of Δa, Δ hue, and ΔE mainly describe the samples at the end of the storage.

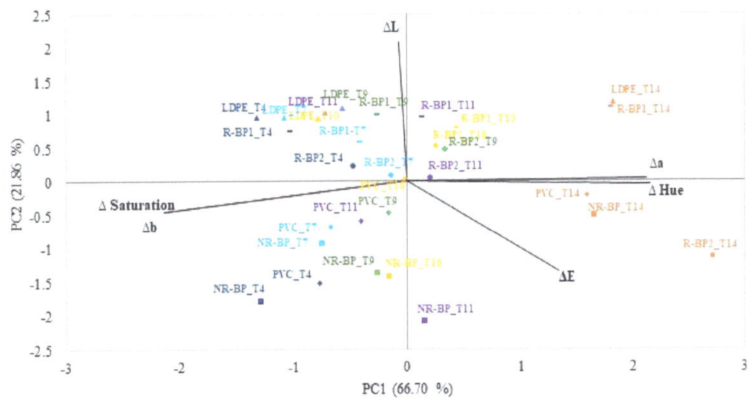


**Figure 1.** Biplot from the principal component analysis applied to interaction of the cling film and the storage time in relation to color parameters (ΔE, ΔL, Δa, Δb, Δ hue, and Δ saturation). Different symbols identify different cling film materials, while different colors identify different storage times.

### 3.2. *Effect of the Cling Films on Cheese Aroma Profile over Storage Time*

The biplot of the PCA applied to the e-nose data is shown in Figure 2. The two first principal components, PC1 and PC2, accounted for 88.03% of the total variance at 70.76% and 17.27%, respectively. The first PC discriminates samples according to their storage time, outlining the trend in the aroma profile of wrapped cheese samples. The aromatic profiles of the cheese slices packaged in the different cling films appear to have a similar evolution over time. During the first control times (T4 and T7), the cheese wrapped with R-BP1 seemed to have a slower decay than the cheeses packaged with the other cling films, while at the end of the storage period (14 days), the cheeses packaged with R-BP1 cling film seemed to have a slightly faster decay, similar to the cheeses packaged with NR-BP. At day 14 of storage, the cheeses wrapped with R-BP2 showed a comparable development of the aromatic profile to the cheeses wrapped with the commercial reference cling films PVC and LDPE. These results could be due to the fact that these two cling films have higher WVTR values, compared to both fossil-based plastics cling films (PVC and LDPE), and higher $O_2$TR values compared to PVC cling film. Thus, lower water vapor and oxygen permeability barriers could be responsible of a lower cheese sensory shelf life [45]. A high degree of oxygen diffusion from the surrounding atmosphere in cheeses and thus increased oxygen in packages can in fact cause rancidity and other oxidative degradation reactions in lipid-containing foods, leading not only to a possible risk of reduced quality and safety, but also to the production and spread of unpleasant and undesirable odors that can spread outside the package as well [46,47]. The similar oxygen permeability values, instead, of the R-BP2 film with the LDPE cling film could confirm the greater similarity in the development of the aroma profile at the end of storage.

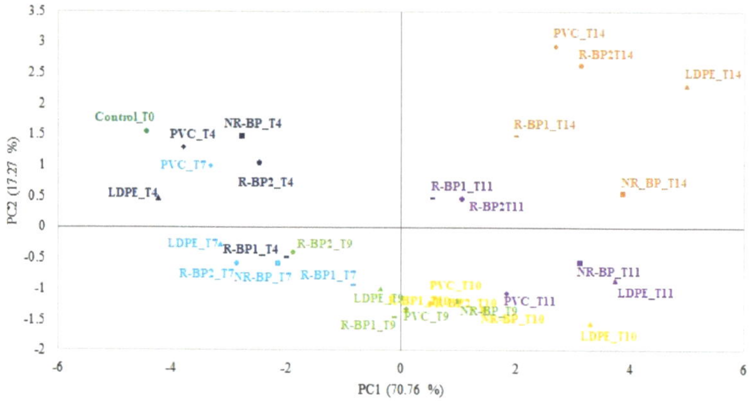


**Figure 2.** Biplot from the principal component analysis applied to e-nose responses. Different colors identify different storage times.

### 3.3. *Effect of the Cling Films on Acceptability for Cheese Appearance and Odor over Storage Time*

The mean values of the scores provided by all subjects regarding the acceptability evaluation of appearance and odor of the slices of cheese packaged into the five cling films are reported in Table 4. In terms of appearance, both cling film materials and different storage times had a significant effect on the subjects' judgement. In fact, subjects preferred cheese wrapped in R-BP1 and R-BP2 cling films that resulted in significant difference ($p$ = 0.005) with the cheese wrapped in NR-BP film. Slices of cheese wrapped into PVC and LDPE films were not significantly different from the cheese wrapped with the other three types of cling films. Regarding storage times, the subjects distinguished between the two storage times and preferred cheese stored for 7 days ($p$ = 0.026). Conversely, in terms of odor, there was not a significant difference between the two storage times. However, concerning the effect of material on odor acceptability, significant differences were found ($p$ = 0.0003). Subjects showed a higher acceptability for R-BP1 cling film packaged cheese, while the lowest mean score was achieved for cheese packaged in NR-BP film. R-BP2 packaged cheese was not found to be significantly different from either cheese wrapped with R-BP1 or cheese wrapped with PVC and LDPE, but there was a significant difference with the NR-BP packaged cheese.

**Table 4.** Acceptability mean values for cheese appearance and odor (n = 56).

| Variables | Appearance | Odor |
|---|---|---|
| *Cling film* | | |
| R-BP1 | **5.7** $^{a}$ | **5.3** $^{a}$ |
| R-BP2 | **5.6** $^{a}$ | **5.1** $^{ab}$ |
| NR-BP | **5.1** $^{b}$ | **4.6** $^{c}$ |
| PVC | **5.4** $^{ab}$ | **4.6** $^{bc}$ |
| LDPE | **5.6** $^{ab}$ | **4.7** $^{bc}$ |
| $p$ | **0.005** | **0.0003** |
| *Storage time* | | |

**Table 4.** *Cont.*

| Variables | Appearance | Odor |
|---|---|---|
| T7 | **5.6** [a] | 4.9 [a] |
| T10 | **5.3** [b] | 4.8 [a] |
| *p* | **0.026** | 0.776 |

Mean values in bold and different lowercase letters within a column indicate statistically significant differences in terms of appearance and odor acceptability (Tukey's HSD test, $p < 0.05$).

The preference map obtained from the PCA applied to the appearance acceptability data expressed by the subjects during the sensory evaluation is shown in Figure 3. The total variance explained for the samples based on the first two significant dimensions was 38.83%, with PC1 and PC2 accounting for 21.18% and 17.65%, respectively. Samples were distributed along the first component PC1 mainly according to the type of cling film, showing a clear difference between cheeses wrapped with R-BP1 and R-BP2 from the samples wrapped with the other types (NR-BP, PVC, and LDPE). Instead, samples tended to be distributed along the second component PC2 as a function of storage time. According to time, there was a similarity between cheeses packaged with R-BP1 and R-BP2 films, and it was evident that cheeses packaged with R-BP1 and R-BP2, even with longer storage times, were appreciated more than cheeses with shorter storage times packaged into NR-BP, PVC, and LDPE cling films by a higher proportion of subjects (identified by the vectors in Figure 3). In fact, the visual distribution of the subjects' acceptability seems to lean more towards cheeses packaged with R-BP1 and R-BP2 films and stored for 7 and 10 days.

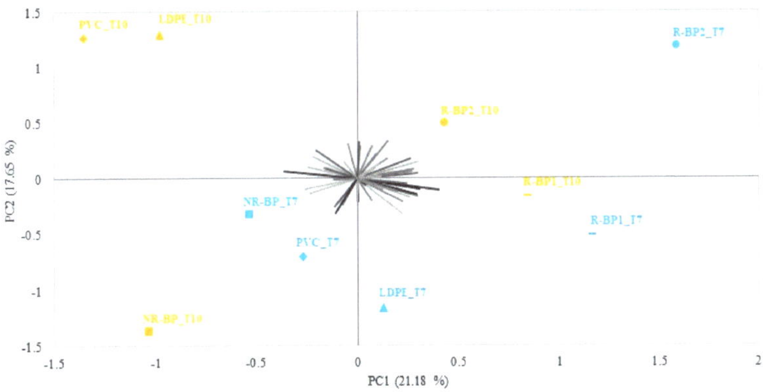


**Figure 3.** Biplot from the principal component analysis applied to acceptability data for cheese appearance (n = 56). Different symbols identify the two different storage times (T7 and T10).

The results obtained from the sensory tests, according to which cheese packaged with the R-BP1 and R-BP2 coating films exhibited higher acceptability in terms of appearance, are partially in disagreement with the results of the instrumental tests applied in the present study, which would instead seem to show a higher decay of some color parameters. However, the sensory tests showed that cheese packed with the NR-BP cling film was less liked by consumers in terms of appearance, which could validate the results obtained from the instrumental tests, which showed that cheese packed with NR-BP differs the most from the control.

The total variance explained by the preference map obtained for the evaluation of the odor of the stored cheese by PCA applied to the acceptability data expressed by the subjects during the sensory evaluation was 40.2%, with PC1 and PC2 accounting for 25.17% and 15.25%, respectively (Figure 4). Even this preference map seems to confirm the subjects' preferences for cheese stored with the cling films R-BP1 and R-BP2. Indeed, the visual distribution of acceptability response is more oriented to the slices of cheese wrapped into R-BP1 and R-BP2 cling films, also indicating in the case of odor acceptability, a similarity of appreciation between these two cling films, whereas these two samples were quite clearly distinguished from the cheese stored in the LDPE, PVC, and NR-BP cling films. This partially disagrees with previous studies according to which foods packaged with biodegradable films were less appreciated than those packaged with conventional packaging. In fact, cheese packaged with the R-BP1 cling film was more appreciated in terms of odor by the participants than cheese packaged with the others cling films, except for R-BP2. That is the case of the work of Pluta-Kubica et al. [37], in which the odor of the cheese wrapped in the active biodegradable film was less appreciated by the panelists than that of the control, possibly as a consequence of the decrease in water content following a much higher WVTR of the biodegradable film. Amjadi et al. [48] also observed lower hedonic odor scores and overall acceptance during storage of white cheese wrapped in biodegradable films containing zinc oxide nanoparticles.

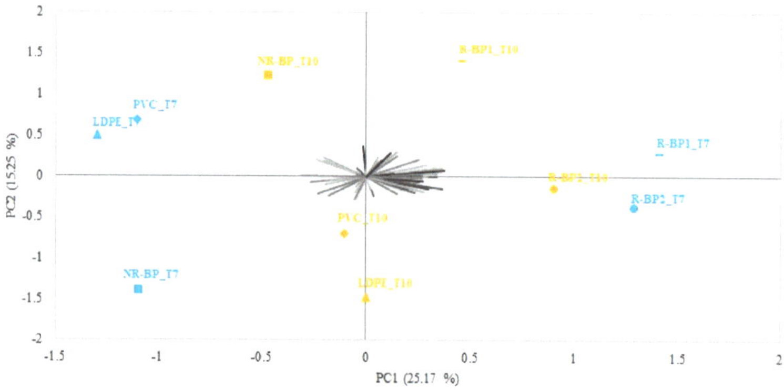


**Figure 4.** Biplot from the principal component analysis applied to acceptability data for cheese odor (n = 56). Different symbols identify the two different storage times (T7 and T10).

The fact that this study only examined appearance and odor variation during the storage period of packaged products emerged as a limitation, since the variations in terms of flavor and texture that food products may be subjected to during storage could not be identified [49–53]. Therefore, future research could envisage tests including all sensory modalities. Moreover, other analyses (e.g., weight loss monitoring, fat oxidation analyses, microbiological analyses, pH measurements, and rheological analyses) on wrapped cheese would be helpful in order to provide a more comprehensive evaluation of the preservation performance offered by the films analyzed. It could also be interesting in the future to analyze the effects of these cling films on other types of cheese (e.g., hard cheese), dairy products, and food products, such as fruit and vegetables, in order to assess whether the films may have different effects when in contact with different food matrices.

## 4. Conclusions

This study examined the potential of biobased and biodegradable cling films derived from renewable and circular sources in preserving cheese appearance and odor, comparing them with commercially cling films derived from non-renewable sources and with fossil-based plastic cling films, in order to reduce the future production and consumption of disposable food contact materials that have a negative impact on the environment. Based on the determination of cheese color variation and odor profile development, the results showed a different protecting performance of the packaging materials as a function of storage time and cling film type. Specifically, according to the results obtained from instrumental analysis, cheeses packaged in films derived from renewable and nonrenewable sources were found to be more prone to color variation than cheeses packaged in PVC and LDPE, probably due to different water vapor and oxygen transmission rate values.

However, when analyzing the variation over time, comparable performances were found between R-BP1 and R-BP2 films and LDPE films, while biodegradable film derived from nonrenewable sources showed more similarities with PVC film. Regarding odor evolution, although similarities were found between the biodegradable and synthetic films, the samples packed with NR-BP and R-BP1 seemed to have slightly faster decay in the last days of storage, unlike the PVC-packed cheese, which showed a slower decay. Overall, this study shows how the biobased and biodegradable cling films tested, obtained from renewable and circular sources, have a potential to be comparable solutions to conventional plastic materials. Therefore, future research could be addressed to improving these innovative films able to offer multiple benefits for sustainable development.

**Author Contributions:** Conceptualization, C.C. and L.T.; methodology, C.C. and L.T.; software, C.C.; formal analysis, C.C.; investigation, C.C. and L.T.; resources, L.T.; data curation, C.C.; writing—original draft preparation, C.C.; writing—review and editing, L.T.; visualization, C.C. and L.T.; supervision, L.T.; project administration, L.T.; funding acquisition, L.T. All authors have read and agreed to the published version of the manuscript.

**Funding:** This work was supported by the PRIME (Processi e pRodotti Innovativi di chiMica vErde) project funded by the POR FESR 2014/2020 Programme, Asse I—Azione I.1b.2.2 Regione Piemonte, within the "Piattaforma Tecnologica per la Bioeconomia" (Grant. N. 234, 2019, April 19).

**Institutional Review Board Statement:** The study was conducted according to the guidelines of the Declaration of Helsinki and approved by the Ethics Committee of the University of Gastronomic Sciences (Ethics Committee Proceeding n. 2022.01).

**Informed Consent Statement:** Informed consent was obtained from all of the subjects involved in the study.

**Data Availability Statement:** The data presented in this study are available on request from the corresponding author. The data are not publicly available due to privacy.

**Acknowledgments:** The authors are thankful to: Tiziana Milizia and Alessandro D'Elicio, coordinators of the PRIME project, for providing the cling film materials and data on their physical-chemical properties, as well as to all the individuals who participated in the study.

**Conflicts of Interest:** The authors declare no conflict of interest. The funder had no role in the design of the study; in the collection, analyses, or interpretation of data; in the writing of the manuscript; or in the decision to publish the results.

## References

1. Popović, S.Z.; Lazić, V.L.; Hromiš, N.M.; Šuput, D.Z.; Bulut, S.N. Biopolymer Packaging Materials for Food Shelf-Life Prolongation. In *Biopolymers for Food Design*; Elsevier: Amsterdam, The Netherlands, 2018; pp. 223–277. ISBN 9780128115015.
2. Cazón, P.; Velazquez, G.; Ramírez, J.A.; Vázquez, M. Polysaccharide-Based Films and Coatings for Food Packaging: A Review. *Food Hydrocoll.* **2017**, *68*, 136–148. [CrossRef]
3. Wang, H.; Wang, L. Developing a Bio-Based Packaging Film from Soya by-Products Incorporated with Valonea Tannin. *J. Clean. Prod.* **2017**, *143*, 624–633. [CrossRef]
4. Huang, T.; Qian, Y.; Wei, J.; Zhou, C. Polymeric Antimicrobial Food Packaging and Its Applications. *Polymers* **2019**, *11*, 560. [CrossRef]
5. Chirilli, C.; Molino, M.; Torri, L. Consumers' Awareness, Behavior and Expectations for Food Packaging Environmental Sustainability: Influence of Socio-Demographic Characteristics. *Foods* **2022**, *11*, 2388. [CrossRef]
6. Rajvanshi, J.; Sogani, M.; Kumar, A.; Arora, S.; Syed, Z.; Sonu, K.; Gupta, N.S.; Kalra, A. Perceiving Biobased Plastics as an Alternative and Innovative Solution to Combat Plastic Pollution for a Circular Economy. *Sci. Total Environ.* **2023**, *874*, 162441. [CrossRef] [PubMed]
7. Versino, F.; Ortega, F.; Monroy, Y.; Rivero, S.; López, O.V.; García, M.A. Sustainable and Bio-Based Food Packaging: A Review on Past and Current Design Innovations. *Foods* **2023**, *12*, 1057. [CrossRef]
8. Khodaei, D.; Álvarez, C.; Mullen, A.M. Biodegradable Packaging Materials from Animal Processing Co-Products and Wastes: An Overview. *Polymers* **2021**, *13*, 2561. [CrossRef] [PubMed]
9. Fortunati, E.; Mazzaglia, A.; Balestra, G.M. Sustainable Control Strategies for Plant Protection and Food Packaging Sectors by Natural Substances and Novel Nanotechnological Approaches. *J. Sci. Food Agric.* **2019**, *99*, 986–1000. [CrossRef]
10. Petronilho, S.; Oliveira, A.; Domingues, M.R.; Nunes, F.M.; Coimbra, M.A.; Gonçalves, I. Hydrophobic Starch-Based Films Using Potato Washing Slurries and Spent Frying Oil. *Foods* **2021**, *10*, 2897. [CrossRef]

11. Mai, S.; Ma, Y.; Liu, H.; Li, C.; Song, Y.; Hu, K.; Chen, X.; Chen, Y.; Zou, W. Dynamic Modulation of $SO_2$ Atmosphere for Enhanced Fresh-Keeping of Grapes Using a Novel Starch-Based Biodegradable Foam Packaging. *Foods* **2023**, *12*, 2222. [CrossRef]
12. Reichert, C.L.; Bugnicourt, E.; Coltelli, M.B.; Cinelli, P.; Lazzeri, A.; Canesi, I.; Braca, F.; Martínez, B.M.; Alonso, R.; Agostinis, L.; et al. Bio-Based Packaging: Materials, Modifications, Industrial Applications and Sustainability. *Polymers* **2020**, *12*, 1558. [CrossRef] [PubMed]
13. Lee, D.S.; Yam, K.L.; Piergiovanni, L. *Food Packaging Science and Technology*, 1st ed.; CRC Press: Boca Raton, FL, USA, 2008; Available online: https://doi-org.bibliopass.unito.it/10.1201/9781439894071 (accessed on 3 August 2023).
14. Licciardello, F. Packaging, Blessing in Disguise. Review on Its Diverse Contribution to Food Sustainability. *Trends Food Sci. Technol.* **2017**, *65*, 32–39. [CrossRef]
15. Giannakourou, M.C.; Tsironi, T.N. Application of Processing and Packaging Hurdles for Fresh-Cut Fruits and Vegetables Preservation. *Foods* **2021**, *10*, 830. [CrossRef] [PubMed]
16. Barlow, C.Y.; Morgan, D.C. Polymer Film Packaging for Food: An Environmental Assessment. *Resour. Conserv. Recycl.* **2013**, *78*, 74–80. [CrossRef]
17. Rubilar, J.F.; Cruz, R.M.S.; Silva, H.D.; Vicente, A.A.; Khmelinskii, I.; Vieira, M.C. Physico-Mechanical Properties of Chitosan Films with Carvacrol and Grape Seed Extract. *J. Food Eng.* **2013**, *115*, 466–474. [CrossRef]
18. Elsabee, M.Z.; Abdou, E.S. Chitosan Based Edible Films and Coatings: A Review. *Mater. Sci. Eng. C* **2013**, *33*, 1819–1841. [CrossRef]
19. Malhotra, B.; Keshawani, A.; Kharkwal, H. Natural polymer based cling films for food packaging. *Int. J. Pharm. Pharm. Sci.* **2015**, *7*, 10–18.
20. Piscopo, A.; Zappia, A.; De Bruno, A.; Pozzo, S.; Limbo, S.; Piergiovanni, L.; Poiana, M. Use of Biodegradable Materials as Alternative Packaging of Typical Calabrian Provola Cheese. *Food Packag. Shelf Life* **2019**, *21*, 100351. [CrossRef]
21. Vimala Bharathi, S.K.; Maria Leena, M.; Moses, J.A.; Anandharamakrishnan, C. Zein-Based Anti-Browning Cling Wraps for Fresh-Cut Apple Slices. *Int. J. Food Sci. Technol.* **2020**, *55*, 1238–1245. [CrossRef]
22. Dinnella, C.; Torri, L.; Caporale, G.; Monteleone, E. An Exploratory Study of Sensory Attributes and Consumer Traits Underlying Liking for and Perceptions of Freshness for Ready to Eat Mixed Salad Leaves in Italy. *Food Res. Int.* **2014**, *59*, 108–116. [CrossRef]
23. Tanner, D. Impacts of Storage on Food Quality. In *Reference Module in Food Science*; Elsevier: Amsterdam, The Netherlands, 2016.
24. Imtiyaz, H.; Soni, P.; Yukongdi, V. Role of Sensory Appeal, Nutritional Quality, Safety, and Health Determinants on Convenience Food Choice in an Academic Environment. *Foods* **2021**, *10*, 345. [CrossRef] [PubMed]
25. Ares, G.; Giménez, A.; Gámbaro, A. Sensory Shelf Life Estimation of Minimally Processed Lettuce Considering Two Stages of Consumers' Decision-Making Process. *Appetite* **2008**, *50*, 529–535. [CrossRef] [PubMed]
26. Aly, S.S.H. Studying the Migration of Plasticizers from Plastic Packaging to Local Processed Cheese. *Middle East J. Appl. Sci.* **2016**, *6*, 957–963.
27. Panseri, S.; Chiesa, L.M.; Zecconi, A.; Soncini, G.; De Noni, I. Determination of Volatile Organic Compounds (VOCs) from Wrapping Films and Wrapped PDO Italian Cheeses by Using HS-SPME and GC/MS. *Molecules* **2014**, *19*, 8707–8724. [CrossRef] [PubMed]
28. Cheng, H.D.; Jiang, X.H.; Sun, Y.; Wang, J. Color Image Segmentation: Advances and Prospects. *Pattern Recognit.* **2000**, *34*, 2259–2281. [CrossRef]
29. Torri, L.; Rinaldi, M.; Chiavaro, E. Electronic Nose Evaluation of Volatile Emission of Chinese Teas: From Leaves to Infusions. *Int. J. Food Sci. Technol.* **2014**, *49*, 1315–1323. [CrossRef]
30. Civille, G.V.; Carr, B.T. Factors Influencing Sensroy Vedicts. In *Sensory Evaluation Techniques*, 5th ed.; CRC Press: Boca Raton, FL, USA, 2015; Chapter 4; pp. 45–50.
31. Zhang, H.; Tang, Z.; Rasco, B.; Tang, J.; Sablani, S.S. Shelf-Life Modeling of Microwave-Assisted Thermal Sterilized Mashed Potato in Polymeric Pouches of Different Gas Barrier Properties. *J. Food Eng.* **2016**, *183*, 65–73. [CrossRef]
32. Skrypec, S.; Doh, H.; Whiteside, W.S. Effect of Oxygen Scavenging Films and Modified Atmosphere on the Color Quality of Hot-Filled Freestone Peach Puree. *J. Food Process. Eng.* **2021**, *44*, e13681. [CrossRef]
33. Chudy, S.; Bilska, A.; Kowalski, R.; Teichert, J. Colour of Milk and Milk Products in CIE Lab Space. *Med. Weter.* **2020**, *76*, 77–81.
34. Mashabela, M.; Mahajan, P.V.; Sivakumar, D. Influence of Different Types of Modified Atmosphere Packaging Films and Storage Time on Quality and Bioactive Compounds in Fresh-Cut Cauliflower. *Food Packag. Shelf Life* **2019**, *22*, 100374. [CrossRef]
35. Sonar, C.R.; Al-Ghamdi, S.; Marti, F.; Tang, J.; Sablani, S.S. Performance Evaluation of Biobased/Biodegradable Films for in-Package Thermal Pasteurization. *Innov. Food Sci. Emerg. Technol.* **2020**, *66*, 102485. [CrossRef]
36. Tarakçı, Z.; Durmuş, Y. Utjecaj Različitih Načina Pakiranja Na Neke Parametre Zrenja Sira Tulum. *Mljekarstvo* **2016**, *66*, 293–303. [CrossRef]
37. Pluta-Kubica, A.; Jamróz, E.; Khachatryan, G.; Florkiewicz, A.; Kopel, P. Application of Furcellaran Nanocomposite Film as Packaging of Cheese. *Polymers* **2021**, *13*, 1428. [CrossRef]
38. Di Marzo, S.; Di Monaco, R.; Cavella, S.; Romano, R.; Borriello, I.; Masi, P. Correlation between Sensory and Instrumental Properties of Canestrato Pugliese Slices Packed in Biodegradable Films. *Trends Food Sci. Technol.* **2006**, *17*, 169–176. [CrossRef]
39. Peixoto, A.M.; Petronilho, S.; Domingues, M.R.; Nunes, F.M.; Lopes, J.; Pettersen, M.K.; Grøvlen, M.S.; Wetterhus, E.M.; Gonçalves, I.; Coimbra, M.A. Potato Chips Byproducts as Feedstocks for Developing Active Starch-Based Films with Potential for Cheese Packaging. *Foods* **2023**, *12*, 1167. [CrossRef] [PubMed]

40. Weber, C.J.; Haugaard, V.; Festersen, R.; Bertelsen, G. Production and Applications of Biobased Packaging Materials for the Food Industry. *Food Addit. Contam.* **2002**, *19*, 172–177. [CrossRef]
41. Kalendova, A.; Dagmar, M.; Francois, G.J.; Miroslav, S. Polymer/Clay Nanocomposites and Their Gas Barrier Properties. *Polym. Compos.* **2013**, *34*, 1418–1424. [CrossRef]
42. Nilsen-Nygaard, J.; Fernández, E.N.; Radusin, T.; Rotabakk, B.T.; Sarfraz, J.; Sharmin, N.; Sivertsvik, M.; Sone, I.; Pettersen, M.K. Current Status of Biobased and Biodegradable Food Packaging Materials: Impact on Food Quality and Effect of Innovative Processing Technologies. *Compr. Rev. Food Sci. Food Saf.* **2021**, *20*, 1333–1380. [CrossRef]
43. Brandelero, R.P.H.; Brandelero, E.M.; de Almeida, F.M. Filmes Biodegradáveis de Amido/PVOH/Alginato Em Sistema de Embalagem Para Alface (*Lactuca sativa* L.) Minimamente Processada. *Cienc. Agrotecnol.* **2016**, *40*, 510–521. [CrossRef]
44. Zhang, H.; Patel, J.; Bhunia, K.; Al-Ghamdi, S.; Sonar, C.R.; Ross, C.F.; Tang, J.; Sablani, S.S. Color, Vitamin C, β-Carotene and Sensory Quality Retention in Microwave-Assisted Thermally Sterilized Sweet Potato Puree: Effects of Polymeric Package Gas Barrier during Storage. *Food Packag. Shelf Life* **2019**, *21*, 100324. [CrossRef]
45. Petersen, K.; Vñggemose Nielsen, P.; Bertelsen, G.; Lawther, M.; Olsen, M.B.; Nilsson, N.H.; Mortensen, G. Potential of Biobased Materials for Food Packaging. *Food Sci. Technol.* **1999**, *10*, 52–68. [CrossRef]
46. Nájera, A.I.; Nieto, S.; Barron, L.J.R.; Albisu, M. A Review of the Preservation of Hard and Semi-Hard Cheeses: Quality and Safety. *Int. J. Environ. Res. Public Health* **2021**, *18*, 9789. [CrossRef] [PubMed]
47. Mokwena, K.K.; Tang, J.; Dunne, C.P.; Yang, T.C.S.; Chow, E. Oxygen Transmission of Multilayer EVOH Films after Microwave Sterilization. *J. Food Eng.* **2009**, *92*, 291–296. [CrossRef]
48. Amjadi, S.; Emaminia, S.; Nazari, M.; Davudian, S.H.; Roufegarinejad, L.; Hamishehkar, H. Application of Reinforced ZnO Nanoparticle-Incorporated Gelatin Bionanocomposite Film with Chitosan Nanofiber for Packaging of Chicken Fillet and Cheese as Food Models. *Food Bioproc. Technol.* **2019**, *12*, 1205–1219. [CrossRef]
49. Feng, H.; Zhang, M.; Liu, P.; Liu, Y.; Zhang, X. Evaluation of IoT-Enabled Monitoring and Electronic Nose Spoilage Detection for Salmon Freshness during Cold Storage. *Foods* **2020**, *9*, 1579. [CrossRef]
50. Jiang, H.; He, Y.; Chen, Q. Qualitative Identification of the Edible Oil Storage Period Using a Homemade Portable Electronic Nose Combined with Multivariate Analysis. *J. Sci. Food Agric.* **2021**, *101*, 3448–3456. [CrossRef]
51. Chen, Y.; Wei, Z.; Zhang, T.; Ng, K.H.; Ye, J.; He, W. Physicochemical, Electronic Nose and Tongue, Sensory Evaluation Determination Combined with Chemometrics to Characterize *Ficus hirta* Vahl. (*Moraceae*) Beer. *J. Food Qual.* **2022**, *2022*, 1–13. [CrossRef]
52. Rasekh, M.; Karami, H. Application of Electronic Nose with Chemometrics Methods to the Detection of Juices Fraud. *J. Food Process. Preserv.* **2021**, *45*, e15432. [CrossRef]
53. Chacón, I.; Roales, J.; Lopes-Costa, T.; Pedrosa, J.M. Analyzing the Organoleptic Quality of Commercial Extra Virgin Olive Oils: IOC Recognized Panel Tests vs. Electronic Nose. *Foods* **2022**, *11*, 1477. [CrossRef]

foods

MDPI

Article

# Application Potential of *Trichoderma* in the Degradation of Phenolic Acid-Modified Chitosan

**Maria Swiontek Brzezinska [1,*], Beata Kaczmarek-Szczepańska [2,*], Grażyna B. Dąbrowska [3], Marta Michalska-Sionkowska [1], Katarzyna Dembińska [1], Agnieszka Richert [3], Marcela Pejchalová [4], Sweta Binod Kumar [1] and Agnieszka Kalwasińska [1]**

[1] Department of Environmental Microbiology and Biotechnology, Faculty of Biological and Veterinary Sciences, Nicolaus Copernicus University in Torun, Lwowska 1, 87-100 Torun, Poland; mms@umk.pl (M.M.-S.); katarzynadembinska6@gmail.com (K.D.); sweta14789@gmail.com (S.B.K.); kala@umk.pl (A.K.)

[2] Department of Biomaterials and Cosmetics Chemistry, Faculty of Chemistry, Nicolaus Copernicus University in Torun, Gagarina 7, 87-100 Torun, Poland

[3] Department of Genetics, Faculty of Biological and Veterinary Sciences, Nicolaus Copernicus University in Torun, Lwowska 1, 87-100 Torun, Poland; browsk@umk.pl (G.B.D.); a.richert@umk.pl (A.R.)

[4] Department of Biological and Biochemical Sciences, Faculty of Chemical Technology, University of Pardubice, Sudentska 573, 53210 Pardubice, Czech Republic; marcela.pejchalova@upce.cz

* Correspondence: swiontek@umk.pl (M.S.B.); beata.kaczmarek@umk.pl (B.K.-S.)

**Abstract:** The aim of the study was to determine the potential use of fungi of the genus *Trichoderma* for the degradation of phenolic acid-modified chitosan in compost. At the same time, the enzymatic activity in the compost was checked after the application of a preparation containing a suspension of the fungi *Trichoderma* (spores concentration $10^5$/mL). The *Trichoderma* strains were characterized by high lipase and aminopeptidase activity, chitinase, and β-1,3-glucanases. *T. atroviride* TN1 and *T. citrinoviride* TN3 metabolized the modified chitosan films best. Biodegradation of modified chitosan films by native microorganisms in the compost was significantly less effective than after the application of a formulation composed of *Trichoderma* TN1 and TN3. Bioaugmentation with a *Trichoderma* preparation had a significant effect on the activity of all enzymes in the compost. The highest oxygen consumption in the presence of chitosan with tannic acid film was found after the application of the consortium of these strains (861 mg $O_2$/kg after 21 days of incubation). Similarly, chitosan with gallic acid and chitosan with ferulic acid were found after the application of the consortium of these strains (849 mg$O_2$/kg and 725 mg $O_2$/kg after 21 days of incubation). The use of the *Trichoderma* consortium significantly increased the chitinase activity. The application of *Trichoderma* also offers many possibilities in sustainable agriculture. *Trichoderma* can not only degrade chitosan films, but also protect plants against fungal pathogens by synthesizing chitinases and $\beta$-1,3 glucanases with antifungal properties.

**Keywords:** chitosan; *Trichoderma*; biodegradation

**Citation:** Swiontek Brzezinska, M.; Kaczmarek-Szczepańska, B.; Dąbrowska, G.B.; Michalska-Sionkowska, M.; Dembińska, K.; Richert, A.; Pejchalová, M.; Kumar, S.B.; Kalwasińska, A. Application Potential of *Trichoderma* in the Degradation of Phenolic Acid-Modified Chitosan. *Foods* **2023**, *12*, 3669. https://doi.org/10.3390/foods12193669

Academic Editor: Sílvia Petronilho

Received: 5 September 2023
Revised: 30 September 2023
Accepted: 3 October 2023
Published: 5 October 2023



## 1. Introduction

Plastics are used in many areas such as packaging, construction materials, and hygiene products due to their chemical resistance. Unfortunately, they are not susceptible to microbial degradation, which makes them a serious environmental pollutant. For this reason, interest in biodegradable polymers has increased [1]. Chitosan obtained in the deacetylation process of chitin [2] has antimicrobial and fungicidal properties and can therefore be applied as a natural biocide in medicine, food production, and agriculture. The biocidal properties of chitosan result from its ability to degrade the cell wall and membrane, bind metal ions, and combine with DNA [3].

Chitosan-based films in particular can be used in the food packaging industry. They are modified just as often with various biocidal substances, e.g., tannic acid [4]. Phenolic substances are synthesized by plants as a variety of secondary metabolites (phytochemicals) known to be involved in defense mechanisms and, over the past few years, some of these molecules have been recognized to have health benefits, including antimicrobial properties [5]. They are capable of forming strong hydrogen bonds with polymers and thus, they are good cross-linkers for proteins and polysaccharides [6]. The presence of phenolic acids in chitosan film increases its stability for food packaging applications. Despite their use, the film polymers should be biodegradable [7,8]. Polymer degradation is a complex process involving physical, chemical, and biological factors [9]. Undoubtedly, microorganisms producing hydrolytic enzymes contribute to their biodegradation [8,10–12]. Fungi, for example, *Pseudozyma japonica* Y7-09 and *Trichoderma viride* 99, degraded polycaprolactone (PCL) [13,14], *Clitocybe* sp. and *Laccaria laccata* degraded polylactide (PLA) and polyethylene terephthalate (PET) [9], and *Aspergillus* degraded polyethylene (PE) [15]. Vivi et al. [16] reported that polycaprolactone (PCL) and polyvinyl chloride (PVC) were degraded by *Chaetomium globosum* ATCC 16021. A sample of each polymer was placed in a Petri dish with complete agar and a second one on the incomplete medium. A spare sample was kept as a control inside an envelope in the dark. Individual samples of polymers (PCL and PVC) were immersed in 25 mL mix suspension with a concentration of $10^6$ spores/mL for 15 s. The samples were analyzed by means of morphological and color changes, mass loss, optical microscopy (OM), and scanning electron microscopy (SEM) after 28 days of culturing. After the incubation period, visual observations of the PCL films showed many micropores and cracks, pigmentation, surface erosion and hyphal adhesion on the sample surfaces, and a mass loss of up to 75%. Also, *Cladosporium* sp., *Mucor* sp., *Penicillium caseicolun*, *Penicillium citrinium*, and *Rhizopus* sp. degraded chitosan-gellan and polylysine-gellan fibers [17]. Degradation of phenolic acid-modified chitosan film can be a problem, as native microorganisms are unable to degrade the polymers. For this reason, it is so important to introduce additional microorganisms to aid in the degradation of the chitosan film. In recent years, microbial formulations have been increasingly developed to accelerate polymers' biodegradation. In the case of chitosan films, it is worthwhile to select chitinolytic microorganisms with multifarious enzymes which are capable of degrading chitin and its derivatives. Therefore, it is be suitable to use the fungi of the genus *Trichoderma*, which are already used in agriculture and known for their chitinolytic properties [18]. *Trichoderma* spp. belong to the Ascomycota, which are the most numerous species of fungi in almost all temperate and tropical soils. They are characterized by rapid growth, abundant sporulation, and the production of several hydrolytic enzymes [11,19].

Our previous reports in the literature indicated the possibility of the practical use of fungi like *Trichoderma* in accelerating the biodegradation of polymeric materials [9,11]. The research conducted by Znajewska et al. [14] has shown that *T. viride* accelerated the degradation of polycaprolactone in soil. Moreover, *T. viride* also showed growth on polymer materials including PE, PCL, PET, PLA, and PHB in in vitro studies [20]. Dąbrowska et al. [11] reported that the *T. viride* strain GZ1 produced hydrophobin proteins in contact with PLA and PET, producing a hydrophobin film that enabled the mycelium to adhere to the surface of the polymer.

Food packaging produced on an industrial scale from materials that are difficult to degrade represents a serious environmental problem. They take many years to decompose. An alternative to current packaging materials could be chitosan films modified with gallic, ferulic, and tannic acids. These materials can be degraded in the environment by native microorganisms. However, not all native microorganisms are able to degrade chitosan films due to their biocidal properties. In order to accelerate biodegradation, a microbiological preparation based on *Trichoderma* has been developed to support their decomposition. Fungi of the *Trichoderma* genus are known especially for the production of antifungal enzymes, which may be an additional feature when applied to the environment.

## 2. Experiments

### 2.1. Isolation and Molecular Identification of Fungal Strains

Saprophytic fungi isolated from oak wood were used in this study. The culture of the fungus stored in microbiological solid potato dextrose agar (Difco, Franklin Lakes, NJ, USA) was transferred to the same medium and incubated at 23 °C for 5 days.

Mycelia fragments were homogenized in liquid nitrogen for molecular identification. The genomic DNA from mycelia was extracted using the Plant & Fungi DNA Purification Kit (EURx, Gdansk, Poland) according to the manufacturer's protocol. The quality and quantity of the isolated DNA were checked spectrophotometrically (Nanodrop) and electrophoretically by separation in a 1% agarose gel with ethidium bromide (BioRad, Hamburg, Germany). The isolated DNA served as a template for the PCR reaction which was run under the following conditions: The amplification reaction used a primer with the sequences (5′-CTTGGTCATTTAGAGGAAGTA-3′ and 5′-TCCTCCGCTTATTGATATGC-3′), named ITS1_f and ITS2, respectively. The genomic region including internal transcribed regions ITS2 (the region between 5.8S rRNA and 28S rRNA genes) and ITS1 (the region between 18S rRNA and 5.8 rRNA genes) was amplified [21,22]. The amplified PCR products were sequenced using the Sanger method (Genomed, Warsaw, Poland). The obtained sequences were compared with those contained in GenBank NCBI (https://blast.ncbi.nlm.nih.gov/Blast.cgi (accessed on 3 October 2022) and MycoBank Polyphasic Identifications Database (http://www.mycobank.org (accessed on 3 October 2022).

### 2.2. Chitosan Films Preparation

Phenolic acids (tannic acid, ferulic acid, and gallic acid) used for chitosan film preparation were acquired from Sigma Aldrich (Saint Louis, MO, USA). Phenolic acids were selected based on our previous research, where those compounds were used as chitosan crosslinkers [23]. Raw compounds such as chitosan and all phenolic acids were dissolved in 0.1 M acetic acid as a solvent to obtain a solution at 1% concentration. Phenolic acids were added to the chitosan (1/10 *w*/*w*%) and mixed on a magnetic stirrer. Mixtures were put on the plastic holder (40 mL per 10 × 10 cm) and left for solvent evaporation in room conditions. This resulted in obtaining thin dry films.

### 2.3. Enzymatic Characteristics of Trichoderma

The study included the determination of hydrolytic enzymes involved in the degradation of organic matter: lipase, aminopeptidase, α, and β-glucosidase. For this purpose, a liquid medium containing (g/L) sucrose 10 and yeast extract 3 was inoculated with fungal spores ($10^5$/mL) and incubated for 4 days at 25 °C. The culture was then centrifuged (10 min, 10,000× *g*) and the supernatant was collected for the determination of the enzyme activity. Subsequently, 2 mL of each supernatant was added to its corresponding substrate (0.2 mL): 4-Methylumbelliferyl butyrate for lipase activity, L-Leucine-7-amido-4-methylcoumarin hydrochloride for aminopeptidase activity, 4-methylumbelliferyl α-D-glucoside, and 4-methylumbelliferyl β-D-glucoside for α and β-glucosidase activity. The final concentration of substrates in the sample was 50 μmol/mL. Incubation was carried out in a thermoblock at 40 °C for one hour. After incubation, the amount of MUF and MCA released was measured using a Hitachi F—2500 spectrophotometer [24–26].

Excitation/emission wavelengths were centered on 318/445 nm for methylumbelliferyl (MUF) and 345/425 for amido-4-methylcoumarinhydrochloride (MCA). The unit of activity (U) of enzymes was taken as μmol of MUF/MCA released per hour.

*Trichoderma* is known for its antifungal activity, therefore the activity of chitinase and β-1,3-glucanase (enzymes having antifungal properties) were also determined. For this purpose, a liquid medium containing (g/L) sucrose 5, yeast extract 1, and colloidal chitin 10 was inoculated with a suspension of fungal spores ($10^5$/mL) and incubated at 25 °C for 4 days. After incubation, the culture was centrifuged (10 min, 10 rpm) and the activity of chitinase and β-1,3-glucanase was determined from the supernatant.

For the measurement of chitinase activity, the procedure mentioned in the above paragraph was used, taking methylumbelliferyl-N-acetyl-β-D-glucosaminide as the substrate [25]. The β-1,3-glucanase activity was determined according to Wu et al. [27] by the spectrophotometric method. In brief, 500 µL of supernatant was mixed with 500 µL of laminarin 0.5% ($v/w$) in 100 mM sodium acetate buffer (pH 5.5). The reaction was allowed to proceed at 50 °C for 60 min after which it was terminated by heating for 5 min at 100 °C. Thereafter, 2 mL of dinitrosalicylate (DNS) 1% ($v/w$) was added to the reaction and the mixture was boiled for 10 min. After cooling the samples, the sugar concentration was measured with a spectrophotometer at 540 nm. The β-1,3-glucanase of activity was expressed as µmoL of glucose per hour.

### 2.4. Biological Activity of Trichoderma in the Presence of Chitosan Films

The biological activity of *Trichoderma* in the presence of the test films was determined using the respirometric method with the OxiTop Control system (WTW, Wroclaw, Poland) according to the modified method of Swiontek Brzezinska et al. [28]. Into the OxiTop measuring dishes, 100 mL of sterile medium containing (g/L) sucrose 5, yeast extract 0.5, and fragments (2 cm × 2 cm) of the test films (total mass 1 g) were placed and inoculated with 0.1 mL of *Trichoderma* suspension (spores concentration $10^5$/mL). The quivers with $CO_2$ absorbent (0.4 g NaOH) were placed in the measuring dishes and the measuring heads were screwed on. Incubation was carried out at 26 °C for 14 days. The control sample contained a polymer-free medium inoculated with the same amount of fungal suspension. The biological activity was expressed as mg of $O_2$/L of culture.

### 2.5. Fourier Transform Infrared Spectroscopy—Attenuated Total Reflectance (FTIR–ATR)

FTIR–ATR analysis was made for each kind of chitosan film modified by phenolic acids. Films before and after degradation were placed under the diamond ATR crystal and scanned with 4 $cm^{-1}$ resolution in the range of 4000–600 $cm^{-1}$ using Nicolet iS10 spectrometer (Thermo Fisher Scientific Inc., Waltham, MA, USA). Each registered spectrum was obtained by averaging 64 scans. Spectra were recorded and graphs were prepared in the Excel program. Only the CTS/GA and CTS/TA films were selected for testing as the other films were broken and were not suitable for further analysis.

### 2.6. Mechanical Properties

The mechanical testing was carried out using a testing machine (Shimadzu EZ-Test EZ-SX, Kyoto, Japan). Samples were inserted between two handles and stretched (5 mm/min). Young's Modulus was calculated from the slope of the stress–strain curve in the linear region with the use of the Trapezium X Texture program. The thickness of the samples was measured using a thickness gauge (Sylvac, Yverdon-les-Bains, Switzerland). Each measurement was carried out in ten repetitions [6].

### 2.7. Scanning Electron Microscopy (SEM)

The morphology of the samples was studied using Scanning Electron Microscope (SEM) (LEO Electron Microscopy, Thornwood, NY, USA). The films were covered with gold before the observation. Scanning electron microscope images were made with a magnification of 1000×.

### 2.8. Biodegradation of Chitosan Films in the Compost after Application of Trichoderma Strains

The biodegradation of the chitosan films in the compost after the application of *Trichoderma* strains was determined using the OxiTop system [29] according to the modified procedure described previously. For this purpose, 100 g of compost was placed in the jar of the OxiTop system, mixed with chitosan film fragments (total mass, 1 g, 2 cm × 2 cm), and inoculated with a 1 mL fungal culture ($10^9$ CFU/mL). Incubation was carried out at 26 °C for 21 days.

The study was carried out in variants: variant 1—biodegradation of chitosan films by native compost microorganisms; variant 2—biodegradation of chitosan film after application of *T. atroviride* TN1; variant 3—biodegradation of chitosan film after application *T. citrinoviride* TN3; variant 4—biodegradation of chitosan film after application of consortium composed of TN1 and TN3. The control sample contained compost without chitosan films and without supplementation with a fungal culture. The susceptibility to biodegradation of chitosan films was determined based on oxygen consumption and expressed as $mgO_2$/kg compost. The physicochemical parameters of the compost are shown in Table 1. The pH of the compost was measured by the potentiometric method using a glass electrode in 1 M KCl. The Tiurin procedure was used to assess total organic carbon (TOC), while total nitrogen (TN) was determined by the Kjeldahl method [30]. Soil humidity was measured with a tensionometer T1-q-30. $P_2O_5$, $K_2O$, Mg, N-$NO_3$, and $NNH_4$ were determined according to the following standards: PN-ISO 10390:1997, PN-R-04022:1996, PN-R-04022:1996/Az1:2002, PN-R-04020:1994/Az1:2004, and PN-R-04028:1997 [31–35]. Biological oxygen consumption (BOD) at 20 °C for 5 days was measured using the OxiTop measurement system [36]. Hydrolase activity was checked using fluorescein diacetate (Sigma-Aldrich) by following Adam & Duncan's [37] method, with a small modification. Microbial abundance was tested using the Koch plate method. Compost samples, after a series of tenfold dilutions, were transferred to nutrient agar (for bacteria), potato glucose agar (for molds), and actinomycetes agar (for actinomycetes).

**Table 1.** Physicochemical properties of the compost. Values are expressed as mean ± SD ($n$ = 3).

| Parameters | Values |
|---|---|
| pH | 7.5 |
| humidity | 200 mbar/hPa |
| TOC (g/kg) | 351 |
| TN (g/kg) | 27 |
| N-$NH_4$ (mg/kg) | 2.72 |
| N-$NO_3$ (mg/kg) | 29.0 |
| $P_2O_5$ (mg/100 g) | 26.2 |
| Mg (mg/100 g) | 7.1 |
| $K_2O$ (mg/100 g) | 27.5 |
| $BOD_5$ ($mgO_2$/kg) | 80 |
| Hydrolase activity (mg/kg) | 14 |
| Heterotrophic bacteria numbers (CFU/g) | $45 \times 10^5$ |
| Actinomycete numbers (CFU/g) | $39 \times 10^4$ |
| Mold numbers (CFU/g) | $23 \times 10^3$ |

### 2.9. Effect of Application of *Trichoderma* on Hydrolytic Enzymes Activity in Compost

Compost samples without chitosan films (control), compost samples containing chitosan films without fungal supplementation, and compost samples with chitosan films after bioaugmentation with a *Trichoderma* preparation were used for the study. To determine the enzymatic activity, 10 g of compost was ten-fold diluted using saline (0.9% NaCl) and shaken for 20 min. The activities of aminopeptidase, lipase, chitinase, α, and β-glucosidase were determined in the diluted samples as described previously (point 2.3).

### 2.10. Statistical Analysis

The statistical analyses of studies in 3.2, 3.3, and 3.8 point were conducted using Past v. 3.08 [38], with three replicates for each type of data. For all datasets, we calculated summary statistics (mean and standard deviation), performed a normality test (Shapiro–Wilk test), and assessed homogeneity of variance (Levene's test). One-way ANOVA was employed to examine differences in enzyme activity and BOD between sample groups with different *Trichoderma* strains. Tukey's pairwise test was applied for post hoc comparisons.

The statistical analysis of the rest of the study was performed using commercial software (SigmaPlot 14.0, Systat Software, San Jose, CA, USA). The Shapiro–Wilk test was used to assess the normal distribution of the data. All the results were calculated as means ± standard deviations (SD) and statistically analyzed using one-way analysis of variance (one-way ANOVA). Multiple comparisons versus the control group between means were performed using the Bonferroni *t*-test with the statistical significance set at $p < 0.05$.

## 3. Results and Discussion

### 3.1. Molecular Identification of *Trichoderma*

Conventional methods for the identification of *Trichoderma* spp. using cultural and morphological methods might provide erroneous results [39]. A better approach is to use molecular phylogenetic characteristics in combination with conventional methods [40]. One of the molecular methods used to identify fungi is rDNA sequencing [41–43] identification which has made it possible to identify fungi belonging to the species of *Trichoderma atroviride* and *Trichoderma citrinoviride*. The resulting sequences designated TN1–3 have been deposited with GeneBank NCBI under accession numbers (TN1—ID OP586643.1; TN2—ID OP586653.1 and TN3—ID: OP586652.1). A comparison of the TN1 sequence with the sequences available from MycoBase showed homologies to *T. atroviride* and *T. citrinoviride*. The TN1 sequence showed homology to *T. atroviride* (NBRC 101776) with 99.82% sequence similarity, TN2 was 99.83% homologous to *T. atroviride* (FMR12650), and TN3 sequence showed 100% similarity to *T. citrinoviride* (DAOM 172792). Identification of fungi is often problematic and it is not always possible to identify them by species [43]. There are various molecular characterization techniques to confirm *Trichoderma* isolates, such as amplification and sequence analysis of the internal transcribed spacer gene ITS 1 and 2 and the translation elongation factor 1-alpha (tef1) encoding gene and BLAST interface in TrichOKEY and TrichoBLAST (https://trichokey.com/index.php/trichoblast (accessed on 3 October 2022) [39]. Dendrogram prepared using nucleotide sequences of the ITS region of the 18S rRNA gene and using Clustal W showed in Figure S1.

### 3.2. Enzymatic Activity of *Trichoderma* Strains

The results of enzymatic activities of three different *Trichoderma* cultures designated as TN1, TN2, and TN3 are presented in Table 2.

**Table 2.** Enzymatic activity of *Trichoderma* strains. Values are expressed as mean ± SD ($n$ = 3). Different letters in the same line indicate statistically significant differences with $p < 0.05$).

| Enzymes Activity (U/h) | *Trichoderma artoviride* TN1 | *Trichoderma artoviride* TN2 | *Trichoderma citrinoviride* TN3 |
|---|---|---|---|
| chitinase | 0.82 ± 0.03 $^{a}$ | 0.4 ± 0.03 $^{c}$ | 0.70 ± 0.04 $^{b}$ |
| β—1,3—glucanase | 1.63 ± 0.02 $^{a}$ | 0.7± 0.02 $^{b}$ | 1.9 ± 0.01 $^{a}$ |
| lipase | 2.23 ± 0.01 $^{a}$ | 0.51 ± 0.20 $^{c}$ | 1.99 ± 0.06 $^{b}$ |
| α-glucosidase | 0.08 ± 0.01 $^{a}$ | 0.06 ± 0.01 $^{b}$ | 0.08 ± 0.04 $^{a}$ |
| β-glucosidse | 0.05 ± 0.01 $^{b}$ | 0.02 ± 0.01 $^{c}$ | 0.09 ± 0.02 $^{a}$ |
| aminopeptidase | 1.7 ± 0.02 $^{b}$ | 1.9 ± 0.01 $^{a}$ | 1.8 ± 0.01 $^{a}$ |

Strain *T. atroviride* TN1 and *T. citrinoviride* TN3 appeared to be the most enzymatically active. All the *Trichoderma* strains were characterized by the highest lipase and aminopeptidase activity. These enzymes are important in the degradation of organic matter, which can be deposited in significant quantities in the environment. All the tested isolates produced chitinases and β-1,3-glucanases, which are known for their fungicidal properties. By producing fungicidal chitinases, *Trichoderma* strains are used for plant protection.

Also, some secondary metabolites and hydrolytic enzymes excreted by *Trichoderma* have been shown to be involved in inhibiting the growth of pathogenic microorganisms and stimulating plant growth [44]. Cherkupally et al. [45] studied the *Trichoderma* genus for activity on amylases, cellulases, peptidases, pectinases, and chitinases. They reported that *T. harzianum*, *T. pseudokoningii*, *T. atroviride*, *T. viride*, *T. virens*, and *T. koningii* had high chitinolytic activity. Similarly, Xia et al. [46] found that *T. viride* showed high chitinase activity. Win et al. [47] isolated *Trichoderma asperellum*, which presented antifungal activity against four species of phytopathogenic fungi from the *Fusarium* genus. In addition, *Trichoderma asperellum* has been shown to secrete mycolytic enzymes (chitinase and β-1,3-glucanase). The number of chitinase genes showed great variation in fungal genomes, ranging from a single gene in *Schizosaccharomyces pombe* to 36 genes in *T. virens* [48].

*Trichoderma* have the ability to detoxify pesticides and herbicides [49]. This is certainly due to the processing of substances by extracellular and intracellular enzymes. According to Abdenaceur et al. [50], *Trichoderma* isolates obtained in the ecosystem niche of olives showed high potential for plant growth promoting biomolecule production. According to Wang et al. [51], *Trichoderma lentiforme* ACCC30425 showed the highest lipase production capacity at high pH. According to Qian et al. [52], the filamentous fungus *Trichoderma reesei* is one of the best-known cellulolytic organisms and the significant producer of cellulases for industrial applications. In addition, aspartic protease P6281 secreted by *Trichoderma harzianum* is important in mycoparasitism on phytopathogenic fungi [53].

### 3.3. Respirometric Activity of Trichoderma in the Presence of Modified Chitosan Films

The tested *Trichoderma* strains degraded chitosan films at different levels (Table 3).

**Table 3.** Biological oxygen demand of *Trichoderma* strains in the presence of chitosan films modified by phenolic acids.

| Materials | *Trichoderma artoviride* TN1 | *Trichoderma artoviride* TN2 | *Trichoderma citrinoviride* TN3 |
|---|---|---|---|
| CTS | * 137 ± 2.2 [a] | 89 ± 3.3 [a] | 110 ± 2.5 [a] |
| CTS/TA | 210 ± 3.1 [b] | 120 ± 3.4 [b] | 178 ± 3.2 [b] |
| CTS/GA | 104 ± 4.0 [a] | 98 ± 1.3 [a] | 117 ± 3.3 [a] |
| CTS/FA | 257 ± 4.2 [c] | 130 ± 3.2 [b] | 220 ± 4.1 [c] |

* Biological activity was expressed in mg $O_2$/L after 8 days of incubation. Values are expressed as mean ± SD ($n$ = 3). Different letters in the columns indicate statistically significant differences with $p < 0.05$).

*T. atroviride* TN1 and *T. citrinoviride* TN3 metabolized the tested films best. At the same time, a statistically significant influence of phenolic acids incorporated into chitosan on the biological activity of *Trichoderma* was observed. CTS/FA films were best used by *Trichoderma* (BOD 130–257 mg $O_2$/L after 8 days). In turn, CTS/GA films were less prone to biodegradation (BOD 98–117 mg $O_2$/L after 8 days).

Chitosan has many intrinsic characteristics including biocompatibility and biodegradability [54]. Furthermore, it shows antimicrobial, antitumor, and antioxidant properties [55]. Inhibition of fungal and bacterial growth depends on the molecular weight of chitosan and its functional groups. Smaller oligomeric chitosan molecules can easily penetrate the cell membrane and inhibit RNA transcription to prevent cell growth [56].

Our previous research [4] showed that chitosan film modified with tannic acid was difficult to degrade by microorganisms. However, there is no doubt that there might be some microorganisms in the environment capable of utilizing chitosan as a source of carbon. Ohkawa et al. [17] studied the degradation of chitosan–gellan (CGF) by seven species of soil filamentous fungi. Biodegradation processes were observed under the microscope and it was revealed that only *P. caseicolum* on the CGF grew along with the breakdown of the fiber matrices. In the biochemical oxygen demand (BOD) test, the degradation of CGF by *A. oryzae* was the highest at 59% among the seven species of tested fungi. *Penicillium caseicolum* and *P. citrinum* also degraded the CGF at the level of 41% and 10%, respectively. Babaee et al. [57] examined the biodegradation of chitosan nanoparticles (CNPs) films using *T. versicolor*. According to these authors, the chitosan films underwent a fast degradation process during storage within the first 30 days and complete degradation after 50–60 days.

*3.4. Fourier Transform Infrared Spectroscopy—Attenuated Total Reflectance (FTIR–ATR)*

The degradation of chitosan films modified by phenolic acids was observed in all studied conditions (Figure 1A), however, it was not controlled by the type of phenolic acid added to chitosan. The degradation peak in the range of 3600–2700 cm$^{-1}$ was observed from amine and hydroxyl groups. A peak was not observed after the degradation. Similarly, a peak at 1025 cm$^{-1}$ from COH and $CH_2OH$ groups was observed before degradation and was not noticed after the treatment with fungi. It suggested that the degradation process broke C-N and C-O bonds. Peaks at 2055 and 1835 cm$^{-1}$ were present after degradation. It proved that the number of C=C and C=O groups had increased. The characteristic peaks of chitosan films modified by tannic acid are shown in Figure 1B. Similar observations were also found for CTS/GA acid films (Figure 1C).

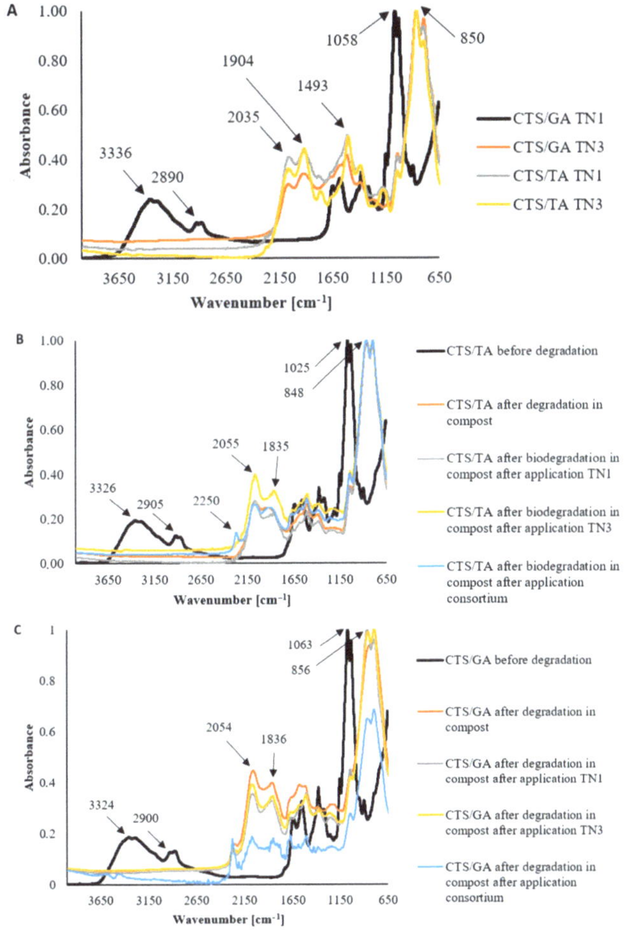


**Figure 1.** The FTIR–ATR spectra of (**A**) chitosan film modified by gallic acid (CTS/GA) and tannic acid (CTS/TA) after degradation of *Trichoderma* strains (TN1 and TN3); chitosan film modified by tannic acid (**B**) and gallic acid (**C**) before and after degradation in compost after application *Trichoderma* strains (TN1 and TN3).

Through previous studies, it has been shown that chitosan can be modified by phenolic acids addition. The FTIR spectra of chitosan-based materials with ferulic acid, gallic acid, and tannic acid were published before by Kaczmarek-Szczepańska et al. [21]. Significant changes in the localization and the presence of characteristic groups on the spectra were not observed. It was assumed that covalent bonds between chitosan and phenolic acids are not formed, and the cross-linking process is based on the hydrogen bond formation between hydrophilic groups.

Stoleru et al. [58] tested the biodegradation of poly(lactic acid)/chitosan by *Phanerochaete chrysosporium*, a white rot fungus. The degradation was followed by infrared spectroscopy (FTIR). Notable structural changes resulted from the fungal biodegradation of PLA/chitosan-based samples which showed the onset of hydrolytic or fungal degradation. Biodegradation by the fungi increased the crystallinity of the samples, which means that degradation occurs primarily in amorphous regions. It should be noted that chitosan in the initial days of incubation may act as an antimicrobial agent and the fungi tested may be less active in degrading chitosan films. Stoleru et al. [58] reported that changes in structure are less pronounced for polylactide/chitosan composite than for uncoated polylactide which undergoes important changes after a short incubation period. Therefore, after degradation in the presence of P. chrysosporium, due to hydrolytic cleavage of polylactide backbone and chitosan deacetylation, both components of the layered composites exhibited structural changes. Our previous research [4] showed that bacteria isolated from chitosan films modified by tannic acid accelerated the degradation of the materials. Before the degradation, films based on chitosan and tannic acid in a ratio of 80/20 had a big peak at 1289 $cm^{-1}$ from the C-O-C group, which corresponded to the backbone of the chitosan polymeric chain. It decreased after biodegradation. This may be due to the fact that the bond is broken by the action of microorganisms.

### 3.5. Mechanical Properties

All tested samples after the degradation process showed greater thickness than before (Table 4).

**Table 4.** The thickness of films based on chitosan (CTS) with tannic acid (TA), ferulic acid (FA), and gallic acid (GA) before and after degradation (significantly different $p < 0.05$ from control: [a] vs. CTS, [b] vs. CTS/TA, [c] vs. CTS/GA).

| Sample | Thickness [mm] | Sample | Thickness [mm] |
|---|---|---|---|
| CTS | 0.090 ± 0.009 | CTS/TA after degradation in compost | 1.073 ± 0.011 [a,b] |
| CTS/TA | 0.131 ± 0.008 [a] | CTS/TA after degradation in compost after application TN1 | 1.038 ± 0.017 [a,b] |
| CTS/FA | 0.100 ± 0.010 | CTS/TA after degradation in compost after application TN3 | 1.098 ± 0.010 [a,b] |
| CTS/GA | 0.080 ± 0.006 | CTS/TA after degradation in compost after application consortium | 1.671 ± 0.013 [a,b] |
| | | CTS/GA after degradation in compost | 1.011 ± 0.008 [c] |
| | | CTS/GA after degradation in compost after application TN1 | 1.045 ± 0.014 [c] |
| | | CTS/GA after degradation in compost after application TN3 | 1.024 ± 0.009 [c] |
| | | CTS/GA after degradation in compost after application consortium | 1.781 ± 0.018 [c] |

As a result of degradation, the film shrinks, thereby allowing the increase in thickness to be observed. Mechanical parameters such as Young Modulus, maximum tensile strength, and elongation at break were determined (Figure 2).

It was noticed that the highest value of Young Modulus and maximum tensile strength were noticed for the material composed of chitosan and ferulic acid (Figure 2). Mathew & Abraham [59] showed that the incorporation of ferulic acid to starch/chitosan films enhances the tensile strength. In terms of gallic acid incorporated into chitosan/gelatin, the chitosan-based films modified by gallic acid showed higher Emod and $\sigma$max values than the films without phenolic acids. Similar results were obtained by Rui et al. [60] where gallic acid addition resulted in the increase in mechanical properties of the fabricated films. No significant differences in the elongation at break values were observed between the studied samples. Qin et al. [61] observed that the addition of tannic acid to chitosan results in an increase in tensile strength as well as the Young Modulus. 0Also, Sun et al. [62] documented that the presence of gallic acid results in a significant increase in the tensile strength of the studied chitosan/gallic acid films. No statistically significant differences in the elongation at break values were observed between the studied samples. Similar results were obtained by Rivero et al. [63] where the elongation at break did not significantly change after the addition of phenolic acids.

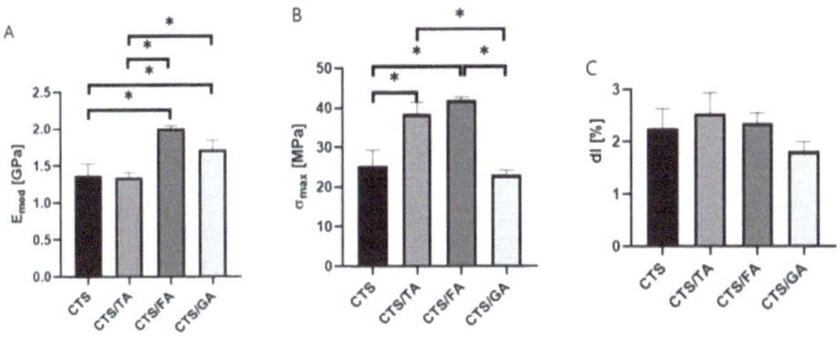


**Figure 2.** The Young Modulus (**A**), maximum tensile strength (**B**), and elongation at break (**C**) of films based on chitosan (CTS) with tannic acid (TA), ferulic acid (FA), and gallic acid (GA) (* significantly different $p < 0.05$).

### 3.6. Scanning Electron Microscopy (SEM)

The images of CTS/TA and CTS/GA before degradation showed the smooth surface of films without any scratches or structures (Figure 3).

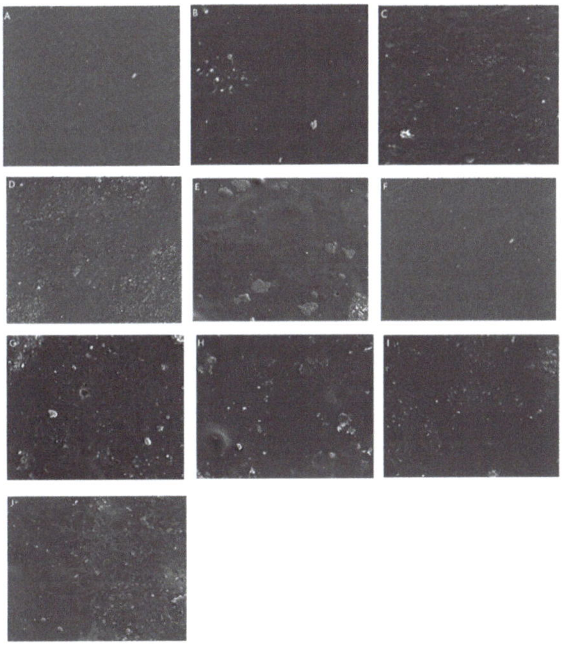

**Figure 3.** The SEM images (1000×) of (**A**) CTS/TA before degradation; (**B**) CTS/TA after degradation in compost; (**C**) CTS/TA after biodegradation in compost after application of TN1; (**D**) CTS/TA after biodegradation in compost after application of TN3; (**E**) CTS/TA after biodegradation in compost after application of consortium; (**F**) CTS/GA before degradation; (**G**) CTS/GA after degradation in compost; (**H**) CTS/GA after biodegradation in compost after application of TN1; (**I**) CTS/GA after biodegradation in compost after application of TN3; (**J**) CTS/GA after biodegradation in compost after application of consortium.

It can be observed that after degradation in the compost, the soil grains are present on the surface of both types of films. For images after biodegradation in the compost after application, the *Trichoderma* application allowed us to observe the presence of fungi-like structures that adhere to the materials' surface. It suggests that the biodegradation process includes the adhesion of fungi to the CTS/TA and CTS/GA surface, both TN1 and TN3 types. Similar structures were observed by Poorna Chandrika et al. [64] on SEM images of chitosan-PEG blended films using *Trichoderma* and by Cheng et al. [65] on carvacrol/sodium alginate films.

### 3.7. Biodegradation of Chitosan Films in the Compost after Application of Trichoderma

The biodegradation of the chitosan film modified by phenolic acids after application of *Trichoderma* is shown in Figure 4.

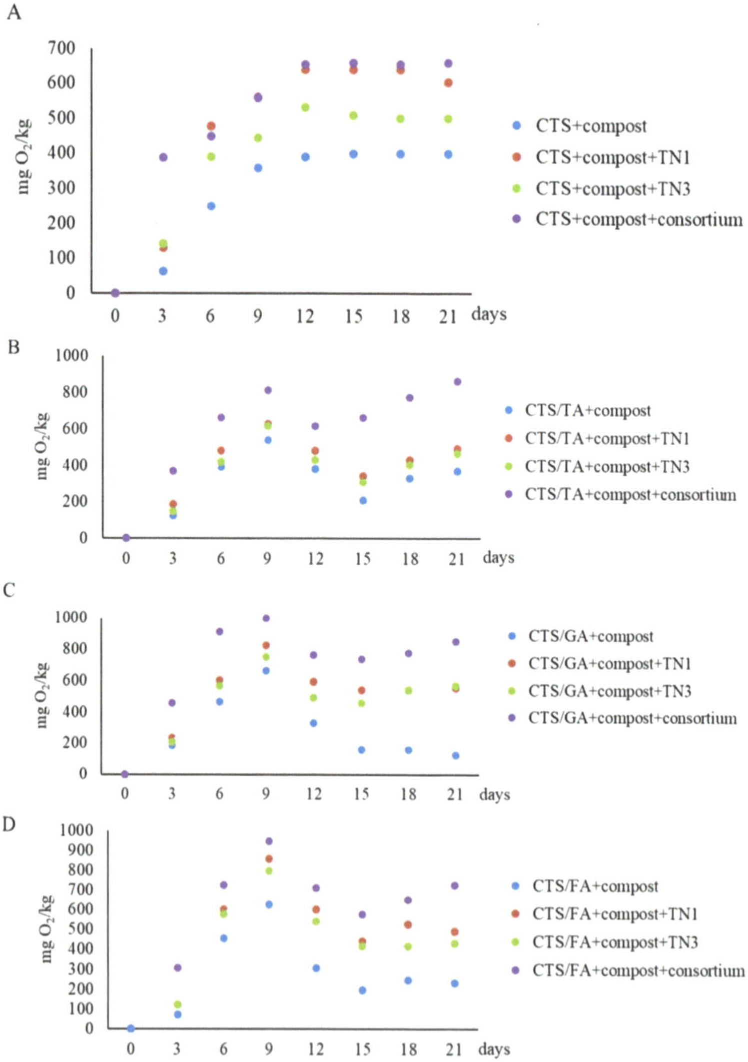


**Figure 4.** The biodegradation of chitosan film modified by phenolic acids after application of *Trichoderma* strains (TN1 and TN3): chitosan (**A**), chitosan modified tannic acid (**B**), chitosan modified gallic acid (**C**), and chitosan modified ferulic acid (**D**).

Overall, the oxygen consumption by microorganisms in the presence of chitosan films modified by phenolic acids was significantly dependent on the type of phenolic acids used, the fungal culture, and the interaction of the two factors (Table 5).

**Table 5.** Two-way ANOVA results on the influence of phenolic compounds, fungal strain/consortium (experiment variant), and their interaction on biological oxygen demand (BOD 21) in compost.

| | Sum of Squares | df | Mean Square | F | *p*-Value |
|---|---|---|---|---|---|
| Phenolic substance | $4.48 \times 10^4$ | 3 | $1.49 \times 10^4$ | $1.25 \times 10^3$ | $3.32 \times 10^{-33}$ |
| Variant | $1.46 \times 10^6$ | 3 | $4.88 \times 10^5$ | $4.08 \times 10^4$ | $2.24 \times 10^{-57}$ |
| Interaction | $2.46 \times 10^5$ | 9 | $2.74 \times 10^4$ | $2.29 \times 10^3$ | $2.55 \times 10^{-42}$ |
| Within | $3.83 \times 10^2$ | 32 | $1.20 \times 10^1$ | | |
| Total | $1.76 \times 10^6$ | 47 | | | |

The application of *T. atroviride* TN1 and *T. citrinoviride* TN3 similarly influenced the degradation of the CTS/TA film. The respiratory activity after 21 days of incubation was 492 $mgO_2/kg$ and 468 $mgO_2/kg$. However, the highest oxygen consumption in the presence of CTS/TA film was found after the application of the consortium of these strains (861 mg $O_2/kg$ after 21 days of incubation). Similarly, chitosan with gallic acid (CTS/GA) was least degraded by the native microorganisms of the compost (Figure 4B). The application of *T. atroviride* TN1 and *T. citrinoviride* TN3 similarly influenced the degradation of the CTS/GA film. The respiratory activity was 553 $mgO_2/kg$ and 566 $mgO_2/kg$, respectively, after 21 days of incubation. However, the highest oxygen consumption in the presence of CTS/GA was found after the application of the consortium of these strains (849 $mgO_2/kg$ after 21 days of incubation). Similar relationships were observed after the introduction of chitosan with ferulic acid (CTS/FA) into the compost after the application of the *Trichoderma* consortium (Figure 4C). The films before and after degradation in Oxi-Top are shown in Figure 5.

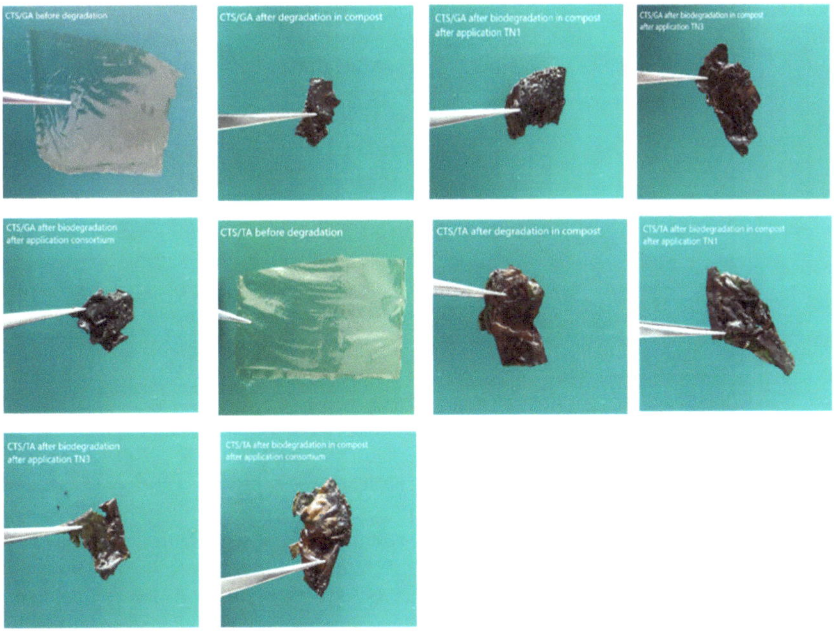


**Figure 5.** The films before and after degradation in OxiTop.

Composting is a significant method of waste management, including polymer films. Altun et al. [66] used controlled composting reactors to study the biodegradation of chitosan films. Microbial diversity was also determined by PCR-denaturing gradient gel electrophoresis (DGGE). The dominant taxonomic groups of microorganisms in composting reactors at the phylum level were Ascomycota and Proteobacteria. Carbon dioxide emissions data showed a high degradation rate of chitosan in reactors inoculated with microorganisms, which were capable of producing chitosanase. The control reactor contained only chitosan films (no microbial inoculum). In comparison, the group I microorganisms (*Bacillus circulans, Bacillus licheniformis, Streptomyces roseolus, Streptomyces zaomyceticus, Penicillium islandicum*, and *Penicillium chrysogenum*) increased the biodegradation rate. On the other hand, the group II inoculant (*Mycoplana dimorpha, Fennellia flavipes*, and *Chaetomium globosum*) reduced the efficiency of chitosan biodegradation. In the biodegradation of chitosan, the appropriate selection of microbial strains for inoculation is important, especially strains producing chitinases of different subclasses. Kaczmarek-Szczepańska et al. [4] reported that the bacteria isolated from the soil degraded the chitosan films modified by tannic acid (50CTS/50TA) better than the bacteria isolated from the compost. Their study also showed that consortia of bacteria (*Acinetobacter calcoaceticus* GTA2, *Bacillus mycoides* GTA7, *Pseudomonas arsenicoxydans* GTA8, and Pseudomonas laurylsulfativorans GTA9) added to soil had a positive effect on the biodegradation of the 50CTS/50TA. Similarly, consortia of bacteria (*Pseudomonas hunanensis* KTA1 and *Pseudomonas psychrophila* KTA3) added to the compost accelerated the degradation of chitosan films modified by tannic acid (50CTS/50TA).

Our study showed that *T. atroviride* metabolized chitosan-based materials, but the process was relatively slow. The chitosan films used for this study were modified with phenolic acids, which had antimicrobial properties. Therefore, it can be speculated that the degradation of such materials might start with the adaptation of microorganisms to the polymers and the production of enzymes such as chitosanase. Znajewska et al. [14] reported that *T. viride* DAR5, *T. viride* 3333, and *T. viride* 154 accelerated polycaprolactone (PCL) degradation in the soil. The greatest degree of polymer degradation expressed in terms of mass loss was shown by *T. viride* 3333 (complete disintegration of the PCL film). The other two strains degraded PCL to a lesser extent (PCL mass loss of 27.85% and 11.58%, respectively). These results were obtained after 6 months of PCL storage in the soil. On the other hand, the studies by Dąbrowska et al. [11] showed that *T. viride* GZ1 is capable of producing hydrolytic enzymes degraded polylactide and polyethylene terephthalate, producing hydrophobin in the first stage of biodegradation.

The introduction of engineered consortia of microorganisms into the environment often has a positive effect not only on the biodegradation processes of organic matter but also improves soil quality and favorably influences plant development. For example, the addition of suitable microorganism-based boosters compensates for stress and reduced plant growth caused by weeds, drought, heavy metals, salt, and other adverse environmental conditions [67].

### 3.8. Effect of Application of *Trichoderma* on Hydrolytic Enzymes Activity in Compost

Extracellular enzymes secreted by microorganisms into the compost significantly affect the degradation of organic matter. The introduction of the *Trichoderma* preparation into the compost further enriched the compost with enzymes capable of degrading macromolecular compounds (Table 6).

Lipase and aminopeptidase were found to be the most active. The greatest changes in enzyme activity were found for chitinases. These enzymes showed the highest activities after the introduction into the compost chitosan films with ferulic acid. The application of the *Trichoderma* consortium significantly increased the activity of chitinases in the compost. In contrast, the activity of glucosidases in the compost was the lowest even after the introduction of chitosan films and *Trichoderma*.

**Table 6.** Effect of chitosan film modified by phenolic acids on changes in soil hydrolytic enzyme activity after *Trichoderma* strains bioaugmentation.

| Variants | Enzyme Activity (U/h) | | | | |
|---|---|---|---|---|---|
| | Chit. | Lip. | α-Gluc. | β-Gluc | Amino. |
| Compost | 4.4 ±0.14 [a] | 19.4 ± 0.1 [a] | 4.2 ±0.13 [a] | 4.2 ± 0.03 [a] | 8.2 ± 0.05 [a] |
| CTS/TA + compost | 3.1 ± 0.03 [b] | 15.6 ± 0.01 [b] | 4.5 ± 0.01 [a] | 4.5 ± 0.01 [a] | 6.8 ± 0.05 [b] |
| CTS/TA + compost + TN1 | 5.8 ± 0.07 [c] | 16.8 ± 0.10 [b] | 4.5 ± 0.01 [a] | 4.3 ± 0.03 [a] | 9.2 ± 0.06 [c] |
| CTS/TA + compost + TN3 | 5.3 ± 0.06 [c] | 17.9 ± 0.07 [b] | 4.4 ± 0.01 [a] | 3.9 ± 0.03 [a] | 9.6 ± 0.02 [c] |
| CTS + compost + TN1 + TN3 | 5.8 ± 0.21 [c] | 34.2 ± 0.31 [c] | 4.6 ± 0.41 [a] | 4.2 ± 0.02 [a] | 9.4 ± 0.06 [c] |
| CTS/GA + compost | 4.0 ± 0.18 [a] | 7.0 ± 0.05 [b] | 3.6 ± 0.01 [a] | 4.1 ± 0.01 [a] | 1.3 ± 0,03 [b] |
| CTS/GA + compost + TN1 | 5.5 ± 0.01 [a] | 13.2 ± 0.12 [c] | 3.7 ± 0.01 [a] | 4.2 ± 0.02 [a] | 5.8 ± 0.01 [c] |
| CTS/GA + compost + TN3 | 6.6 ± 0.03 [b] | 13.4 ± 0.01 [c] | 3.0 ± 0.01 [a] | 4.3 ± 0.02 [a] | 6.2 ± 0.01 [c] |
| CTS/GA + compost + TN1 + TN3 | 11.2 ± 0.12 [c] | 13.6 ± 0.01 [c] | 3.0 ± 0.01 [a] | 3.9 ± 0.01 [a] | 5.4 ± 0.05 [c] |
| CTS/FA + compost | 3.9 ± 0.28 [a] | 19.9 ± 0.08 [a] | 4.7 ± 0.01 [a] | 4.2 ± 0.08 [a] | 3.7 ± 0.02 [b] |
| CTS/FA + compost + TN1 | 25 ± 0.11 [b] | 14.0 ± 0.04 [b] | 4.8 ± 0.01 [b] | 4.4 ± 0.01 [a] | 7.4 ± 0.03 [c] |
| CTS/FA + compost + TN3 | 28.3 ± 0.28 [b] | 23.3 ± 0.01 [b] | 4.6 ± 0.03 [b] | 4.9 ± 0.01 [a] | 6.5 ± 0.03 [c] |
| CTS/FA + compost + N1 + TN3 | 38.5 ± 0.57 [c] | 25.1 ± 0.01 [b] | 4.3 ± 0.02 [b] | 4.6 ± 0.01 [a] | 6.1 ± 0.06 [c] |

Values are expressed as mean ± SD ($n$ = 3). Chit—chitinase, Lip—lipase, α-gluc—α-glucosidase, β-gluc—β-glucosidase, and Amino—aminopeptodase. Different letters in the column indicate statistically significant differences at $p < 0.05$ according to the HDS Tukey's test.

The introduction of materials with antimicrobial properties into the environment often affects the production of extracellular enzymes by microorganisms. And our results also showed that for most enzymes, inhibition of activity was observed after the introduction of chitosan films with ferulic, tannic, and gallic acids. Further, the addition of the *Trichoderma* consortium to the compost improved enzyme synthesis, especially chitinase, lipase, and aminopeptidase. Klose et al. [68] fumigated toxic compounds including methyl bromide, propargyl bromide, chloropicrin, and sandy soil, and then checked the activity of acid phosphatase, arylsulfatase, β-glucosidase, and dehydrogenase. Their tests showed that arylsulfatase (62%), dehydrogenase (35%), acid phosphatase (22%), and β-glucosidase (6%) activities decreased within 90 days after fumigation. Su et al. [69] studied enzyme activity in arsenic-contaminated soils. After inoculation of *Trichoderma asperellum* SM-12F1 into contaminated soil, they observed β-glucosidase (155%), chitinase (211%), and phosphatase (108%) activities in SM soils, while significantly decreased β-glucosidase (81%), phosphatase (54%), aminopeptidase (60%), and catalase (67%) activities in soils. Such a result demonstrates the positive effects of bioaugmentation, which is increasingly practiced. Inoculation of *Trichoderma* into the environment can improve enzymatic activity in the rhizosphere of plants. Yedidia et al. [70] reported that cucumber roots inoculated with *Trichoderma harzianum* T-203 showed higher chitinase, β-1,3-glucanase, cellulase, and peroxidase activities up to 72 h after inoculation compared to roots without *Trichoderma*. Soil enzyme activity is sensitive to environmental changes and has been proposed as an indicator of soil quality due to its relationship with soil biology [71] and rapid responses to changes in management and soil environment [72,73]. Therefore, the introduction of biocidal materials into the environment is associated with the inhibition of enzyme activity. Inoculation with microorganisms with high enzymatic activity can improve soil quality as well as soil enzyme activity.

## 4. Conclusions

Chitosan is known primarily for its antifungal properties. Chitosan-based liquid preparations have been developed for plant protection. Chitosan films are also used in medicine as a dressing material. The introduction of biocidal phenolic acids into chitosan film makes it possible to use it as food packaging to protect against the growth of both bacteria and fungi. The potential of modified chitosan films may be seen in the protection of fruits and vegetables. Such foil can protect them against the development of fungi. However, packaging materials should be degradable in the environment. Studies have shown that chitosan modified with phenolic acid is less readily degraded by native microorganisms. Only after the application of *Trichoderma* does the degradation of chitosan films occur more effectively. At the same time, studies have shown a beneficial effect of the use of *Trichoderma* on the enzymatic activity of compost. Chitinases, the activity of which increased after the introduction of the preparation, can control the development of phytopathogens.

**Supplementary Materials:** The following supporting information can be downloaded at: https://www.mdpi.com/article/10.3390/foods12193669/s1, Figure S1: Dendrogram prepared using nucleotide sequences of the ITS region of the 18S rRNA gene and using Clustal W (https://www.genome.jp/tools-bin/clustalw, accessed on 3 October 2022). Sequences from NCBI's GeneBank were used to analysis: (Acc. no. OP586652.1) *T. citrinoviride*, (OP586653.1) *T. atroviride* G, (OP586643.1), *T. atroviride* isolate Fb1, (AY380909.1) *T. viride* strain ATCC 28038, (X93978.1) *T. viride* strain ATCC 18652, (AF279395.1) *T. citrinoviride* strain ATCC 90178, (JQ745264.1) *T. atroviride* strain ATCC 20476, (AF278796.1) *T. atroviride* strain ATCC 36042, (AF127152.1) *T. atroviride* ATCC 28036, (DQ315445.1) Trichoderma trixiae strain ATCC 32630, (NG_059411.1) *Trichoderma koningii* ATCC 64262, (DQ315434.1) *Trichoderma paraviridescens* strain ATCC 20898, (AF127154.1) *Trichoderma reesei* QM6a ATCC 13631, (AF127145.1) *T. viride* strain Tr48 ATCC 5242, (AY625068.1) *Trichoderma harzianum* strain NRRL 13019.

**Author Contributions:** M.S.B.: Conceptualization, methodology, investigation, project administration, supervision, and writing—original draft. G.B.D.: Investigation and writing—original draft. B.K.-S.: Writing—original draft, investigation, and data curation. M.M.-S.: Investigation. K.D.: Writing—review and editing. A.R.: Writing—original draft. M.P.: Writing—review and editing, and data curation. S.B.K.: Writing—review and editing. A.K.: Writing—review and editing, and data curation. All authors have read and agreed to the published version of the manuscript.

**Funding:** This research was supported by funds provided by Nicolaus Copernicus University (Torun, Poland) to maintain research potential and the Excellence Initiative Research University competition for scientific groups—BIOdegradable PACKaging materials research group no. 4101.00000085 IDUB/Research Group.

**Data Availability Statement:** The data used to support the findings of this study can be made available by the corresponding author upon request.

**Conflicts of Interest:** The funders had no role in the design of the study; in the collection, analyses, or interpretation of data; in the writing of the manuscript; or in the decision to publish the results.

## References

1. Vroman, I.; Tighzert, L. Biodegradable Polymers. *Materials* **2009**, *2*, 307–344. [CrossRef]
2. Priyadarshi, R.; Rhim, J.W. Chitosan-based biodegradable functional films for food packaging applications. *Innov. Food Sci. Emerg. Technol.* **2020**, *62*, 102346. [CrossRef]
3. Wang, W.; Xue, C.; Mao, X. Chitosan: Structural modification, biological activity and application. *Int. J. Biol. Macromol.* **2020**, *164*, 4532–4546. [CrossRef] [PubMed]
4. Kaczmarek-Szczepańska, B.; Michalska-Sionkowska, M.; Mazur, O.; Świątczak, J.; Swiontek Brzezinska, M. The role of microorganisms in biodegradation of chitosan/tannic acid materials. *Int. J. Biol. Macromol.* **2021**, *184*, 584–592. [CrossRef] [PubMed]
5. Borges, A.; Ferreira, C.; Saavedra, M.J.; Simões, M. Antibacterial activity and mode of action of ferulic and gallic acids against pathogenic bacteria. *Microb. Drug Resist.* **2013**, *19*, 256–265. [CrossRef]
6. Kaczmarek-Szczepańska, B.; Wekwejt, M.; Mazur, O.; Zasada, L.; Pałubicka, A.; Olewnik-Kruszkowska, E. The physicochemical and antibacterial properties of chitosan-based materials modified with phenolic acids irradiated by UVC light. *Int. J. Mol. Sci.* **2021**, *22*, 6472. [CrossRef]
7. Richert, A.; Olewnik-Kruszkowska, E.; Dąbrowska, G.B.; Dąbrowski, H.P. The Role of Birch Tar in Changing the Physicochemical and Biocidal Properties of Polylactide-Based Films. *Int. J. Mol. Sci.* **2022**, *23*, 268. [CrossRef]
8. Richert, A.; Kalwasińska, A.; Jankiewicz, U.; Swiontek Brzezinska, M. Effect of birch tar embedded in polylactide on its biodegradation. *Int. J. Biol. Macromol.* **2023**, *239*, 124226. [CrossRef]
9. Janczak, K.; Dąbrowska, G.B.; Raszkowska-Kaczor, A.; Kaczor, D.; Hrynkiewicz, K.; Richert, A. Biodegradation of the plastics PLA and PET in cultivated soil with the participation of microorganisms and plants. *Int. Biodeterior. Biodegrad.* **2020**, *155*, 1050. [CrossRef]
10. Richert, A.; Dąbrowska, G.B. Bacterial biofilm on PLA films and methods of its identification. *Ecol. Quest.* **2020**, *31*, 31–38. [CrossRef]
11. Dąbrowska, G.B.; Garstecka, Z.; Olewnik-Kruszkowska, E.; Szczepańska, G.; Ostrowski, M.; Mierek-Adamska, A. Comparative study of structural changes of polylactide and poly(ethylene terephthalate) in the presence of Trichoderma viride. *Int. J. Mol. Sci.* **2021**, *22*, 3491. [CrossRef] [PubMed]

12. Richert, A.; Dąbrowska, G.B. Enzymatic degradation and biofilm formation during biodegradation of polylactide and polycaprolactone polymers in various environments. *Int. J. Biol. Macromol.* **2021**, *176*, 226–232. [CrossRef] [PubMed]
13. Abdel-Motaal, F.F.; El-Sayed, M.A.; El-Zayat, S.A.; Ito, S.I. Biodegradation of poly (-caprolactone) (PCL) film and foam plastic by Pseudozyma japonica sp. nov., a novel cutinolytic ustilaginomycetous yeast species. *3 Biotech* **2014**, *4*, 507–512. [CrossRef] [PubMed]
14. Znajewska, Z.; Dąbrowska, G.B.; Hrynkiewicz, K.; Janczak, K. Biodegradation of polycaprolactone by Trichoderma viride fungi. *Chem. Ind.* **2018**, *97*, 1676–1679. [CrossRef]
15. Kannahi, M.; Thamizhmarai, T. Biodegradation of plastic by *Aspergillus* sp. *Int. J. Trend Sci. Res. Dev.* **2018**, *2*, 683–690. [CrossRef]
16. Vivi, V.K.; Martins-Franchetti, S.M.; Attili-Angelis, D. Biodegradation of PCL and PVC: Chaetomium globosum (ATCC 16021) activity. *Folia Microbiol.* **2019**, *64*, 1–7. [CrossRef] [PubMed]
17. Ohkawa, K.; Yamada, M.; Nishida, A.; Nishi, N.; Yamamoto, H. Biodegradation of chitosan-gellan and poly (L-lysine)-gellan polyion complex fibers by pure cultures of soil filamentous fungi. *J. Polym. Environ.* **2000**, *8*, 59–66. [CrossRef]
18. Znajewska, Z.; Dąbrowska, G.B.; Narbutt, O. Trichoderma viride stains stimulating the growth and development of winter rapeseed (*Brassica napus* L.). *Prog. Plant Prot.* **2018**, *58*, 264–269. [CrossRef]
19. Harman, G.E.; Howell, C.R.; Viterbo, A.; Chet, I.; Lorito, M. Trichoderma species—Opportunistic, avirulent plant symbionts. *Nat. Rev. Microbiol.* **2004**, *2*, 43–56. [CrossRef]
20. Janczak, K.; Dąbrowska, G.; Znajewska, Z.; Hrynkiewicz, K. Searching for the fungi capable to grow on polymeric materials. *Przemysł Chem.* **2014**, *93*, 1206–1209. [CrossRef]
21. Kaczmarek-Szczepańska, B.; Polkowska, I.; Paździor-Czapula, K.; Nowicka, B.; Gierszewska, M.; Michalska-Sionkowska, M.; Otrocka-Domagała, I. Chitosan/Phenolic Compounds Scaffolds for Connective Tissue Regeneration. *J. Funct. Biomater.* **2023**, *14*, 69. [CrossRef] [PubMed]
22. Vancov, T.; Keen, B. Amplification of soil fungal community DNA using the ITS86F and ITS4 primers. *FEMS Microbiol. Lett.* **2009**, *296*, 91–96. [CrossRef] [PubMed]
23. Raja, H.A.; Miller, A.N.; Pearce, C.J.; Oberlies, N.H. Fungal identification using molecular tools: A primer from the natural products research community. *J. Nat. Prod.* **2017**, *80*, 756–770. [CrossRef] [PubMed]
24. Hoppe, H.G. Significance of exoenzymatic activities in the ecology of brackish water: Measurements by means of methylumbelliferyl-substrates. *Mar. Ecol. Prog. Ser.* **1983**, *11*, 299–308. [CrossRef]
25. Freeman, C.; Liska, G.; Ostle, N.J.; Jones, S.E.; Lock, M.A. The use of fluorogenic substrates for measuring enzyme activity in peatlands. *Plant Soil* **1995**, *175*, 147–152. [CrossRef]
26. Niemi, R.M.; Vepsäläinen, M. Stability of the fluorogenic enzyme substrates and pH optima of enzyme activities in different Finnish soils. *J. Microbiol. Methods* **2005**, *60*, 195–205. [CrossRef] [PubMed]
27. Wu, Q.; Dou, X.; Wang, Q.; Guan, Z.; Cai, Y.; Liao, X. Isolation of β-1,3-Glucanase-Producing Microorganisms from Poria cocos Cultivation Soil via Molecular Biology. *Molecules* **2018**, *23*, 1555. [CrossRef]
28. Swiontek Brzezinska, M.; Jankiewicz, U.; Walczak, M. Biodegradation of chitinous substances and chitinase production by the soil actinomycete Streptomyces rimosus. *Int. Biodeterior. Biodegrad.* **2013**, *84*, 104–110. [CrossRef]
29. Swiontek Brzezinska, M.; Richert, A.; Kalwasińska, A.; Świątczak, J.; Deja-Sikora, E.; Walczak, M.; Michalska-Sionkowska, M.; Piekarska, K.; Kaczmarek-Szczepańska, B. Microbial degradation of polyhydroxybutarate with embedded polyhexamethylene guanidine derivatives. *Int. J. Biol. Macromol.* **2021**, *187*, 309–318. [CrossRef]
30. Bednarek, R.; Dziadowiec, H.; Pokojska, U.; Prusinkiewicz, Z. *Badania Ekologiczno-Gleboznawcze*; Wydawnictwo Naukowe PWN: Warszawa, Poland, 2004; ISBN 83-01-14216-2.
31. *PN-ISO 14235*; Determination of Organic Carbon Content by Oxidation With Dichromate (VI) in Sulphuric Acid Medium. Polish Committee for Standardisation: Warsaw, Poland, 2003. (In Polish)
32. *PN-R-04022*; Chemical and Agricultural Analysis of Soil—Determination of Available Potassium in Mineral Soils. Polish Committee for Standardisation: Warsaw, Poland, 1996. (In Polish)
33. *PN-R-04028*; Soil Chemical and Agricultural Analysis, Sampling and Determination of Nitrogen and Ammonium Ions in Mineral Soils. Polish Committee for Standardisation: Warsaw, Poland, 1997. (In Polish)
34. *PN-R-04022:1996/Az1*; Soil Chemical and Agricultural Analysis -Determination of Available Potassium in Mineral Soils. Polish Committee for Standardisation: Warsaw, Poland, 2002. (In Polish)
35. *PN-R-04020:1994/Az1*; Chemical and Agricultural Analysis of Soil—Determination of Available Magnesium Content. Polish Committee for Standardisation: Warsaw, Poland, 2004. (In Polish)
36. Platen, H.; Wirtz, A. The measurement of respiration activity soil with a respirometric method (measuring system Oxi Top Control). In *WTW/OxiTopR/Appliance*; Fachbereich KMUB, Umwelt-und Hygienetechnik und Zentrum f¸r Umwelttechnologie: Giessen, Germany, 1999.
37. Adam, G.; Duncan, H. Development of a sensitive and rapid method for the measurement of total microbial activity using fluorescein diacetate (FDA) in a range of soils. *Soil Biol. Biochem.* **2001**, *33*, 943–951. [CrossRef]
38. Hammer, Ř.; Harper, D.A.T.; Ryan, P.D. PAST: Paleontological Statistics Software Package for Education and Data Analysis. *Palaeontol. Electron.* **2001**, *4*, 9.
39. Savitha, M.J.; Sriram, S. Morphological and molecular identification of Trichoderma isolates with biocontrol potential against Phytophthora blight in red pepper. *Pest Manag. Hortic. Ecosyst.* **2015**, *21*, 194–202.

40. Soesanto, L.; Sri-Utami, D.; Rahayuniati, R.F. Morphological characteristics of four Trichoderma isolates and two endophytic Fusarium isolates. *Can. J. Sci. Ind. Res.* **2011**, *2*, 294–306.
41. Choi, I.-Y.; Hong, S.-B.; Yadav, M.C. Molecular and morphological characterization of green mold, Trichoderma spp. isolated from oyster mushrooms. *Mycobiology* **2003**, *31*, 74–80. [CrossRef]
42. Narayan, K.; Kotasthane, A. Genetic relatedness among Trichoderma isolates inhibiting a pathogenic fungi Rhizoctonia solani. *J. Biotechnol.* **2006**, *5*, 580–584.
43. Lücking, R.; Aime, M.C.; Robbertse, B.; Miller, A.N.; Ariyawansa, H.A.; Aoki, T.; Cardinali, G.; Crous, P.W.; Druzhinina, I.S.; Geiser, D.M.; et al. Unambiguous identification of fungi: Where do we stand and how accurate and precise is fungal DNA barcoding? *IMA Fungus* **2020**, *11*, 14. [CrossRef] [PubMed]
44. Contreras-Cornejo, H.A.; López-Bucio, J.S.; Méndez-Bravo, A.; Macías-Rodríguez, L.; Ramos-Vega, M.; Guevara-García, A.A.; López-Bucio, J. Mitogen-Activated Protein Kinase 6 and Ethylene and Auxin Signaling Pathways Are Involved in Arabidopsis Root-System Architecture Alterations by Trichoderma atroviride. *Mol. Plant Microbe Interact.* **2015**, *28*, 701–710. [CrossRef]
45. Cherkupally, R.; Amballa, H.; Bhoomi, N.R. In vitro screening for enzymatic activity of Trichoderma species for biocontrol potential. *Ann. Plant Sci.* **2017**, *6*, 1784–1789. [CrossRef]
46. Xia, W.; Liu, P.; Liu, J. Advance in chitosan hydrolysis by non-specific cellulases. *Bioresour. Technol.* **2008**, *99*, 6751–6762. [CrossRef]
47. Win, T.T.; Bo, B.; Malec, P.; Khan, S.; Fu, P. Newly isolated strain of Trichoderma asperellum from disease suppressive soil is a potential bio-control agent to suppress Fusarium soil borne fungal phytopathogens. *J. Plant Pathol.* **2021**, *103*, 549–561. [CrossRef]
48. Gruber, S.; Seidl-Seiboth, V. Self versus non-self: Fungal cell wall degradation in Trichoderma. *Microbiology* **2021**, *158*, 26–34. [CrossRef] [PubMed]
49. Vázquez, M.B.; Barrera, V.; Bianchinotti, V. Molecular identification of three isolates of Trichoderma harzianum isolated from agricultural soils in Argentina, and their abilities to detoxify in vitro metsulfuron methyl. *Botany* **2015**, *93*, 793–800. [CrossRef]
50. Abdenaceur, R.; Farida, B.T.; Mourad, D.; Rima, H.; Zahia, O.; Fatma, S.H. Effective biofertilizer *Trichoderma* spp. isolates with enzymatic activity and metabolites enhancing plant growth. *Int. Microbiol.* **2022**, *25*, 817–829. [CrossRef]
51. Wang, Y.; Ma, R.; Li, S.; Gong, M.; Yao, B.; Bai, Y.; Gu, J. An alkaline and surfactant-tolerant lipase from Trichoderma lentiforme ACCC30425 with high application potential in the detergent industry. *AMB Expres* **2018**, *8*, 95. [CrossRef] [PubMed]
52. Qian, Y.; Zhong, L.; Sun, Y.; Sun, N.; Zhang, L.; Liu, W.; Zhong, Y. Enhancement of cellulase production in Trichoderma reesei via disruption of multiple protease genes identified by comparative secretomics. *Front. Microbiol.* **2019**, *10*, 2784. [CrossRef]
53. Deng, J.J.; Huang, W.Q.; Li, Z.W.; Lu, D.L.; Zhang, Y.; Luo, X.C. Biocontrol activity of recombinant aspartic protease from Trichoderma harzianum against pathogenic fungi. *Enzym. Microb. Technol.* **2018**, *112*, 35–42. [CrossRef] [PubMed]
54. Aranaz, I.; Mengibar, M.; Harris, R.; Panos, I.; Miralles, B.; Acosta, N.; Gemma, G.; Heras, A. Functional characterization of chitin and chitosan. *Curr. Chem. Biol.* **2009**, *3*, 203–230. [CrossRef]
55. Islam, S.; Bhuiyan, M.R.; Islam, M.N. Chitin and chitosan: Structure, properties and applications in biomedical engineering. *J. Polym. Environ.* **2017**, *25*, 854–866. [CrossRef]
56. Klaykruayat, B.; Siralertmukul, K.; Srikulkit, K. Chemical modification of chitosan with cationic hyperbranched dendritic polyamidoamine and its antimicrobial activity on cotton fabric. *Carbohydr. Polym.* **2010**, *80*, 197–207. [CrossRef]
57. Babaee, M.; Garavand, F.; Rehman, A.; Jafarazadeh, S.; Amini, E.; Cacciotti, I. Biodegradability, physical, mechanical and antimicrobial attributes of starch nanocomposites containing chitosan nanoparticles. *Int. J. Biol. Macromol.* **2022**, *195*, 49–58. [CrossRef]
58. Stoleru, E.; Hitruc, E.G.; Vasile, C.; Oprică, L. Biodegradation of poly (lactic acid)/chitosan stratified composites in presence of the Phanerochaete chrysosporium fungus. *Polym. Degrad. Stab.* **2017**, *143*, 118–129. [CrossRef]
59. Mathew, S.; Abraham, T.E. Characterisation of ferulic acid incorporated starch–chitosan blend films. *Food Hydrocoll.* **2008**, *22*, 826–835. [CrossRef]
60. Rui, L.; Xie, M.; Hu, B.; Zhou, L.; Yin, D.; Zeng, X. A comparative study on chitosan/gelatin composite films with conjugated or incorporated gallic acid. *Carbohydr. Polym.* **2017**, *173*, 473–481. [CrossRef]
61. Qin, Z.; Huang, Y.; Xiao, S.; Zhang, H.; Lu, Y.; Xu, K. Preparation and characterization of high mechanical strength chitosan/oxidized tannic acid composite film with Schiff base and hydrogen bond crosslinking. *Int. J. Mol. Sci.* **2022**, *23*, 9284. [CrossRef] [PubMed]
62. Sun, X.; Wang, Z.; Kadouh, H.; Zhou, K. The antimicrobial, mechanical, physical and structural properties of chitosan–gallic acid films. *LWT-Food Sci. Technol.* **2014**, *57*, 83–89. [CrossRef]
63. Rivero, S.; Garcia, M.A.; Pinotti, A. Physical and chemical treatments on chitosan matrix to modify film properties and kinetics of biodegradation. *J. Mater. Phys. Chem.* **2013**, *1*, 51–57.
64. Poorna Chandrika, K.S.V.; Prasad, R.D.; Godbole, V. Development of chitosan-PEG blended films using Trichoderma: Enhancement of antimicrobial activity and seed quality. *Int. J. Biol. Macromol.* **2019**, *126*, 282–290. [CrossRef] [PubMed]
65. Cheng, M.; Wang, J.; Zhang, R.; Kong, R.; Lu, W.; Wang, X. Characterization and application of the microencapsulated carvacrol/sodium alginate films as food packaging materials. *Int. J. Biol. Macromol.* **2019**, *141*, 259–267. [CrossRef]
66. Altun, E.; Çelik, E.; Ersan, H.Y. Tailoring the microbial community for improving the biodegradation of chitosan films in composting environment. *J. Polym. Environ.* **2020**, *28*, 1548–1559. [CrossRef]
67. Ahmad, R.; Jilan, G.; Arshad, M.; Zahir, Z.A.; Khalid, A. Bio-conversion of organic wastes for their recycling in agriculture: An overview of perspectives and prospects. *Ann. Microbiol.* **2007**, *57*, 471–479. [CrossRef]

68. Klose, S.; Acosta-Martínez, V.; Ajwa, H.A. Microbial community composition and enzyme activities in a sandy loam soil after fumigation with methyl bromide or alternative biocides. *Soil Biol. Biochem.* **2006**, *38*, 1243–1254. [CrossRef]
69. Su, S.; Zeng, X.; Bai, L.; Williams, P.N.; Wang, Y.; Zhang, L.; Wu, C. Inoculating chlamydospores of Trichoderma asperellum SM-12F1 changes arsenic availability and enzyme activity in soils and improves water spinach growth. *Chemosphere* **2017**, *175*, 497–504. [CrossRef] [PubMed]
70. Yedidia, I.; Benhamou, N.; Kapulnik, Y.; Chet, I. Induction and accumulation of PR proteins activityduring early stages of root colonizationby the mycoparasite Trichoderma harzianum strain T-203. *Plant Physiol. Biochem.* **2000**, *38*, 863–873. [CrossRef]
71. Henry, H.A. Reprint of "Soil extracellular enzyme dynamics in a changing climate". *Soil Biol. Biochem.* **2013**, *56*, 53–59. [CrossRef]
72. García-Ruiz, R.; Ochoa, V.; Vinegla, B.; Hinojosa, M.B.; Pena-Santiago, R.; Liébanas, G.; Linares, J.C.; Carreira, J.A. Soil enzymes, nematode community and selected physico-chemical properties as soil quality indicators in organic and conventional olive oil farming: Influence of seasonality and site features. *Appl. Soil Ecol.* **2009**, *41*, 305–314. [CrossRef]
73. Burke, D.J.; Weintraub, M.N.; Hewins, C.R.; Kalisz, S. Relationship between soil enzyme activities, nutrient cycling and soil fungal communities in a northern hardwood forest. *Soil Biol. Biochem.* **2011**, *43*, 795–803. [CrossRef]

foods

Article

# Dynamic Modulation of $SO_2$ Atmosphere for Enhanced Fresh-Keeping of Grapes Using a Novel Starch-Based Biodegradable Foam Packaging

**Shihua Mai [1,†], Yue Ma [1,†], Hongsheng Liu [1,2], Chao Li [1], Yuqing Song [1], Kaizhen Hu [1], Xinyan Chen [1], Ying Chen [3,4] and Wei Zou [1,*]**

1 School of Food Science and Engineering, South China University of Technology, Guangzhou 510641, China; femaish@mail.scut.edu.cn (S.M.); feyuem@mail.scut.edu.cn (Y.M.); liuhongsheng@scut.edu.cn (H.L.); fe201920124894@mail.scut.edu.cn (C.L.); 202121028199@mail.scut.edu.cn (Y.S.); mehukaizhen@mail.scut.edu.cn (K.H.); 202130310114@mail.scut.edu.cn (X.C.)
2 Sino-Singapore International Joint Research Institute, Guangzhou 510663, China
3 School of Food Science and Engineering, Yangzhou University, Yangzhou 225127, China; 008265@yzu.edu.cn
4 Department of Food Science and Technology, National University of Singapore, Science Drive 2, Singapore 117542, Singapore
* Correspondence: wei.zou@uq.net.au; Tel.: +86-20-87113845
† These authors contributed equally to this work.

**Abstract:** To improve the fresh-keeping of highly perishable fruits with high commercial value, a novel starch-based foam packaging material was developed in this study. The foam incorporated the antiseptic ingredient $Na_2S_2O_5$, which chemically interacted with environmental moisture to release $SO_2$ as an antifungal agent. Scanning electron microscopy (SEM), moisture absorption and mechanical measurements were used to characterize the unique sandwich-like inner structure of the foam which allowed for the modulable release of $SO_2$. The starch-based foam exhibited sufficient resilience (~100%) to provide ideal cushioning to prevent physical damage to fresh fruits during transportation. When 25 $g/m^2$ of $Na_2S_2O_5$ was applied, the foam stably released over 100 ppm $SO_2$ and demonstrated satisfactory antifungal performance (inhibition over 60%) in terms of maintaining the appearance and nutritional values (such as soluble solids 14 vs. 11%, total acidity 0.45 vs. 0.30%, and Vitamin C 3.4 vs. 2.5 mg/100 g) of fresh grapes during a 21 day storage period. Additionally, the residual $SO_2$ (14 mg/kg) also meets the safety limits (<30 mg/kg). These research findings suggest great potential for the utilization of this novel foam in the food industry.

**Keywords:** starch-based foam package; sandwich-like inner structure; cushioning; antisepsis; fresh-keeping

**Citation:** Mai, S.; Ma, Y.; Liu, H.; Li, C.; Song, Y.; Hu, K.; Chen, X.; Chen, Y.; Zou, W. Dynamic Modulation of $SO_2$ Atmosphere for Enhanced Fresh-Keeping of Grapes Using a Novel Starch-Based Biodegradable Foam Packaging. *Foods* **2023**, *12*, 2222. https://doi.org/10.3390/foods12112222

Academic Editors: Mario M. Martinez and Sílvia Petronilho

Received: 26 April 2023
Revised: 29 May 2023
Accepted: 30 May 2023
Published: 31 May 2023



## 1. Introduction

Foam is a commonly used packaging material for fruit preservation and protection. Foam boxes, typically made from polystyrene (PS), can provide partial protection against bruising caused by external physical damage. However, they are not effective in delaying the natural decay process caused by microbial metabolism. For example, grapes, being a typical type of berry fruit, are highly susceptible to the gray mold pathogen Botrytis cinerea Pers [1–3]. The growth of this pathogen depends on various factors such as temperature and humidity during storage [4].

To minimize the economic loss of post-harvest fruit, $Na_2S_2O_5$, a fungicide, has been used to chemically react with moisture from the fruit surface or surroundings, leading to the release of sulfur dioxide ($SO_2$) to effectively kill harmful microbes within the packaging box [5–7]. However, this approach has limitations: traditional packaging materials cannot achieve long-term release of fungicides and non-biodegradable petrochemical-based foam poses a significant burden when disposed into the environment.

Starch foam has recently gained attention as a substitute for petrochemical-based products due to its abundant and renewable nature, low cost, and biodegradability. It has been particularly used in loose-fill packaging applications [8]. Additionally, starch foam is known for its native hydrophilicity, which makes it favorable for moisture absorption and retention [9,10]. This property has traditionally been utilized in modified atmosphere packaging, with the expectation of achieving a long-term antifungal effect on susceptible berry fruits.

To achieve this goal, a novel strategy was proposed in this study, involving the development of a starch-based foam with a sandwich-like structure that incorporates pre-loaded $Na_2S_2O_5$ powder between two starch foam sheets. It is expected that this structure will enable the foam to absorb moisture from fruit respiration or the external environment, and then chemically react with $Na_2S_2O_5$ to release $SO_2$ as an antifungal ingredient in a modulable manner. Previous research has indicated the potential of utilizing composite packaging materials to modify the environmental atmosphere and improve the preservation of fruits. For instance, the effectiveness of aloe vera composite packaging has been demonstrated in suppressing the growth of aerobic mesophilic bacteria, yeast, and mold. This ultimately helps to preserve the essential nutrients and minimize loss during the fruit's shelf life [11]. Multiple techniques, including SEM observation, moisture absorption, and mechanical measurements, were employed to characterize the unique functional structure and release characteristics of $SO_2$. The antimicrobial effects and real performance of this novel starch-based biodegradable foam on fresh grapes were also assessed. It is hypothesized that this innovative foam could serve as a long-term effective fresh-keeping shield for susceptible perishable fruits with high commercial value.

## 2. Materials and Methods

### 2.1. Materials and Reagents

In this work, the starch-based foam was prepared using hydroxypropyl cornstarch with about 25% amylose content, which was provided by Derui Starch Co., (Luohe, China) and the initial water content was about 13%. Polyvinyl alcohol (078-20) and glycerol were obtained from Sinopec Petrochemical Co., Ltd. (Shanghai, China). Calcium carbonate was purchased from Yifeng Co. (Guangzhou, China). $Na_2S_2O_5$ was supplied by Aladdin Chemical Co., (Shanghai, China). All chemicals used were of analytical grade. The Botrytis cinerea was from BeNa Culture Collection Co., Ltd. (Xinyang, China), and Potato Dextrose Agar (PDA) medium was from BeNa Culture Collection Co., Ltd. (Xinyang, China).

'Kyoho' grapes were purchased from local fruit market (Guangzhou, China). The preservation experiment was performed in 5.5 L airtight containers at room temperature (about 25 °C).

### 2.2. Starch-Based Foam Preparation

Starch-based foam was produced by two-step extrusions according to our previous method [12]. Firstly, hydroxypropyl cornstarch (80% $w/w$), glycerol (5% $w/w$), polyvinyl alcohol (PVA) (5% $w/w$), fiber (4% $w/w$), and water (6% $w/w$) were homogeneously mixed by a high-speed mixer (Shende SRZ-500, Zhangjiagang, China), and then were processed into pellets using a twin-screw co-rotating extruder (Keya, Nanjing, China) at a speed of 120 rpm. The temperature profile of the barrels was set at 30, 30, 80, 110, 110, 110, 80, and 80 °C, respectively. Subsequently, the pellets were dried for a constant moisture content of 16% and fed into a single-screw extruder (Tongjia, Jining, China) at a speed of 180 rpm. The temperature profile of the barrels was of 100, 180, 180 °C, and a sheet die with 400 tmm width was heated to 200 °C. After it, the foams were cut into sheets with a size of 4 cm × 3 cm.

In this study, the starch-based foam with dynamic modulation of $SO_2$ atmosphere was newly designed and the preparation procedure was shown in Figure 1. Initially, $Na_2S_2O_5$ was evenly spread and loaded between two starch foam sheets, which were subsequently pressed together by using a hot-pressing method. In this case, a little amount of water was used as an adhesive and sprinkled on the narrow inner edge of the sheets, improving starch foam sticking without interference with the chemical reaction. In order to ensure that the residual concentration of $SO_2$ remains below the safe limit of 30 mg/kg, the concentration range of $Na_2S_2O_5$ was carefully selected through trial testing to meet the necessary safety requirements. The objective of the study is to achieve a $Na_2S_2O_5$ concentration that is as low as possible while still providing effective antiseptic performance against microbial decay in grapes. This approach will also help reduce the manufacturing cost of the novel starch-based foam packaging. Thus, different dosages of $Na_2S_2O_5$ (5, 15 and 25 g/m$^2$) were used. The control group was also prepared without the addition of $Na_2S_2O_5$. In the study, $Na_2S_2O_5$ was manually cast into the starch-based pads. However, it is important to note that for practical applications, the process should be automated using a machine to ensure the most even distribution possible.

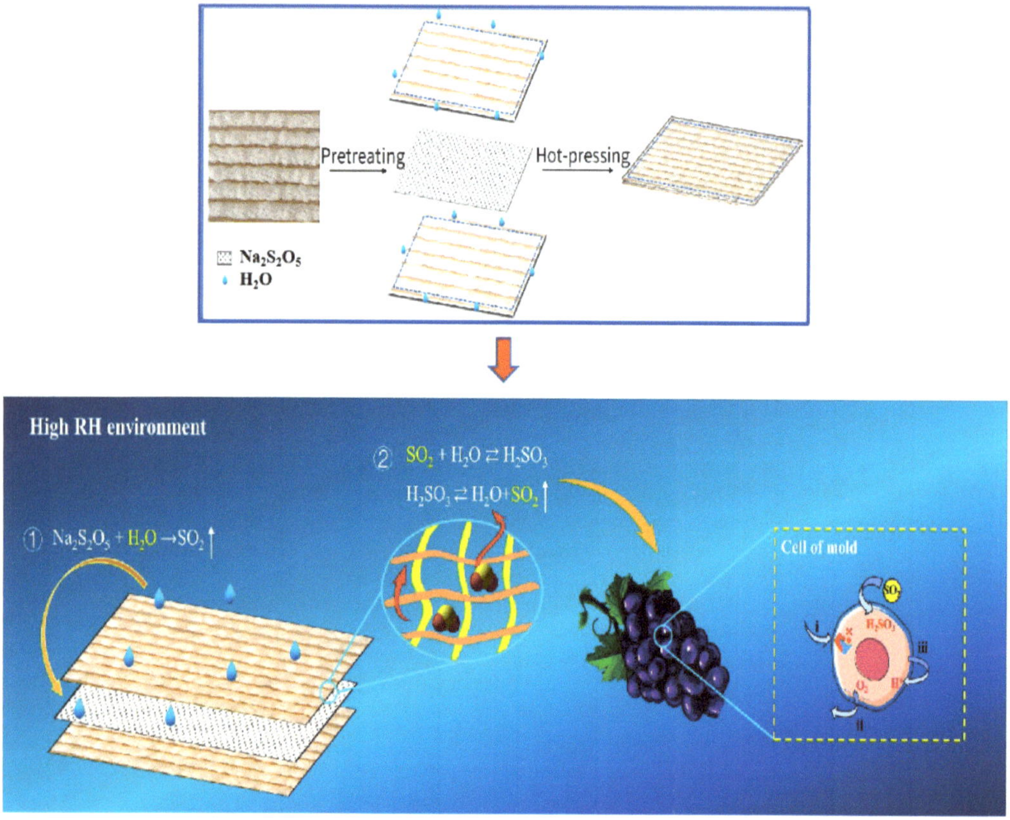


**Figure 1.** An illustration of starch-based foams pre-loaded with $Na_2S_2O_5$ and its functions.

## *2.3. Scanning Electron Microscope (SEM)*

SEM (EVO18, Zeiss, Oberkochen, Germany) was used to observe the raw materials and the cross-section of starch-based foams. A lower voltage of 10 kV was used during the test in order not to damage the cell structures. The samples were coated with gold before measurement [13].

## *2.4. Moisture Absorption*

The objective of the starch-based foam package is to preserve and transport the locally grown "Kyoho grapes" in Guangzhou, China. These fresh grapes hold significant commercial value, but they typically ripen off the vine in August when the local humidity reaches high levels, often around 90–100% according to climate records released by NASA (Figure S1). Due to the grape granules' vulnerability to rupture under humidity exceeding 95% [14], the main focus of the study is to investigate the efficacy of the starch-based foam package in effectively preserving the local "Kyoho grapes" within such extremely humid conditions. Foam sheets were first dried in an oven at 40 °C for 12 h and kept in the controlled containers with relative humidity (RH) of 86%, 93% and 98%, which were adjusted with saturated KCl, $Na_2SO_4$ and $K_2SO_4$ solutions, respectively. Absorbed moisture was calculated by the weight difference [15,16]. The absorption kinetics was investigated with a mathematical model proposed by Peleg (1988) [17], which was calculated using Equation (1):

$$M_{(t)} = M_0 + \left(\frac{t}{K_1 + K_2 t}\right), \quad (1)$$

where $M_{(t)}$ is the moisture at different time, $M_0$ is the initial moisture, $t$ is the incubation time, $K_1$ is the Peleg flux constant, while $K_2$ is Peleg capacity constant [10].

### 2.5. Compressive Strength, Resilience, and Recovery

The mechanical properties of the starch-based foams were measured by a universal testing machine (Instron, Model 5566, Norwood, MA, USA) after storage at the containers with different levels of RH for 1 week. The samples were fastened and compressed by a metal plate at a rate of 5 mm/min to a depth of 50% of its height. The detailed measurement was performed according to our previous study [12].

### 2.6. Artificial Vibrating Test

The packages were designed with and without starch-based foams, respectively, and then they were applied for grape preservation. Grapes were preserved by a one-day vibration testbed (M/MN100, Mooney Laboratory Equipment Co., Ltd., Dongguan, China) at a frequency of 2 Hz to simulate vibrating during transportation [18].

### 2.7. Measuring $SO_2$ Content

The yield of $SO_2$ was measured using a handheld gas analyzer [19] (Shenzhen Fukaite Technology Co., Ltd., Shenzhen, China). In the measurement, the foams loaded with $Na_2S_2O_5$ were kept in an airtight container with a hole of 0.5 mm in diameter. Humidity of the container was adjusted according to the experimental requirement. The release of $SO_2$ under room humidity of 86, 93, and 98% were recorded and calculated according to the growing indication of h, respectively. The indication was separated by 24 h.

### 2.8. Antimicrobial Analysis

Antimicrobial activity was determined experimentally according to a previous method [4,20]. The PDA medium plates were inoculated with Botrytis cinerea in different containers. Medium plates were used 3 times as replicates and the whole trial was repeated twice. The plates were incubated at 25 °C and radial fungal growth (mm) of Botrytis cinerea was measured after 7 days of incubation. The percentage of a reduction in colony diameter (CD) was calculated according to Equation (2):

$$\mathrm{CD}(\%) = \left(\frac{dc - dt}{dc}\right) \times 100, \quad (2)$$

where $dc$ and $dt$ are the average diameter of fungal growth in the control and treated samples, respectively.

### 2.9. Measuring Total Soluble Solids (TSS)

TSS value affects the taste as it indicates the level of sweetness of the fruit [5]. For the measurement of TSS, berries were randomly taken from the same bunch of grapes and squeezed into juice [21,22], and then measured with a digital hand-held refractometer

(Shanghai Xinrui Electric Co., Ltd., Shanghai, China). Each replicate consisted of 3 grapes and every grape was measured three times.

*2.10. Measuring Titratable Acid (TA)*

TA measures the total acid concentration present in food items and in this work was determined according to the reference [23]. Grapes (10.0 g) were ground in a mortar and transferred into a 100 mL volumetric flask, kept for 30 min, and then were filtered. After it, 20 mL supernatants were transferred into a conical flask and 0.1 mL 1% (m/m) phenolphthalein solution was added, and then titrated by 0.01 mol/L sodium hydroxide (NaOH). The used NaOH solution was recorded to calculate TA. Each titration was repeated three times.

*2.11. Measuring Vitamin C*

Vitamin C was measured by the 2,6-dichloroindophenoltitration titration method as described in the previous references [24,25]. Grapes (10 g) were put into a mortar, and 50 mL of 20 g/L of the oxalic acid solution was poured and instantly mixed into a homogeneous solution. The homogenate was then fixed the volume to 100 mL with 20 g/L of the oxalic acid solution and shaken well, filtered 3 times. The trial was carried out in triplicate and the level of VC was calculated in 100 mg/g of fresh sample.

*2.12. Measuring $SO_2$ Residues*

The residual $SO_2$ was quantitatively measured by a distillation method according to the standard of GB 5009.34-2016 (China). Briefly, 5.0 g grapes were accurately weighed and $SO_2$ is extracted by distillation in the iodine measuring bottle. Then titrating with iodine standard solution and recording the volume. The $SO_2$ residues were calculated according to Equation (3)

$$\mathrm{X(g/kg)} = \frac{(V - V_0) \times 0.032 \times c \times 1000}{m}, \quad (3)$$

where $V$ and $V_0$ are the volume of iodine standard solution consumed for titration of grape samples and the blank group, respectively, $c$ is the concentration of iodine standard solution, and m is the mass of grape samples.

*2.13. Statistical Analysis*

The experiment was arranged based on a completely randomized design with three replicates. One-way ANOVA was used to analyze the differences, and results with $p < 0.05$ were considered as significantly different.

## 3. Results

*3.1. Starch-Based Foam's Morphological Inner Structure*

Figure 2 shows the ideal appearance of the starch-based foam sheet as a packaging material, with a smooth surface and regular wave-shape. In comparison, the cross-section of the foam sheet shows a high degree of porous inner cells under SEM [10,13], indicating excellent expansion necessary for ensuring adequate buffering and moisture absorption [13,26]. Due to the hydrophilic starch substrate, the interconnecting air cells gradually shrank or collapsed as the surrounding humidity increased from 60–98%, as shown in Figure 3.

*3.2. Starch-Based Foam's Moisture Absorption Behavior*

To better understand the moisture absorption properties of the starch-based foam, it was subjected to room humidity of 86, 93, and 98%, as shown in Figure 4. The absorption process can be divided into two phases: an initial fast phase (from ~7 to 15% $w/w$) within the first three to five days, followed by a slower phase that gradually reached equilibrium by the end of the 21-day assay period. A higher room humidity (98 vs. 86%) resulted in faster absorption rates and higher maximal moisture absorption content (25 vs. 15% $w/w$), as confirmed by the Peleg kinetic model characterization of higher $K_1$ and lower $K_2$ values

(Table 1). These findings suggest that water absorption occurs at a faster rate and larger capacity under higher environmental humidity, consistent with previous research [15,16,27].

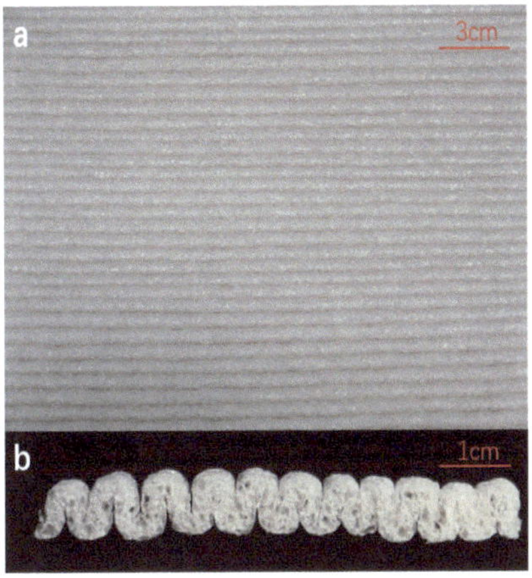


**Figure 2.** Photograph of surface (**a**) and cross-section (**b**) of starch-based foam.

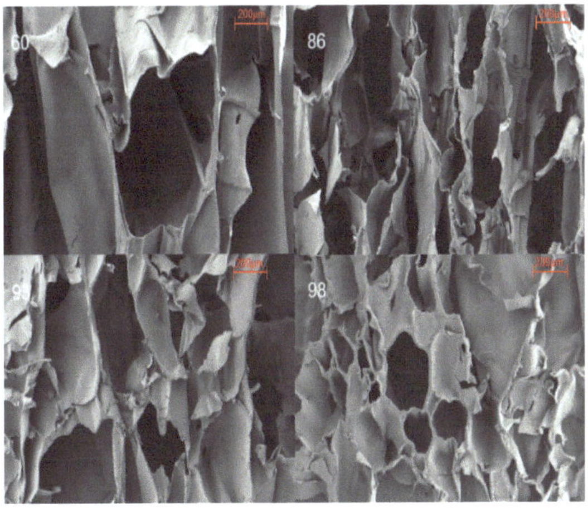


**Figure 3.** Inner-cell structure of starch-based foam under humidity levels of 60%, 86%, 93%, and 98%.

**Table 1.** Compressive properties of foam sheets at different humidity levels.

| Humidity Levels | First Compressive Strength (Kpa) | Second Compressive Strength (Kpa) | Resilience (%) | Recovery (%) |
|---|---|---|---|---|
| Control (60%) | 345.09 ± 3.46 [a] | 359.45 ± 7.40 [a] | 104.15 ± 1.10 [b] | 90.21 ± 1.43 [a] |
| 86% | 80.08 ± 4.85 [b] | 83.73 ± 4.26 [b] | 104.63 ± 1.46 [b] | 84.72 ± 1.37 [b] |
| 93% | 28.42 ± 1.34 [c] | 31.11 ± 1.23 [c] | 109.51 ± 1.50 [a] | 78.36 ± 2.06 [c] |
| 98% | 15.09 ± 0.62 [d] | 15.97 ± 0.57 [d] | 105.82 ± 0.96 [b] | 70.19 ± 2.48 [d] |

One-way ANOVA method used; difference significant (in terms of a, b, c and d) at the 0.05 level.

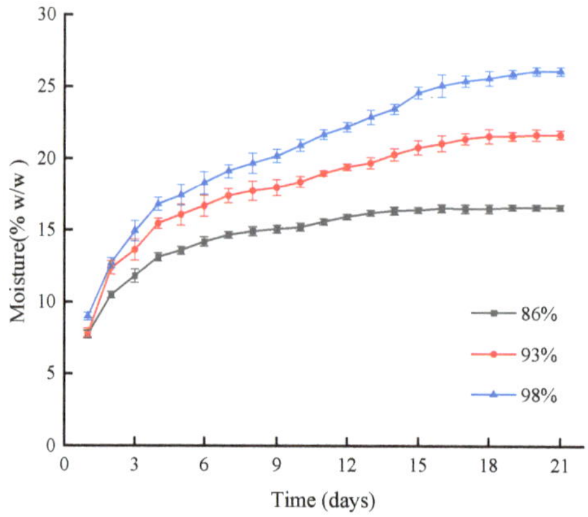


**Figure 4.** Moisture percentage of starch-based foam over time at humidity levels of 60%, 86%, 93%, and 98%.

### 3.3. Starch-Based Foam's Mechanical Properties

The mechanical properties of starch foam, such as compressive strength and recovery ratio, depend on its inner loose-fill structure, which can vary under different environmental humidity conditions. Table 2 shows that the compressive strength of the foam was high at around 350–360 kPa under 60% humidity but dropped to only 80 kPa with an increase of humidity to 86%. With further increases in humidity, the compressive strength of the foam continued to decline, reaching only 15 kPa under 98% humidity. Such significant weakening of mechanical performance is disadvantageous for packaging materials and can be attributed to the plasticization effect caused by the absorption of more water molecules into the starch-based foam framework at higher environmental humidity.

**Table 2.** Peleg model parameters of starch foams at humidity levels of 86%, 93%, and 98%.

| Humidity Levels | $K_1$ | $K_2$ |
|---|---|---|
| 86% | 64.17 | 1.88 |
| 93% | 74.29 | 1.42 |
| 98% | 85.50 | 1.15 |

In comparison, the Resilience Percentage of the starch-based foam remained almost unaltered (at ~105%) regardless of the increase of humidity from 60 to 98%. Additionally, the Recovery Percentage showed only a slight decrease from ~90 to 70%. This suggests that the starch-based foam can still play an ideal cushioning role as a fresh-package material, particularly under high humid conditions. The compression test is conducted to simulate the physical conditions experienced by the starch-based foam package during transportation. Its purpose is to verify whether the foam can withstand the considerable external pressure exerted on the package during fruit transportation. This test aims to ensure the reliability of the foam in protecting the fruits from physical damage prior to the onset of microbial decay.

All these mechanical properties can help maximally reduce the physical damage to the susceptible fruits during transportation. In the study, fresh grapes were packed using starch-based foam and subjected to a one-day assessment at 25 °C under artificial vibrating conditions without the application of the $Na_2S_2O_5$ as a preservative (Figure 5). The control group exhibited several fallen granules, evident bruising, and fermentation smells (see $a_1$ and $a_2$). In contrast, the grapes packed inside the starch-based foam remained almost fresh without any visible physical damage (see $b_1$ and $b_2$). The starch-based foam was thus confirmed to play a satisfactory cushioning role in protecting susceptible fresh fruits during transportation [18,28,29].

**Figure 5.** Fresh grapes with and without the starch-based foam package (**a$_1$**,**b$_1$**), and their respective appearance after one-day artificial vibrating assessment (**a$_2$**,**b$_2$**). No $Na_2S_2O_5$ as preservative was applied in the test.

### 3.4. Starch-Based Foam Preloaded with $Na_2S_2O_5$ to Prevent Microbial Decay

To better preserve fresh fruits and prevent microbial decay, $Na_2S_2O_5$ powder (interacting with environmental moisture for releasing the antiseptic $SO_2$ atmosphere) is often applied in food preservation [30]. However, adding $Na_2S_2O_5$ powder without any packaging is not an appropriate strategy, as it can result in the excessive and uncontrollable release of $SO_2$ (red lines in Figure 6a–c). For instance, $SO_2$ content released from $Na_2S_2O_5$ (5 g/m$^2$) under typical high humidity (98%) was initially high within the first three days but quickly declined afterwards (Figure 6c). In comparison, when pre-loaded into the hydrophilic starch substrate of the foam package with its unique sandwich-like inner structure, $Na_2S_2O_5$ powder can release a more controlled amount of $SO_2$ content in two phases during the storage assay. This is due to moisture being absorbed into the starch substrate, enabling a more controlled release of $SO_2$.

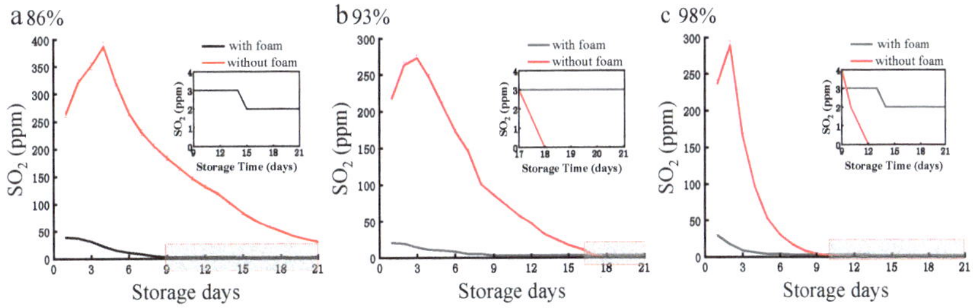


**Figure 6.** Time-dependent $SO_2$ concentration released from $Na_2S_2O_5$ (5 g/m$^2$) with and without starch-based package foam at humidity levels of 86% (**a**), 93% (**b**), and 98% (**c**) respectively.

The release of $SO_2$ was within the range of antiseptic and food safety requirements, and pre-loading $Na_2S_2O_5$ powder at 5 g/m$^2$ was sufficient to generate enough safe content of $SO_2$ atmosphere. Higher content (15 or 25 g/m$^2$) resulted in excessive accumulation of $SO_2$ atmosphere, which may incur safety risks (over 30 mg/kg *Codex Alimentarius International Food Standards*). Figure 7 aims to illustrate the starch-based foam package can release relatively low but stable concentration of $SO_2$ during the 21-day trail. The safe level of $SO_2$ often refer to its residual content on grapes, which is clearly indicated as 30 mg/kg in last Figure and well discussed below in Section 3.5.

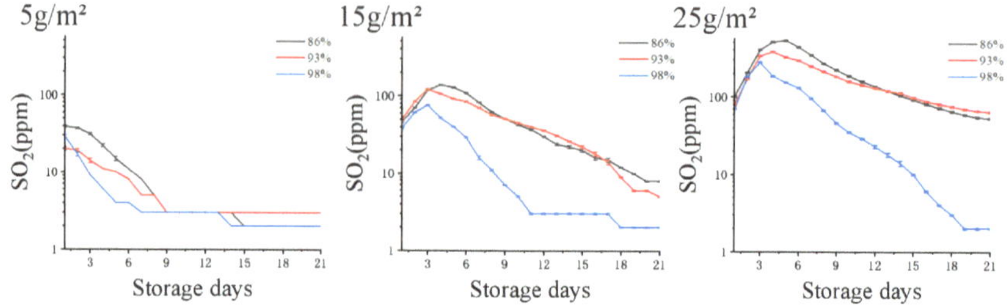


**Figure 7.** Time-dependent $SO_2$ concentration released from starch-based foam package with varying $Na_2S_2O_5$ amounts (5, 15, and 25 g/m$^2$) at humidity levels of 86%, 93%, and 98%.

The release of $SO_2$ must comply with both antiseptic and food safety requirements. As shown in Figure 7, using 5 g/m$^2$ of pre-loading $Na_2S_2O_5$ powder may not be sufficient to generate a safe level of $SO_2$ atmosphere compared to higher contents (15 or 25 g/m$^2$) over a 21 day storage period, which may result in inadequate protection against microbial decay. On the other hand, excessive usage of $Na_2S_2O_5$ poses a food safety risk as it may leave residual $SO_2$ content on the fresh fruits above the threshold limit [31]. Both of these important aspects will be addressed in detail in the following sections.

A seven-day assay has been widely recognized in previous studies as an effective method to evaluate the efficacy of inhibiting microbial growth [4,20,32]. In this study, the same assay duration was employed to investigate whether the $Na_2S_2O_5$ preloaded starch-based foam could effectively inhibit the growth of Botrytis cinerea, the main culprit behind the microbial decay of fresh grapes. During a seven-day assay, the $Na_2S_2O_5$ preloaded starch-based foam was tested for effectiveness against the typical Botrytis cinerea (the main cause of fresh grapes' microbial decay). The antimicrobial effectiveness was evaluated through the quantification of inhibition percentages (Figure 8a,b) and illustrated in the study [32]. The results showed that the antimicrobial performance was highly dependent on the concentration of antiseptic chemicals, with an inhibition percentage of above 50% observed when the $Na_2S_2O_5$ concentration was over 25 g/m$^2$. Additionally, the study found that the release of $SO_2$ from $Na_2S_2O_5$ was facilitated by the presence of more moisture within the enclosure. This was evidenced by a significant increase in the antimicrobial inhibition percentage when the humidity level was set to 93% compared to 86%. However, the excellent antimicrobial performance decreased when the humidity level exceeded 98%. Because this humid condition can promote the growth of various microbial communities, which may compromise the effectiveness of the preservation method.

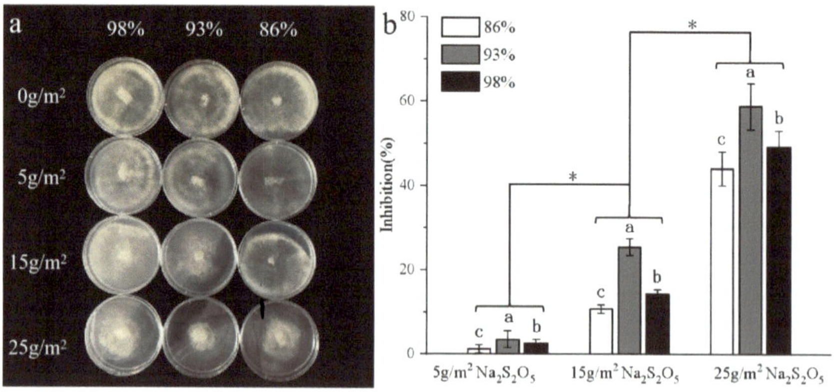


**Figure 8.** Antimicrobial activity (**a**) and the inhibition rate (**b**) of the starch-based foam with 0, 5, 15, and 25 g/m$^2$ of $Na_2S_2O_5$ at humidity levels of 86%, 93%, and 98%. One-way ANOVA method used; difference significant (in terms of a, b, c and *) at the 0.05 level.

### 3.5. The Fresh-Keeping Performance of Starch-Based Foam Preloaded with $Na_2S_2O_5$

The use of antiseptic $Na_2S_2O_5$ powder preloaded into starch-based foam has been shown to effectively preserve fresh grapes susceptible to Botrytis cinerea (Figure 9) [7,33,34]. The starch-based foam alone resulted in rapid visible decay of grapes from the sixth day of storage, but even a small addition of $Na_2S_2O_5$ (just 5 g/m$^2$) helped maintain the grapes' fresh appearance for up to 18 days. Further increasing the amount of $Na_2S_2O_5$ to 15–25 g/m$^2$ preserved the fresh appearance of the grapes throughout the entire 21 day storage period. Total soluble solids, acidity, and vitamin C are recognized as the three key nutritional parameters of grapes, which tend to decrease due to microbial metabolism during storage [35]. The release of $SO_2$ atmosphere from $Na_2S_2O_5$ has been previously validated as an effective method to inhibit microbial growth and preserve the nutritional quality of grapes during storage [36] (Figure 10a–c). Because an appropriate content of $Na_2S_2O_5$ (from 5 to 25 g/m$^2$) was found to better preserve the total soluble solids, acidity, and Vitamin C, which would otherwise be quickly lost during storage. However, considering the common safety threshold of $SO_2$ residues (30 mg/kg, U.S. Federal Registry, 1986), the results indicate that adding $Na_2S_2O_5$ powder at ~25 g/m$^2$ is ideal (only ~14 mg/kg of $SO_2$ residues) for achieving a good balance between fresh-keeping effectiveness and food safety.

All these results indicate a significant improvement in the fresh-keeping ability of the starch-based foam packaging when the appropriate amount of antiseptic $Na_2S_2O_5$ is included.

**Figure 9.** Appearance of fresh grapes packed with starch-based foam with varying $Na_2S_2O_5$ amounts (5, 15, and 25 g/m$^2$) over a 21-day period.

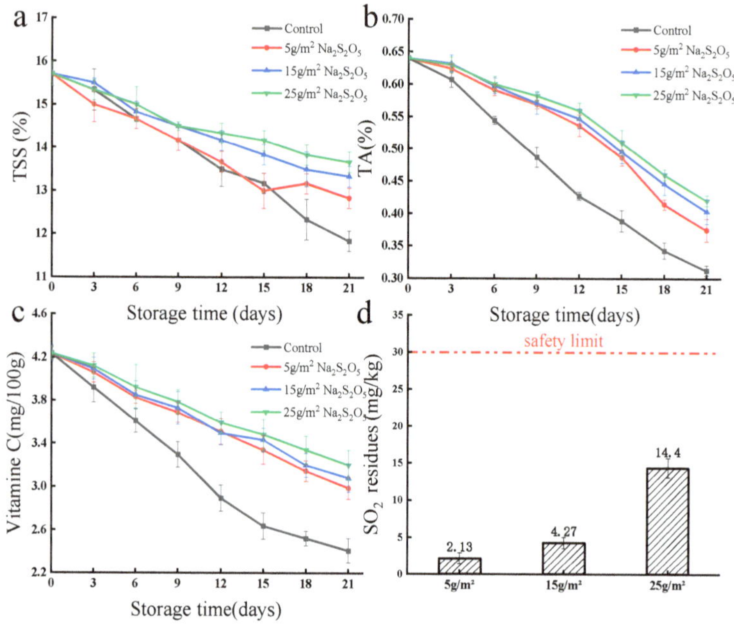


**Figure 10.** Changes in total soluble solids (TSS, (**a**)), acidity (TA, (**b**)), Vitamin C content (**c**), and $SO_2$ residual concentration (**d**) of fresh grapes packed with starch-based foam containing varying amounts of $Na_2S_2O_5$ (5, 15, and 25 $g/m^2$) over a 21-day period.

## 4. Conclusions

A novel starch-based foam was successfully developed in this study as a packaging material for fresh fruits, featuring a unique sandwich-like inner structure. This foam not only provided sufficient resilience to protect the fruits from physical damage during transportation but also facilitated the pre-loaded $Na_2S_2O_5$ powder to chemically interact with environmental moisture, allowing for modulable release of the antiseptic $SO_2$ atmosphere. As a result, the starch-based foam effectively preserved the appearance and key nutrients (such as soluble solids, acidity, and Vitamin C) of fresh grapes during a 21-day storage period, indicating its great potential for fresh-keeping of perishable fruits with high commercial value in the food industry.

In summary, the research successfully validates the hypothesis by developing a novel starch-based foam packaging that effectively preserves highly perishable fresh grapes of significant commercial value. However, there are several limitations that should be addressed in future studies. Firstly, there is a need for characterizing the grapes' microbial stability and sensory properties during storage in the starch-based foam package. Furthermore, the applicability of the starch-based foam packaging in dry regions with lower humidity needs to be explored. Additionally, its potential for wider application to other perishable fruits beyond grapes should be investigated. Further applied research is necessary to optimize the dosage and distribution of $Na_2S_2O_5$ to meet specific requirements and enhance the overall performance of the packaging system.

**Supplementary Materials:** The following supporting information can be downloaded at: https://www.mdpi.com/article/10.3390/foods12112222/s1, Figure S1: The humidity record from July to September in Guangzhou, China (NASA, 2022).

**Author Contributions:** Conceptualization, S.M., Y.M., C.L. and H.L.; methodology, S.M., Y.M. and C.L.; software, S.M., Y.M. and C.L.; validation, S.M., Y.M. and C.L.; formal analysis, S.M., Y.M., C.L. and W.Z.; investigation, S.M., Y.M., C.L., W.Z., X.C., Y.C., Y.S. and K.H.; resources, H.L.; data curation, S.M., Y.M. and C.L.; writing—original draft preparation, S.M., Y.M., C.L., X.C., Y.C., Y.S. and K.H.; writing—review and editing, W.Z.; visualization, S.M., Y.M. and C.L.; supervision, H.L.; project administration, H.L.; funding acquisition, H.L. All authors have read and agreed to the published version of the manuscript.

**Funding:** This research was funded by the National Natural Science Foundation of China (32272340) and the 111 project (B17018).

**Data Availability Statement:** Data is contained within the article (or Supplementary Material).

**Acknowledgments:** The modified starch foam was kindly supported by Guangzhou Green-Fill Material Technology Co., Ltd. W.Z. acknowledges the support from Guangdong Jiashili Food Group Co., Ltd.

**Conflicts of Interest:** The authors declare no conflict of interest.

## References

1. Chen, Y.H.; Li, Y.F.; Wei, H.; Li, X.X.; Zheng, H.T.; Dong, X.Y.; Xu, T.F.; Meng, J.F. Inhibition efficiency of wood vinegar on grey mould of table grapes. *Food Biosci.* **2020**, *38*, 100755. [CrossRef]
2. Morales-Cedeno, L.R.; Orozco-Mosqueda, M.D.; Loeza-Lara, P.D.; Parra-Cota, F.I.; de los Santos-Villalobos, S.; Santoyo, G. Plant growth-promoting bacterial endophytes as biocontrol agents of pre- and post-harvest diseases: Fundamentals, methods of application and future perspectives. *Microbiol. Res.* **2021**, *242*, 126612. [CrossRef]
3. Wang, K.T.; Liao, Y.X.; Kan, J.Q.; Han, L.; Zheng, Y.H. Response of direct or priming defense against botrytis cinerea to methyl jasmonate treatment at different concentrations in grape berries. *Int. J. Food Microbiol.* **2015**, *194*, 32–39. [CrossRef]
4. Youssef, K.; de Oliveira, A.G.; Tischer, C.A.; Hussain, I.; Roberto, S.R. Synergistic effect of a novel chitosanisilica nanocomposites-based formulation against gray mold of table grapes and its possible mode of action. *Int. J. Biol. Macromol.* **2019**, *141*, 247–258. [CrossRef]
5. Ahmed, S.; Roberto, S.R.; Domingues, A.R.; Shahab, M.; Chaves, O.J.; Sumida, C.H.; de Souza, R.T. Effects of different sulfur dioxide pads on botrytis mold in 'italia' table grapes under cold storage. *Horticulturae* **2018**, *4*, 29. [CrossRef]
6. Pretel, M.T.; Martinez-Madrid, M.C.; Martinez, J.R.; Carreno, J.C.; Romojaro, F. Prolonged storage of 'aledo' table grapes in a slightly co2 enriched atmosphere in combination with generators of so2. *LWT-Food Sci. Technol.* **2006**, *39*, 1109–1116. [CrossRef]
7. Zutahy, Y.; Lichter, A.; Kaplunov, T.; Lurie, S. Extended storage of 'red globe' grapes in modified $SO_2$ generating pads. *Postharvest Biol. Technol.* **2008**, *50*, 12–17. [CrossRef]
8. Tapia-Blacido, D.R.; Aguilar, G.J.; de Andrade, M.T.; Rodrigues, M.F.; Guareschi-Martins, F.C. Trends and challenges of starch-based foams for use as food packaging and food container. *Trends Food Sci. Technol.* **2022**, *119*, 257–271. [CrossRef]
9. Machado, C.M.; Benelli, P.; Tessaro, I.C. Constrained mixture design to optimize formulation and performance of foams based on cassava starch and peanut skin. *J. Polym. Environ.* **2019**, *27*, 2224–2238. [CrossRef]
10. Bergel, B.F.; da Luz, L.M.; Santana, R.M.C. Effect of poly(lactic acid) coating on mechanical and physical properties of thermoplastic starch foams from potato starch. *Prog. Org. Coat.* **2018**, *118*, 91–96. [CrossRef]
11. Esmaeili, Y.; Zamindar, N.; Paidari, S.; Ibrahim, S.A.; Nafchi, A.M. The synergistic effects of aloe vera gel and modified atmosphere packaging on the quality of strawberry fruit. *J. Food Process. Preserv.* **2021**, *45*, e16003. [CrossRef]
12. Duan, Q.F.; Zhu, Z.Y.; Chen, Y.; Liu, H.S.; Yang, M.; Chen, L.; Yu, L. Starch-based foams nucleated and reinforced by polysaccharide-based crystals. *ACS Sustain. Chem. Eng.* **2022**, *10*, 2169–2179. [CrossRef]
13. Meng, L.; Liu, H.; Yu, L.; Duan, Q.; Chen, L.; Liu, F.; Shao, Z.; Shi, K.; Lin, X. How water acting as both blowing agent and plasticizer affect on starch-based foam. *Ind. Crops Prod.* **2019**, *134*, 43–49. [CrossRef]
14. Vilela Pinto, J.A.; Weber Schorr, M.R.; Thewes, F.R.; Ceconi, D.L.; Both, V.; Brackmann, A.; Fronza, D. Relative humidity during cold storage on postharvest quality of 'niagara rosada' table grapes. *Cienc. Rural* **2015**, *45*, 386–391. [CrossRef]
15. Mali, S.; Debiagi, F.; Grossmann, M.V.E.; Yamashita, F. Starch, sugarcane bagasse fibre, and polyvinyl alcohol effects on extruded foam properties: A mixture design approach. *Ind. Crops Prod.* **2010**, *32*, 353–359. [CrossRef]
16. Glenn, G.M.; Orts, W.J.; Nobes, G.A.R. Starch, fiber and caco3 effects on the physical properties of foams made by a baking process. *Ind. Crops Prod.* **2001**, *14*, 201–212. [CrossRef]
17. Peleg, M. An empirical model for the description of moisture sorption curves. *J. Food Sci.* **1988**, *53*, 1216–1217. [CrossRef]
18. Zhang, Y.; Li, C.; Fu, X.; Ma, N.; Bao, X.; Liu, H. Characterization of a novel starch-based foam with a tunable release of oxygen. *Food Chem.* **2022**, *389*, 133062. [CrossRef]
19. Lichter, A.; Zutahy, Y.; Kaplunov, T.; Lurie, S. Evaluation of table grape storage in boxes with sulfur dioxide-releasing pads with either an internal plastic liner or external wrap. *Horttechnology* **2008**, *18*, 206–214. [CrossRef]
20. Ji, D.C.; Chen, T.; Ma, D.Y.; Liu, J.L.; Xu, Y.; Tian, S.P. Inhibitory effects of methyl thujate on mycelial growth of botrytis cinerea and possible mechanisms. *Postharvest Biol. Technol.* **2018**, *142*, 46–54. [CrossRef]

21. Shao, Y.; Wang, K.; Xuan, G.; Gao, C.; Hu, Z. Soluble solids content monitoring for shelf-life assessment of table grapes coated with chitosan using hyperspectral imaging. *Infrared Phys. Technol.* **2021**, *115*, 103725. [CrossRef]
22. Bangar, S.P.; Whiteside, W.S.D.; Ozogul, F.; Dunno, K.D.; Cavender, G.A.; Dawson, P. Development of starch-based films reinforced with cellulosic nanocrystals and essential oil to extend the shelf life of red grapes. *Food Biosci.* **2022**, *47*, 101621. [CrossRef]
23. Chen, R.C.; Wu, P.W.; Cao, D.Y.; Tian, H.Q.; Chen, C.K.; Zhu, B.Z. Edible coatings inhibit the postharvest berry abscission of table grapes caused by sulfur dioxide during storage. *Postharvest Biol. Technol.* **2019**, *152*, 1–8. [CrossRef]
24. Nia, A.E.; Taghipour, S.; Siahmansour, S. Pre-harvest application of chitosan and postharvest aloe vera gel coating enhances quality of table grape (*Vitis vinifera* l. Cv. 'Yaghouti') during postharvest period. *Food Chem.* **2021**, *347*, 11.
25. Xu, B.; Wu, S. Preservation of mango fruit quality using fucoidan coatings. *LWT* **2021**, *143*, 111150. [CrossRef]
26. Tianyu, J.; Qingfei, D.; Jian, Z.; Hongsheng, L.; Long, Y. Starch-based biodegradable materials: Challenges and opportunities. *Adv. Ind. Eng. Polym. Res.* **2020**, *3*, 8–18.
27. Cha, J.Y.; Chung, D.S.; Seib, P.A.; Flores, R.A.; Hanna, M.A. Physical properties of starch-based foams as affected by extrusion temperature and moisture content. *Ind. Crops Prod.* **2001**, *14*, 23–30. [CrossRef]
28. Finkenstadt, V.L.; Felker, F.C.; Fanta, G.F.; Kenar, J.A.; Selling, G.W.; Hornback, K.J.; Fisk, D.L. Extruded foams prepared from high amylose starch with sodium stearate to form amylose inclusion complexes. *J. Appl. Polym. Sci.* **2016**, *133*, 10. [CrossRef]
29. Benezet, J.C.; Stanojlovic-Davidovic, A.; Bergeret, A.; Ferry, L.; Crespy, A. Mechanical and physical properties of expanded starch, reinforced by natural fibres. *Ind. Crops Prod.* **2012**, *37*, 435–440. [CrossRef]
30. Zhang, Z.Q.; Liu, T.; Xu, Y.; Chen, Y.; Chen, T.; Li, B.Q.; Tian, S.P. Sodium pyrosulfite inhibits the pathogenicity of botrytis cinerea by interfering with antioxidant system and sulfur metabolism pathway. *Postharvest Biol. Technol.* **2022**, *189*, 9. [CrossRef]
31. Chen, X.Y.; Zhu, Z.Q.; Zhang, X.S.; Antoce, A.O.; Mu, W.S. Modeling the microbiological shelf life of table grapes and evaluating the effects of constant concentrations of sulfur dioxide. *J. Food Process. Preserv.* **2017**, *41*, e13058. [CrossRef]
32. Pan, Q.; Zhou, C.; Yang, Z.; He, Z.; Wang, C.; Liu, Y.; Song, S.; Gu, H.; Hong, K.; Yu, L.; et al. Preparation and characterization of chitosan derivatives modified with quaternary ammonium salt and quaternary phosphate salt and its effect on tropical fruit preservation. *Food Chem.* **2022**, *387*, 132878. [CrossRef] [PubMed]
33. Youssef, K.; Chaves, O.J.; Muhlbeier, D.T.; Roberto, S.R. Sulphur dioxide pads can reduce gray mold while maintaining the quality of clamshell-packaged 'brs nubia' seeded table grapes grown under protected cultivation. *Horticulturae* **2020**, *6*, 20. [CrossRef]
34. Artes-Hernandez, F.; Tomas-Barberan, F.A.; Artes, F. Modified atmosphere packaging preserves quality of so2-free 'superior seedless' table grapes. *Postharvest Biol. Technol.* **2006**, *39*, 146–154. [CrossRef]
35. Chaves, O.J.; Youssef, K.; Koyama, R.; Ahmed, S.; Dominguez, A.R.; Muhlbeier, D.T.; Roberto, S.R. Control of gray mold on clamshell-packaged 'benitaka' table grapes using sulphur dioxide pads and perforated liners. *Pathogens* **2019**, *8*, 14.
36. Luo, P.; Li, F.P.; Liu, H.H.; Yang, X.M.; Duan, Z.H. Effect of fucoidan-based edible coating on antioxidant degradation kinetics in strawberry fruit during cold storage. *J. Food Process. Preserv.* **2020**, *44*, e14381. [CrossRef]

foods

MDPI

Article

# Antioxidant Capacity and Phenolics Profile of Portuguese Traditional Cultivars of Apples and Pears and Their By-Products: On the Way to Newer Applications

**João David Teixeira [1,2], Ana Rita Soares Mateus [1,2,3,4], Claudia Sanchez [5,6], Pier Parpot [7,8], Carina Almeida [1,9,10,*] and Ana Sanches Silva [1,2,3,11,*]**

1 National Institute for Agrarian and Veterinary Research (INIAV), I.P., Rua dos Lágidos, Lugar da Madalena, 4485-655 Vila do Conde, Portugal; david.teixeira@iniav.pt (J.D.T.); anarsmateus@hotmail.com (A.R.S.M.)
2 Center for Study in Animal Science (CECA), ICETA, University of Porto, 4050-453 Porto, Portugal
3 Faculty of Pharmacy, University of Coimbra, Polo III, Azinhaga de Stª Comba, 3000-548 Coimbra, Portugal
4 REQUIMTE/LAVQ, R. D. Manuel II, Apartado 55142, 4501-401 Porto, Portugal
5 National Institute for Agricultural and Veterinary Research (INIAV), I.P., 2460-059 Alcobaça, Portugal; claudia.sanchez@iniav.pt
6 GREEN-IT—Bioresources for Sustainability R&D Unit, ITQB NOVA, 2780-157 Oeiras, Portugal
7 Department of Chemistry, University of Minho, 4704-553 Braga, Portugal; parpot@quimica.uminho.pt
8 Center of Biological Chemistry, University of Minho, 4704-553 Braga, Portugal
9 LEPABE—Laboratory for Process Engineering, Environment, Biotechnology and Energy, Faculty of Engineering, University of Porto, Rua Dr. Roberto Frias, 4200-465 Porto, Portugal
10 AliCE—Associate Laboratory in Chemical Engineering, Faculty of Engineering, University of Porto, Rua Dr. Roberto Frias, 4200-465 Porto, Portugal
11 Associate Laboratory for Animal and Veterinary Sciences (Al4AnimalS), 1300-477 Lisbon, Portugal
* Correspondence: carina.almeida@iniav.pt (C.A.); ana.silva@iniav.pt (A.S.S.)

**Citation:** Teixeira, J.D.; Soares Mateus, A.R.; Sanchez, C.; Parpot, P.; Almeida, C.; Sanches Silva, A. Antioxidant Capacity and Phenolics Profile of Portuguese Traditional Cultivars of Apples and Pears and Their By-Products: On the Way to Newer Applications. *Foods* **2023**, *12*, 1537. https://doi.org/10.3390/foods12071537

Academic Editors: Sílvia Petronilho and Mario M. Martinez

Received: 10 March 2023
Revised: 28 March 2023
Accepted: 29 March 2023
Published: 5 April 2023



**Abstract:** Pears (*Pyrus communis* L.) and apples (*Malus domestica* Borkh.) are two of the most popular fruits worldwide. The phenolic compounds they offer are associated with human health benefits due to their antioxidant properties. Since these fruits' by-products are not yet fully exploited, it is important to characterize them, especially in terms of their antioxidant properties. The aim of this study was to determine the antioxidant properties of old traditional cultivars, six regional pear cultivars and five regional apple cultivars grown in the Alcobaça region (Portugal). Antioxidant capacity assays were used to evaluate the antioxidant properties. Generally, the antioxidant capacity, total phenolics content (TPC), and total flavonoids content (TFC) of fruit byproducts (both seeds and peels) were higher than the corresponding mesocarp, indicating their potential as sources of beneficial antioxidant compounds. Moreover, a UHPLC-ToF-MS method was optimized and validated in order to quantify 21 distinct phenolics in these fruit samples. The analytical method's suitability for quantifying phenolic compounds was demonstrated by an evaluation of linearity, limit of detection, limit of quantification, precision and accuracy. This method was used to determine the phenolic composition of samples of regional (local) cultivars. The phenolics in the fruit samples with the highest concentrations were phlorizin and chlorogenic acid. Principal component analysis (PCA) was used to separate distinct fruit species while emphasizing their similarities and differences.

**Keywords:** ultra-high performance liquid chromatography; time-of-flight mass spectrometry; phenolics; apples; pears; by-products; antioxidant capacity; fructose; principal component analysis

## 1. Introduction

It is well established that fruits are a very important part of the human diet since they provide numerous nutrients such as minerals and dietary fiber, as well as bioactive compounds [1,2]. One of the most produced fruits worldwide is the apple (*Malus domestica*), mainly in the Asian continent, with the largest orchard fields being located in China [3,4].

There are over 7000 cultivars of apples identified in the world, each with a unique set of characteristics comprising size, color, firmness, shape, texture, flavor (including sweet, sour, and bitter sensations), juiciness, aroma, and nutritional value [5]. These characteristics influence the quality of the fruit and the products they originate [6]. Pears (*Pyrus communis* L.) are also planted in abundance, and like apples, they are mainly produced in China, with over 3000 cultivars identified in this country's repertoire [7].

Both of these fruits are known for being major sources of bioactive compounds, such as phenolic compounds [8,9]. These are considered responsible for desirable health properties such as anti-inflammatory, antibacterial and antioxidant activities [9,10], which are proven to have a significant role in the prevention of diseases such as cancer [11], diabetes [12], age-related functional decline [13], cardiovascular and neurodegenerative diseases such as Alzheimer's [14–16].

Over the past few years, consumers have become more informed, and they have changed their buying and consumption habits [17]. In this line, people are developing an interest in 'natural' foods/products and favor these over processed products [18,19]. With the previous set of characteristics in mind, the food industry aims to develop their products, incorporating ingredients that are more close to 'natural' foods, rich in bioactive compounds, convenient, and largely accepted by their organoleptic properties, making fruits among the most suitable foods for this goal [20].

Since the production of fruit-based products has grown in that time period, one of the challenges that the food industry faces is the amount of waste being generated from the production of fruit-based products. This poses a major concern regarding environmental pollution [21] because large quantities of by-products are discarded without any treatment. However, these by-products (seeds and peels), are a great source of bioactive compounds (such as phenolics) [22] with interesting biological properties that can have a role on the prevention and/or treatment of emerging diseases [23].

Biodiversity is declining at an alarming rate [24,25], endangering the conservation of various species. This crisis requires careful consideration and setting of objectives in order to preserve biodiversity in the world [26,27]. One way of tackling this problem is to find uses for lesser-known species, making them targets of interest for the food industry.

Therefore, the objective of this study is to determine the antioxidant properties of five apple cultivars (*Pêro de Borbela, Pardo Lindo, Repinau, Pêro Coimbra* and *Noiva*) and six pear cultivars (*Bela-Feia, Torres Novas, Carapinheira, Carapinheira Roxa, Lambe-os-Dedos* and *Amorim*), from the Alcobaça (Portugal) region to characterize Portuguese traditional cultivars regarding these fruits' bioactive composition, for which there is a lack of information. Moreover, the evaluation of the fructose content of both apple and pear cultivars was performed to conclude their potential to substitute sugar in food formulations. Moreover, in accordance with international guidelines, an analytical method was optimized and validated for the detection of 21 individual phenolic compounds in fruits using UHPLC-ToF-MS. PCA was used as a valid tool to distinguish between fruit species/cultivars and fruits parts. These data will allow to preserve fruits' biodiversity and might increase the interest of both fruit growers and food industry by the exploration of some of these cultivars.

## 2. Materials and Methods

### *2.1. Cultivars under Study*

Five apple and six pear cultivars from the Alcobaça region (Portugal) were collected from Vieira de Natividade Fruit Research Station (Alcobaça) of the National Institute of Agrarian and Veterinary Research (INIAV), between July and November 2021 and July and October 2022. The apple regional cultivars included were *Pardo Lindo, Repinau, Pêro de Borbela, Pêro Coimbra,* and *Noiva*, and the pear regional cultivars were *Carapinheira, Carapinheira Roxa, Amorim, Lambe-os-Dedos, Bela-Feia* and *Torres Novas*. Photographs of the Portuguese traditional cultivars can be found in this publication—see Supplementary Materials, Figure S1.

## 2.2. Sample Preparation

Ten to twelve fruits from each cultivar were used in this study. Each fruit was separated into three sections: peels, seeds and pulp (mesocarp). The three sections were individually homogenized with a grinder homogenizer (Ultra Turrax® T25, Janke and Kunkel IKA, Stavfen, Germany). Two grams of the homogenized sample was weighted into a 50 mL Falcon tube and mixed with 20 mL of the selected solvent (two separate extracts were prepared—one using water as the solvent and the other using 95% ethanol) on an Ultra-TURRAX homogenizer for 3 min. The remainder of the homogenates were stored at $<-20$ °C.

For the UHPLC phenolic compounds analysis, a solid–liquid extraction methodology was used. For this purpose, two grams of each sample was added to 10 mL of MeOH:$H_2O$:formic acid (49.95:49.95:0.10 $v/v/v$) and further sonicated at room temperature for 10 min. After this, the solution was agitated for 15 min in a horizontal shaker and centrifuged at 2250× *g* for 10 min at 20 °C. The supernatant was removed to another Falcon tube, and the extract was repeated with another 10 mL of the solvent. The second extract was merged with the first one.

## 2.3. Chemicals and Reagents

The phenolic compound standards (4-hydroxybenzoic acid, apigenin, caffeic acid, catechin, chlorogenic acid, epicatechin, eriodictyol, gallic acid, luteolin, naringenin, o-coumaric acid, *p*-coumaric acid, phloridzin, quercetin, quercetin-3-B-D-glucoside, quercitrin, rutin, sinapic acid, syringic acid, trans-ferrulic acid, and vanillic acid), (±)-6-hydroxy-2,5,7,8-tetramethylchromane-2-carboxylic acid (Trolox), 2,2-diphenyl-1-picrylhydrazyl (DPPH), Folin-Cioucalteu reagent, sodium carbonate, sodium nitrite, aluminum chloride, sodium hydroxide, β-Carotene, chloroform, Tween® 40, linoleic acid, fructose, resorcinol, thiourea, and formic acid were purchased from Sigma Aldrich. Ethanol, methanol, acetic acid, and hydrochloric acid were purchased from Honeywell.

## 2.4. Evaluation of the Dry Mass Percentage

Dry mass content (%) was measured for each sample in a hot air conventional oven (Memmert, Nurenberg, Germany). Fruit samples of approximately 10 g, cut into slices, were dried at 70 ± 1 °C until they reached a constant mass, according to AOAC [28]. All determinations were made in triplicate.

## 2.5. Antioxidant Activity Assays

### 2.5.1. DPPH Radical Scavenging Assay

The DPPH radical scavenging assay was employed as described by Martins et al. [29]. Briefly, 50 µL of sample was mixed with 2 mL of the DPPH radical solution (14.2 µg/mL) in a 15 mL Falcon tube, kept in the dark for 30 min and the absorbance was read in a spectrophotometer (UV-Vis Spectrophotometer (U-2810, Hitachi, Digilab, Sydney, NSW, Australia) at 515 nm. A calibration curve ($y = 0.8457x - 3.2621$, $r^2 = 0.9980$) was drawn up using different concentrations of Trolox (6-hydroxy-2,5,7,8-tetramethylchroman-2-carboxylic acid), with a working range of 5–100 µg/mL. Results were expressed as µg Trolox equivalent (TE)/g of fruit. The inhibiting percentage was calculated according to the following formula:

$$Inhibiting\ percentage\ (\%) = \left(\frac{Abs\ control - Abs\ sample}{Abs\ control}\right) \times 100, \quad (1)$$

where *Abs control* is the absorbance of the control and *Abs sample* corresponds to the absorbance of the sample. All experiments were conducted using duplicates.

### 2.5.2. β-Carotene Bleaching Assay

This method was carried out following the procedure of Miller [30]. In this method, 1 mL of a β-carotene solution (0.2 mg/mL) in chloroform was mixed with 20 mg of linoleic acid and 200 mg of Tween® 40 emulsifier. The chloroform was evaporated on a rotary evaporator at 40–45 °C. Then, 100 mL of ultrapure MilliQ® water, which was vortexed for 30 min, was added, and thoroughly agitated, until an emulsion was formed. Finally, to 200 µL of each sample, 5 mL of the β-carotene emulsion was added and held for two h at 55 °C in a water bath. Both the samples and the control absorbance were read in a spectrophotometer at 470 nm at the conclusion of this time frame. The antioxidant activity coefficient (*AAC*) was calculated resorting to the following equation:

$$AAC = \left( \frac{Abs\ sample - A2\ control}{A0\ control - A2\ control} \right) \times 100, \quad (2)$$

where *A0 control* is the absorbance of the control at the initial time, $t = 0$ min, *A2 control* is the absorbance of the control after 120 min at 50 °C, and *Abs sample* is the absorbance of the sample after it has also been subjected to 50 °C for 120 min in a water bath. All experiments were conducted using duplicates.

### 2.6. Total Phenolics Content (TPC) Assay

In order to estimate the total phenolics content of the fruit extracts, the method by Erkan et al. [31] was applied. This method consists of mixing 1 mL of sample with 7.5 mL of Folin–Cioucalteu reagent (10% $v/v$) and letting it react for 5 min, before adding 7.5 mL of $Na_2CO_3$ (60 mg/mL) and reading the absorbance, after 120 min at 725 nm. A calibration curve ($y = 0.0065x - 0.0057$, $r^2 = 0.9997$) was drawn up using different concentrations of gallic acid with a working range of 5–150 µg/mL. Results were expressed as µg Gallic acid equivalents (GAE)/g of fruit. All experiments were conducted using duplicates.

### 2.7. Total Flavonoids Content (TFC) Assay

The total flavonoid content method was performed according to Barbosa et al. [32]. In this assay, 1 mL of sample was mixed with 4 mL of ultrapure water and 0.3 mL of sodium nitrite (50 mg/mL). After 5 min, 0.6 mL of aluminum chloride (100 mg/mL) was added to the mixture and after another 6 min, 2 mL of sodium hydroxide (40 mg/m) and 2.1 mL of ultrapure water were added. The absorbance of samples was then read in a spectrophotometer at 510 nm. A calibration curve ($y = 0.0017x + 0.0165$, $r^2 = 0.9986$) was drawn up using different concentrations of Epicatechin, with a working range of 5–150 µg/mL. Results were expressed as µg Epicatechin equivalents (EE)/g of fruit. All experiments were conducted using duplicates.

### 2.8. Fructose Content Assay

The total amount of fructose present in the samples was assessed using the method described by Ashwell [33]. For the performance of this method, resorcinol reagent was prepared by mixing 1 g of resorcinol with 250 mg of thiourea in 100 mL of glacial acetic acid. Then, 2 mL of sample was mixed with 1 mL of the resorcinol reagent, and 7 mL of a diluted hydrochloric acid solution was added to the mixture. The sample mixtures were kept in a water bath at 80 °C for 10 min. At the end of this period, all samples were cooled under tap water for 5 min and read in a spectrophotometer at 520 nm. A calibration curve ($y = 0.0016x + 0.0524$, $r^2 = 0.9995$) was drawn up using different concentrations of fructose, with a working range of 5–500 µg/mL. Results were expressed as mg fructose/g of fruit. All experiments were conducted using duplicates.

### 2.9. UHPLC-ToF-MS Conditions

The detection and quantification of the phenolic compounds were performed using a Nexera X2 Shimadzu UHPLC coupled with a 5600+ ToF-MS detector (SCIEX, Foster City, CA, USA) equipped with a Turbo Ion Spray electrospray ionization source working in positive mode (ESI+). Regarding the analytical column, a Acquity UPLC BEH C18 (2.1 mm × 100 mm, 1.7 µm) was used. The column temperature was kept at 20 °C, and the autosampler was maintained at 20 °C. The chromatographic separation took place in gradient mode using an aqueous solution of 0.1% formic acid (eluent A) and methanol with

0.1% formic acid (eluent B) as the mobile phase. The injection volume for both standards and samples was 20 μL. The selected gradient program was the following: 0–0.5 min kept at 90% [A]; 0.5–8 min from 90% to 20% [A] and kept until the end of the run at 20% [A], completing a total run time of 8.1 min. Using the Analyst® TF 1.7 software (SCIEX, Foster City, CA, USA) and the following parameters for mass spectrometry, the acquisition was carried out in full scan from 100 to 750 Da: ion source voltage of 5500 V; source temperature of 575 °C; curtain gas (CUR) of 30 psi; Gas 1 and Gas 2 of 55 psi; and declustering potential (DP) of 100 V. To provide accurate mass resolution, the ToF-MS detector was calibrated every 7 injections in the method's mass range.

PeakView™ 2.2 and MultiQuant™ 3.0 software (SCIEX, Foster City, CA, USA) were used for phenolic compound identification and data processing. PeakView™ 2.2 software automatically presents the isotope match. Two parameters and their accompanying equations (Equations (3) and (4)) were employed for phenolic compound identification: (1) maximum retention time deviation (ΔRRT) of 0.1 min (Equation (3)); and (2) exact mass deviation (m) with a tolerance of 5 ppm (Equation (4)).

$$\text{RRT} = \left(\frac{\text{RRT spiked samples} - \text{RRT standard}}{\text{RRT standard}}\right) \times 100 \quad (3)$$

$$\Delta\text{m (ppm)} = \left(\frac{\text{Exact mass} - \text{Detected mass}}{\text{Exact mass}}\right) \times 10^6 \quad (4)$$

where RRT is the relative retention time and Δm is the exact mass deviation.

### 2.10. Validation of the UHPLC-ToF-MS Method

The method was validated by the determination of the working range, linearity, limit of detection (LOD), limit of quantification (LOQ), precision and accuracy (through recovery assays). LOD and LOQ were determined as the concentration that originates a signal-to-noise ratio (S/N) ≥3 and ≥10, respectively. According to Directive 96/23/EC, the trueness of measurements can be assessed through the recovery of additions of known amounts of the analytes to a blank matrix, when certified reference materials are not available.

### 2.11. Principal Component Analysis (PCA)

Principal component analysis, classification and regression by machine learning were carried out using Python (Python 3).

### 2.12. Statistical Analysis

To evaluate the effect of the harvest year, cultivar, and part of the fruit on the antioxidant properties and phenolic profile of the samples a two-way analysis of variance (ANOVA) was performed. Significance was defined at least at $p < 0.05$. All data analyses were carried out using Microsoft Excel 365 equipped with Analysis ToolPak.

## 3. Results and Discussion

### 3.1. Dry Mass Percentage Assay

Measurement of the dry mass content led to the information presented on Table 1. The observed percentages are similar to the ones reported by Instituto Nacional de Saúde Dr Ricardo Jorge (INSA), 17.1% for apple [34] and 14.9% for the mean of five cultivars of pear [35]. The old traditional Portuguese cultivars of apples showed dry mass contents between 18.3 and 22.8% while the pear cultivars exhibited values held between 16.7 and 20.7%.

**Table 1.** Dry mass percentage of apple and pear cultivars. Results are expressed in means ± standard deviation.

| Fruit | Cultivar | Dry Mass Percentage |
|---|---|---|
| Apple | *Pêro Coimbra* | 18.8 ± 0.4 |
| | *Repinau* | 20.8 ± 0.4 |
| | *Pardo Lindo* | 22.8 ± 0.5 |
| | *Pêro de Borbela* | 18.6 ± 0.1 |
| | *Noiva* | 18.3 ± 0.7 |
| Pear | *Bela-Feia* | 20.2 ± 1.4 |
| | *Torres Novas* | 17.3 ± 1.0 |
| | *Carapinheira Roxa* | 20.5 ± 0.3 |
| | *Lambe-os-Dedos* | 16.7 ± 1.0 |
| | *Amorim* | 20.7 ± 1.5 |
| | *Carapinheira* | 19.7 ± 0.5 |

## 3.2. Antioxidant Capacity Assays

Four tests were conducted to determine the antioxidant properties of the fruit cultivars. The DPPH radical scavenging assay and the β-carotene bleaching assay were used to measure the antioxidant capacity. The other two tests evaluated the content of compounds that contribute to the antioxidant properties. These tests consisted of the total phenolic content (TPC) assay and the total flavonoid content (TFC) assay. The tests were repeated for two harvests of each cultivar, 2021 and 2022, thus analyzing if the time frame can be considered a factor that originates differences in the amount of phenolic compounds present in the fruit. Other possible factors are the edaphoclimatic conditions, the cultivar itself, or the pesticide/fertilizer appliance.

Tables 2 and 3 show the inhibiting percentage (IP) and AAC obtained by all the cultivars of apples and pears, respectively.

**Table 2.** Antioxidant capacity of different portions of apple cultivars using two methods: DPPH radical scavenging assay and the β-carotene bleaching assay. Results are expressed as means ± standard deviation.

| Cultivar | Portion | Inhibition Percentage (%) | | μg Trolox Equivalents (TE)/g FW | | Antioxidant Activity Coefficient (AAC) | |
|---|---|---|---|---|---|---|---|
| | | 2021 | 2022 | 2021 | 2022 | 2021 | 2022 |
| *Pardo Lindo* | Peel | $9.22^{a} \pm 0.15$ | $7.69^{i} \pm 0.87$ | $148^{a} \pm 1.81$ | $130^{i} \pm 10.3$ | $124^{A} \pm 10.9$ | $134^{AG} \pm 2.21$ |
| | Seed | $5.10^{b} \pm 0.46$ | $4.15^{g} \pm 0.22$ | $98.8^{b} \pm 5.44$ | $87.7^{g} \pm 2.57$ | $62.7^{B} \pm 10.9$ | $88.9^{C} \pm 4.41$ |
| | Pulp | $0.06^{c} \pm 0.55$ | $0.62^{c} \pm 0.00$ | $39.3^{c} \pm 6.47$ | $45.8^{c} \pm 0.00$ | $76.8^{BC} \pm 9.05$ | $64.0^{B} \pm 4.41$ |
| *Repinau* | Peel | $14.5^{d} \pm 0.00$ | $12.5^{g} \pm 0.65$ | $210^{d} \pm 0.00$ | $186^{g} \pm 7.72$ | $89.6^{C} \pm 12.7$ | $88.9^{C} \pm 4.41$ |
| | Seed | $5.42^{b} \pm 0.31$ | $6.46^{b} \pm 0.44$ | $103^{b} \pm 3.63$ | $115^{b} \pm 5.15$ | $42.3^{D} \pm 10.9$ | $37.4^{D} \pm 2.21$ |
| | Pulp | $0.13^{c} \pm 0.46$ | $1.85^{c} \pm 0.00$ | $40.1^{c} \pm 5.39$ | $60.4^{c} \pm 0.00$ | $82.0^{CE} \pm 9.05$ | $51.5^{BD} \pm 4.41$ |
| *Pêro Coimbra* | Peel | $19.2^{e} \pm 0.15$ | $24.7^{d} \pm 0.39$ | $266^{e} \pm 1.82$ | $331^{d} \pm 4.65$ | $127^{AF} \pm 10.9$ | $142^{FG} \pm 4.41$ |
| | Seed | $5.21^{b} \pm 0.00$ | $6.67^{bg} \pm 0.79$ | $100^{b} \pm 0.00$ | $117^{g} \pm 9.29$ | $147^{G} \pm 10.9$ | $150^{G} \pm 2.21$ |
| | Pulp | $0.00^{c} \pm 0.00$ | $1.39^{c} \pm 0.39$ | $33.4^{c} \pm 7.25$ | $55.0^{c} \pm 4.65$ | $85.8^{C} \pm 14.5$ | $70.2^{BE} \pm 0.00$ |
| *Pêro de Borbela* | Peel | $27.8^{f} \pm 0.61$ | $39.7^{j} \pm 0.87$ | $367^{f} \pm 7.25$ | $508^{j} \pm 10.3$ | $143^{FG} \pm 9.05$ | $245^{J} \pm 4.41$ |
| | Seed | $6.72^{g} \pm 0.92$ | $14.6^{d} \pm 1.09$ | $118^{g} \pm 10.9$ | $211^{d} \pm 12.9$ | $109^{A} \pm 7.24$ | $130^{AG} \pm 4.41$ |
| | Pulp | $0.98^{c} \pm 0.46$ | $1.38^{c} \pm 0.22$ | $50.1^{c} \pm 5.44$ | $54.9^{c} \pm 2.57$ | $71.7^{BE} \pm 5.43$ | $87.4^{C} \pm 6.62$ |
| *Noiva* | Peel | $53.4^{h} \pm 1.00$ | $51.5^{k} \pm 1.96$ | $670^{h} \pm 11.9$ | $648^{k} \pm 23.2$ | $283^{H} \pm 3.62$ | $270^{H} \pm 8.83$ |
| | Seed | $13.3^{d} \pm 0.18$ | $11.5^{d} \pm 0.65$ | $196^{d} \pm 2.16$ | $175^{d} \pm 7.72$ | $196^{I} \pm 7.24$ | $122^{A} \pm 6.62$ |
| | Pulp | $3.93^{b} \pm 0.36$ | $6.15^{g} \pm 0.44$ | $85.1^{b} \pm 4.31$ | $111^{g} \pm 5.15$ | $76.8^{BC} \pm 9.05$ | $65.5^{BE} \pm 6.62$ |

Different letters (a–k) indicate statistically significant differences among IP of 2021 and 2022 or among μg Trolox Equivalents values of 2021 and 2022 ($p \leq 0.05$). Different letters (A–J) indicate statistically significant differences among AAC of 2021 and 2022 ($p \leq 0.05$).

**Table 3.** Antioxidant capacity of different portions of pear cultivars using two methods: DPPH radical scavenging assay and the β-carotene bleaching assay. Results are expressed in means ± standard deviation.

| Cultivar | Portion | Inhibition Percentage (%) | | μg Trolox Equivalent (TE)/g FW | | Antioxidant Activity Coefficient (AAC) | |
|---|---|---|---|---|---|---|---|
| | | 2021 | 2022 | 2021 | 2022 | 2021 | 2022 |
| *Bela-Feia* | Peel | 13.2 [a] ± 0.31 | 15.8 [h] ± 0.89 | 195 [a] ± 3.63 | 225 [h] ± 10.5 | 161 [A] ± 9.05 | 178 [A] ± 11.0 |
| | Seed | 11.9 [a] ± 0.31 | 12.8 [a] ± 0.30 | 180 [a] ± 3.63 | 190 [a] ± 3.51 | 122 [B] ± 7.24 | 90.5 [C] ± 6.62 |
| | Pulp | 1.63 [b] ± 0.15 | 3.15 [g] ± 0.30 | 57.8 [b] ± 1.81 | 75.8 [g] ± 3.51 | 83.2 [C] ± 3.62 | 68.6 [CD] ± 6.62 |
| *Torres Novas* | Peel | 7.37 [c] ± 0.00 | 8.19 [c] ± 0.30 | 126 [c] ± 0.00 | 135 [c] ± 3.51 | 88.4 [C] ± 3.62 | 112 [B] ± 11.0 |
| | Seed | 8.03 [c] ± 0.31 | 7.98 [c] ± 0.59 | 133 [c] ± 3.63 | 133 [c] ± 7.03 | 85.8 [C] ± 10.9 | 76.4 [CE] ± 8.83 |
| | Pulp | 0.00 [d] ± 0.00 | 0.00 [d] ± 0.00 | 24.5 [d] ± 5.44 | 46.0 [b] ± 3.51 | 64.0 [D] ± 9.05 | 57.7 [D] ± 4.41 |
| *Carapinheira Roxa* | Peel | 12.6 [a] ± 0.00 | 15.8 [h] ± 1.49 | 187 [a] ± 0.00 | 225 [h] ± 17.6 | 96.0 [CE] ± 7.24 | 109 [BE] ± 11.0 |
| | Seed | 7.48 [c] ± 0.15 | 7.77 [c] ± 0.30 | 127 [c] ± 1.81 | 130 [c] ± 3.51 | 60.2 [D] ± 10.9 | 49.9 [D] ± 2.20 |
| | Pulp | 0.11 [e] ± 0.46 | 3.15 [g] ± 0.30 | 39.9 [e] ± 5.44 | 75.8 [g] ± 3.51 | 52.5 [DF] ± 10.9 | 35.9 [F] ± 4.41 |
| *Lambe-os-Dedos* | Peel | 9.98 [f] ± 0.61 | 12.2 [a] ± 0.59 | 157 [f] ± 7.25 | 183 [a] ± 7.03 | 129 [B] ± 7.24 | 139 [B] ± 0.00 |
| | Seed | 3.47 [g] ± 0.31 | 9.03 [cf] ± 0.89 | 79.6 [g] ± 3.63 | 145 [cf] ± 10.54 | 39.7 [F] ± 10.9 | 67.1 [CD] ± 8.83 |
| | Pulp | 0.22 [e] ± 0.61 | 3.57 [g] ± 0.89 | 41.1 [e] ± 7.25 | 80.8 [g] ± 10.54 | 60.2 [D] ± 7.24 | 62.4 [D] ± 3.54 |
| *Amorim* | Peel | 15.4 [h] ± 1.31 | 13.7 [ah] ± 1.89 | 221 [h] ± 3.63 | 200 [ah] ± 10.54 | 216 [G] ± 7.24 | 228 [G] ± 2.21 |
| | Seed | 12.9 [a] ± 0.15 | 14.5 [ah] ± 0.89 | 191 [a] ± 1.81 | 210 [ah] ± 10.54 | 152 [A] ± 10.9 | 175 [A] ± 6.62 |
| | Pulp | 8.57 [c] ± 0.46 | 8.19 [c] ± 0.89 | 140 [c] ± 5.44 | 135 [c] ± 10.54 | 76.8 [CD] ± 9.05 | 98.3 [C] ± 4.41 |
| *Carapinheira* | Peel | 14.8 [h] ± 0.00 | - * | 213 [h] ± 0.00 | - * | 163 [A] ± 7.24 | - * |
| | Seed | 21.4 [i] ± 0.15 | - * | 291 [i] ± 1.81 | - * | 165 [A] ± 7.24 | - * |
| | Pulp | 0.65 [b] ± 0.31 | - * | 46.3 [b] ± 3.63 | - * | 41.0 [F] ± 9.05 | - * |

* The authors did not had access to the 2022 harvest of the *Carapinheira* pear cultivar. Different letters (a–k) indicate statistically significant differences among IP of 2021 and 2022 or among μg Trolox Equivalents values of 2021 and 2022 harvest years ($p \leq 0.05$). Different letters (A–J) indicate statistically significant differences among AAC of 2021 and 2022 harvest years ($p \leq 0.05$).

The *Noiva* cultivar presented the highest IP, in the peel portion (53.4 ± 1.00 and 51.5 ± 1.96% in 2021 and 2022, respectively), followed by the *Pêro de Borbela* (27.8 ± 0.61 and 39.7 ± 0.87% in 2021 and 2022, respectively). However, the seed portion of the 2022 harvest of the *Pêro de Borbela* cultivar presented a higher IP (14.6 ± 1.09%) than the ones presented by both harvests of the *Noiva* cultivar (13.3 ± 0.18 and 11.5 ± 0.65%). One common point among all the apple cultivars is that the by-products present a higher IP than the pulp, in which the maximum registered IP among edible portions was 6.15 ± 0.44% for the *Noiva* cultivar. According to the DPPH radical scavenging inhibition assay, the *Noiva* and *Pêro de Borbela* cultivars stand out among all the apple cultivars that were examined in terms of their antioxidant capacity, mostly in the by-products. These findings are comparable to those published by Moni Bottu et al. [10] who found that using a similar solvent ($MeOH{:}H_2O$ 80:20) and reporting the results in fresh weight (FW), the tested apple cultivars, *Bittersweet* and *Jonagold*, obtained an IP of 14.4 ± 1.8% and 13.4 ± 3.1%, respectively. These results, however, represent the whole fruit instead of a portion of it. The same problem was registered in the work of Mignard et al. [5], where among all the 155 analyzed cultivars, the minimum and maximum relative antioxidant content identified were 1.7 and 44.6 mg Trolox/100 g FW. In the present work, if we convert our results into mg Trolox/100 g of FW, we obtain values ranging from 3.34 to 64.70, meaning that the studied cultivars possess similar antioxidant properties to those studied by the authors.

The *Noiva* cultivar excelled all other cultivars regarding the AAC value registered, in both by-products (283 ± 3.62 and 270 ± 8.83 in the peel and 196 ± 7.24 and 122 ± 6.62 in the seed) but not in the pulp. The second highest AAC was registered in the peel of the *Pêro de Borbela* cultivar (245 ± 4.41). All the AAC values for pulp were very similar to each other, with the results being held between 51.5 ± 4.41 (*Repinau*) and 87.4 ± 6.64 (*Pêro de Borbela*). These findings are in agreement with the results reported by Sara et al. [36], and in some cases, such as in the *Noiva* by-products, our results showed higher antioxidant capacity than the cultivars presented by other authors.

Regarding the pear cultivars, the *Amorim* and *Bela-Feia*, as well as the *Carapinheira* (of which we were only able to obtain data from the 2021 harvest) cultivars can be highlighted regarding the IP they present in the seed portion, 15.4 ± 1.31 and 13.7 ± 1.89%, 13.2 ± 0.31 and 15.8 ± 0.89%, and 14.8 ± 0.00%, respectively. In contrast to the apple cultivars, pear seeds present, in some cases, significantly higher IP than the peel portion of the same fruit. The *Carapinheira* cultivar is the greatest example of this, showing an IP of 21.4 ± 0.15%, but the *Amorim* 2022 and *Torres Novas* 2021 harvests present the same profile regarding this parameter. The IP on the pulps of pear cultivars is also interesting to analyze because according to the DPPH radical assay, the mesocarp of all the samples under study has a similar level of antioxidant capacity, nevertheless with significant differences. The *Amorim* pear cultivar outperforms all other cultivars, registering an IP of 8.57 ± 0.46 and 8.19 ± 0.89%, over 200 percent higher than all the other cultivars. The *Torres Novas* cultivar has not shown any IP across the two harvests. Kolniak-Ostek et al. [9] studied the antioxidant activities of different portions of the *Radona* pear cultivar, reporting DPPH radical scavenging assay values of 1210.1 ± 17.8, 1632.7 ± 18.5, and 426.0 ± 5.9 μmol TE/100 g for the peels, seeds and pulp, respectively. These can be considered very similar to the results we obtained—the maximum value in the same unit of measurement was 1438.7 ± 25.5 μmol TE/100 g DW (*Amorim* Peel) In another study, Guan et al. [37], measured the antioxidant activity of 22 Asian pear cultivars, including *Niitaka* and *Xuehua*, and found IP's between 7.08% and 31.5%.

In general, the by-products present significantly higher AAC values than the pulp. Once again, the *Amorim* pear cultivar outperformed all the other cultivars, registering an AAC of 216 ± 7.24 and 228 ± 2.21 in the peel portion. However, the *Lambe-os-Dedos* cultivar shows a value of 60.2 ± 7.24 in the pulp, representing a higher AAC than the seed portion (39.7 ± 10.9). No pulp was able to reach AAC values close to those presented by the peel portion. This is in line with the research of Oaldje-Pavlovic et al. [38], where the authors indicate that pear peel extracts have more potent antioxidant activity than pear pulp extracts.

Koleva et al. [39] reported that the β-carotene bleaching assay is restricted to less polar compounds, whereas the DPPH radical scavenging assay is unaffected by sample polarity. This could be an indicator of why we achieved better results regarding the measured antioxidant activity with the DPPH scavenging assay than with the β-carotene bleaching assay.

Tables 4 and 5 present the TPC (in μg of Gallic Acid Equivalents (GAE) per gram FW) and TFC (in μg of Epicatechin Equivalents (EE) per gram of FW) found for the different cultivars of apples and pears, respectively.

**Table 4.** Total Phenolics Content (TPC) and Total Flavonoids Content (TFC) of different portions of apple cultivars. Results are expressed in means ± standard deviation.

| Cultivar | Portion | Total Phenolics Content (μg GAE/g FW) | | Total Flavonoids Content (μg EE/g FW) | |
|---|---|---|---|---|---|
| | | 2021 | 2022 | 2021 | 2022 |
| *Pardo Lindo* | Peel | 846 $^{a}$ ± 13.1 | 748 $^{j}$ ± 2.19 | 295 $^{A}$ ± 0.00 | 427 $^{I}$ ± 12.4 |
| | Seed | 587 $^{b}$ ± 12.0 | 298 $^{e}$ ± 1.09 | 211 $^{B}$ ± 4.14 | 246 $^{AB}$ ± 12.4 |
| | Pulp | 366 $^{c}$ ± 1.09 | 149 $^{k}$ ± 2.19 | 43.7 $^{C}$ ± 0.00 | 35.0 $^{C}$ ± 4.14 |
| *Repinau* | Peel | 962 $^{d}$ ± 21.9 | 744 $^{j}$ ± 2.19 | 562 $^{D}$ ± 12.4 | 579 $^{D}$ ± 20.7 |
| | Seed | 598 $^{b}$ ± 19.7 | 276 $^{e}$ ± 2.19 | 262 $^{A}$ ± 8.28 | 225 $^{B}$ ± 16.6 |
| | Pulp | 326 $^{ce}$ ± 3.28 | 127 $^{k}$ ± 2.19 | 37.9 $^{C}$ ± 0.00 | 37.9 $^{C}$ ± 0.00 |
| *Pêro Coimbra* | Peel | 1207 $^{f}$ ± 27.3 | 1146 $^{l}$ ± 49.2 | 749 $^{E}$ ± 12.4 | 667 $^{J}$ ± 29.0 |
| | Seed | 623 $^{b}$ ± 13.1 | 875 $^{a}$ ± 8.74 | 360 $^{F}$ ± 0.00 | 237 $^{B}$ ± 24.8 |
| | Pulp | 291 $^{e}$ ± 0.00 | 276 $^{e}$ ± 26.2 | 70.1 $^{C}$ ± 4.14 | 17.4 $^{C}$ ± 12.4 |
| *Pêro de Borbela* | Peel | 1424 $^{g}$ ± 43.7 | 1651 $^{m}$ ± 43.7 | 778 $^{G}$ ± 12.4 | 1205 $^{K}$ ± 20.7 |
| | Seed | 627 $^{b}$ ± 5.46 | 700 $^{i}$ ± 2.19 | 228 $^{B}$ ± 4.14 | 497 $^{L}$ ± 4.14 |
| | Pulp | 294 $^{e}$ ± 0.00 | 310 $^{e}$ ± 12.0 | 37.9 $^{C}$ ± 0.00 | 64.2 $^{C}$ ± 4.14 |
| *Noiva* | Peel | 1964 $^{h}$ ± 0.00 | 1801 $^{n}$ ± 38.2 | 1284 $^{H}$ ± 0.00 | 1389 $^{M}$ ± 24.8 |
| | Seed | 690 $^{i}$ ± 2.19 | 627 $^{b}$ ± 3.28 | 357 $^{F}$ ± 4.14 | 313 $^{A}$ ± 16.6 |
| | Pulp | 261 $^{e}$ ± 1.09 | 275 $^{e}$ ± 1.09 | 64.2 $^{C}$ ± 4.14 | 84.7 $^{C}$ ± 8.28 |

Different letters (a–n) indicate statistically significant differences among Total Phenolics of 2021 and 2022 harvest years ($p \leq 0.05$). Different letters (A–M) indicate statistically significant differences among Total Flavonoids of 2021 and 2022 harvest years ($p \leq 0.05$).

**Table 5.** TPC and TFC of different portions of pear cultivars. Results are expressed in means ± standard deviation.

| Cultivar | Portion | Total Phenolics Content (µg GAE/g) | | Total Flavonoids Content (µg EE/g) | |
|---|---|---|---|---|---|
| | | 2021 | 2022 | 2021 | 2022 |
| *Bela-Feia* | Peel | $948^{a} \pm 25.1$ | $1027^{a} \pm 74.3$ | $389^{A} \pm 8.28$ | $418^{A} \pm 24.8$ |
| | Seed | $778^{b} \pm 24.0$ | $1399^{e} \pm 238$ | $155^{B} \pm 8.28$ | $196^{B} \pm 41.4$ |
| | Pulp | $339^{c} \pm 2.19$ | $199^{g} \pm 73.2$ | $11.6^{C} \pm 20.7$ | $5.70^{C} \pm 12.4$ |
| *Torres Novas* | Peel | $862^{a} \pm 12.0$ | $783^{b} \pm 18.6$ | $339^{D} \pm 12.4$ | $430^{A} \pm 57.9$ |
| | Seed | $632^{d} \pm 7.65$ | $474^{d} \pm 9.83$ | $216^{E} \pm 4.14$ | $208^{E} \pm 24.8$ |
| | Pulp | $280^{c} \pm 6.56$ | $187^{g} \pm 39.3$ | $0.00^{C} \pm 0.00$ | $0.00^{C} \pm 0.00$ |
| *Carapinheira Roxa* | Peel | $991^{a} \pm 32.8$ | $1139^{f} \pm 44.8$ | $257^{E} \pm 12.4$ | $319^{DF} \pm 8.28$ |
| | Seed | $484^{d} \pm 13.1$ | $500^{d} \pm 50.3$ | $111^{B} \pm 12.4$ | $114^{B} \pm 16.6$ |
| | Pulp | $306^{c} \pm 1.09$ | $237^{c} \pm 21.9$ | $0.00^{C} \pm 0.00$ | $0.00^{C} \pm 0.00$ |
| *Lambe-os-Dedos* | Peel | $731^{bd} \pm 19.7$ | $1064^{af} \pm 44.8$ | $292^{F} \pm 12.4$ | $354^{AD} \pm 24.8$ |
| | Seed | $466^{c} \pm 17.5$ | $511^{d} \pm 108$ | $126^{B} \pm 0.00$ | $137^{B} \pm 8.28$ |
| | Pulp | $299^{c} \pm 7.65$ | $170^{g} \pm 30.6$ | $0.00^{C} \pm 0.00$ | $0.00^{C} \pm 0.00$ |
| *Amorim* | Peel | $1409^{e} \pm 40.4$ | $1095^{f} \pm 43.7$ | $547^{G} \pm 33.1$ | $541^{G} \pm 41.4$ |
| | Seed | $865^{ab} \pm 25.1$ | $500^{d} \pm 80.9$ | $237^{E} \pm 16.6$ | $158^{B} \pm 12.4$ |
| | Pulp | $367^{c} \pm 4.37$ | $197^{g} \pm 29.5$ | $23.3^{C} \pm 4.14$ | $26.2^{C} \pm 8.28$ |
| *Carapinheira* | Peel | $1102^{f} \pm 35.0$ | - * | $371^{A} \pm 24.8$ | - * |
| | Seed | $1164^{f} \pm 39.3$ | - * | $448^{D} \pm 8.28$ | - * |
| | Pulp | $376^{c} \pm 4.37$ | - * | $61.3^{H} \pm 0.00$ | - * |

* The authors did not had access to the 2022 harvest of the *Carapinheira* pear cultivar. Different letters (a–g) indicate statistically significant differences among Total Phenolics of 2021 and 2022 harvest years ($p \leq 0.05$). Different letters (A–G) indicate statistically significant differences among Total Flavonoids of 2021 and 2022 harvest years ($p \leq 0.05$).

The TPC and TFC assays allow to conclude there is a separation between the content in phenolics and flavonoids in the three portions of the apple fruits. In all the cultivars, the peels presented a higher level of both classes of compounds, followed by the seeds and finally the pulp. Among all the by-products, the peel of the *Noiva* cultivar revealed the highest content of phenolics (1964 ± 0.00 µg GAE/g), followed by the *Pêro de Borbela* peels (1651 ± 43.7 µg GAE/g) and *Pêro Coimbra* peels (1207 ± 27.3 µg GAE/g). The other two cultivars presented similar TPC, although their content is significatively different. The seeds portion of the regional cultivars presents the same profile as the peel portions; however, the TPC values are much closer to each other across the cultivars. In this portion of the fruit, the *Pêro Coimbra* cultivar registered the highest TPC (875 ± 8.74 µg GAE/g). Across the two years, the TPC and TFC values of all the portions are significantly different, meaning that the edaphoclimatic conditions may have an important role in the amount of phenolic compounds that are produced by each cultivar. In 2012, Ceymann et al. [40] measured the polyphenol profile of 104 European apple cultivars, reporting TPC values ranging from 520 to 3790 µg GAE/g FW, meaning that the regional cultivars we assessed in this study have similar phenolic content to those analyzed by the authors.

Regarding the flavonoid content, the *Noiva* cultivar has the highest TFC in the peels (1390 ± 24.8 µg EE/g), although the *Pêro de Borbela* cultivar has higher amount in the seeds (497 ± 4.14 µg EE/g). Statistical analysis indicated there are significant differences in the TFC in the pulp portion. Gulsunoglu et al., while studying the effect of fermentation with a specific fungus on the enhancement of the phenolic compound concentration in apples [41], reported a control TFC value of <1500 µg CE/g DW (catechin equivalents per gram of dry weight), and Mignard et al. [5] reported values ranging from 7 to 1421 µg CE/g FW when assessing the antioxidant traits of 155 apple cultivars. These authors, however, used catechin as the standard for the TFC assay, thus providing the results in catechin equivalents rather than epicatechin equivalents, making the comparison with our work more challenging.

Concerning pear cultivars, the *Carapinheira* cultivar constitutes an exception to the portions´ profile displayed throughout the samples. The TPC and TFC values are significantly greater in the seeds' portion (1164 ± 39.3 μg GAE/g and 448 ± 8.28 μg EE/g, respectively) than in the peels´ portion (1102 ± 35.0 μg GAE/g and 371 ± 24.8 μg EE/g). Among all the studied samples, this is the only cultivar in which this is registered. However, *Carapinheira* is not the pear cultivar that presents the highest TPC in the peels (the *Amorim* cultivar has a TPC of 1409 ± 40.4 μg GAE/g). A TPC value of 701 ± 7.94 μg GAE/g for the whole fruit was published by Sinha et al. [42] when studying the effects of the preservation of antioxidant properties in pears under cold storage, although in whole fruit. Furthermore, Zhou et al. [43] reported a TPC of 207 ± 13.7 μg GAE/g in the pulp of *Packham* pears. This represents a value held between the TPC values interval that the pulp of the assessed cultivars achieved in this study (170 ± 30.6–376 ± 4.37 μg GAE/g).

Regarding the TFC, the *Torres Novas*, *Carapinheira Roxa* and *Lambe-os-Dedos*, did not register any content of flavonoids, in both harvests, in the pulps. Nevertheless, the other three cultivars presented TFC values ranging from 5.7 ± 12.4 to 61.3 ± 0.00 μg EE/g. Guan et al. [37] reported the total flavonoid content in the pulp of 22 Asian cultivars of pear, with contents held between 23.10 and 104.28 μg quercetin equivalents/g. In the by-products, the TFC was significantly higher. The peels of the *Amorim* cultivar presented the highest TFC, at 547 ± 33.1 μg EE/g. However, Wu et al. [44] reported TFC´s from 3124.6 to 6216.6 μg rutin equivalents/g of fruit. Because the total flavonoid content is expressed in different units in different scientific studies, it is very difficult to compare the results. The presentation of phenolic compounds in the scientific literature would benefit greatly from standardization or, at the very least, more harmonization [45].

### 3.3. Fructose Content Assay

Table 6 displays the results for the total fructose content of tested apple and pear samples.

**Table 6.** Total fructose content of different portions of apple and pear cultivars. Results are expressed in means ± standard deviation.

| Apple Cultivar | Portion | Total Fructose Content (mg Fructose/g FW) | | Pear Cultivar | Portion | Total Fructose Content (mg Fructose/g FW) | |
|---|---|---|---|---|---|---|---|
| | | 2021 | 2022 | | | 2021 | 2022 |
| | Peel | 95.0 $^{a}$ ± 4.61 | 96.0 $^{a}$ ± 4.17 | | Peel | 48.3 $^{A}$ ± 3.95 | 47.7 $^{A}$ ± 2.20 |
| *Pardo Lindo* | Seed | 84.6 $^{b}$ ± 5.71 | 82.8 $^{b}$ ± 3.95 | *Bela-Feia* | Seed | 57.6 $^{AB}$ ± 4.83 | 62.4 $^{BC}$ ± 5.05 |
| | Pulp | 107 $^{a}$ ± 5.05 | 110 $^{a}$ ± 2.85 | | Pulp | 56.5 $^{AB}$ ± 5.93 | 52.4 $^{A}$ ± 4.39 |
| | Peel | 89.6 $^{a}$ ± 4.83 | 87.4 $^{c}$ ± 3.51 | | Peel | 57.3 $^{AB}$ ± 4.39 | 60.0 $^{BC}$ ± 3.73 |
| *Repinau* | Seed | 71.8 $^{a}$ ± 5.05 | 74.6 $^{a}$ ± 3.73 | *Torres Novas* | Seed | 68.8 $^{C}$ ± 4.83 | 66.9 $^{BC}$ ± 3.95 |
| | Pulp | 88.5 $^{a}$ ± 4.61 | 89.0 $^{ac}$ ± 5.27 | | Pulp | 44.3 $^{A}$ ± 3.07 | 39.2 $^{A}$ ± 4.61 |
| | Peel | 97.7 $^{a}$ ± 3.95 | 101 $^{a}$ ± 4.61 | | Peel | 60.4 $^{BC}$ ± 4.39 | 62.6 $^{BC}$ ± 4.39 |
| *Pêro Coimbra* | Seed | 94.0 $^{a}$ ± 6.15 | 92.4 $^{a}$ ± 2.20 | *Carapinheira Roxa* | Seed | 66.0 $^{BC}$ ± 3.51 | 66.0 $^{BC}$ ± 5.71 |
| | Pulp | 95.8 $^{c}$ ± 5.71 | 97.5 $^{c}$ ± 2.41 | | Pulp | 63.7 $^{BC}$ ± 5.05 | 57.6 $^{AB}$ ± 3.95 |
| | Peel | 89.6 $^{a}$ ± 3.95 | 84.2 $^{a}$ ± 4.75 | | Peel | 48.9 $^{A}$ ± 4.39 | 51.9 $^{A}$ ± 5.05 |
| *Pêro de Borbela* | Seed | 75.2 $^{b}$ ± 5.49 | 82.8 $^{a}$ ± 6.19 | *Lambe-os-Dedos* | Seed | 49.3 $^{A}$ ± 4.39 | 51.4 $^{A}$ ± 6.15 |
| | Pulp | 107 $^{c}$ ± 4.17 | 102 $^{c}$ ± 2.55 | | Pulp | 57.9 $^{A}$ ± 5.27 | 56.2 $^{A}$ ± 5.49 |
| | Peel | 73.6 $^{b}$ ± 5.05 | 79.0 $^{b}$ ± 3.75 | | Peel | 69.9 $^{C}$ ± 3.73 | 75.2 $^{D}$ ± 2.85 |
| *Noiva* | Seed | 71.3 $^{b}$ ± 5.71 | 82.5 $^{a}$ ± 3.99 | *Amorim* | Seed | 87.6 $^{D}$ ± 2.41 | 88.4 $^{D}$ ± 3.51 |
| | Pulp | 85.6 $^{a}$ ± 6.59 | 94.9 $^{a}$ ± 4.92 | | Pulp | 51.1 $^{A}$ ± 2.20 | 61.4 $^{BC}$ ± 2.63 |
| | | | | | Peel | 46.8 $^{A}$ ± 3.95 | - * |
| | | | | *Carapinheira* | Seed | 51.3 $^{A}$ ± 4.61 | - * |
| | | | | | Pulp | 68.8 $^{C}$ ± 4.39 | - * |

* The authors did not had access to the 2022 harvest of the *Carapinheira* pear cultivar. Different letters (a–c) indicate statistically significant differences among Total Fructose content of apple cultivars of 2021 and 2022 harvest years ($p \leq 0.05$). Different letters (A–D) indicate statistically significant differences among Total Fructose content of pear cultivars of 2021 and 2022 harvest years ($p \leq 0.05$).

The analysis of the fructose content shows that both the pulp and the by-products possess a very similar level of this sugar. However, the pulps of the *Pardo Lindo*, *Pêro de Borbela* and *Repinau* present the three highest fructose contents among all the portions and cultivars (110 ± 2.85 and 107 ± 5.05, 107 ± 4.17 and 102 ± 2.55, 101 ± 4.61 and 97.7 ± 3.95 mg/g, respectively). It also allows us to say that apple cultivars have a higher content than pear cultivars in all the portions, reporting values held between 71.3 ± 5.71 and 110 ± 2.85 mg/g, except for the *Amorim* pear seed (87.6 ± 2.41 and 88.4 ± 3.51 mg/g FW). The lowest levels of fructose in the pear cultivars, were registered by the pulp of the *Torres Novas* cultivar (39.2 ± 4.61 and 44.3 ± 3.07 mg/g), followed by the peels of the *Carapinheira* and *Bela-Feia* cultivars (46.8 ± 3.95, 47.7 ± 2.20 and 48.3 ± 3.95 mg/g, respectively).

### 3.4. Validation of the Analytical Method

Linearity was assessed using solvent calibration curves in different ranges for various phenolic compounds. Calibration curves with at least five calibration points were plotted for the studied phenolic compounds. The determination coefficients ($r^2$) were always higher than 0.9905 (except for 4-hydroxibenzoic acid, gallic acid, vanillic acid), highlighting that the selected ranges are suitable to the quantification of the selected 21 phenolic compounds. Table 7 presents the results of linear range, LOQ, LOD, retention time, and recovery (at two spiking levels) for the analyzed phenolic compounds. Recovery between 80 and 120% is considered acceptable, however, if the recovery percentage is over 70%, results can be considered satisfactory. In this perspective, the recovery of the different phenolics (70.2–97.3%) is acceptable, and the extraction process is suitable for the extraction of most of the phenolic compounds from apple and pear fruit matrices.

**Table 7.** Linearity, sensitivity and recovery percentages for the UHPLC-ToF-MS method. (n.d. = not determined).

| Standard | rt (min) | Equation | $r^2$ | Recovery Percentage | | Linear Range (μg/mL) | LOQ (μg/g) | LOD (μg/g) |
|---|---|---|---|---|---|---|---|---|
| | | | | Spiking Level 0.1 (mg/100 g) | Spiking Level 1.0 (mg/100 g) | | | |
| 4-Hydroxybenzoic Acid | 3.34 | $y = 329{,}563x + 21{,}605$ | 0.9805 | 92.0 | 97.3 | 0.25–10 | 2.5 | 1.0 |
| Apigenin | 6.21 | $y = 4 \times 10^7 + 89{,}331$ | 0.9993 | 85.1 | 77.1 | 0.0025–5 | 0.025 | 0.01 |
| Caffeic Acid | 3.52 | $y = 385{,}020x + 7627$ | 0.9960 | 95.8 | 87.9 | 0.5–10 | 5.0 | 2.5 |
| Catechin | 3.34 | $y = 3 \times 10^6 + 105{,}538$ | 0.9933 | 75.2 | 76.9 | 0.25–10 | 2.5 | 1.0 |
| Chlorogenic Acid | 3.23 | $y = 393{,}929x + 7739.1$ | 0.9929 | n.d. | 78.3 | 0.25–10 | 2.5 | 1.0 |
| Epicatechin | 3.59 | $y = 3 \times 10^6 + 53{,}261$ | 0.9935 | 72.4 | 83.3 | 0.1–10 | 1.0 | 0.5 |
| Eriodyctiol | 5.53 | $y = 8 \times 10^6 + 27{,}573$ | 0.9986 | 84.3 | 87.8 | 0.005–10 | 0.05 | 0.025 |
| Gallic Acid | 1.18 | $y = 90{,}742x + 6145.5$ | 0.9814 | 82.5 | 87.7 | 0.01–10 | 0.1 | 0.05 |
| Luteolin | 5.71 | $y = 2 \times 10^7 + 288{,}365$ | 0.9937 | 87.0 | 75.2 | 0.025–5 | 0.25 | 0.1 |
| Naringenin | 6.06 | $y = 5 \times 10^6 + 5148.3$ | 0.9994 | 80.6 | 87.0 | 0.25–10 | 2.5 | 1.0 |
| *o*-Coumaric Acid | 4.86 | $y = 246{,}146x + 1871.2$ | 0.9905 | 85.6 | 88.2 | 0.5–10 | 5.0 | 2.5 |
| *p*-Coumaric Acid | 4.11 | $y = 261{,}740x + 10{,}462$ | 0.9923 | 81.5 | 87.5 | 0.5–10 | 5.0 | 2.5 |
| Phlorizin | 4.88 | $y = 49{,}825x + 1778.4$ | 0.9915 | 75.4 | 85.8 | 0.025–5 | 0.25 | 0.1 |
| Quercetin | 5.79 | $y = 4 \times 10^6 + 5067.2$ | 0.9983 | 84.4 | 86.7 | 0.025–10 | 0.25 | 0.1 |
| Quercetin-3-B-D-Glucoside | 4.32 | $y = 3 \times 10^6 + 4105.4$ | 0.9969 | 94.0 | 88.1 | 0.025–10 | 0.25 | 0.1 |
| Quercitrin | 4.62 | $y = 1 \times 10^6 + 1287.9$ | 0.9992 | 77.5 | 85.4 | 0.025–10 | 0.25 | 0.1 |
| Rutin | 4.13 | $y = 1 \times 10^6 + 3288.2$ | 0.9946 | 80.5 | 87.1 | 0.25–10 | 2.5 | 1.0 |
| Sinapic Acid | 4.14 | $y = 415{,}150x + 11{,}493$ | 0.9960 | 82.2 | 86.5 | 0.25–5 | 2.5 | 1.0 |
| Syringic Acid | 3.48 | $y = 492{,}820x + 5813.3$ | 0.9957 | 70.2 | 88.5 | 1.0–10 | 10.0 | 5.0 |
| *trans*-Ferulic Acid | 4.24 | $y = 418{,}556x + 1616.9$ | 0.9953 | n.d. | 82.2 | 1.0–10 | 10.0 | 5.0 |
| Vanillic Acid | 3.49 | $y = 287{,}981x + 31{,}116$ | 0.9853 | 76.0 | 88.9 | 1.0–10 | 10.0 | 5.0 |

Table 8 displays the results of repeatability and inter-day precision for the various phenolic components in a blank apple sample spiked at two different concentration levels. The method's repeatability was assessed using the relative standard deviation ($RSD_r$) for all phenolics, using the same sample and operator in a short period of time. The relative standard deviation ($RSD_R$) was used to assess the method's intra-day precision over three separate analysis days, two different concentration levels, and multiple operators.

**Table 8.** Results of the relative standard deviation repeatability ($RSD_r$) and relative standard deviation of inter-day precision ($RSD_R$) at different spiking levels.

| Standard | Spiking Level | $RSD_r$ (%) | $RSD_R$ (%) | Standard | Spiking Level | $RSD_r$ (%) | $RSD_R$ (%) |
|---|---|---|---|---|---|---|---|
| 4-Hydroxybenzoic Acid | 0.1 mg/100 g | 2.50<br>1.63<br>1.53 | 0.78 | *p*-Coumaric Acid | 0.1 mg/100 g | 10.1<br>4.37<br>6.08 | 10.2 |
| | 1.0 mg/100 g | 11.0<br>4.97<br>2.65 | 3.03 | | 1.0 mg/100 g | 3.33<br>2.74<br>2.88 | 4.39 |
| Apigenin | 0.1 mg/100 g | 2.18<br>1.55<br>1.38 | 7.35 | Phlorizin | 0.1 mg/100 g | 4.34<br>2.23<br>5.89 | 5.93 |
| | 1.0 mg/100 g | 2.62<br>1.32<br>0.74 | 4.10 | | 1.0 mg/100 g | 4.72<br>1.69<br>1.12 | 6.34 |
| Caffeic Acid | 0.1 mg/100 g | 12.2<br>7.87<br>8.88 | 23.8 | Quercetin | 0.1 mg/100 g | 5.71<br>3.74<br>4.08 | 3.73 |
| | 1.0 mg/100 g | 3.60<br>3.03<br>3.53 | 4.89 | | 1.0 mg/100 g | 2.96<br>2.84<br>1.47 | 4.61 |
| Catechin | 0.1 mg/100 g | 4.01<br>1.56<br>1.99 | 3.82 | Quercetin-3-B-D-Glucoside | 0.1 mg/100 g | 5.52<br>2.84<br>3.78 | 4.38 |
| | 1.0 mg/100 g | 6.39<br>1.35<br>2.61 | 5.37 | | 1.0 mg/100 g | 3.40<br>2.11<br>1.16 | 4.79 |
| Chlorogenic Acid | 0.1 mg/100 g | 6.33<br>1.27<br>2.40 | 4.52 | Quercitrin | 0.1 mg/100 g | 8.51<br>1.94<br>3.97 | 13.3 |
| | 1.0 mg/100 g | 3.97<br>1.73<br>1.51 | 6.56 | | 1.0 mg/100 g | 1.83<br>1.49<br>1.05 | 5.37 |
| Epicatechin | 0.1 mg/100 g | 2.78<br>2.04<br>1.74 | 5.39 | Rutin | 0.1 mg/100 g | 7.65<br>4.60<br>4.73 | 5.88 |
| | 1.0 mg/100 g | 5.48<br>6.60<br>1.70 | 3.49 | | 1.0 mg/100 g | 4.52<br>2.88<br>2.35 | 4.35 |
| Eriodyctiol | 0.1 mg/100 g | 6.38<br>2.43<br>2.21 | 5.17 | Sinapic Acid | 0.1 mg/100 g | 6.97<br>5.07<br>2.98 | 4.42 |
| | 1.0 mg/100 g | 4.30<br>2.40<br>1.62 | 5.50 | | 1.0 mg/100 g | 3.56<br>5.47<br>2.21 | 6.91 |
| Gallic Acid | 0.1 mg/100 g | 2.10<br>7.18<br>6.25 | 14.2 | Syringic Acid | 0.1 mg/100 g | 5.90<br>2.50<br>2.25 | 8.78 |
| | 1.0 mg/100 g | 10.8<br>16.6<br>6.48 | 10.2 | | 1.0 mg/100 g | 3.40<br>2.00<br>1.69 | 5.62 |
| Luteolin | 0.1 mg/100 g | 3.51<br>1.49<br>1.90 | 7.54 | *trans*-Ferulic Acid | 0.1 mg/100 g | 11.4<br>6.42<br>5.67 | 8.93 |
| | 1.0 mg/100 g | 3.91<br>1.35<br>1.77 | 5.99 | | 1.0 mg/100 g | 4.67<br>1.33<br>3.15 | 4.20 |
| Naringenin | 0.1 mg/100 g | 22.5<br>14.9<br>12.9 | 20.1 | Vanillic Acid | 0.1 mg/100 g | 12.3<br>7.28<br>10.6 | 9.26 |
| | 1.0 mg/100 g | 9.74<br>3.08<br>1.27 | 7.61 | | 1.0 mg/100 g | 3.37<br>3.37<br>1.69 | 5.51 |
| *o*-Coumaric Acid | 0.1 mg/100 g | 5.93<br>6.69<br>5.93 | 6.81 | | | | |
| | 1.0 mg/100 g | 2.32<br>2.05<br>2.27 | 6.07 | | | | |

The values obtained for repeatability are acceptable, ranging from 0.74 to 22.5%, referring to Apigenin and Naringenin, respectively. The reported values for the inter-day precision can also be considered acceptable even though the highest $RSD_R$ was 23.8% (caffeic acid).

### 3.5. Quantification of Individual Phenolic Compounds

A total of 18 different phenolic compounds were identified in apple samples, but only 16 were quantified. In the pear samples, 19 phenolics were detected, of which 18 were quantified. Tables 9 and 10 summarize the findings from these samples. As may be seen, chlorogenic acid is the most abundant phenolic in both fruits, followed by phlorizin and epicatechin. In the pear cultivars, caffeic acid is very predominant in the samples, while not present at all in the apple cultivars. Syringic acid was another phenolic compound strictly detected in the pear samples. From the 21 analyzed phenolics, apigenin and luteolin were not quantified in any of the fruit cultivars. Apigenin was detected in apples, but at levels lower than LOQ (0.0025 μg/mL). As far as the authors are aware, this is the first study to report the presence of Apigenin and Luteolin in apples and pears, although at trace levels, and quantify 4-Hydroxybenzoic acid, Eriodyctiol and Naringenin in both fruits.

Regarding the apple portions, by-products present much higher phenolic concentration when compared to the edible part of the same cultivar. These results are in agreement with the findings of Savikin et al. [46] in which the authors highlight that higher contents were found in the peel of four apple cultivars. There is no direct relation between the part of the apple fruit and the content of phenolic compounds, among by-products, since in two of the cultivars the total content was higher on seeds and in the other three cultivars, on peels. The highest phenolic content was found in the seeds of the *Pêro de Borbela* cultivar, accumulating a total of 782.6 μg/g FW, most of which were represented by phlorizin (318.7 μg/g FW), followed by chlorogenic acid (293.5 μg/g FW) and epicatechin (59.31 μg/g FW). Phlorizin has been reported to have a variety of bioactivities. In fact, phlorizin is currently extracted from some apple pomace to be used preventing obesity, and suppressing bacterial development [47,48]. These phlorizin levels are higher than those published by Karaman et al. [49] and Mihailovic et al. [50], at −23 to 159 μg/g FW and 207 μg/g FW, respectively. Chlorogenic acid levels were also higher than others previously reported in the literature. Bílková et al. [51] studied the benefits of ultra-low oxygen conditions in long-term storage and reported a chlorogenic acid content of 1.4 to 99.6 μg/g FW. The authors reported that the analysis was made just after harvesting of the fruits, which makes the results very comparable to ours. In our case, we divided the fruits in three portions, while Bílková et al. used the whole fruit. Raudone et al. [52] also revealed the phenolic content of six apple cultivars, where the chlorogenic acid levels ranged between 250 and 2000 μg/g of dry fruit. If we take into consideration the dry mass percentage of our cultivars (Table 1), we can conclude that the studied cultivars contained between 153.2 and 675.3 μg/g of dry fruit of chlorogenic acid, representing similar levels, although lower than those mentioned by the previous authors.

Regarding pear cultivars, the distribution of the phenolics among the fruit parts was very well outlined, the peels were the part where the major concentration of phenolics was found, followed by the seeds and finally the pulp. Cui et al. [53] reported the same profile for 14 Asian pear cultivars when analyzing one of the major phenolics in pears, arbutin, and Brahem et al. [54] revealed higher content of phenolics in the peel of 19 pear cultivars, in comparison to the pulp, supporting our finding.

As mentioned before, caffeic and syringic acid were quantified in the pear cultivars, and not on the apple cultivars. Caffeic acid was determined to be one of the most present phenolics in the pear samples, especially in the peels portion, where the concentration of this phenolic compound was up to 27.09 μg/g FW. This represents much higher levels of caffeic acid than the ones published by Colaric et al. [55] in *Williams* pear whole fruit (0.86 μg/g FW). The major phenolic compound in pear fruits, is however, chlorogenic acid, ranging from 21.4 to 340 μg/g FW, across the different portions. Salta et al. [56] studied the phenolic composition of five pear cultivars, including *Rocha* pear. The authors presented chlorogenic acid levels of 43–79 μg/g FW for the other four cultivars and 624 μg/g FW for the *Rocha* pear (whole fruit), meaning that our samples were very close to most of the studied cultivars.

**Table 9.** Phenolic compounds (µg/g FW) in different parts of apple regional cultivars (P: Peel, S: Seed, M: Mesocarp (pulp)).

| Phenolic Compound | *Pardo Lindo* | | | *Repinau* | | | *Pêro Coimbra* | | | *Pêro de Borbela* | | | *Noiva* | | |
|---|---|---|---|---|---|---|---|---|---|---|---|---|---|---|---|
| | **P** | **S** | **M** | **P** | **S** | **M** | **P** | **S** | **M** | **P** | **S** | **M** | **P** | **S** | **M** |
| 4-Hydroxybenzoic Acid | 2.226 | - | - | 2.112 | <LOQ | - | 12.37 | 1.025 | - | 40.98 | 27.10 | 1.119 | 36.35 | 8.274 | 1.688 |
| Apigenin | <LOQ | <LOQ | - | - | <LOQ | - | - | <LOQ | - | - | - | - | - | <LOQ | - |
| Caffeic Acid | - | - | - | - | - | - | - | - | - | - | - | - | - | - | - |
| Catechin | - | - | - | 3.536 | - | - | - | - | - | 4.508 | 5.389 | - | 5.986 | 0.612 | - |
| Chlorogenic Acid | 34.92 | 9.955 | 3.189 | 1.062 | 6.155 | 0.610 | 65.84 | - | 9.003 | 43.07 | 293.5 | 10.42 | 123.6 | 183.9 | 55.21 |
| Epicatechin | <LOQ | 4.590 | <LOQ | 4.495 | 0.824 | <LOQ | 25.84 | - | <LOQ | 96.11 | 59.31 | 3.095 | 80.70 | 16.37 | 3.745 |
| Eriodyctiol | 0.010 | 0.014 | - | - | 0.050 | - | 0.010 | <LOQ | <LOQ | - | - | - | - | - | - |
| Gallic Acid | 5.445 | 4.126 | 3.781 | 4.785 | 5.535 | 4.079 | 2.756 | 1.867 | 2.320 | 4.530 | 5.608 | 4.539 | 6.876 | 6.715 | 7.857 |
| Luteolin | <LOQ | <LOQ | - | - | - | - | <LOQ | - | <LOQ | - | - | - | - | - | - |
| Naringenin | 0.078 | 0.052 | - | 0.074 | 0.078 | - | 0.346 | 0.158 | 0.033 | - | - | - | - | - | - |
| o-Coumaric Acid | 6.099 | 5.904 | - | - | 12.61 | - | 2.504 | - | - | 16.16 | 23.67 | 0.634 | 7.350 | 6.440 | - |
| *p*-Coumaric Acid | <LOQ | <LOQ | - | - | 3.394 | - | 2.368 | - | - | 15.46 | 10.46 | - | 6.158 | 1.908 | - |
| Phlorizin | 11.36 | 33.36 | - | 2.704 | 58.82 | <LOQ | 23.11 | 13.37 | 1.364 | 137.8 | 318.7 | 9.778 | 67.26 | 86.40 | 4.165 |
| Quercetin | 2.479 | - | - | - | 0.096 | - | 0.514 | - | - | 2.295 | 0.254 | - | 0.180 | <LOQ | - |
| Quercetin-3-B-D-Glucoside | 4.647 | 0.991 | - | 23.96 | 1.741 | - | 49.69 | 1.708 | 0.117 | 154.4 | 20.27 | 0.293 | 9.352 | 0.923 | 0.064 |
| Quercitrin | 16.08 | 2.323 | 0.476 | 20.27 | 2.702 | - | 59.52 | 9.973 | 1.251 | - | 8.793 | 2.914 | 1.540 | 0.859 | 0.406 |
| Rutin | 0.595 | 0.093 | - | 2.154 | 0.099 | - | 4.122 | - | - | 87.24 | 4.053 | - | 3.707 | 0.376 | - |
| Sinapic Acid | - | - | - | - | - | - | - | - | - | - | - | - | - | - | - |
| Syringic Acid | - | - | - | - | - | - | - | - | - | - | - | - | - | - | - |
| trans-Ferulic Acid | 1.090 | <LOQ | - | <LOQ | 1.516 | - | 1.108 | 3.829 | <LOQ | <LOQ | - | - | - | - | - |
| Vanillic Acid | 8.997 | 4.092 | <LOQ | 6.863 | 4.989 | 1.427 | 9.957 | 5.739 | 2.357 | 9.181 | 5.495 | <LOQ | 6.153 | 5.587 | 2.546 |
| **SUM** | **94.02** | **65.50** | **7.447** | **72.01** | **98.61** | **6.116** | **260.1** | **37.67** | **16.45** | **611.8** | **782.6** | **32.79** | **355.2** | **318.4** | **75.68** |

**Table 10.** Phenolic compounds (µg/g FW) in different parts of pear regional cultivars (P: Peel, S: Seed, M: Mesocarp (pulp)).

| Phenolic Compound | *Bela Feia* | | | *Torres Novas* | | | *Carapinheira Roxa* | | | *Lambe-os-Dedos* | | | *Amorim* | | | *Carapinheira* | | |
|---|---|---|---|---|---|---|---|---|---|---|---|---|---|---|---|---|---|---|
| | P | S | M | P | S | M | P | S | M | P | S | M | P | S | M | P | S | M |
| 4-Hydroxybenzoic Acid | 4.679 | 4.151 | - | 2.193 | - | - | 16.59 | 1.894 | - | 17.05 | 8.045 | - | 13.65 | 4.042 | - | 16.10 | 13.57 | - |
| Apigenin | - | - | - | - | - | - | - | - | - | - | - | - | - | - | - | - | - | - |
| Caffeic Acid | 14.53 | - | - | 23.84 | 9.508 | 3.994 | 17.27 | 3.835 | - | 22.80 | - | - | 27.09 | 8.777 | - | 15.80 | 5.894 | 2.574 |
| Catechin | 0.620 | 1.224 | - | - | 2.530 | - | 4.263 | 1.992 | - | 2.781 | 2.173 | - | 2.358 | 0.263 | - | 1.782 | 2.579 | <LOQ |
| Chlorogenic Acid | 46.36 | 21.40 | - | 103.0 | 123.1 | 33.29 | 277.5 | 32.83 | 45.17 | 118.0 | 118.3 | 65.12 | 63.23 | 66.87 | 75.13 | 340.2 | 156.7 | 21.95 |
| Epicatechin | 8.953 | 8.782 | - | 0.190 | 5.100 | - | 36.85 | 4.567 | 0.683 | 37.30 | 14.90 | 0.192 | 29.61 | 9.053 | 0.312 | 30.87 | 28.21 | 0.760 |
| Eriodyctiol | 0.095 | - | - | 0.036 | 0.059 | - | - | - | - | <LOQ | - | - | <LOQ | - | - | <LOQ | - | - |
| Gallic Acid | - | 1.708 | - | - | - | - | 2.836 | - | 3.037 | 2.131 | - | 5.802 | - | - | - | 2.803 | 4.454 | 3.408 |
| Luteolin | - | - | - | - | <LOQ | - | - | - | - | <LOQ | - | - | - | - | - | - | - | - |
| Naringenin | 0.329 | 0.028 | - | 0.192 | 0.112 | - | 0.110 | <LOQ | <LOQ | 0.122 | <LOQ | - | 0.130 | <LOQ | - | 0.062 | <LOQ | - |
| o-Coumaric Acid | - | - | - | 1.015 | 0.748 | - | - | - | - | - | - | - | - | - | - | - | - | - |
| *p*-Coumaric Acid | 2.469 | - | - | - | 4.067 | - | - | - | - | 0.971 | - | - | - | - | - | - | - | - |
| Phlorizin | - | - | 2.605 | - | - | - | 2.139 | - | - | - | - | - | - | - | 0.792 | 4.445 | 5.753 | 4.904 |
| Quercetin | 0.861 | 0.109 | - | - | 0.978 | - | 0.826 | 0.191 | - | 0.350 | <LOQ | - | 0.530 | - | - | 0.508 | 0.044 | - |
| Quercetin-3-B-D-Glucoside | 14.68 | 1.085 | <LOQ | 22.27 | 1.406 | 0.028 | 5.580 | 0.531 | 0.066 | 4.751 | 0.770 | 0.052 | 7.868 | 0.290 | <LOQ | 19.76 | 1.296 | <LOQ |
| Quercitrin | 1.432 | 0.176 | - | 5.333 | 0.403 | - | 0.915 | 0.179 | - | 1.264 | 0.330 | - | 3.100 | 0.152 | - | 4.801 | 0.488 | 0.212 |
| Rutin | 0.307 | - | - | 32.60 | 0.762 | - | 1.611 | 0.060 | - | - | 0.163 | - | 0.453 | 0.124 | - | 59.61 | 4.979 | - |
| Sinapic Acid | - | - | - | - | - | - | - | - | - | - | - | - | 1.078 | 0.612 | - | - | - | - |
| Syringic Acid | 2.198 | 1.438 | 1.309 | - | 1.322 | 0.519 | 7.897 | 3.917 | 2.080 | 5.167 | 6.036 | 3.348 | 4.297 | 5.761 | 3.472 | 1.805 | 1.307 | - |
| trans-Ferulic Acid | 1.351 | 1.533 | <LOQ | - | 1.680 | - | 1.936 | <LOQ | - | 1.659 | 2.932 | 1.669 | 1.575 | 3.921 | 1.092 | - | <LOQ | - |
| Vanillic Acid | 4.486 | - | 3.482 | 13.37 | - | 5.012 | 25.69 | 7.677 | 9.662 | 15.39 | 29.49 | 22.67 | 11.65 | 69.36 | 24.85 | 18.88 | 13.47 | 5.571 |
| **SUM** | **103.3** | **41.64** | **7.397** | **204.0** | **151.7** | **42.84** | **402.0** | **57.67** | **60.69** | **229.7** | **183.1** | **98.85** | **166.6** | **169.2** | **105.7** | **517.4** | **238.8** | **39.37** |

### 3.6. Results of the Principal Component Analysis

The principal component analysis (PCA) is a mathematical process that employs an orthogonal transformation to turn a set of potentially correlated observations into a set of values of linearly uncorrelated variables known as principal components. The major goal is to reduce the number of dimensions while maintaining the contribution of all starting variables in order to recognize visual patterns. The correlation plot (Figure 1) shows that the fructose content is independent from all other parameters and that the TPC, TFC and DPPH radical inhibition are strongly correlated.

PCA uses eigenvectors to transform the original variables into new ones called principal components, which are determined by assigning a coefficient to each original variable proportional to their contribution to this transformation (rotation) in order to maximize the variances of the first few components. The biplot graph (Figure 2), which projects the samples (scores) and the variables (loadings) onto a two- or three-dimensional graph, demonstrates first how the samples are related to one another and second how each variable affects each sample.

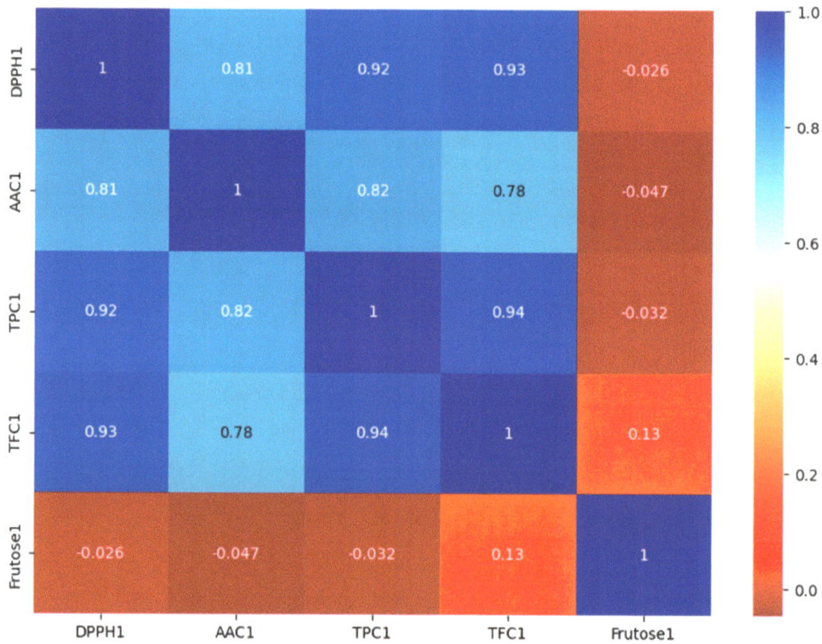


**Figure 1.** Correlation plot among the studied parameters.

These two PCs explain 95% of the observed variation in the original data. It can be noticed that the peel of the *Noiva* variety has a very distinct set of characteristics compared to the other portions. We can also see that the fructose content is the main differentiator between apple and pear portions. TFC, TPC and DPPH radical inhibition are the main factors that contribute to the separation of the pulp and by-products in the biplot chart. This graph confirms also the correlation between these last variables.

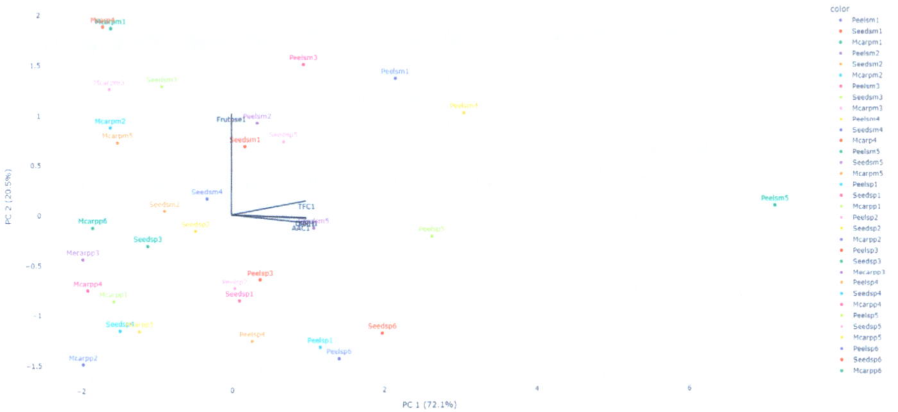


**Figure 2.** Biplot PC2 vs. PC1 for different apple and peer varieties and for the different parts of the fruits. (Peelsm1: Peel of *Pêro Coimbra* Apple, Seedsm1: Seeds of *Pêro Coimbra* Apple, Mcarpm1: Pulp of *Pêro Coimbra* Apple, Peelsm2: Peel of *Pardo Lindo* Apple, Seedsm2: Seeds of *Pardo Lindo* Apple, Mcarpm2: Pulp of *Pardo Lindo* Apple, Peelsm3: Peel of *Repinau* Apple, Seedsm3: Seeds of *Repinau* Apple, Mcarpm3: Pulp of *Repinau* Apple, Peelsm4: Peel of *Pêro de Borbela* Apple, Seedsm4: Seeds of *Pêro de Borbela* Apple, Mcarpm4: Pulp of *Pêro de Borbela* Apple, Peelsm5: Peel of *Noiva* Apple, Seedsm5: Seeds of *Noiva* Apple, Mcarpm5: Pulp of *Noiva* Apple, Peelsp1: Peel of *Bela-Feia* Pear, Seedsp1: Seeds of *Bela-Feia* Pear, Mcarpp1: Pulp of *Bela-Feia* Pear, Peelsp2: Peel of *Torres Novas* Pear, Seedsp2: Seeds of *Torres Novas* Pear, Mcarpp2: Pulp of *Torres Novas* Pear, Peelsp3: Peel of *Carapinheira Roxa* Pear, Seedsp3: Seeds of *Carapinheira Roxa* Pear, Mcarpp3: Pulp of *Carapinheira Roxa* Pear, Peelsp4: Peel of *Lambe-os-Dedos* Pear, Seedsp4: Seeds of *Lambe-os-Dedos* Pear, Mcarpp4: Pulp of *Lambe-os-Dedos* Pear, Peelsp5: Peel of *Amorim* Pear, Seedsp5: Seeds of *Amorim* Pear, Mcarpp5: Pulp of *Amorim* Pear, Peelsp6: Peel of *Carapinheira* Pear, Seedsp6: Seeds of *Carapinheira* Pear, Mcarpp6: Pulp of *Carapinheira* Pear).

### 3.6.1. Principal Component Analysis

PCA performed for the individual apple peels, which explains 99% of the variation of the original data (Figure 3), shows that the *Pêro Coimbra*, *Repinau* and *Pêro de Borbela* peels are the most similar among the group. *Noiva* and *Pardo Lindo* peels are detached from the others, meaning that they possess the most singular profile. The fructose content affects strongly the peels of Pêro Coimbra and Repinau apple varieties, while TFC has higher effect on Peel of Pêro de Borbela specie. The graph shows also that the fructose content and the parameter AAC are inversely proportional.

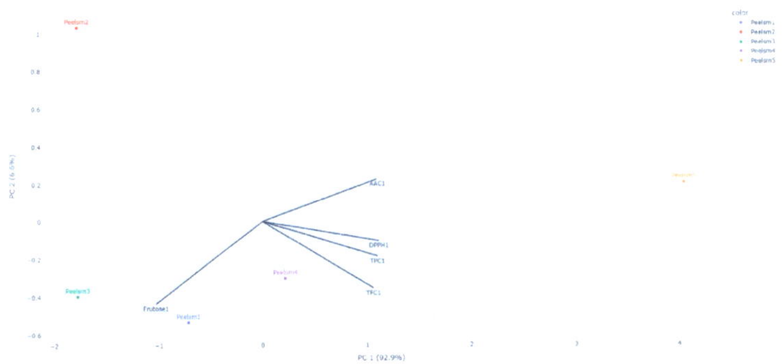

**Figure 3.** Biplot PC2 vs. PC1 for the peel moiety of the different apple varieties. (Peelsm1: Peel of *Pêro Coimbra* Apple, Peelsm2: Peel of *Pardo Lindo* Apple, Peelsm3: Peel of *Repinau* Apple, Peelsm4: Peel of *Pêro de Borbela* Apple, Peelsm5: Peel of *Noiva* Apple).

3.6.2. Classification with Machine Learning Algorithms

The availability of multivariate data matrices makes it feasible to classify individuals and rapidly extract the most crucial information from data using statistical and mathematical methods. In order to place new, unknown samples in one of the recognized classes based on their measurement pattern, supervised pattern recognition techniques use information about the samples' membership in a given group (class or category). In this case, there were two types of samples: apples and pears. Supervised pattern recognition requires a training set of samples that fall into recognized categories in order to create a model for the classification of the data. The decision tree (DT) classifier was used, and the data were divided into a training (60%) and a test group (40). The scores for the training and the test groups were, respectively, 0.9474 and 0.8571, indicating that the algorithm provides a satisfactory classification. The confusion matrix which shows how often a sample that belongs to a given class was classified as belonging to the other, were given in Figure 4. for the training and the test group. The results show that only a few percentages of the samples are incorrectly classified.

The same principle was applied to the samples' membership to one of three portions (peels, seeds and pulp). The scores corresponding to the training and the test groups were, respectively, 0.9474 and 0.7857. Although the results are less satisfactory than those found for the previous case, the score corresponding to the test group indicating a relatively good model fit remains acceptable. The confusion matrix given in Figure 5 shows that the highest number of false positives corresponds to the second class.

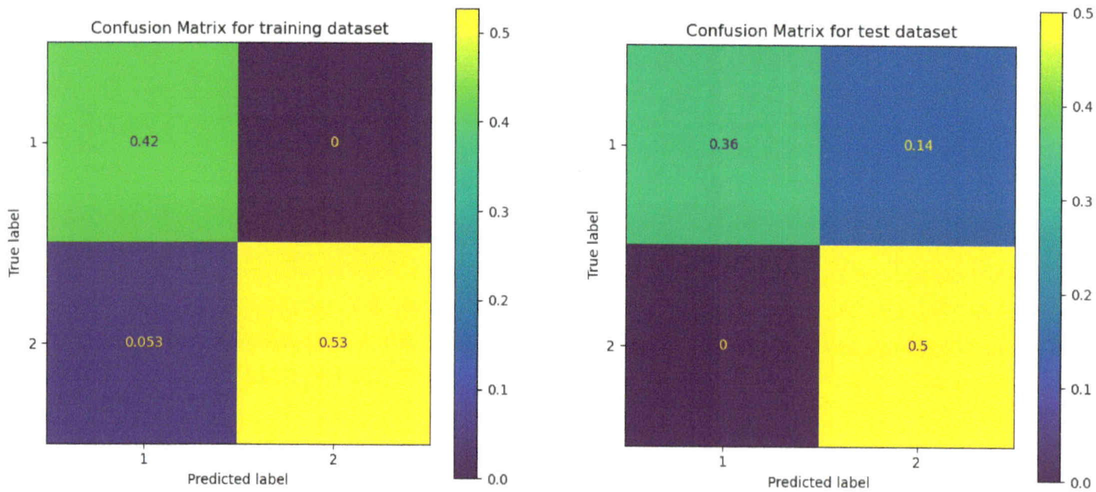


**Figure 4.** Confusion matrix of the training group (at the (**left**)) and the test group (at the (**right**)).

Another use for machine learning could be the prediction of how one of the variables is going to behave when the rest of them were known. In this study, the TPC of different parts of distinct cultivars of apples and pears was assessed using Linear Regression regressor. The results (Figure 6) indicate that the regression model was able to predict the total phenolics content of the specified sample with a coefficient of determination of 0.93 for the training group and 0.73 for the test group.

The regression can be also applied to predict TFC. The scores for training and test groups in this case are, respectively, 0.8754 and 0.8491.

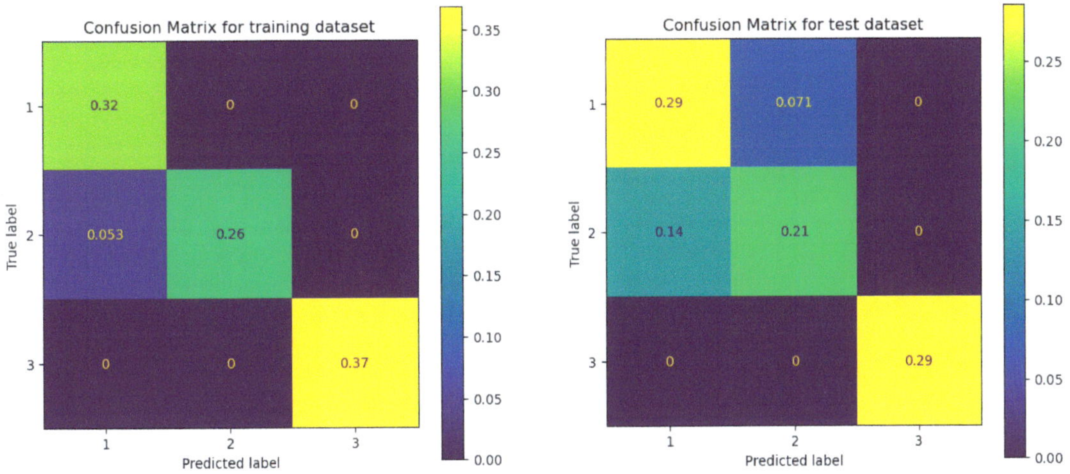


**Figure 5.** Confusion matrix of the training and test groups, regarding the samples' membership to one of three portions.

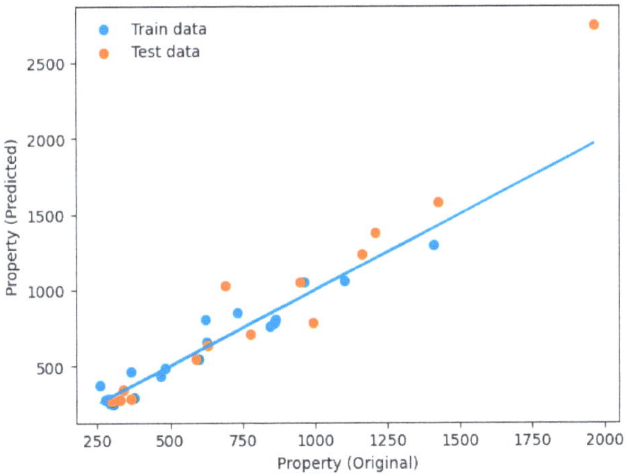


**Figure 6.** Representation of the results of the regression model created to predict the TPC in different fruit parts.

## 4. Conclusions

Fruit consumption is essential for a well-balanced diet, healthy habits, and illness prevention. In fact, in order to achieve these nutritional and health benefits, The World Health Organization (WHO) suggests eating at least 400 g daily, more or less five pieces [57]. Such information even originated popular sayings as: "One apple a day, keeps the doctor away". Yet, since only around 70% of the fruit is edible, there are a lot of by-products produced from these fruits. Reusing fruit by-products is both necessary and advantageous because they contain a significant amount of phenolic compounds. Thus, this generates increasing interest from the food industry because it turns the otherwise considered waste, into material with the potential to be used with other applications such as functional foods and/or active packaging, towards a circular economy. Another important aspect to consider from this study is the need to guarantee that biodiversity continues to exist. If we can prove that regional, i.e., less well-known and less-produced cultivars of fruits, have attributes that will increase their production at a large scale, we can predict that these fruits will generate a large amounts of waste (by-products).

This study successfully established an analytical method for the determination of phenolic compounds in fruit pulps and by-products using ultra-high performance liquid chromatography coupled with time-of-flight mass spectrometry. The method was applied to thirty-three samples of regional fruits. The limits of quantification obtained were all lower than 10 μg/g in the case of phenolic acids and 2.5 μg/g in the case of other phenolics. Recovery tests showed values between 70 and 96% for the lower fortification level (0.1 mg/100 g) and 75 and 97% for the higher fortification level (1.0 mg/100 g).

The antioxidant capacity of the regional cultivars of apples and pears was evaluated and compared. In general, apple cultivars present a higher content of total phenolics and total flavonoids, as well as a higher antioxidant capacity.

The multivariate analysis was used in order to classify different apple and pear varieties and separate them into several groups. Another goal was to study the effect of different parameters on the definition of chemical properties of the samples. PCA was performed separately for the whole fruit samples and for the apples alone with total cumulative variances of, respectively, 93 and 99% corresponding to the first two PCs. These results demonstrate that multivariate analysis, with a particular emphasis on PCA, can be a useful technique for separating various fruit species while highlighting their similarities and differences. Supervised machine learning algorithms were also used for classify and predict the value of a given parameters with satisfactory outcome. In this context, the regression model was able to predict the total phenolics content of the specified sample with a coefficient of determination of 0.73 for the test group.

Although a very descriptive analysis of the antioxidant properties and total phenolics, flavonoids and fructose content were performed, in the future, it would be interesting to evaluate also possible contaminants of these samples such as mycotoxins and pesticides residues, to ensure food safety when reusing the fruits by-products.

**Supplementary Materials:** The following supporting information can be downloaded at: https://www.mdpi.com/article/10.3390/foods12071537/s1, Figure S1: Photos of the traditional cultivars of apples and pears selected for the study.

**Author Contributions:** Conceptualization, J.D.T., C.S., C.A. and A.S.S.; Data curation, J.D.T. and C.S.; Formal analysis, J.D.T. and A.R.S.M.; Funding acquisition, C.A. and A.S.S.; Investigation, A.R.S.M., C.S. and P.P.; Methodology, A.R.S.M. and P.P.; Project administration, C.A.; Resources, J.D.T.; Software, P.P.; Supervision, P.P., C.A. and A.S.S.; Validation, J.D.T. and P.P.; Writing—original draft, J.D.T.; Writing—review and editing, A.R.S.M., C.S., P.P., C.A. and A.S.S. All authors have read and agreed to the published version of the manuscript.

**Funding:** This research was funded by the clabel+ project: Innovative Natural, Nutritious and Consumer Oriented "Clean Label" Foods with the reference POCI-01-0247-FEDER-046080 financed by the Competitiveness and Internationalization Thematic Operational Programme (PO CI), under the COMPETE2020, PORTUGAL2020 Partnership Agreement, through the co-financing of European Regional Development Fund (FEDER). J.D. Teixeira is grateful for his fellowship in the frame of clabel+ project. The work was supported by UIDB/00211/2020, LA/P/0045/2020 (ALiCE), UIDB/00511/2020 and UIDP/00511/2020 (LEPABE), UIDB/04551/2020 (GREEN-IT—Bioresources for Sustainability) funded by national funds through FCT/MCTES, UIDB/04469/2020 (CEB) and by LA/P/0029/2020 (LABBELS), Centre of Chemistry (UID/QUI/0686/2020) and project BioTecNorte (operation NORTE-01-0145-FEDER-000004), supported by the Northern Portugal Regional Operational Programme (NORTE 2020), under the Portugal 2020 Partnership Agreement, through the European Regional Development Fund (ERDF).

**Data Availability Statement:** Data sharing is not applicable to this article.

**Acknowledgments:** We thank to Patricia Vicente, from INIAV/Alcobaça, for technical support in the field.

**Conflicts of Interest:** The authors declare no conflict of interest.

## References


1. Harris, J.; van Zonneveld, M.; Achigan-Dako, E.G.; Bajwa, B.; Brouwer, I.D.; Choudhury, D.; de Jager, I.; de Steenhuijsen Piters, B.; Dulloo, M.E.; Guarino, L.; et al. Fruit and Vegetable Biodiversity for Nutritionally Diverse Diets: Challenges, Opportunities, and Knowledge Gaps. *Glob. Food Secur.* **2022**, *33*, 100618. [CrossRef]
2. Peng, F.; Ren, X.; Du, B.; Niu, K.; Yu, Z.; Yang, Y. Insoluble Dietary Fiber of Pear Fruit Pomace (Pyrus Ussuriensis Maxim) Consumption Ameliorates Alterations of the Obesity-Related Features and Gut Microbiota Caused by High-Fat Diet. *J. Funct. Foods* **2022**, *99*, 105354. [CrossRef]
3. Ma, Q.; Bi, J.; Yi, J.; Wu, X.; Li, X.; Zhao, Y. Stability of Phenolic Compounds and Drying Characteristics of Apple Peel as Affected by Three Drying Treatments. *Food Sci. Hum. Wellness* **2021**, *10*, 174–182. [CrossRef]
4. Home | Food and Agriculture Organization of the United Nations. Available online: https://www.fao.org/home/en (accessed on 10 March 2023).
5. Mignard, P.; Beguería, S.; Reig, G.; Font i Forcada, C.; Moreno, M.A. Genetic Origin and Climate Determine Fruit Quality and Antioxidant Traits on Apple (Malus × Domestica Borkh). *Sci. Hortic.* **2021**, *285*, 110142. [CrossRef]
6. Tao, H.; Sun, H.; Wang, Y.; Wang, X.; Guo, Y. Effects of Water Stress on Quality and Sugar Metabolism in 'Gala' Apple Fruit. *Hortic. Plant J.* **2023**, *9*, 60–72. [CrossRef]
7. Zhang, Y.; Cao, Y.; Huo, H.; Xu, J.; Tian, L.; Dong, X.; Qi, D.; Liu, C. An Assessment of the Genetic Diversity of Pear (*Pyrus* L.) Germplasm Resources Based on the Fruit Phenotypic Traits. *J. Integr. Agric.* **2022**, *21*, 2275–2290. [CrossRef]
8. Salazar-Orbea, G.L.; García-Villalba, R.; Bernal, M.J.; Hernández, A.; Tomás-Barberán, F.A.; Sánchez-Siles, L.M. Stability of Phenolic Compounds in Apple and Strawberry: Effect of Different Processing Techniques in Industrial Set Up. *Food Chem.* **2023**, *401*, 134099. [CrossRef]
9. Kolniak-Ostek, J. Chemical Composition and Antioxidant Capacity of Different Anatomical Parts of Pear (*Pyrus communis* L.). *Food Chem.* **2016**, *203*, 491–497. [CrossRef] [PubMed]
10. Moni Bottu, H.; Mero, A.; Husanu, E.; Tavernier, S.; Pomelli, C.S.; Dewaele, A.; Bernaert, N.; Guazzelli, L.; Brennan, L. The Ability of Deep Eutectic Solvent Systems to Extract Bioactive Compounds from Apple Pomace. *Food Chem.* **2022**, *386*, 132717. [CrossRef]
11. Wei, L.; Liao, Z.; Ma, H.; Wei, J.; Peng, C. Antioxidant Properties, Anti- SARS-CoV-2 Study, Collagenase and Elastase Inhibition Effects, Anti-Human Lung Cancer Potential of Some Phenolic Compounds. *J. Indian Chem. Soc.* **2022**, *99*, 100416. [CrossRef]
12. Adelusi, T.I.; Boyenle, I.D.; Tolulope, A.; Adebisi, J.; Fatoki, J.O.; Ukachi, C.D.; Oyedele, A.-Q.K.; Ayoola, A.M.; Timothy, A.A. GCMS Fingerprints and Phenolic Extracts of Allium Sativum Inhibit Key Enzymes Associated with Type 2 Diabetes. *J. Taibah Univ. Med. Sci.* **2023**, *18*, 337–347. [CrossRef]
13. Zhang, T.; Wei, X.; Miao, Z.; Hassan, H.; Song, Y.; Fan, M. Screening for Antioxidant and Antibacterial Activities of Phenolics from Golden Delicious Apple Pomace. *Chem. Cent. J.* **2016**, *10*, 47. [CrossRef]
14. Ali, A.; Cottrell, J.J.; Dunshea, F.R. Identification and Characterization of Anthocyanins and Non-Anthocyanin Phenolics from Australian Native Fruits and Their Antioxidant, Antidiabetic, and Anti-Alzheimer Potential. *Food Res. Int.* **2022**, *162*, 111951. [CrossRef]
15. Vetrani, C. (Poly)Phenols and Cardiovascular Diseases_ Looking in to Move Forward. *J. Funct. Foods* **2020**, *71*, 104013. [CrossRef]
16. Fatima, M.T.; Bhat, A.A.; Nisar, S.; Fakhro, K.A.; Al-Shabeeb Akil, A.S. The Role of Dietary Antioxidants in Type 2 Diabetes and Neurodegenerative Disorders: An Assessment of the Benefit Profile. *Heliyon* **2023**, *9*, e12698. [CrossRef] [PubMed]
17. Mesías, F.J.; Martín, A.; Hernández, A. Consumers' Growing Appetite for Natural Foods: Perceptions towards the Use of Natural Preservatives in Fresh Fruit. *Food Res. Int.* **2021**, *150*, 110749. [CrossRef]
18. Talwar, S.; Kaur, P.; Kumar, S.; Hossain, M.; Dhir, A. What Determines a Positive Attitude towards Natural Food Products? An Expectancy Theory Approach. *J. Clean. Prod.* **2021**, *327*, 129204. [CrossRef]
19. Lavilla, M.; Gayán, E. Consumer Acceptance and Marketing of Foods Processed Through Emerging Technologies. In *Innovative Technologies for Food Preservation*; Elsevier: Amsterdam, The Netherlands, 2018; pp. 233–253. ISBN 978-0-12-811031-7.
20. Bruno de Sousa Sabino, L.; Leônia da Costa Gonzaga, M.; de Siqueira Oliveira, L.; Souza Gomes Duarte, A.; Alexandre e Silva, L.M.; Sousa de Brito, E.; Wilane de Figueiredo, R.; Morais Ribeiro da Silva, L.; Machado de Sousa, P.H. Polysaccharides from Acerola, Cashew Apple, Pineapple, Mango and Passion Fruit Co-Products: Structure, Cytotoxicity and Gastroprotective Effects. *Bioact. Carbohydr. Diet. Fibre* **2020**, *24*, 100228. [CrossRef]
21. Sette, P.; Fernandez, A.; Soria, J.; Rodriguez, R.; Salvatori, D.; Mazza, G. Integral Valorization of Fruit Waste from Wine and Cider Industries. *J. Clean. Prod.* **2020**, *242*, 118486. [CrossRef]
22. Kowalska, H.; Czajkowska, K.; Cichowska, J.; Lenart, A. What's New in Biopotential of Fruit and Vegetable by-Products Applied in the Food Processing Industry. *Trends Food Sci. Technol.* **2017**, *67*, 150–159. [CrossRef]
23. Antonio, A.S.; Wiedemann, L.S.M.; Galante, E.B.F.; Guimarães, A.C.; Matharu, A.S.; Veiga-Junior, V.F. Efficacy and Sustainability of Natural Products in COVID-19 Treatment Development: Opportunities and Challenges in Using Agro-Industrial Waste from Citrus and Apple. *Heliyon* **2021**, *7*, e07816. [CrossRef] [PubMed]
24. Díaz, S.; Zafra-Calvo, N.; Purvis, A.; Verburg, P.H.; Obura, D.; Leadley, P.; Chaplin-Kramer, R.; De Meester, L.; Dulloo, E.; Martín-López, B.; et al. Set Ambitious Goals for Biodiversity and Sustainability. *Science* **2020**, *370*, 411–413. [CrossRef] [PubMed]
25. Zhang, L.; Yang, L.; Chapman, C.A.; Peres, C.A.; Lee, T.M.; Fan, P.-F. Growing Disparity in Global Conservation Research Capacity and Its Impact on Biodiversity Conservation. *One Earth* **2023**, *6*, 147–157. [CrossRef]

26. Macfarlane, N.B.W.; Adams, J.; Bennett, E.L.; Brooks, T.M.; Delborne, J.A.; Eggermont, H.; Endy, D.; Esvelt, K.M.; Kolodziejczyk, B.; Kuiken, T.; et al. Direct and Indirect Impacts of Synthetic Biology on Biodiversity Conservation. *iScience* **2022**, *25*, 105423. [CrossRef]
27. Cortés-Capano, G.; Hausmann, A.; Di Minin, E.; Kortetmäki, T. Ethics in Biodiversity Conservation: The Meaning and Importance of Pluralism. *Biol. Conserv.* **2022**, *275*, 109759. [CrossRef]
28. Horwitz, W. (Ed.) *Official Methods of Analysis of AOAC International*, 18th ed.; AOAC International: Gaithersburg, MD, USA, 2006; ISBN 978-0-935584-77-6.
29. Martins, C.; Vilarinho, F.; Sanches Silva, A.; Andrade, M.; Machado, A.V.; Castilho, M.C.; Sá, A.; Cunha, A.; Vaz, M.F.; Ramos, F. Active Polylactic Acid Film Incorporated with Green Tea Extract: Development, Characterization and Effectiveness. *Ind. Crops Prod.* **2018**, *123*, 100–110. [CrossRef]
30. Miller, H.E. A Simplified Method for the Evaluation of Antioxidants. *J. Am. Oil Chem. Soc.* **1971**, *48*, 91. [CrossRef]
31. Erkan, N.; Ayranci, G.; Ayranci, E. Antioxidant Activities of Rosemary (Rosmarinus officinalis L.) Extract, Blackseed (*Nigella sativa* L.) Essential Oil, Carnosic Acid, Rosmarinic Acid and Sesamol. *Food Chem.* **2008**, *110*, 76–82. [CrossRef]
32. Barbosa, C.H.; Andrade, M.A.; Séndon, R.; Silva, A.S.; Ramos, F.; Vilarinho, F.; Khwaldia, K.; Barbosa-Pereira, L. Industrial Fruits By-Products and Their Antioxidant Profile: Can They Be Exploited for Industrial Food Applications? *Foods* **2021**, *10*, 272. [CrossRef]
33. Ashwell, G. Colorimetric Analysis of Sugars. *Methods Enzym.* **1957**, *3*, 73–105. [CrossRef]
34. INSA_pt. Available online: http://portfir.insa.pt/foodcomp/food?21438 (accessed on 10 March 2023).
35. INSA_pt. Available online: http://portfir.insa.pt/foodcomp/food?22189 (accessed on 10 March 2023).
36. Sara, J.; Abdolrasoul, E.; Ahmad, K.; Hossain, B. Evaluation of Antioxidant Activity of Malus Domestica Fruit Extract from Kashan Area. *Afr. J. Agric. Res.* **2015**, *10*, 2136–2140. [CrossRef]
37. Guan, J. Chapter 17—How Cultivars Influence Fruit Composition: Total Phenols, Flavonoids Contents, and Antioxidant Activity in the Pulp of Selected Asian Pears. In *Processing and Impact on Active Components in Food*; Academic Press: Cambridge, MA, USA, 2015; p. 7.
38. Oaldje-Pavlovic, M.; Alimpic-Aradski, A.; Savic, A.; Jankovic, S.; Milutinovic, M.; Marin, P.; Duletic-Lausevic, S. Traditional Varieties and Wild Pear from Serbia: A Link among Antioxidant, Antidiabetic and Cytotoxic Activities of Fruit Peel and Flesh. *Bot. Serb.* **2021**, *45*, 203–213. [CrossRef]
39. Koleva, I.I.; van Beek, T.A.; Linssen, J.P.H.; Groot, A.D.; Evstatieva, L.N. Screening of Plant Extracts for Antioxidant Activity: A Comparative Study on Three Testing Methods. *Phytochem. Anal.* **2002**, *13*, 8–17. [CrossRef]
40. Ceymann, M.; Arrigoni, E.; Schärer, H.; Bozzi Nising, A.; Hurrell, R.F. Identification of Apples Rich in Health-Promoting Flavan-3-Ols and Phenolic Acids by Measuring the Polyphenol Profile. *J. Food Compos. Anal.* **2012**, *26*, 128–135. [CrossRef]
41. Gulsunoglu, Z.; Purves, R.; Karbancioglu-Guler, F.; Kilic-Akyilmaz, M. Enhancement of Phenolic Antioxidants in Industrial Apple Waste by Fermentation with *Aspergillus* spp. *Biocatal. Agric. Biotechnol.* **2020**, *25*, 101562. [CrossRef]
42. Sinha, A.; Gill, P.P.S.; Jawandha, S.K.; Kaur, P.; Grewal, S.K. Chitosan-Enriched Salicylic Acid Coatings Preserves Antioxidant Properties and Alleviates Internal Browning of Pear Fruit under Cold Storage and Supermarket Conditions. *Postharvest Biol. Technol.* **2021**, *182*, 111721. [CrossRef]
43. Zhou, L.; Liu, W.; Stockmann, R.; Terefe, N.S. Effect of Citric Acid and High Pressure Thermal Processing on Enzyme Activity and Related Quality Attributes of Pear Puree. *Innov. Food Sci. Emerg. Technol.* **2018**, *45*, 196–207. [CrossRef]
44. Wu, C.-S.; Gao, Q.-H.; Guo, X.-D.; Yu, J.-G.; Wang, M. Effect of Ripening Stage on Physicochemical Properties and Antioxidant Profiles of a Promising Table Fruit 'Pear-Jujube' (Zizyphus Jujuba Mill.). *Sci. Hortic.* **2012**, *148*, 177–184. [CrossRef]
45. Costa, D.C.; Costa, H.S.; Albuquerque, T.G.; Ramos, F.; Castilho, M.C.; Sanches-Silva, A. Advances in Phenolic Compounds Analysis of Aromatic Plants and Their Potential Applications. *Trends Food Sci. Technol.* **2015**, *45*, 336–354. [CrossRef]
46. Šavikin, K.; Živković, J.; Zdunić, G.; Gođevac, D.; Đorđević, B.; Dojčinović, B.; Đorđević, N. Phenolic and Mineral Profiles of Four Balkan Indigenous Apple Cultivars Monitored at Two Different Maturity Stages. *J. Food Compos. Anal.* **2014**, *35*, 101–111. [CrossRef]
47. El-Hawary, S.S.; Hammam, W.E.; El-Mahdy El-Tantawi, M.; Yassin, N.A.Z.; Kirollos, F.N.; Abdelhameed, M.F.; Abdelfattah, M.A.O.; Wink, M.; Sobeh, M. Apple Leaves and Their Major Secondary Metabolite Phlorizin Exhibit Distinct Neuroprotective Activities: Evidence from in Vivo and in Silico Studies. *Arab. J. Chem.* **2021**, *14*, 103188. [CrossRef]
48. Wang, X.; Wang, J.; Wei, L.; Hu, C.Y.; Deng, H.; Guo, Y.; Meng, Y.H. Apple Phlorizin Oxidation Product 2 Inhibits Proliferation and Differentiation of 3T3-L1 Preadipocytes. *J. Funct. Foods* **2019**, *62*, 103525. [CrossRef]
49. Karaman, Ş.; Tütem, E.; Başkan, K.S.; Apak, R. Comparison of Antioxidant Capacity and Phenolic Composition of Peel and Flesh of Some Apple Varieties: Antioxidant Capacity and Phenolic Content of Apples. *J. Sci. Food Agric.* **2013**, *93*, 867–875. [CrossRef]
50. Mihailović, N.R.; Mihailović, V.B.; Kreft, S.; Ćirić, A.R.; Joksović, L.G.; Đurđević, P.T. Analysis of Phenolics in the Peel and Pulp of Wild Apples (Malus sylvestris (L.) Mill.). *J. Food Compos. Anal.* **2018**, *67*, 1–9. [CrossRef]
51. Bílková, A.; Baďurová, K.; Svobodová, P.; Vávra, R.; Jakubec, P.; Chocholouš, P.; Švec, F.; Sklenářová, H. Content of Major Phenolic Compounds in Apples: Benefits of Ultra-Low Oxygen Conditions in Long-Term Storage. *J. Food Compos. Anal.* **2020**, *92*, 103587. [CrossRef]
52. Raudone, L.; Raudonis, R.; Liaudanskas, M.; Janulis, V.; Viskelis, P. Phenolic Antioxidant Profiles in the Whole Fruit, Flesh and Peel of Apple Cultivars Grown in Lithuania. *Sci. Hortic.* **2017**, *216*, 186–192. [CrossRef]

53. Cui, T.; Nakamura, K.; Ma, L.; Li, J.-Z.; Kayahara, H. Analyses of Arbutin and Chlorogenic Acid, the Major Phenolic Constituents in Oriental Pear. *J. Agric. Food Chem.* **2005**, *53*, 3882–3887. [CrossRef]
54. Brahem, M.; Renard, C.M.G.C.; Eder, S.; Loonis, M.; Ouni, R.; Mars, M.; Le Bourvellec, C. Characterization and Quantification of Fruit Phenolic Compounds of European and Tunisian Pear Cultivars. *Food Res. Int.* **2017**, *95*, 125–133. [CrossRef] [PubMed]
55. Colaric, M.; Stampar, F.; Solar, A.; Hudina, M. Influence of Branch Bending on Sugar, Organic Acid and Phenolic Content in Fruits of 'Williams' Pears (Pyrus communis L.). *J. Sci. Food Agric.* **2006**, *86*, 2463–2467. [CrossRef]
56. Salta, J.; Martins, A.; Santos, R.G.; Neng, N.R.; Nogueira, J.M.F.; Justino, J.; Rauter, A.P. Phenolic Composition and Antioxidant Activity of Rocha Pear and Other Pear Cultivars—A Comparative Study. *J. Funct. Foods* **2010**, *2*, 153–157. [CrossRef]
57. Weltgesundheitsorganisation; FAO (Eds.) *Diet, Nutrition, and the Prevention of Chronic Diseases: Report of a WHO-FAO Expert Consultation*; Joint WHO-FAO Expert Consultation on Diet, Nutrition, and the Prevention of Chronic Diseases, 2002, Geneva, Switzerland; WHO Technical Report Series; World Health Organization: Geneva, Switzerland, 2003; ISBN 978-92-4-120916-8.

*foods*

MDPI

*Article*

# Pomegranate Peel Powder: In Vitro Efficacy and Application to Contaminated Liquid Foods

**Valentina Lacivita [1], Adriana Lordi [1], Tamara Posati [2], Roberto Zamboni [2], Matteo Alessandro Del Nobile [1,*] and Amalia Conte [1]**

1 Department of Agricultural Sciences, Food and the Environment, University of Foggia, Via Napoli, 71122 Foggia, Italy
2 Consiglio Nazionale delle Ricerche, Istituto per la Sintesi Organica e la Fotoreattività (CNR-ISOF), Via Piero Gobetti 101, 40129 Bologna, Italy
* Correspondence: matteo.delnobile@unifg.it

**Abstract:** In this study the recycling of pomegranate peel powder (PPP) was proposed. In particular, the use of powder loaded in a silk fibroin polymeric matrix to create an active pad was tested. For the sake of comparison, the powder alone was also analysed. Both powder and active pad efficacy was assessed in two different food systems, soymilk (rich in proteins), preliminarily contaminated with *Pseudomonas* spp. and yeasts, and apple juice (rich in carbohydrates), preliminarily contaminated with *Alyciclobacillus acidoterrestris*. Three different concentrations of powder alone and powder in the pad were tested (5%, 7.5% and 10% $w/v$) in both types of beverages. To assess a possible dependence of the efficacy on the powder granulometry, different powder sizes were preliminarily analysed on *Pseudomonas* spp. and yeasts using an in vitro test. PPP was effective on both *Pseudomonas* spp. and yeasts. No significant differences appeared among the tested granulometries and therefore in the subsequent tests powder with an average diameter of 250 μm was used. Results recorded with soymilk and apple juice were different. When applied to the soymilk, the activity of PPP in the pad was less effective than that recorded when the powder was directly added to the beverage. With the two highest powder concentrations directly added to food, more than four log cycle reductions in *Pseudomonas* spp. and yeast cells were recorded, compared to soymilk without any powder. Compared to the control sample, all the soymilk samples either with PPP or with the active pad showed a delayed microbial and fungal growth. When applied to apple juice, both powder and pad were effective at completely inhibiting the proliferation of *A. acidoterrestris* (<$10^2$ CFU/g).

**Keywords:** pomegranate peel; fruit by-products; soymilk; apple juice; natural additives

**Citation:** Lacivita, V.; Lordi, A.; Posati, T.; Zamboni, R.; Del Nobile, M.A.; Conte, A. Pomegranate Peel Powder: In Vitro Efficacy and Application to Contaminated Liquid Foods. *Foods* **2023**, *12*, 1173. https://doi.org/10.3390/foods12061173

Academic Editors: Sílvia Petronilho and Mario M. Martinez

Received: 6 February 2023
Revised: 28 February 2023
Accepted: 7 March 2023
Published: 10 March 2023



## 1. Introduction

In the last years, food waste has become a global problem with serious implications on economic aspects and on the environment, and above all on climate, soil, water resources and biodiversity [1]. Therefore, waste reduction should become a priority. To date, reuse within the human food chain seems to be the best option [2]. Most food by-products come from fruit and vegetable processing in terms of seeds, peel, and leaves. These by-products, being rich in polyphenols and flavonoids, are considered natural antimicrobial and antioxidant agents [3], and therefore they are prized for their nutritional properties [4]. For this reason, by-products, or their extracts, can be recycled and reused as raw materials, to produce fortified foods [5,6] or to prolong fresh food shelf life [7–10].

Among the various by-products, pomegranate peels and their extracts gained great attention as they turned out to be powerful antimicrobial and antioxidant sources of compounds, such as tannins and anthocyanins [11]. The pomegranate fruit consists of 50% peel. A recent estimate considered the pomegranate peel world production to be about 1.9 million metric tons per year [12]. Due to the abundant amount, several uses of residues from processing of pomegranate were evaluated. As recently reviewed by El Barnossi et al. [12], these by-products may be used directly or after treatments as animal feed, bio-fertilizer, bio-adsorbent, production of biogas, bioethanol, biohydrogen and value-added products such as essential oil, enzymes, food, medical and cosmetic products.

In the food sector, Incoronato et al. [13] designed a fortified pancake with all parts of pomegranate (juice and by-products), Bourekoua et al. [14] evaluated the effect of the addition of pomegranate seed powder on physical, sensorial and antioxidant properties of gluten-free bread. Another interesting case-study concerns the possibility of extending the shelf life of fresh fish burgers fortified with whole pomegranate juice, peel, and seeds [15]. The use of pomegranate peel powder up to 3% was also proven to be effective as a natural preservative in high quality beef sausages during storage at 4 °C [16].

From the literature, it emerges that the use of by-products as sources of bioactive compounds to develop films with antimicrobial and/or antioxidant properties is also very common [17,18]. For example, pomegranate components have been used for the development of active polymeric systems [19]. Mushtaq et al. [20] made a zein-based film by incorporating pomegranate peel extracts in the polymeric matrix. Bertolo et al. [21] functionalized chitosan/gelatin-based materials by pomegranate peel extract. Chitosan-based coatings with incorporated pomegranate peel extracts have been applied to rainbow trout [22] and Pacific white shrimp [23]. Giannelli et al. [24] recently developed an active pad based on silk fibroin and pomegranate peel powder, with the potential to be adopted for food packaging. It is abundantly recognized that for regulatory reasons, delivery systems of active compounds are advantageous. A sustained release of phenolics over time is observed, compared to dipping foods in phenolic solution or adding a phenolic solution to the food before packaging [25]. In addition, biopolymers from biowastes offer great opportunities to decrease the devastating overuse of plastic-based packaging [18]. Starting from results of Giannelli et al. [24], in the current study, the active pad made up of silk fibroin and pomegranate peel powder was applied to contaminated fresh liquid food. The active pad was developed by considering that silk fibroin acted as a reservoir of antimicrobial phenolics contained in the pomegranate peel, which reached the food mainly through diffusion. In our active pad it was expected that release rate of peel phenolics over time would be adequate to maintain a sufficient concentration to inhibit growth of unwanted microorganisms and yeasts in two types of perishable food beverages.

Milk-based drinks and fresh fruit juice are beverages prone to microbial deterioration during storage, as well as to chemical, physical and sensorial changes [26–28]. Alternative preservation techniques to thermal processing have been proposed for fresh beverages to satisfy consumer demand for more fresh-tasting juice; however, these solutions are still limited and require further optimization to find a large-scale application [29]. For example, juices were treated by high pressure [30,31], also combined with ultrasound and pulsed electric field [32], by microwave heating and thermo-sonication [29,33,34], by cold atmospheric pressure plasma [35] and by carbonation [36].

In view of the world production of a plentiful amount of waste generated by pomegranate fruit and considering that these wastes present risks to the environment and human health and also taking into account the increase in their economic benefits, the pomegranate by-product valorisation through an active pad development could be very promising. Therefore, from the perspective of taking advantage of by-products and reduction of waste, the aim of this study was to develop an active pad loaded with pomegranate peel powder. For comparison, the peel powder alone was also tested. A preliminary in vitro test was made to assess differences among different powder granulometries. Once the proper powder size was selected, three concentrations of pomegranate peel powder, alone and in the pad were, respectively, tested against *Pseudomonas* spp. and yeasts, previously inoculated in soymilk, and against *A. acidoterrestris* previously inoculated in apple juice. The ability of the powder alone and of the active pad in improving the microbial stability of the two previously inoculated fresh food beverages was assessed during an appropriate storage period.

## 2. Materials and Methods

### 2.1. Pomegranate Peel Powder Preparation and In Vitro Efficacy

The pomegranate peel powder (PPP) was obtained according to the study of Giannelli et al. [24] from the pomegranate fruit of the Wonderful variety, kindly supplied by A.P.O. Association (Foggia, Italy). Briefly, pomegranates were carefully washed, rinsed and cut to separate the peel from the arils. The by-products were dried in a food dehydrator (at 38 °C for 48 h) up to a final humidity of 8.77%. After drying, the by-products were ground into a powder in a lab blender and then sieved. Different granulometries were obtained, 100, 150, 250, 400 and 500 μm, using correct sieves (ENCO srl, Spinea, Italy). The PPP with five different sizes, was tested against *Pseudomonas* spp. and yeasts, as typical food spoilage organisms. In particular, *Pseudomonas putida* and *Pseudomonas fluorescens* isolated from spoiled mozzarella cheese, and yeasts isolated from red grape marc, were used. The two strains of *Pseudomonas* spp. were maintained in Plate Count broth (PC, Oxoid, Milan, Italy) at −20 °C with the addition of 30% of glycerol as stock cultures. Prior to the antimicrobial tests, exponentially growing cultures were obtained by allowing each strain to grow in appropriate broth at 25 °C for 24 h. Then, a cocktail of the two strains was prepared by mixing 1% of each culture. The cocktail of *Pseudomonas* spp. and yeasts at a concentration of $10^3$ CFU/mL were both prepared by diluting growing cultures with sterile saline solution (9 g/L NaCl). For testing the powder efficacy, each granulometry of PPP (5% $w/w$) was placed in 20 tubes containing 10 g of PC broth (10 tubes for *Pseudomonas* spp. and 10 for yeasts). All tubes were incubated at 25 °C for 72 h. At different incubation times (4, 24, 48 and 72 h) aliquots of 1 mL were taken from each tube for microbiological analyses carried out by plate count (incubation at 25° C for 48 h). All analyses were performed twice on two different samples. In all the samples the measurement of pH, conducted in triplicate, was performed on the first homogenized dilution of the samples, by a pH meter (Crison, Barcelona, Spain) with an accuracy of 0.01 pH, previously calibrated with 4.01 and 7.00 pH buffer solutions.

### 2.2. Pad Development

Pads based on silk fibroin, with and without pomegranate peel powder (PPP), were prepared according to the study of Giannelli et al. [24]. To extract fibroin, *Bombyx mori* silkworm white cocoons (CREA, Padua, Italy) were used. The silk fibroin extraction was carried out according to details also reported in the abovementioned article of Giannelli et al. [24]. The active pad, at a ratio 30:70 by mass (Silk fibroin:PPP), was prepared by blending a 6% $w/v$ of water silk fibroin solution containing 20% $w/w$ of glycerol vs. silk fibroin to induce water insolubility [37], with a suitable amount of pomegranate powder. The active pad was obtained by using a support of polydimethylsiloxane (5 × 4 cm) as substrate. Five mL of the obtained blend were drop-casted, left to dry at room temperature, and peeled off from the substrate. As control, pads of the same dimensions without any powder were also prepared.

### 2.3. Scanning Electron Microscopy (SEM)

SEM analysis was performed with a Zeiss EVO LS 10 LaB6 scanning electron microscope. (U.S. Headquarters, New York, NY, USA). The PPP (2–3 mg) and the obtained silk fibroin-PPP pad (about 0.5 cm × 0.5 cm) samples were deposited on standard aluminium substrate and observed at an acceleration voltage of 5 kV and a working distance of 5 mm.

### 2.4. PPP and Active Pad Efficacy in Previously Inoculated Soy Milk and Apple Juice

The PPP, the control pad and the active pads were tested in two different liquid foods: the soymilk, which is rich in proteins, and the apple juice, which is rich in carbohydrates. The soymilk and apple juice compositions reported on the packaging of each product are listed in Table 1.

**Table 1.** Composition of soy milk and apple juice.

| Food Matrix | Soy Milk | Apple Juice |
|---|---|---|
| Fat | 1.7 g/100 mL | - |
| Carbohydrates | 2.9 g/100 mL | 11 g/100 mL |
| Proteins | 3 g/100 mL | - |
| Fibres | 0.2 g/100 mL | - |
| Salt | 0.10 g/100 mL | - |
| Vitamin B2 | 0.21 mg/100 mL | - |
| Vitamin B12 | 0.38 g/100 mL | - |
| Vitamin D | 1.5 µg/100 mL | - |
| Calcium | 120 µg/100 mL | - |

To simulate contaminated soy milk, the samples were inoculated with *Pseudomonas* spp. (mix of *P. putida* and *P. fluorescens*) and yeasts. The inoculation ($10^3$ CFU/mL) was obtained by diluting growing cultures of *Pseudomonas* spp. and yeasts in soymilk. Different concentrations of PPP (granulometry 250 µm) were tested, specifically, 5%, 7.5% and 10% ($w/v$). The same PPP concentrations in the liquid food were used when the active pad was tested. Therefore, the grams of pad necessary to reach the same powder amount in the milk were calculated keeping in mind that the concentration of PPP in the active pad is 30:70 by mass (silk fibroin:PPP). Specifically, 0.8 g of pad corresponded to 5% of PPP in the liquid food, 1.2 g corresponded to 7.5%, whereas 1.6 g corresponded to 10%. In each tube, soymilk (10 g), with a proper amount of either PPP or active pad, were mixed and then stored at 4 °C. For the control sample, the tubes were inoculated and stored without adding any powder/pad. At different storage times aliquots of 1 mL were taken from each tube for microbiological analyses (plate count technique at 25 °C for 48 h). All analyses were performed twice on two different samples.

The PPP and the active pad efficacy were also tested in apple juice contaminated by *A. acidoterrestris*. The *A. acidoterrestris* (DSM 3922) was provided by DSMZ (Germany). It was revitalized in Malt Extract Broth (MEB, Oxoid, Milan, Italy) at 44 °C for 48 h. The inoculum in apple juice was prepared by diluting the overnight culture to approximately $10^3$ CFU/mL. In each inoculated tube containing 10 g apple juice, 5%, 7.5% and 10% ($w/v$) of PPP (granulometry 250 µm) and 0.8 g, 1.2 g or 1.6 g of the active pad were added, respectively. For the control samples, some tubes were inoculated and stored without adding any powder/pad. All the apple juice samples were stored at 37 °C for two weeks. During storage, an aliquot of 1 mL was taken from each apple juice, diluted and then 0.1 mL was spread on Malt Extract Agar (MEA, Oxoid, Milan, Italy). For *A. acidoterrestris* count, the plates were incubated at 44 °C for 48 h. The pad in both soymilk and apple juice was placed on the bottom of the container and the liquid product was then added. The pad did not float in the liquid food, it did not disintegrate or break. In Figure 1 the original pad developed in the current study and its future potential application in beverage bottle is represented.

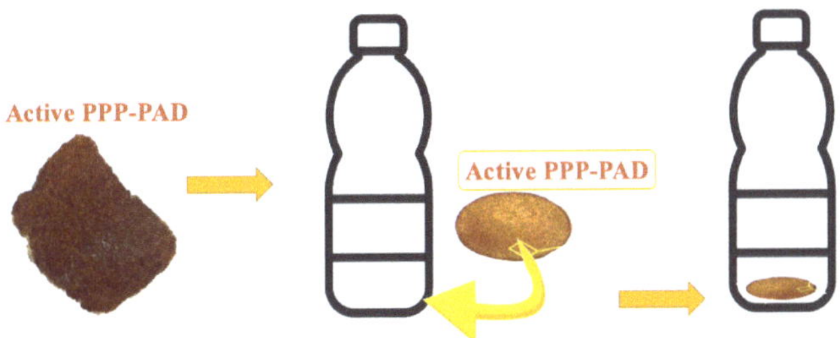


**Figure 1.** Pad developed in the current study and its future application in a beverage bottle.

In the broths and in all the soymilk and apple juice samples the measurement of pH was carried out as reported in the previous Section 2.1.

### 2.5. Data Elaboration and Statistical Analyses

All the experimental data were compared by one way ANOVA and Duncan's multiple range test with the option of homogeneous groups to highlight significant differences among them ($p < 0.05$). STATISTICA 7.1 for Windows (StatSoft, Inc., Tulsa, OK, USA) was the software adopted.

## 3. Results and Discussion

### 3.1. In Vitro Test: Influence of Granulometry on PPP Efficacy

PPP at different sizes was preliminarily tested on *Pseudomonas* spp. and yeasts under in vitro conditions to assess if any dependence between efficacy and powder granulometry existed. Results recorded with powder in contact with the selected microbial and fungal species are reported in Figure 2a,b. As can be seen, Figure 2a highlights a great efficacy of PPP on the two selected *Pseudomonas* strains. In fact, in the control sample the microorganisms grew very rapidly and within one day reached $10^8$ CFU/mL, whereas in the samples with PPP the microbial proliferation was substantially inhibited and remained below $10^4$ CFU/mL during the entire observation period. As abundantly recognized, foods contaminated by *Pseudomonas* spp. become unacceptable when the viable cell concentration exceeds $10^6$ CFU/mL, and therefore data recorded assessed an important efficacy of PPP against this spoilage group. Any significant differences among samples ($p < 0.05$) were observed. In Figure 2b the trend of yeasts is reported. In this case the control sample reached the undesired yeast concentration of $10^5$ CFU/mL within about the first 20 h of monitoring, whereas it took about the double time to reach the same fungal concentration for all the samples enriched with the powder. As can be inferred from the data shown in the figure, no significant differences among sizes can be highlighted ($p < 0.05$). Considering the recognized properties of peel fruit on yeasts and various gram-negative and gram-positive bacteria yet reported in the literature, our results are in line with other scientific data [38–41]. The peel effects can be ascribed to phenolic compounds, especially anthocyanins and tannins [38]. In a previous study, we also found high contents of polyphenols and flavonoids in the same variety of pomegranate peel [13]. Recently, Xiang et al. [39] reviewed the bioactivity of pomegranate peel extract, highlighting that it could contain up to more than 20 polyphenol compounds with a variety of biological activities, such as antioxidative, antitumour, anti-inflammatory, neuroprotection, anti-viral, and anti-bacterial.

The most accredited mechanism of action of pomegranate peel is due to the position of the hydroxyl groups (OH) in the aromatic ring of polyphenols. These groups interact with microbial cell membranes, provoking loss of cellular components and damaging metabolic processes [38,42].

As regards pH, Tables 2 and 3 report data of broths inoculated with *Pseudomonas* spp. and yeasts, respectively. Apart from some significant differences among samples ($p < 0.05$), the most striking feature of the data is the fact that pH of the control samples in both cases ranged between 6 and 7, whereas for all the samples with PPP pH values around 4 were measured. This finding is not surprising considering that pH of PPP is less than 5 [12]. The same pH reduction was also found when pomegranate peel was added to brine of fresh-cut fruit, to a pancake formulation and to a fish-based burger [13,15,43]. Considering that low pH generally favours yeast growth [44], the effects of PPP recorded in the current study against yeasts is very promising.

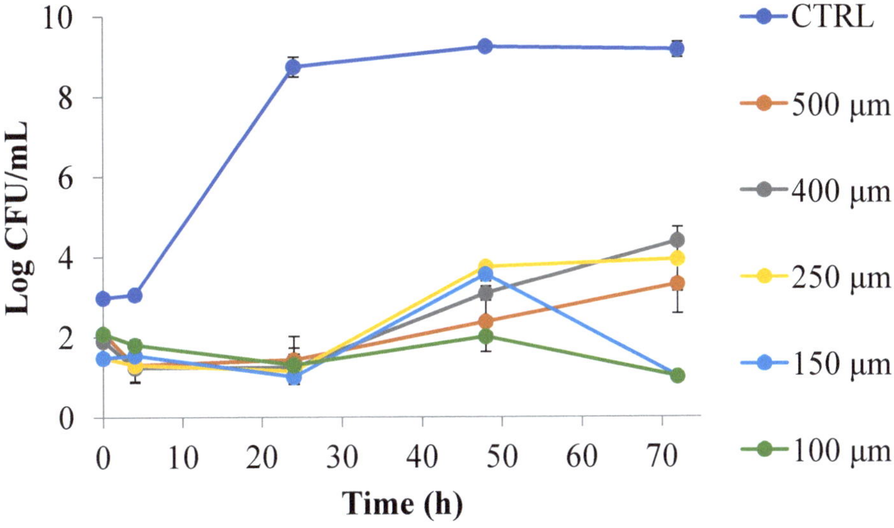


(a)

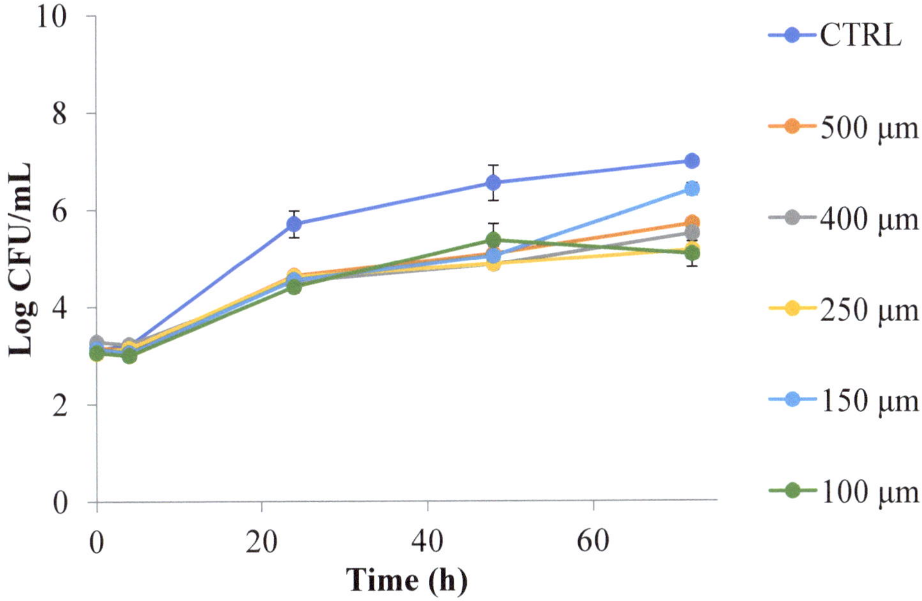


(b)

**Figure 2.** *Pseudomonas* spp. (**a**) and yeasts (**b**) in PC broth with (5%) and without PPP with granulometry from 500 to 100 µm.

**Table 2.** Values of pH for broths inoculated with *Pseudomonas* spp., with and without PPP.

| Time (h) | CTRL | 500 μm | 400 μm | 250 μm | 150 μm | 100 μm |
|---|---|---|---|---|---|---|
| 0 | $6.89 \pm 0.01\ ^{d}{}_{E}$ | $3.98 \pm 0.00\ ^{a}{}_{B}$ | $4.02 \pm 0.01\ ^{a,b}{}_{C}$ | $3.89 \pm 0.02\ ^{a}{}_{B}$ | $3.94 \pm 0.01\ ^{a}{}_{A}$ | $4.07 \pm 0.01\ ^{c}{}_{D}$ |
| 4 | $7.01 \pm 0.04\ ^{d}{}_{E}$ | $4.19 \pm 0.01\ ^{d}{}_{D}$ | $4.07 \pm 0.01\ ^{c}{}_{C}$ | $4.04 \pm 0.01\ ^{b}{}_{B,C}$ | $3.97 \pm 0.01\ ^{a}{}_{B}$ | $3.81 \pm 0.12\ ^{a}{}_{A}$ |
| 24 | $6.03 \pm 0.02\ ^{a}{}_{D}$ | $4.06 \pm 0.03\ ^{b,c}{}_{C}$ | $4.00 \pm 0.02\ ^{a}{}_{B}$ | $4.02 \pm 0.01\ ^{b}{}_{B}$ | $3.94 \pm 0.01\ ^{a}{}_{A}$ | $3.96 \pm 0.02\ ^{b}{}_{A}$ |
| 48 | $6.41 \pm 0.04\ ^{b}{}_{D}$ | $4.04 \pm 0.01\ ^{b}{}_{C}$ | $4.04 \pm 0.01\ ^{b}{}_{C}$ | $4.04 \pm 0.00\ ^{b}{}_{C}$ | $3.82 \pm 0.03\ ^{a}{}_{A}$ | $3.92 \pm 0.00\ ^{b}{}_{B}$ |
| 72 | $6.56 \pm 0.05\ ^{c}{}_{B}$ | $4.08 \pm 0.02\ ^{c}{}_{A}$ | $4.10 \pm 0.02\ ^{d}{}_{A}$ | $4.08 \pm 0.03\ ^{c}{}_{A}$ | $3.74 \pm 0.45\ ^{a}{}_{A}$ | $3.97 \pm 0.01\ ^{b}{}_{A}$ |

Data are reported as mean ± SD. Data in the same column with different superscript letters are statistically different ($p < 0.05$). Data in the row with different subscript letters are statistically different ($p < 0.05$).

**Table 3.** Values of pH for broths inoculated with yeasts, with and without PPP.

| Time (h) | CTRL | 500 μm | 400 μm | 250 μm | 150 μm | 100 μm |
|---|---|---|---|---|---|---|
| 0 | $6.97 \pm 0.00\ ^{d}{}_{D}$ | $4.13 \pm 0.01\ ^{c}{}_{C}$ | $4.09 \pm 0.00\ ^{c}{}_{B}$ | $4.10 \pm 0.00\ ^{b}{}_{B}$ | $4.05 \pm 0.00\ ^{c}{}_{A}$ | $4.04 \pm 0.01\ ^{c}{}_{A}$ |
| 4 | $6.85 \pm 0.02\ ^{c}{}_{D}$ | $4.05 \pm 0.00\ ^{b}{}_{C}$ | $4.06 \pm 0.02\ ^{b,c}{}_{C}$ | $4.07 \pm 0.01\ ^{a,b}{}_{C}$ | $3.87 \pm 0.03\ ^{a}{}_{B}$ | $3.83 \pm 0.02\ ^{a}{}_{A}$ |
| 24 | $6.82 \pm 0.11\ ^{c}{}_{C}$ | $4.12 \pm 0.00\ ^{c}{}_{B}$ | $4.16 \pm 0.01\ ^{d}{}_{B}$ | $4.19 \pm 0.07\ ^{c}{}_{B}$ | $3.98 \pm 0.02\ ^{b}{}_{A}$ | $3.88 \pm 0.05\ ^{a}{}_{A}$ |
| 48 | $6.48 \pm 0.02\ ^{b}{}_{E}$ | $4.13 \pm 0.01\ ^{c}{}_{D}$ | $3.94 \pm 0.01\ ^{a}{}_{B}$ | $4.01 \pm 0.01\ ^{a}{}_{C}$ | $3.99 \pm 0.01\ ^{b}{}_{C}$ | $3.86 \pm 0.02\ ^{a}{}_{A}$ |
| 72 | $6.25 \pm 0.02\ ^{a}{}_{C}$ | $4.03 \pm 0.02\ ^{a}{}_{B}$ | $4.04 \pm 0.03\ ^{b}{}_{B}$ | $4.05 \pm 0.01\ ^{a,b}{}_{B}$ | $3.97 \pm 0.01\ ^{b}{}_{A}$ | $3.99 \pm 0.01\ ^{b}{}_{A}$ |

Data are reported as mean ± SD. Data in the same column with different superscript letters are statistically different ($p < 0.05$). Data in the row with different subscript letters are statistically different ($p < 0.05$).

Our positive preliminary results, in line with other findings from the literature, gave us the possibility to select PPP at a size of 250 μm for the subsequent experiments, because no significant differences were recorded among the various sizes ($p > 0.05$) and because 250 μm was one of the most abundant amounts of powder recorded during sifting.

### 3.2. SEM Images of Pad

The surface morphology of developed silk fibroin-PPP pad was investigated by SEM and compared with that of pristine PPP (Figure 3). As shown in Figure 3, the PPP sample was characterized by the presence of isolated grains (B and b), while the pad showed a more compact and continuous structure (A and a), due to the presence of silk fibroin protein that, as previously demonstrated in the work of Giannelli et al. [24], acts as glue for pomegranate powder particles. Moreover, in detail, no cracks in the pad surface areas were detected (A and a), but many white dots probably related to the insoluble aggregates of pomegranate peel particles embedded in the silk fibroin matrix were observed. This kind of morphology was already observed for other fruit peels incorporated in polymeric matrices as chitosan or gelatin [45].

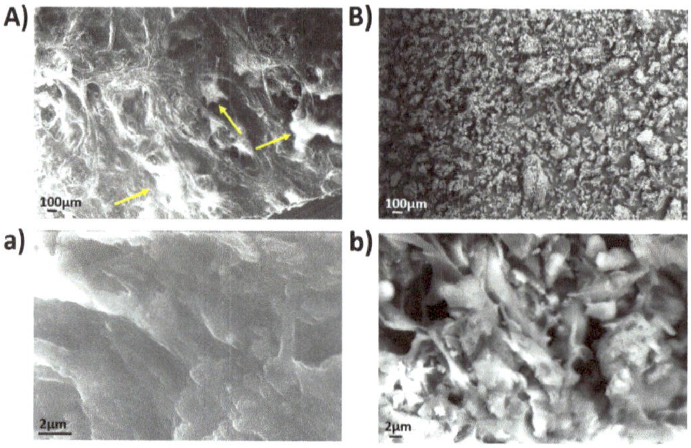


**Figure 3.** SEM images of the silk fibroin-based pad with PPP at different magnifications (**A**,**a**). SEM images of PPP at different magnifications (**B**,**b**).

### 3.3. PPP Efficacy in Previously Inoculated Soymilk and Apple Juice

Data reported in Figure 4a,b show the trend of PPP efficacy in the soymilk inoculated with *Pseudomonas* spp. and yeasts, respectively, during about two weeks of storage under refrigerated temperature. As can be observed, Figure 4a highlights a control sample where *Pseudomonas* spp. rapidly grew and reached a high concentration within a few days. Samples of soymilk containing the fruit peel presented delayed growth, with lag phases that are more marked as the concentration of peel increased. The maximum viable cell concentration in the stationary phase was also different among samples. In fact, the sample with 5% $w/v$ PPP reached about $10^7$ CFU/mL at the end of the observation period, whereas the viable cell concentration of the two samples with 7.5 and 10% $w/v$ PPP remained around $10^5$ CFU/mL and therefore never reached completely unacceptable levels of *Pseudomonas* spp. proliferation. A similar situation can be observed for yeasts. As can be seen in Figure 4b, the control soymilk reached undesired fungal proliferation levels above $10^5$ CFU/mL within 4 days of storage. Patrignani et al. [44] considered a concentration equal to six log CFU/mL as the threshold for yeast acceptability in fruit juice. On the contrary, when added at 5% $w/v$, the PPP delayed yeast growth in soymilk and when added at 7.5 or 10% $w/v$ it completely inhibited fungal proliferation. Therefore, results in the inoculated soymilk enriched with PPP demonstrate that the peel could be advantageously used in the beverage to control undesired spoilage.

Despite the effects recorded with PPP tested in the liquid foods, some differences can be seen in results of the in vitro test. In particular, PPP tested in the PC broth was more effective than in the inoculated soymilk. This experimental evidence can be ascribed to more than one reason: the binding of the active compounds of peel to food components, the protection of microorganisms in the real food matrix and the physiological state of microorganisms under refrigeration conditions, compared to temperatures optimal for microbial growth [46]. As regards interactions, many authors studied phenols and food constituents and observed that higher concentrations than the in vitro bactericidal concentration were necessary to get a bacteriostatic effect in food models. Literature data assessed that the loss of susceptibility of most fungi and bacteria is due to interactions of active phenolics with main food constituents, protein, and fat, at the expense of their interactions with unwanted microbial or fungal cells [47,48]. It is also true that unwanted microorganisms are protected by a "layer" of food constituents limiting the direct contact of antimicrobial phenolics with their cell envelope [49,50].

The pH values are in line with data recorded in the in vitro test, with control samples ranging between 6.4 and 6.8 and active samples ranging between 4.4 and 4.8 during the entire storage period.

Data in Figure 5 report results of *A. acidoterrestris* viable cell concentration evolution in samples of apple juice, with and without PPP. This figure clearly shows the great effect of PPP on the spoiling *A. acidoterrestris* and highlights that even the smallest concentration of PPP (5% $w/v$) was enough to completely provoke the microbial cell death. The initial concentration of the microorganism in the juice samples was around $10^4$ CFU/mL, in the control juice the concentration slightly increased, while in the three juices with PPP it rapidly declined under any microbial detection level. No differences appeared among the three active samples ($p > 0.05$). Thus, PPP at tested concentrations is effective at controlling proliferation of cells of *A. acidoterrestris* in apple juice.

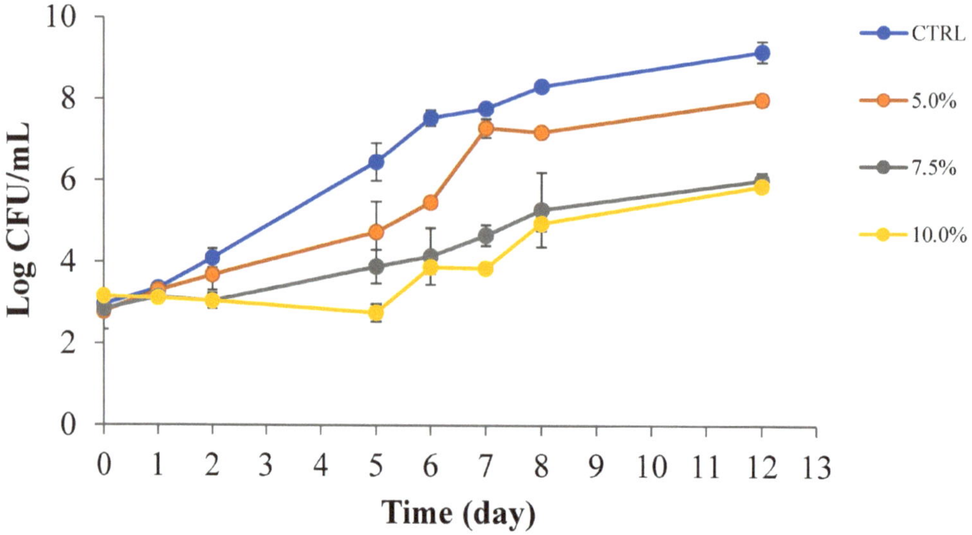


(**a**)

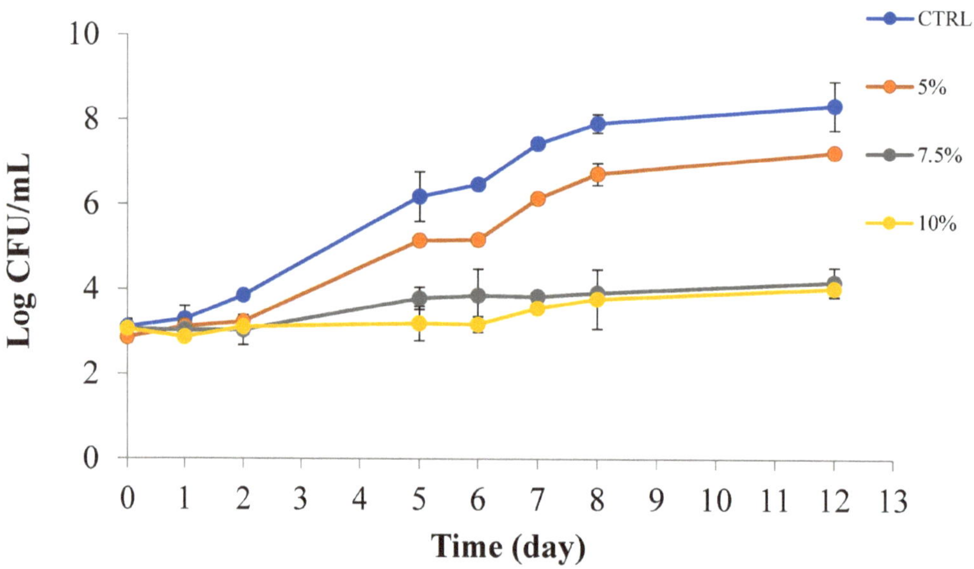


(**b**)

**Figure 4.** *Pseudomonas* spp. (**a**) and yeasts (**b**) in soymilk with PPP at 0, 5, 7.5 and 10% (*w*/*v*).

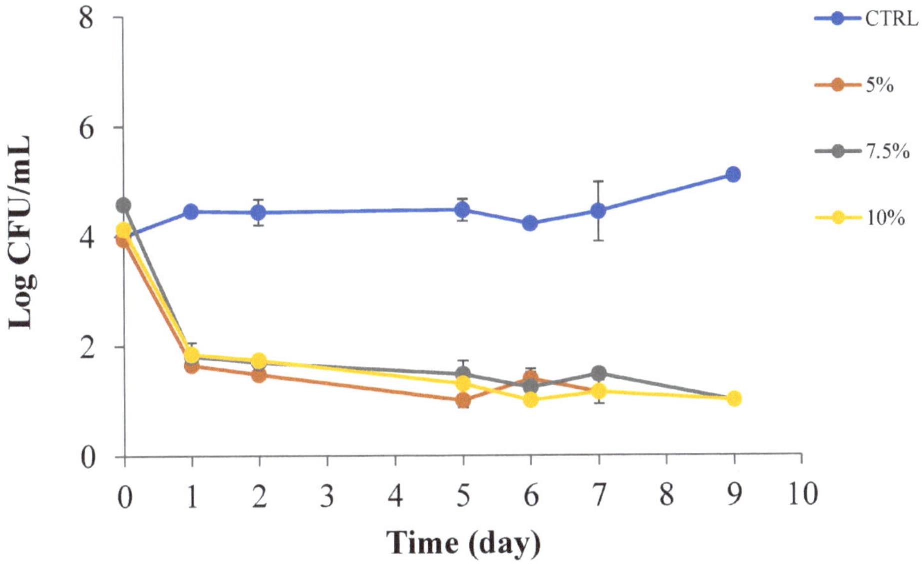


**Figure 5.** *A. acidoterrestris* in apple juice with PPP at 0, 5, 7.5 and 10% ($w/v$).

The great efficacy of PPP in the juice compared to the effects recorded by using the same peel concentration in the inoculated soymilk could be ascribed to different food component interactions. In fact, studies regarding the effect of proteins on the antimicrobial activity of phenolics exist and demonstrate that the antibacterial activity reduction in phenolics in the presence of proteins is correlated with their affinity for proteins [25], which in our case were abundant in the soymilk and absent in the apple juice. Without any doubt, polyphenols also interact with carbohydrates that mainly characterize our apple juice but the studies on these types of interactions are scarce compared to those related to proteins, and therefore, in the absence of such data, the efficiency of PPP in preserving the apple juice can hardly be further discussed [46].

Table 4 reports data of pH in apple juice. As can be seen, pH was around 3.5 in all the samples and for the entire storage period, regardless of the PPP addition. This is because apple juice is an acidic food matrix, and the addition of PPP did not provoke any further pH reduction.

**Table 4.** Values of pH for apple juice inoculated with *A. acidoterrestris*, with and without PPP.

| Time (Day) | CTRL | 5% | 7.5% | 10% |
|---|---|---|---|---|
| 0 | $3.52 \pm 0.01\ ^{b}{}_{A}$ | $3.54 \pm 0.01\ ^{c}{}_{A}$ | $3.52 \pm 0.01\ ^{b}{}_{A}$ | $3.54 \pm 0.01\ ^{c}{}_{A}$ |
| 1 | $3.46 \pm 0.02\ ^{a}{}_{A}$ | $3.46 \pm 0.02\ ^{b}{}_{A}$ | $3.48 \pm 0.01\ ^{c}{}_{A,B}$ | $3.50 \pm 0.00\ ^{b}{}_{B}$ |
| 2 | $3.55 \pm 0.01\ ^{c}{}_{A}$ | $3.58 \pm 0.01\ ^{d}{}_{B}$ | $3.56 \pm 0.01\ ^{d}{}_{A}$ | $3.55 \pm 0.00\ ^{c}{}_{A}$ |
| 5 | $3.46 \pm 0.01\ ^{a}{}_{A}$ | $3.53 \pm 0.00\ ^{c}{}_{C}$ | $3.52 \pm 0.01\ ^{b}{}_{B,C}$ | $3.51 \pm 0.01\ ^{b}{}_{B}$ |
| 6 | $3.45 \pm 0.00\ ^{a}{}_{A,B}$ | $3.46 \pm 0.01\ ^{b}{}_{B}$ | $3.44 \pm 0.01\ ^{a}{}_{A}$ | $3.46 \pm 0.01\ ^{a}{}_{B}$ |
| 7 | $3.51 \pm 0.01\ ^{b}{}_{C}$ | $3.42 \pm 0.02\ ^{a}{}_{A}$ | $3.44 \pm 0.01\ ^{a}{}_{A,B}$ | $3.46 \pm 0.01\ ^{a}{}_{B}$ |
| 9 | $3.51 \pm 0.01\ ^{b}{}_{B}$ | $3.41 \pm 0.01\ ^{a}{}_{A}$ | $3.42 \pm 0.02\ ^{a}{}_{A}$ | $3.44 \pm 0.02\ ^{a}{}_{A}$ |

Data are reported as mean ± SD. Data in the same column with different superscript letters are statistically different ($p < 0.05$). Data in the row with different subscript letters are statistically different ($p < 0.05$).

### 3.4. Active Pad Efficacy in Previously Inoculated Soymilk and Apple Juice

Figure 6a,b shows the trends of microbial and fungal proliferation in the soy-based beverage, while Figure 7 highlights the evolution of *A. acidoterrestris* in apple juice. As can be seen in Figure 6, the microbial and fungal proliferations were not completely affected by the presence of the active pad, even though a certain inhibition can be observed with the two highest PPP concentrations in the polymeric matrix. In fact, looking at Figure 6a it is possible to infer that both the control samples and the soymilk with the active pad 5% *w/v* PPP presents similar behaviour to *Pseudomonas* spp. which reached undesired levels of $10^6$ CFU/mL after 2.5 and 3.5 days, respectively. In the two samples with 7.5 and 10% *w/v* PPP, the same undesired microbial growth was reached after more than 4.5 days, thus suggesting that higher concentrations than 5% *w/v* were suitable to release enough active compounds in the beverage to slow down microbial proliferation. A similar trend can be inferred in Figure 6b for yeasts. The two control samples and the active pad with the lowest PPP concentration became unacceptable for yeast growth within 3 days of storage at 4 °C, whereas the other two active soymilks remained acceptable for more time, about 4 days. These antimicrobial and antifungal effects confirmed data in the literature dealing with other bio-based polymeric matrices containing pomegranate peel or peel extracts, tested against various spoilage and pathogenic strains [51–54]. Kharchoufi et al. [55] showed good bacteriostatic effect by using different concentrations of PPP extract in chitosan film. These authors demonstrated, by an antifungal assay, that the bioactive coatings with the addition of the peel extract, significantly inhibited the growth of *Penicillium digitatum* by producing an inhibition halo around the experimental bioactive film disks. Direct comparisons between our developed active pad in soymilk with other delivery systems also based on pomegranate peel can be hard to extrapolate due to different polymeric matrixes used, different concentrations of peel powder tested, and different food items placed in contact [56]. In addition, it is worth considering that the amount of antimicrobial phenolics released is controlled by their partition equilibrium between the reservoir material and the food matrix in direct contact, and by kinetics of migration of phenolics in food contact material, and in food, respectively [46].

Compared to the powder itself, the pad was less effective which can be explained by more than one reason. The process of producing the active pad may have reduced the concentration of the antimicrobial phenolics present in the powder and consequently in the active pad. The release of the antimicrobial phenolics is hindered by the presence of the polymeric matrix, therefore, active compounds release rate from the active pad is slowed down compared to the powder itself. Last but not least, part of the antimicrobial phenolics could be absorbed by the polymeric matrix, therefore, if compared to the powder itself the amount of antimicrobial phenolics in the liquid food is diminished.

The pH values of control soymilk and of soymilk with the control pad slightly ranged around 7 during the entire monitoring period. On the other hand, lower pH values around 5.9 were recorded in the samples when the respective three active pads were added to the beverage. Considering the low pH of pomegranate peel [12], and its release from the polymeric matrix, it was not surprising that a pH reduction (from about 5.9 to about 4.9) over time was recorded in the liquid food loaded with all the three pads.

As regards the behaviour of *A. acidoterrestris* in apple juice reported in Figure 7, a great inhibition effect of every active pad was recorded, as occurred with the simple PPP reported in Figure 5. In this case it is possible to observe a starting viable cell concentration around $10^4$ CFU/mL and two completely different microbial evolutions trends. In fact, the microbial cells proliferated in the two control juice samples but had a decreasing trend in the three samples containing the active pad, where the growth was completely inhibited. All the considerations reported in the previous paragraph dealing with the active pad efficacy in soymilk, are still valid in apple juice but we observed a more marked effect on *A. acidoterrestris* because the biomolecules' concentration in this case was enough to inhibit the bacteria and therefore powder itself, or powder in the pad, recorded the same antimicrobial activity.

The values of pH in all the juice samples for the entire observation period were 3.5, as reported in Table 4 for the samples with powder.

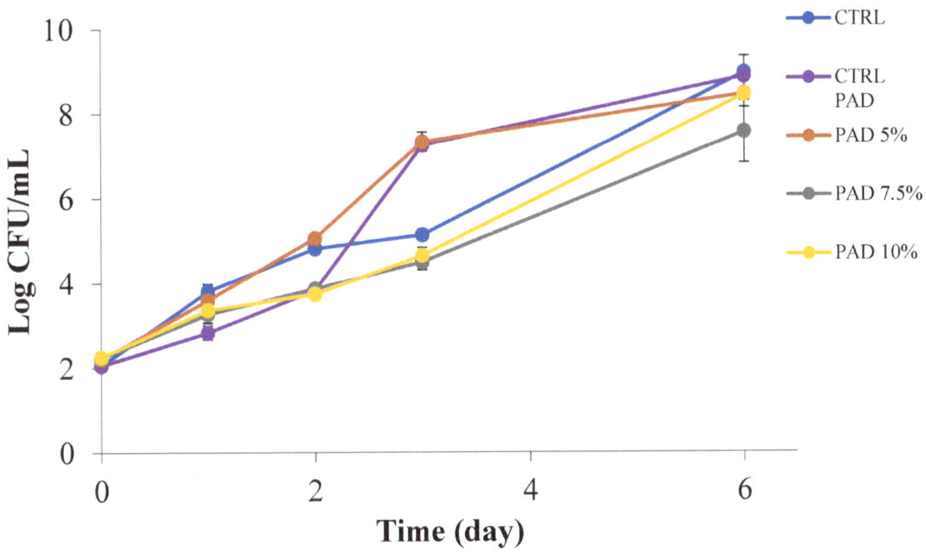


(a)

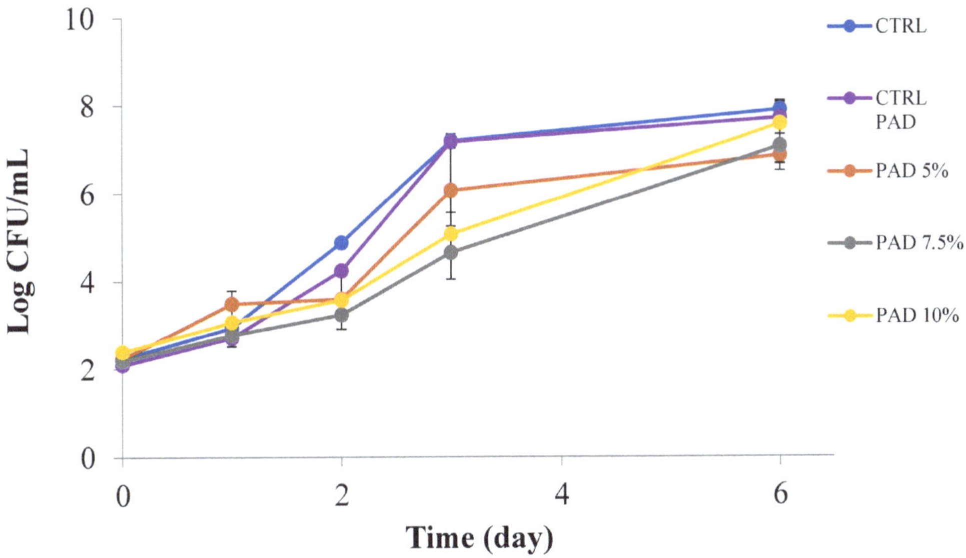


(b)

**Figure 6.** *Pseudomonas* spp. (**a**) and yeasts (**b**) in soymilk with and without pads at 0, 5, 7.5 and 10% (*w*/*v*).

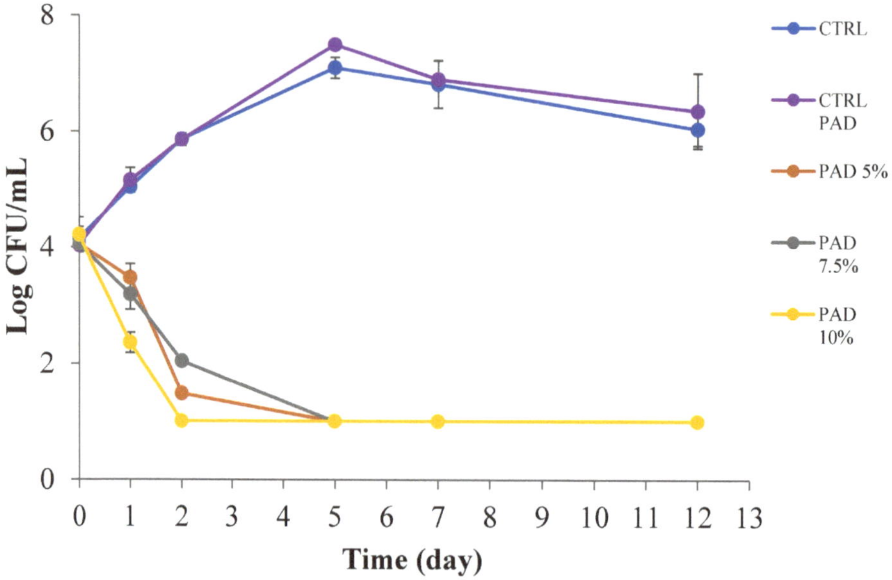


**Figure 7.** *A. acidoterrestris* in apple juice with and without pads at 0, 5, 7.5 and 10% (*w*/*v*).

## 4. Conclusions

In this study pomegranate peel powder was tested against food spoiling microbial and fungal species. The powder itself, and its inclusion in a bio-based polymeric matrix were both considered. In vitro and in situ tests were carried out. Data recorded in vitro proved the effects of PPP on *Pseudomonas* spp. and yeasts, regardless of the powder granulometry. When the powder was tested in soymilk, a certain reduction in the efficacy was recorded, compared to the in vitro tests. This is most probably due to possible interactions between phenolic compounds and food components, in particular proteins. When the powder was tested against *A. acidoterrestris* in apple juice, a complete inhibition was found, probably due to the different food compositions of juice compared to soymilk, the former being constituted solely of carbohydrates. The same PPP concentrations were also tested in silk fibroin-based pads, also using soymilk and apple juice as liquid food systems. In this case also, the results were more interesting for the apple juice because the growth of the *A. acidoterrestris* was completely inhibited by the presence of the active pad, regardless of the PPP concentration. In the soymilk, only the two highest concentrations in the active pad exerted a certain bacteriostatic effect on microbial and fungal proliferation. These preliminary findings gave us interesting information on the powder alone, and on the delivery system based on silk fibroin and pomegranate peel applied to contaminated foods. Considering the obvious advantages in using bio-based polymeric systems that release active compounds, compared to direct addition of these molecules to food, further research could still be carried out to explore concentrations of PPP in the pad able to better preserve food from spoilage. In addition, the interesting experimental evidence recorded with the active pad in apple juice against *A. acidoterrestris* suggested that PPP in the polymeric matrix could be further reduced below 5% *w*/*v*, without affecting the antimicrobial efficacy. Considering that pads containing pomegranate by-products could better prevent sensory alterations in the food product, that are inevitable when the peel is directly added to the food matrix, pad optimization in terms of bacterial and fungal efficacy, would also require study of the preservation of food sensory properties.

**Author Contributions:** Formal analysis, V.L. and A.L.; Conceptualization, Methodology, Data curation, Writing—original draft, T.P. and R.Z.; Conceptualization, Validation, Editing, M.A.D.N.; Conceptualization, Validation, Writing—review and editing, Supervision: A.C. All authors have read and agreed to the published version of the manuscript.

**Funding:** This research received no external funding.

**Institutional Review Board Statement:** Not applicable.

**Informed Consent Statement:** Not applicable.

**Data Availability Statement:** Data will be made available on request.

**Conflicts of Interest:** The authors declare that they have no known competing financial interest or personal relationship that could have appeared to influence the work reported in this paper.

## References

1. Sala, S.; Anton, A.; McLaren, S.J.; Notarnicola, B.; Saouter, E.; Sonesson, U. In quest of reducing the environmental impacts of food production and consumptiom. *J. Clean. Prod.* **2017**, *140*, 387–398. [CrossRef]
2. Vagsholm, I.; Arzoomand, N.S.; Boqvist, S. Food Security, Safety, and Sustainability-Getting the Trade-Offs Right. *Front. Sustain. Food Syst.* **2020**, *4*, 16. [CrossRef]
3. Hygreeva, D.; Pandey, M.; Radhakrishna, K. Potential applications of plant-based derivatives as fat replacers, antioxidants and antimicrobials in fresh and processed meat products. *Meat Sci.* **2014**, *98*, 47–57. [CrossRef] [PubMed]
4. Abid, Y.; Azabou, S.; Jridi, M.; Khemakhem, I.; Bouaziz, M.; Attia, H. Storage stability of traditional Tunisian butter enriched with antioxidant extract from tomato processing by-products. *Food Chem.* **2017**, 233, 476–482. [CrossRef] [PubMed]
5. Arjeh, E.; Akhavan, H.R.; Barzegar, M.; Carbonell-Barrachina, A.A. Bio-active compounds and functional properties of pistachio hull: A review. *Trends Food Sci. Technol.* **2020**, *97*, 55–64. [CrossRef]
6. Helkar, P.B.; Sahoo, A.K.; Patil, N.J. Review: Food Industry By-Products used as a Functional Food Ingredients. *Int. J. Waste Res.* **2016**, *6*, 248.
7. Kim, J.H.; Hong, W.S.; Oh, S.W. Effect of layer-by-layer antimicrobial edible coating of alginate and chitosan with grapefruit seed extract for shelf-life extension of shrimp (*Litopenaeus vannamei*) stored at 4 °C. *Int. J. Biol. Macromol.* **2018**, *120*, 1468–1473. [CrossRef] [PubMed]
8. Angiolillo, L.; Conte, A.; Del Nobile, M.A. A new method to bio-preserve sea bass fillets. *Int. J. Food Microbiol.* **2018**, *271*, 60–66. [CrossRef]
9. Nardella, S.; Conte, A.; del Nobile, M.A. State-of-Art on the Recycling of By-Products from Fruits and Vegetables of Mediterranean Countries to Prolong Food Shelf Life. *Foods* **2022**, *11*, 665. [CrossRef]
10. Panza, O.; Conte, A.; Del Nobile, M.A. Recycling of fig peels to enhance the quality of handmade pasta. *LWT* **2022**, *168*, 113872. [CrossRef]
11. Bonesi, M.; Tundis, R.; Sicari, V.; Loizzo, M.R. The juice of pomegranate (*Punica granatum* L.): Recent studies on its bioactivities. In *Quality Control in the Beverage Industry*; Grumezescu, A.M., Holban, A.M., Eds.; Academic Press: New York, NY, USA, 2019; Volume 17, pp. 459–489.
12. El Barnossi, A.; Moussaid, F.; Housseini, A.I. Tangerine, banana and pomegranate peels valorisation for sustainable environment: A review. *Biotechnol. Rep.* **2021**, *29*, e00574. [CrossRef]
13. Incoronato, A.L.; Cedola, A.; Conte, A.; Del Nobile, M.A. Juice and by-products from pomegranate to enrich pancake: Characterization and shelf-life evaluation. *Int. J. Food Sci. Technol.* **2021**, *56*, 2886–2894. [CrossRef]
14. Bourekoua, H.; Rózyło, R.; Gawlik-Dziki, U.; Benatallah, L.; Zidoune, M.N.; Dziki, D. Pomegranate seed powder as a functional component of gluten-free bread (Physical, sensorial and antioxidant evaluation). *Food Sci. Technol.* **2018**, *53*, 1906–1913. [CrossRef]
15. Panza, O.; Conte, A.; Del Nobile, M.A. Zero-Waste Approach Applied to Pomegranates for prolonging Fish Burger Shelf Life. *Foods* **2022**, *11*, 551. [CrossRef]
16. El-Nashi, H.B.; Fattah, A.F.A.K.A.; Rahman, N.R.A.; El Razik, M.A. Quality characteristics of beef sausage containing pomegranate peels during refrigerated storage. *Ann. Agric. Sci.* **2015**, *60*, 403–412. [CrossRef]
17. Dilucia, F.; Lacivita, V.; Conte, A.; Del Nobile, M.A. Sustainable use of fruit and vegetable by-products to enhance food packaging performance. *Foods* **2020**, *9*, 857. [CrossRef] [PubMed]
18. Hamed, I.; Jakobsen, A.N.; Lerfall, J. Sustainable edible packaging systems based on active compounds from food processing by-products: A review. *Compr. Rev. Food Sci. Food Saf.* **2022**, *21*, 198–226. [CrossRef]
19. Kandylis, P.; Kokkinomagoulos, E. Food Applications and Potential Health Benefits of Pomegranate and its Derivatives. *Foods* **2020**, *9*, 122. [CrossRef]
20. Mushtaq, M.; Gani, A.; Gani, A.; Punoo, H.A.; Masoodi, F.A. Use of pomegranate peel extract incorporated zein film with improved properties for prolonged shelf life of fresh Himalayan cheese (Kalari/kradi). *Innov. Food Sci. Emerg. Technol.* **2018**, *48*, 25–32. [CrossRef]

21. Bertolo, M.R.; Martins, V.C.; Horn, M.M.; Brenelli, L.B.; Plepis, A.M. Rheological and antioxidant properties of chitosan/gelatin-based materials functionalized by pomegranate peel extract. *Carbohydr. Polym.* **2020**, *228*, 115386. [CrossRef]
22. Berizi, E.; Hosseinzadeh, S.; Shekarforoush, S.S.; Barbieri, G. Microbial, chemical, textural and sensory properties of coated rainbow trout by chitosan combined with pomegranate peel extract during frozen storage. *Int. J. Biol. Macromol.* **2018**, *106*, 1004–1013. [CrossRef] [PubMed]
23. Yuan, G.; Lv, H.; Tang, W.; Zhang, X.; Sun, H. Effect of chitosan coating combined with pomegranate peel extract on the quality of Pacific white shrimp during iced storage. *Food Contam.* **2016**, *59*, 818–823. [CrossRef]
24. Giannelli, M.; Lacivita, V.; Posati, T.; Aluigi, A.; Conte, A.; Zamboni, R.; Del Nobile, M.A. Silk Fibroin and Pomegranate By-Products to Develop Sustainable Active Pad for Food Packaging Applications. *Foods* **2021**, *10*, 2921. [CrossRef] [PubMed]
25. Bouhanna, I.; Boussaa, A.; Boumaza, A.; Rigano, D.; Maisto, M.; Basile, A.; Rollini, M.; Limbo, S.; Idoui, T. Characterization and antibacterial activity of gelatin-based film incorporated with *Arbutus unedo* L. fruit extract on *Sardina pilchardus*. *J. Food Process. Preserv.* **2021**, *45*, 15424. [CrossRef]
26. Bayati, M.; Tavakoli, M.M.; Ebrahimi, S.N.; Aliahmadi, A.; Rezadoost, H. Optimization of effective parameters in cold pasteurization of pomegranate juice by response surface methodology and evaluation of physicochemical characteristics. *LWT* **2021**, *147*, 111679. [CrossRef]
27. Kaddumukasa, P.; Imathiu, S.; Mathara, J.; Nakavuma, J. Influence of physicochemical parameters on storage stability: Microbiological quality of fresh unpasteurized fruit juices. *Food Sci. Nutr.* **2017**, *5*, 1098–1105. [CrossRef]
28. Koutchma, T.; Popovic, V.; Ros-Polski, V.; Popielarz, A. Effects of ultraviolet light and high-pressure processing on quality and health-related constituents of fresh juice products. *Comp. Rev. Food Sci. Food Saf.* **2016**, *15*, 844–867. [CrossRef]
29. Anaya-Esparza, L.M.; Velázquez-Estrada, R.M.; Roig, A.X.; García-Galindo, H.S.; Sayago-Ayerdi, S.G.; Montalvo-González, E. Thermo-sonication: An alternative processing for fruit and vegetable juices. *Trends Food Sci. Technol.* **2017**, *61*, 26–37. [CrossRef]
30. Szczepanska, J.; Skyapska, S.; Polaska, M.; Marszalek, K. High pressure homogenization with a cooling circulating system: The effect on physiochemical and rheological properties, enzymes, and carotenoid profile of carrot juice. *Food Chem.* **2022**, *370*, 131023. [CrossRef]
31. Wu, W.; Xiao, G.; Yu, Y.; Xu, Y.; Wu, J.; Peng, J.; Li, L. Effects of high pressure and thermal processing on quality properties and volatile compounds of pineapple fruit juice. *Food Contam.* **2021**, *130*, 108293. [CrossRef]
32. Yildiz, S.; Pokhrel, P.R.; Unluturk, S.; Barbosa-Canovas, G.V. Shelf life extension of strawberry juice by equivalent ultrasound, high pressure, and pulsed electric fields processes. *Food Res. Int.* **2021**, *140*, 110040. [CrossRef]
33. Adulvitayakorn, S.; Azhari, S.H.; Hasan, H. The effects of conventional thermal, microwave heating, and thermosonication treatments on the quality of sugarcane juice. *J. Food Proc. Preserv.* **2020**, *44*, e14322. [CrossRef]
34. Dhumal, S.S.; Karale, A.R.; More, T.A.; Nimbalkar, C.A. Preparation of Pomegranate Juice Concentrate by Various Heating Methods and Appraisal of Its Physicochemical Characteristics. *Int. Soc. Hortic. Sci.* **2015**, *1089*, 473–484. [CrossRef]
35. Starek, A.; Sagan, A.; Andrejko, D.; Chudzik, B.; Kobus, Z.; Kwiatowski, M.; Terebun, P.; Pawlat, J. Possibility to extend the shelf life of NFC tomato juice using cold atmospheric pressure plasma. *Sci. Rep.* **2020**, *10*, 20959. [CrossRef]
36. Park, S.J.; Nurika, I.; Suhartini, S.; Cho, W.H.; Moon, K.D.; Jung, H.J. Carbonation of not from concentrate apple juice positively impacts shelf-life. *LWT* **2020**, *134*, 110128. [CrossRef]
37. Lu, S.; Wang, X.; Lu, Q.; Zhang, X.; Kluge, J.A.; Uppal, N.; Omenetto, F.; Kaplan, D.L. Insoluble and Flexible Silk Films Containing Glycerol. *Biomacromolecules* **2010**, *11*, 143–150. [CrossRef]
38. Andrade, M.A.; Lima, V.; Sanches Silva, A.; Vilarinho, F.; Castilho, M.C.; Khwaldia, K.; Ramos, F. Pomegranate and grape by-products and their active compounds: Are they a valuable source for food applications? *Trends Food Sci. Technol.* **2019**, *86*, 68–84. [CrossRef]
39. Xiang, Q.; Li, M.; Wen, J.; Ren, F.; Yang, Z.; Jiang, X.; Chen, Y. The bioactivity and applications of pomegranate peel extract: A review. *J. Food Biochem.* **2022**, *46*, e14105. [CrossRef] [PubMed]
40. Chen, J.; Liao, C.; Ouyang, X.; Kahramanoglu, I.; Gan, Y.; Li, M. Antimicrobial Activity of Pomegranate Peel and Its Applications on Food Preservation. *J. Food Qual.* **2020**, *2020*, 8850339. [CrossRef]
41. Singh, B.; Singh, J.P.; Kaur, A.; Singh, N. Antimicrobial potential of pomegranate peel: A review. *Int. J. Food Sci. Technol.* **2019**, *54*, 959–965. [CrossRef]
42. Alexandre, E.M.C.; Silva, S.; Santos, S.A.O.; Silvestre, A.J.D.; Duarte, M.F.; Saraiva, J.A.; Pintado, M. Antimicrobial activity of pomegranate peel extracts performed by high pressure and enzymatic assisted extraction. *Food Res. Int.* **2019**, *115*, 167–176. [CrossRef] [PubMed]
43. Lacivita, V.; Incoronato, A.L.; Conte, A.; Del Nobile, M.A. Pomegranate Peel Powder as a Food Preservative in Fruit Salad: A Sustainable Approach. *Foods* **2021**, *10*, 1359. [CrossRef] [PubMed]
44. Patrignani, F.; Vannini, L.; Kamder, S.L.S.; Lanciotti, R.; Guerzoni, M.E. Potential of high-pressure homogenization to inactivate *Zygosaccharomyces bailii* in fruit juices. *J. Food Sci.* **2010**, *75*, 116–120. [CrossRef] [PubMed]
45. Riaz, A.; Lei, S.; Akhtar, H.M.; Wan, P.; Chen, D.; Jabbar, S.; Abid, M.; Hashim, M.M.; Zeng, X. Preparation and characterization of chitosan-based antimicrobial active food packaging film incorporated with apple peel polyphenols. *Int. J. Biol. Macromol.* **2018**, *114*, 547–555. [CrossRef]
46. Oulahal, N.; Degraeve, P. Phenolic-Rich Plant Extracts with Antimicrobial Activity: An Alternative to Food Preservatives and Biocides? *Front. Microbiol.* **2022**, *12*, 753518. [CrossRef] [PubMed]

47. Miceli, A.; Aleo, A.; Corona, O.; Sardina, M.T.; Mammina, C.; Settanni, L. Antibacterial activity of Borago officinalis and Brassica juncea aqueous extracts evaluated in vitro and in situ using different food model systems. *Food Contam.* **2014**, *40*, 157–164. [CrossRef]
48. Xu, X.; Ou, Z.M.; Wu, C.D. Growth media affect assessment of antimicrobial activity of plant-derived polyphenols. *BioMed Res. Int.* **2018**, *2018*, 8308640. [CrossRef]
49. Boziaris, I.S.; Proestos, C.; Kapsokefalou, M.; Komaitis, M. Antimicrobial effect of Filipendula ulmaria plant extract against selected foodborne pathogenic and spoilage bacteria in laboratory media, fish flesh and fish roe product. *Food Technol. Biotechnol.* **2011**, *49*, 263.
50. da Cruz Cabral, L.; Fernández Pinto, V.; Patriarca, A. Application of plant derived compounds to control fungal spoilage and mycotoxin production in foods. *Int. J. Food Microbiol.* **2013**, *166*, 1–14. [CrossRef]
51. Ali, A.; Chen, Y.; Liu, H.; Yu, L.; Baloch, Z.; Khalid, S.; Zhu, J.; Chen, L. Starch-based antimicrobial films functionalized by pomegranate peel. *Int. J. Biol. Macromol.* **2019**, *129*, 1120–1126. [CrossRef]
52. Moghadam, M.; Salami, M.; Mohammadian, M.; Khodadadi, M.; Emam-Djomeh, Z. Development of antioxidant edible films based on mung bean protein enriched with pomegranate peel. *Food Hydroc.* **2020**, *104*, 105735. [CrossRef]
53. Hanani, Z.A.; Aelma Husna, A.B.; Nurul Syahida, S.; Nor Khaizura, M.A.; Jamilah, B. Effect of different fruit peels on the functional properties of gelatin/polyethylene bilayer films for active packaging. *Food Pack. Shelf Life* **2018**, *18*, 201–211. [CrossRef]
54. Hanani, Z.N.; Yee, F.C.; Nor-Khaizura, M.A.R. Effect of pomegranate (*Punica granatum* L.) peel powder on the antioxidant and antimicrobial properties of fish gelatin films as active packaging. *Food Hydroc.* **2019**, *89*, 253–259. [CrossRef]
55. Kharchoufi, S.; Parafati, L.; Licciardello, F.; Muratore, G.; Hamdi, M.; Cirvilleri, G.; Restuccia, C. Edible coatings incorporating pomegranate peel extract and biocontrol yeast to reduce *Penicillium digitatum* postharvest decay of oranges. *Food Microbiol.* **2018**, *74*, 107–112. [CrossRef] [PubMed]
56. Das, A.K.; Nanda, P.K.; Chowdhury, N.R.; Dandapat, P.; Gagaoua, M.; Chauhan, P.; Pateiro, M.; Lorenzo, J.M. Application of Pomegranate by-Products in Muscle Foods: Oxidative Indices, Colour Stability, Shelf Life and Health Benefits. *Molecules* **2021**, *26*, 467. [CrossRef]

foods

MDPI

*Article*

# Potato Chips Byproducts as Feedstocks for Developing Active Starch-Based Films with Potential for Cheese Packaging

**Ana M. Peixoto [1], Sílvia Petronilho [1,2,*], M. Rosário Domingues [1,3,4], Fernando M. Nunes [2], Joana Lopes [1,5], Marit Kvalvåg Pettersen [6], Magnhild S. Grøvlen [6], Elin M. Wetterhus [6], Idalina Gonçalves [5] and Manuel A. Coimbra [1]**

[1] LAQV-REQUIMTE—Associated Laboratory for Green Chemistry of the Network of Chemistry and Technology, Department of Chemistry, Campus Universitário de Santiago, University of Aveiro, 3810-193 Aveiro, Portugal; mac@ua.pt (M.A.C.)
[2] Chemistry Research Centre-Vila Real, Department of Chemistry, University of Trás os-Montes and Alto Douro, Quinta de Prados, 5001-801 Vila Real, Portugal
[3] Mass Spectrometry Centre, Department of Chemistry, Campus Universitário de Santiago, University of Aveiro, 3810-193 Aveiro, Portugal
[4] CESAM—Centre for Environmental and Marine Studies, Department of Chemistry, Campus Universitário de Santiago, University of Aveiro, 3810-193 Aveiro, Portugal
[5] CICECO—Aveiro Institute of Materials, Department of Materials and Ceramic Engineering, Campus Universitário de Santiago, University of Aveiro, 3810-193 Aveiro, Portugal
[6] Nofima—Norwegian Institute of Food, Fisheries and Aquaculture Research, NO-1431 Ås, Norway
* Correspondence: silviapetronilho@ua.pt

**Abstract:** The potato chip industry generates brownish frying residues, which are usually landfilled. While spent frying oil has value as biodiesel, the defatted brownish water-soluble extract (BrE) does not yet have an application. In this work, it was hypothesized that BrE can be a source of compounds for active packaging. BrE is composed of carbohydrates (66.9%), protein (5.7%), and a small amount of phenolics and esterified fatty acids. When incorporated into starch-based formulations and casted, BrE at 5%, 10%, and 15% *w*/*w* (dry starch weight) conferred a yellowish coloration while maintaining the transparency of neat films. The BrE increased the films' traction resistance, elasticity, and antioxidant activity while decreasing their hydrophilicity. Furthermore, starch/15% BrE-based films showed diminished water vapor and good UV-light barrier properties. Their contact with sliced cheese did not change the products' hardness during storage (14 days). Weight loss of the cheese was observed after 7 days of storage, stabilizing at 6.52%, contrary to the cheese packed in polyamide (PA)/polyethylene (PE), already used in food packaging. The cheese packed in the starch/15% BrE-based films showed a significant yellowish darkening and lower content of volatile oxidation products compared to the PA/PE. Therefore, BrE revealed to have compounds with the potential to tune the performance of starch-based films for food packaging.

**Keywords:** potato frying residue; melanoidins; starch films; active packaging; volatile oxidation products; circular economy

**Citation:** Peixoto, A.M.; Petronilho, S.; Domingues, M.R.; Nunes, F.M.; Lopes, J.; Pettersen, M.K.; Grøvlen, M.S.; Wetterhus, E.M.; Gonçalves, I.; Coimbra, M.A. Potato Chips Byproducts as Feedstocks for Developing Active Starch-Based Films with Potential for Cheese Packaging. *Foods* **2023**, *12*, 1167. https://doi.org/10.3390/foods12061167

Academic Editor: Lubomir Lapcik

Received: 24 January 2023
Revised: 24 February 2023
Accepted: 7 March 2023
Published: 9 March 2023



## 1. Introduction

Potato (*Solanum tuberosum* L.) is one of the most widely cultivated and consumed carbohydrate-rich food crop worldwide, with a global production of ca. 359 million tons in 2020 [1]. Although they can be eaten in different ways, potato chips are one of the most widely appreciated and consumed potato product due to the fact of their characteristic salty flavor and crispy texture [2,3]. Their industrial frying process involves multiple steps, including washing, peeling, slicing, and frying in fat or vegetable oil at high temperatures [2]. Several tons of potato byproducts are generated, including washing slurries, peels, spent frying oil, and a brownish frying residue present in the frying oil. This residue is composed of burned potatoes accumulated at the bottom of noncontinuous friers, usually used in small and medium industries and household and restaurant kitchens. The valorization of these byproducts as raw materials to create added-value products is under research, promoting a more natural, low-cost, and sustainable production while encouraging a circular economy [4].

Potato washing slurries have been used to recover starch to develop bioplastics with physicochemical and barrier properties competitive with the ones provided by commercial potato starch [5]. To improve the elastic and hydrophobic properties of starch-based films, spent frying oil [6] and waxes recovered from potato peels [5] have been used. In addition, potato peel phenolic extracts [7] have demonstrated the potential to improve the antioxidant activity and UV-light barrier properties of starch-based films. While the frying oil is channeled for biodiesel production [8], at present, the defatted brownish water-soluble extract (BrE), a putative source of high molecular weight nitrogenous, brown-colored compounds (i.e., melanoidins), is a disposable industrial byproduct that originates environmental issues.

Due to the frying conditions of potatoes (high-temperature and anhydrous environment), melanoidins can be formed during nonenzymatic browning reactions (i.e., Maillard reaction) involving compounds with amino and carbonyl groups [2,9]. Moreover, the fatty acids of triacylglycerides present in the frying oil may be incorporated into the potato chips' melanoidins' structure, similar to what has been observed for other organic acids in coffee melanoidins [10]. Despite these assumptions, the chemical composition of BrE remains unknown. Since melanoidins from thermally processed food have shown bioactive potential [11,12], they should be exploited for different applications, such as in the development of active bio-based materials with potential for food packaging, similar to what has been conducted with potato peels and starch [7].

In this work, it was hypothesized that BrE, when gelatinized with starch recovered from potato washing slurries and casted, can give rise to bioplastics that can be used as active packaging to preserve sliced cheese characteristics, a foodstuff susceptible to lipid oxidation. For this, a hot-water extraction was applied to the defatted brownish frying residue. The influence of BrE on the optical, mechanical, wettability, water solubility, water vapor transmission rate, and active (UV-protective and antioxidant) properties of potato starch-based films was evaluated. The suitability of using starch/BrE-based films as active packaging was assessed in sliced cheese.

## 2. Materials and Methods

### 2.1. Materials

The potato washing slurries and potato chip frying residue were supplied by the A Saloinha Lda. company (Mafra, Portugal). Starch (25% amylose, 59–71 °C gelatinization temperature, and 12.5 J/g enthalpy) was recovered from the lyophilized slurries [5]. The brownish water-soluble extract (BrE, Figure S1) was obtained after defatting (spent frying oil recovered by Soxhlet extraction with a chloroform/methanol (2:1, *v*/*v*) mixture), hot-water extraction (80 °C, 1 h, constant stirring), centrifugation (24,000× *g*, 4 °C, 20 min), and freeze-drying. The BrE was stored in a desiccator containing phosphorus pentoxide until further characterization and usage in the films' production. Glycerol was purchased from Scharlab S.L. (Barcelona, Spain), and the sodium azide, chloroform and methanol were acquired from Sigma-Aldrich (Lisbon, Portugal). All reagents were of analytical grade and used without purification.

### 2.2. Fractionation of Potato Chip Brownish Water-Soluble Extract (BrE)

The BrE was submitted to a fractionation procedure by ultrafiltration (UF) using the Millipore Labscale™ TFF System equipped with a 500 mL reservoir [13]. Five cut-off membranes were used (100, 50, 30, 10, and 5 kDa from Merck Millipore Pellicon XL 50 cassettes with a flat plate format) at room temperature (20 °C ± 2 °C) and working between 10 to 30 psi transmembrane pressures. The separation efficiency was controlled by measuring the permeate conductivity (until it presented a value below 10 μS/cm) and determining the UV–Vis absorption (420 nm and 405 nm to monitor the presence of brown-colored compounds and 325 nm and 280 nm for phenolics and proteins). The ultrafiltration process was repeated for

3 cycles for each membrane (Figure S2). At the end of the procedure, 6 fractions corresponding to the >100 kDa, 50–100 kDa, 30–50 kDa, 10–30 kDa, and 5–10 kDa retentates, and the <5 kDa permeate were recovered. Figure 1 illustrates the flowchart to obtain each fraction according to the relative molecular weight. All UF fractions were frozen, freeze-dried, and stored under an anhydrous atmosphere until further characterization.

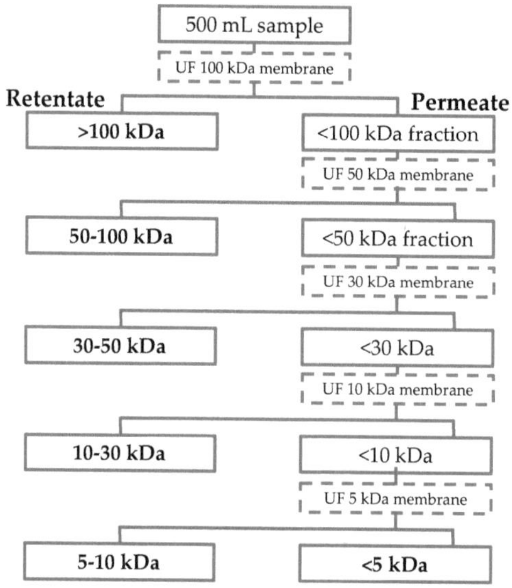


**Figure 1.** Scheme of the ultrafiltration process used for the BrE fractionation using 100, 50, 30, 10, and 5 kDa cut-off membranes.

*2.3. Characterization of Potato Chip Brownish Water-Soluble Extract (BrE)*

The sugar composition and content of the BrE and correspondent UF fractions were determined, in triplicate, by gas chromatography with flame-ionization detection (GC-FID) as alditol acetates. The 2-deoxyglucose was used as the internal standard [14].

The protein content of the BrE and UF fractions was estimated, in triplicate, by determining the total nitrogen in a Truspec 630-200-200 elemental analyzer (St. Joseph, Berrien, MI, USA) with a thermal conductivity detector (TDC). The nitrogen content was converted into protein content (%, *w*/*w* extract) using the 6.25 factor [15].

The total phenolic content of the BrE and UF fractions was determined, in triplicate, using the Folin–Ciocalteu method, where the absorbance of the samples was measured at 750 nm. A calibration curve of gallic acid (50–250 μg/mL) was used, and the results were expressed as μg gallic acid equivalents/mg of extract (μg GAE/mg) [12].

The esterified fatty acids (EFAs) content was determined by GC-FID, in triplicate, as fatty acid methyl esters (FAMEs) after the alkaline-catalyzed transesterification of the esterified fatty acids of the sample. For this, heptadecanoate methyl ester (0.03 mg/mL prepared in n-hexane) and KOH methanolic solution (2M) were used [6]. A DB-FFAP column (30 m × 0.32 mm and 0.25 μm of film thickness, J&W Scientific Inc., Folsom, CA, USA) was used. The compounds were identified by comparing their retention times with those of a commercial FAME mixture ($C_8$–$C_{24}$).

The presence of melanoidins was assessed by determining the specific extinction coefficient ($K_{mix}$) and the melanoidin browning index (MBI). The $K_{mix}$ was determined at 405 nm ($K_{mix\ 405nm}$) to determine the relative content of the brown-colored compounds in each UF fraction [14,16]. Briefly, for each sample, different solutions were prepared in distilled water (0.1 to 1 mg/mL), and the absorbance was measured, in triplicate, at room temperature (20 °C ± 2 °C). The MBI was calculated by dividing the $K_{mix\ 405nm}$ value by the relative content of the unknown material estimated by the difference in the known one (sugars and protein) [12,17].

The antioxidant activity of each sample was determined, in triplicate, using the ABTS (2,2′-azino-bis(3-ethylbenzothiazoline-6-sulphonic acid)) assay [18] and expressed as $IC_{50}$ (minimal concentration of extract required for 50% inhibition of $ABTS^{\bullet+}$). Briefly, 50 µL of each sample was added to 250 µL of 7 mM ABTS solution. The absorbance was read after 20 min at 734 nm using a microplate spectrophotometer (BioTech™, Eon™, Richmond Scientific, Lancashire, Great Britain). Water was used as a blank, ABTS solution as the control, and ascorbic acid as the standard [12].

### 2.4. *Production of Starch/BrE-Based Films by Solvent Casting*

The starch/BrE-based films were produced by solvent casting following a previously described method [6], with the addition of the BrE at different proportions (5%, 10%, and 15% *w*/*w*, starch dry weight). Briefly, a water dispersion of freeze-dried potato starch (40 g/L) was prepared, and 12 g/L of glycerol was added as well as BrE in different proportions (2 g/L, 4 g/L, and 6 g/L. Then, each dispersion was gelatinized at 95 °C ± 0.1 for 30 min with constant stirring (ca. 200 rpm). The potato starch-based films without BrE (neat films) were used as a control.

### 2.5. *Characterization of the Starch/BrE-Based Films*

#### 2.5.1. Chromatic Properties

The chromatic properties of the films were assessed by tristimulus colorimetry (CIELab). The CIELab coordinates $L^*$ (luminosity), $b^*$ (yellow/blue), and $a^*$ (red/green) were determined using a CR-400 Chroma Meter. The total color difference ($\Delta E$) was also assessed [6].

#### 2.5.2. Thickness and Mechanical Properties

The mechanical properties of the films were determined using a texture analyzer (model TA. Hdi, Stable Micro Systems, Surrey, UK). Each film formulation was cut (12 strips for each film sample, each one a size of 1 cm × 9 cm), and the thickness was measured in 3 different strip points with a digital micrometer (ca. 0.001 mm accuracy; Mitutoyo Corporation, Kanagawa, Japan). During the uniaxial tensile tests till film failure, the tensile strength (MPa), elongation at break (%), and Young's modulus (Mpa) values were determined from the stress–strain curves provided by Exponent software [6].

#### 2.5.3. Wettability

The wettability of each film was determined at room temperature (20 °C ± 2 °C) through static water contact angles (WCAs) using a tensiometer (Attention Theta by Biolin Scientific, Madrid, Spain) fitted with an automatic image capture system (One Attension). Each formulation was cut in strips (1 cm × 6 cm) and tested in triplicate for both film surfaces: one exposed to air (film upper surface) and one in contact with a plexiglass plate (film's lower surface). Briefly, 3 µL of ultrapure water were dispensed on the film strips' surfaces, and the corresponding WCAs (at least 6 droplet images along each strip) were measured using One Attention software following the Young–Laplace method [6].

#### 2.5.4. Water Solubility

To determine the films' solubility in water, squares (4 $cm^2$) of each film formulation were cut, weighed, and immersed into distilled water containing sodium azide (0.02% *v*/*v*) at room temperature (20 °C ± 2 °C) at 80 rpm for 14 days. The non-solubilized films were dried (105 °C, overnight), cooled down to room temperature, and weighed. The solubility of each sample was evaluated as a weight loss percentage, in triplicate [6]. The possible migration of the BrE components to the water during the solubility assays was evaluated by UV–Vis (200 to 700 nm), in triplicate, at the end of the assay.

#### 2.5.5. Water Vapor Transmission Rate

The water vapor transmission rate (WVTR) of the films (28 mm diameter per sample) was assessed, in triplicate, using test dishes and a humidity chamber (53% relative humidity) maintained at 23 ± 2 °C and with ca. 160 m/min of air velocity [6].

#### 2.5.6. UV-Protective and Antioxidant Activity

The UV-protective capacity of the films was monitored through a UV–visible spectrophotometer (Shimadzu UV 1280) from 200 to 500 nm. For each spectrum, a representative average of 3 scans was obtained using a data pitch of 0.5 nm, a bandwidth of 2.0 nm, and a scanning speed of 100 nm/min [19].

The antioxidant activity of the films was determined by an adaptation of the ABTS method, as already described [19]. Briefly, each film square (4 $cm^2$) was placed in 1.5 mL $ABTS^{\bullet+}$ solution (7 mM ABTS in 2.45 mM potassium persulfate) and left to react under dark conditions (20 °C ± 2 °C, 80 rpm). The absorbance at 734 nm was measured for 4 h. An $ABTS^{\bullet+}$ solution without film was used as a blank. The antioxidant activity was determined, in triplicate, as the percentage of $ABTS^{\bullet+}$ inhibition.

### *2.6. Sliced Cheese Package with Starch/15% BrE-Based Films*

A commercial sliced cheese (Norvegia® Original, semi-hard matured cheese, 27% fat, valid until 22 November 2022, storage 0–4 °C, L6 17.27, TINE SA, Norway) was packed using the starch/15% BrE-based films. A total of 4 cheese slices were sealed with the films (31 cm × 21 cm) using a sealing machine (Packer®, Packer Poly Sealer, King's Lynn, Norfolk, UK). The packed samples were stored at 4 °C at 80% relative humidity for 14 days. A commercially available plastic of 90 µm polyamide/polyethylene (70 µm PA/20 µm PE, Allfo Vakuumverpackungen, Waltenhofen, Germany) was used as a control. After 7 and 14 days, the cheese samples were evaluated in terms of weight loss, the color CIELab coordinates $L^*$, $a^*$, and $b^*$ were determined using a CR-400 Chroma Meter (Konica Minolta, Inc, NJ, USA), as well as the texture (texture analyzer model TA. Hdi, Stable Micro Systems equipped with a 6 mm cylinder stainless probe for puncture tests [20]). The volatile profile of the cheese samples was analyzed by headspace-gas chromatography-mass spectrometry (HS-GC-MS) [21], where ethyl heptanoate (~98% purity, Fluka, Steinheim, Germany) was used as the internal standard, and the results are expressed as µg ethyl heptanoate equivalents (eq.)/g of the sample. The results are reported as the average of three independent replicates per each sampling moment.

### *2.7. Statistical Analysis*

The results were statistically evaluated using the Student's $t$-test with a significance level of 95% and $p < 0.05$, using the "$t$-test" tool in Excel 2016. Moreover, for multiple comparison analysis, one-way ANOVA with 95% probability level was used using GraphPad Prism version 8 for Windows (trial version GraphPad software, San Diego, CA, USA).

## 3. Results and Discussion

The defatted potato chip sample used in this work corresponded to 58% $w/w$ of the potato chips frying residue supplied by the industry (Figure S1) [6]. From this, the brownish hot water-soluble extract (BrE) corresponded to 57.2% $w/w$ concerning the defatted residue (Table 1). The BrE was fractionated by ultrafiltration (UF) and characterized in terms of carbohydrates, protein, lipids, and phenolic content. Moreover, the presence of melanoidins in the UF fractions was estimated, as well as their antioxidant activity.

**Table 1.** Yield and chemical composition of the BrE and its UF fractions.

| Sample | Yield * | Ara | Gal | Glc | Total Sugars | Total Protein | Total Phenolics | Total EFA | $K_{mix,405\ nm}$ | MBI |
|---|---|---|---|---|---|---|---|---|---|---|
| | % (*w*/*w*) | mol% | | | % (*w*/*w*) | | | | (L/g/cm) | |
| BrE | 57.2 | 0.95 | 5.20 | 93.9 | 66.9 | 5.70 | 0.40 | 0.10 | n.d. | n.d. |
| >100 kDa | 49.3 | 1.19 | 6.06 | 92.7 | 89.4 | 7.60 | 0.90 | 0.11 | 0.27 | n.d. |
| 50–100 kDa | 3.30 | 2.62 | 12.02 | 85.4 | 29.4 | 17.4 | 7.20 | 0.16 | 0.69 | 1.48 |
| <5 kDa | 39.9 | 4.50 | 12.43 | 83.1 | 37.1 | 16.4 | 1.10 | 0.10 | 0.14 | 0.31 |

Ara—arabinose; Gal—galactose; Glc—glucose; EFA—esterified fatty acid; n.d.—not determined. * The BrE yield was expressed as the % of the defatted brownish frying residue, and the yield of the different molecular weight fractions is expressed as the % of the BrE.

### 3.1. Characterization of the Brownish Water-Soluble Extract (BrE)

The BrE was mainly composed of carbohydrates (66.9% *w*/*w*), with glucose, the structural unit of potato starch, as the major constituent (93.9 mol%). Galactose and arabinose, probably from arabinogalactan proteins [22], were also determined in this brownish extract, but in smaller amounts (5.20 mol% and 0.95 mol%, respectively). The BrE also presented protein (5.70% *w*/*w*) and a small amount of phenolics and lipids (ca. 0.40% and ca. 0.10% *w*/*w*, respectively).

The BrE was fractionated by ultrafiltration using five cut-off membranes, followed by freeze-drying. A total of six fractions, corresponding to the >100 kDa, 50–100 kDa, 30–50 kDa, 10–30 kDa, and 5–10 kDa retentates and to the <5 kDa permeate, were obtained. Since the yield of the fractions 30–50 kDa, 10–30 kDa, and 5–10 kDa only represented a total of ca. 3% *w*/*w* of BrE (Figure S3), only the fractions with a higher yield (>100 kDa, 50–100 kDa, and <5 kDa) were chemically characterized (Table 1).

The major fraction (>100 kDa, 49.3% of BrE weight) was very rich in carbohydrates (89.4%), mainly glucose from starch, with a small portion of galactose and arabinose (Table 1). Protein accounted for 7.6%. The second largest fraction (<5 kDa, 39.9% of BrE weight) consisted of 37.1% sugars, with 10.0% accounting for free sugars and 16.4% protein. The presence of minerals can explain the remaining material [23]. Although not relevant in terms of yield (3.30% of BrE weight), the 50–100 kDa fraction was also different from the others due to the fact of its intense brown color, measured at 405 nm ($K_{mix,405\ nm}$ = 0.69 L/g/cm). Carbohydrates accounted for 29.4%, the protein was 17.4%, and the phenolics were 7.20%. Esterified fatty acids accounted for only 0.16%, mainly constituted by oleic (53.7%) and palmitic (36.3%) acid residues derived from spent frying oils due to the fact of their similar proportions [6]. These values allowed to estimate the amount of unknown material of 53.20%, corresponding to a melanoidin browning index (MBI) of 1.48. This value is not as high as the MBI value of 2.44 determined for the 30–100 kDa fraction of instant coffee [17]. In instant coffee, the >100 kDa fraction presented the highest MBI (3.31), not verified for BrE fraction > 100 kDa. This can be explained by the high content of starch in the potato present in the >100 kDa fraction and the high content of protein in the 50–100 kDa fraction, which should have promoted the Maillard reactions with carbohydrates [10,12]. In addition, a significantly higher antioxidant activity was determined in 50–100 kDa fraction ($IC_{50}$ = 0.28 mg/mL, Figure 2), which can have a contribution of melanoidins [11], although the most abundant fractions also present a relevant antioxidant activity. This value was consistent with the $IC_{50}$ value determined for chicory melanoidins ($IC_{50}$ = 0.28 mg/mL) and higher than the one for instant coffee melanoidins ($IC_{50}$ = 0.08 mg/mL) [12]. Due to the low yield of the 50–100 kDa fraction, the unfractionated BrE was used for the preparation of potato starch-based films.

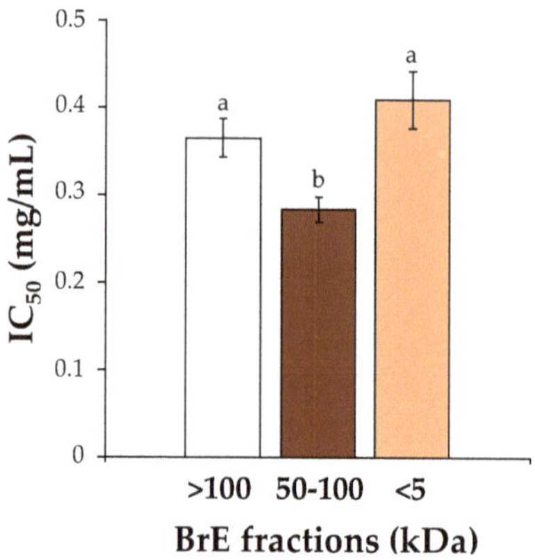


**Figure 2.** Antioxidant activity, expressed as $IC_{50}$ values, determined for each BrE fraction. Different colors represent the chromatic tone of each fraction. The values that are significantly different ($p < 0.05$) are represented by different lowercase letters.

## 3.2. *Characterization of the Potato Starch/BrE-Based Films*

### 3.2.1. Chromatic Properties

The neat films exhibited $L^*$, $a^*$, and $b^*$ values of 93.1, 1.73, and −1.20, respectively (Table S1), following the trend reported for films produced from starch recovered from the potato chip industry [6]. The $L^*$ values significantly decreased from 93.1 (neat films) to a minimum of 89.7 for films containing 15% BrE (Figure 3, Table S1), indicating that the incorporation of the extract decreased the films' lightness. In addition, the $b^*$ (yellow–blue) values significantly increased from −1.20 (neat films) to 1.39 (5% BrE), 4.00 (10% BrE), and 6.31 (15% BrE), thus revealing that the incorporation of the BrE conferred a yellowish coloration to the neat films, this being even more evident for higher BrE dosages. The calculated total color difference (ΔE) values corroborated these chromatic changes, varying from 3.21, 5.94, and 8.26 in films with 5%, 10, and 15% BrE, respectively (Table S1). Thus, the high the BrE amount, the higher the color difference of the materials. These differences can be related to the presence of melanoidins in the 50–100 kDa fraction (Table 1). This trend observed for BrE was found in potato starch-based films containing coffee silverskin, a coffee roasting industry byproduct [19]. Nevertheless, as revealed in the images, the typical transparency of the neat films was preserved for all of the BrE dosages (Figure 3). Moreover, similar $a^*$ values were observed among all of the films (Table S1), meaning that the red–green coordinate was kept constant.

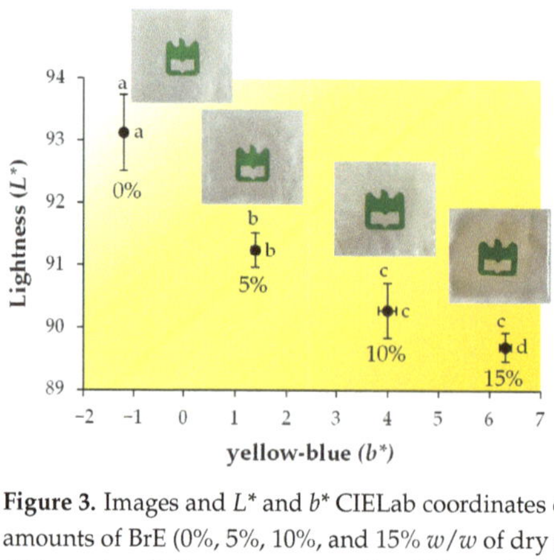


**Figure 3.** Images and $L^*$ and $b^*$ CIELab coordinates of potato starch-based films containing different amounts of BrE (0%, 5%, 10%, and 15% $w/w$ of dry starch weight). The values that are significantly different ($p < 0.05$) are represented by different lowercase letters.

3.2.2. Thickness and Mechanical Properties

The neat films had a thickness of ca. 58 μm and a low elongation at break (ca. 6%) and tensile strength (ca. 11 MPa), although showing a high Young's modulus (ca. 1078 MPa) (Figure 4). These values are consistent with the brittle and rigid character of potato starch-based films [5,6,19].

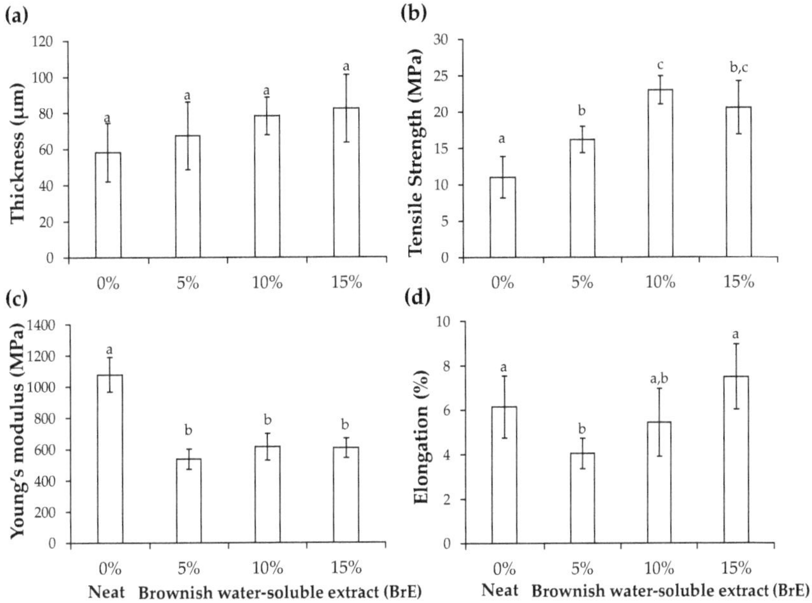


**Figure 4.** Thickness (**a**) and mechanical properties (tensile strength (**b**); Young's modulus (**c**); elongation at break (**d**)) of potato starch-based films containing different BrE amounts (0%, 5%, 10%, and 15% *w*/*w* of dry starch weight). The values that are significantly different ($p < 0.05$) are represented by different lowercase letters.

Compared to the neat films, although no significant differences were observed in thickness (Figure 4a), the incorporation of BrE influenced the mechanical performance of the films. The BrE significantly increased the films' tensile strength from ca. 11 MPa (neat films) to ca. 16 MPa, 23 MPa, and 21 MPa in films with 5%, 10%, and 15% of BrE, respectively (Figure 4b), improving the traction resistance of the materials. Additionally, BrE significantly decreased the Young's modulus from ca. 1078 MPa (neat film) to values lower than ca. 615 MPa (Figure 4c), revealing that their elasticity increased with the incorporation of BrE, independently of its concentration. This effect can be related to the presence of BrE compounds that decrease the cohesion forces within the starch molecules, promoting a discontinuity in the polymeric matrix, similar to the trend observed with starch-based films containing spent frying oil-derived fatty acids [6]. In addition, the incorporation of 5% BrE decreased the elongation at break of the films from ca. 6% (neat film) to ca. 4%, thus decreasing their plasticity. However, for 15% BrE, the elongation at break increased to ca. 8% (Figure 4d). This increase may be justified by the presence of a higher amount of low molecular weight compounds (39.9% of <5 kDa materials, Figure S3), such as phenolic compounds and small peptides (Table 1), which may promote a plasticizing effect due to the decrease in the hydrogen bonding within the starch network. This effect follows the trend already observed for starch-based films containing phenolic-rich extracts of potato peels [7] and hibiscus [24], as well as films with coffee silverskin [19]. Hence, according to the dosage, the BrE increased starch-based films' resistance to break and elasticity, promoting their flexibility.

#### 3.2.3. Wettability

The neat films exhibited water contact angles (WCAs) below ca. 45° at both surfaces, which highlights their hydrophilic character and homogeneous composition, following the trend that has been reported for potato starch-based materials [6,19]. When compared to the neat films, the incorporation of 5, 10, and 15% BrE increased the WCAs on both film surfaces, from ca. 45° (neat films) to ca. 58°, 59°, and 62° at the upper side and from ca. 38° (neat films) to ca. 68°, 58°, and 66° at the lower side, respectively. However, these values were neither statistically different among the three BrE dosages used nor among the two films' surfaces (Figure 5). Therefore, BrE allowed for increasing the water tolerance of potato starch/BrE-based films. Since BrE was mainly composed of carbohydrates 66.9% *w*/*w* (Table 1), their hydroxyl groups may justify all of the WCA values below the hydrophobicity benchmark of 90°. Moreover, the presence of the 50–100 kDa brown compounds and phenolics (Table 1), may have contributed to the increase in the WCA of films containing BrE, similar to the trend observed for starch-based films containing coffee silverskin [19] and potato peel phenolic-rich extract [7]. In addition, it can be suggested that BrE can contain frying oil-derived fatty acids incorporated into the melanoidins structure, similar to what was observed for other organic acids in coffee melanoidins [10], although in insufficient amounts to promote hydrophobicity to the materials.

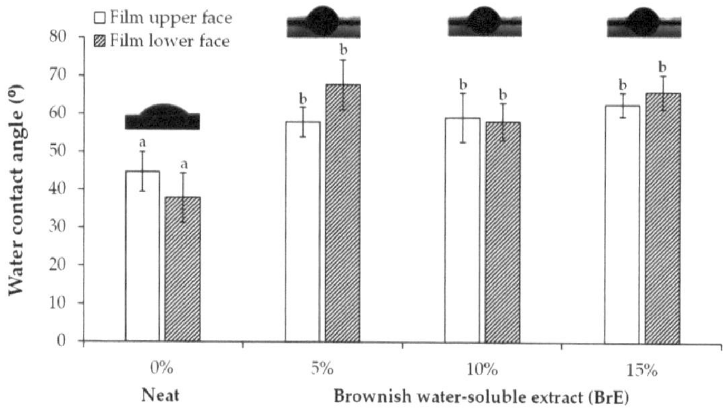


**Figure 5.** Water contact angle (WCAs) of the upper and "lower" surfaces of potato starch-based films containing different amounts of BrE (0%, 5%, 10%, and 15% *w*/*w* of dry starch weight). The values that are significantly different ($p < 0.05$) are represented by different lowercase letters.

#### 3.2.4. Solubility and Water Vapor Permeability

When immersed in distilled water, the neat films (0% BrE, Figure 6a) lost their native weight in ca. 21%, which is in line with the literature on potato starch-based films [6]. This weight loss has been related to the high-water solubility of glycerol [19], the plasticizing agent. However, as a peak at 260 nm was observed in the aqueous medium of the neat films (Table S2), their weight loss can also have the contribution of leaching of the protein present in the recovered potato starch [25]. With the incorporation of BrE, the films' weight loss significantly decreased for all BrE dosages studied (Figure 6a). These results revealed that BrE hinders the film hydration and leaching of water-soluble compounds. When compared to the aqueous medium of the neat films, a significant increase in the absorbance values at 260 nm was determined for all formulations, although not significant among the tested BrE dosages (Table S2). This indicates that the films' weight loss can be related to the solubilization of the BrE components, possibly protein, as observed for the neat films. This suggests that the crescent incorporation of other BrE components, such as melanoidins (Table 1), avoids the glycerol bonding with water molecules and its release from the starch matrix, the efficiency of which varies linearly with the BrE concentration.

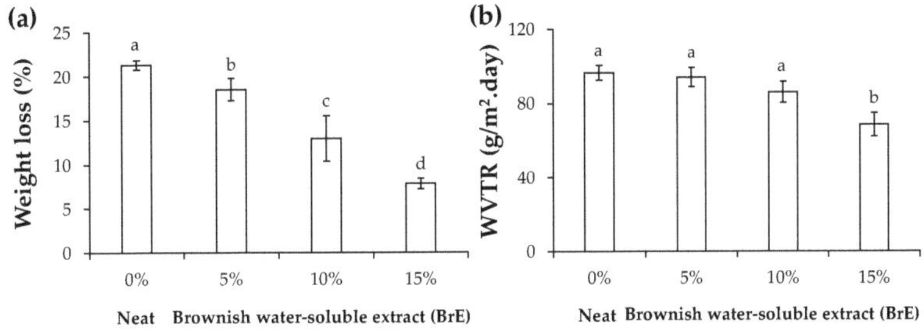


**Figure 6.** Weight loss (**a**) and water vapor transmission rate (WVTR) (**b**) of starch-based films containing different amounts of BrE (0%, 5%, 10%, and 15% *w*/*w* of dry starch weight). The values that are significantly different ($p < 0.05$) are represented by different lowercase letters.

The WVTR of the potato starch-based films was ca. 97 $g/m^2 \cdot day$, which is in line with the literature [19], as amylose and amylopectin from potato starch and glycerol are responsible for the passage of water vapor through the films. The incorporation of 15% BrE allowed to significantly decrease the WVTR values of the neat films to 68 $g/m^2 \cdot day$. Although the other BrE dosages also tended to decrease the WVTR of the films, the values were not statistically different when compared to the neat films (Figure 6b). The highest WVTR decrease observed in films containing 15% BrE reinforces the hypothesis that BrE melanoidins (50–100 kDa, Table 1) had hydrophobic features promoted by the incorporation of frying oil-derived fatty acids, although in an insufficient amount to promote the full water vapor barrier of the materials. A similar trend was observed in starch-based films containing spent frying oil-derived fatty acids [6].

3.2.5. UV-Protective and Antioxidant Properties

From the UV–visible absorption range of 200 nm to 500 nm, two major bands at 250 nm and 340 nm were determined for films containing BrE, while no absorption was observed for the neat films (Figure 7a). This allowed for the conclusion that the BrE brown-colored components conferred a UV-protective ability, especially to the films containing 15% BrE, which had a higher melanoidins proportion ($K_{mix,\,405\,nm}$ = 0.69 L/g/cm and MBI = 1.48, Table 1). In addition, the major band was registered in the UV-C region at 250 nm, the intensity of which significantly increased with the BrE dosage. A less intense band in the UV-A region, at 340 nm, was also determined, being similar for all tested BrE dosages. The protection against UV radiation may be explained by the presence of melanoidins and phenolic compounds in the 50–100 kDa BrE fraction (Table 1). Phenolic compounds from coffee silverskin have been related to the UV absorption properties of potato starch-based films [19]. This property is relevant for the application of starch/BrE-based films in food packaging for preventing food oxidation promoted by UV radiation [26].

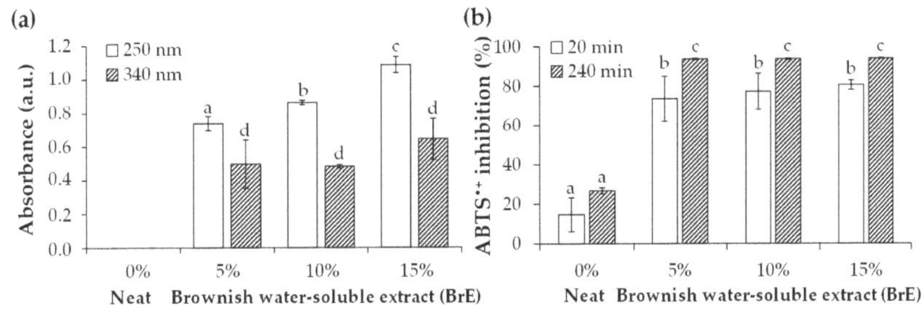


**Figure 7.** UV–visible absorption data at 250 nm and 340 nm (**a**) and antioxidant activity (by $ABTS^{\bullet+}$ inhibition) (**b**) of potato starch-based films containing different amounts of BrE (0%, 5%, 10%, and 15% *w*/*w* of dry starch weight). The values that are significantly different ($p < 0.05$) are represented by different lowercase letters.

The neat films exhibited a percentage of ABTS$^{\bullet+}$ inhibition of ca. 15% and 26% after 20 min and 240 min (Figure 7b). This minimal ABTS$^{\bullet+}$ inhibition can be explained by the presence of the phenolic compounds in the starch used to produce the films [7,27]. However, this inhibition greatly increased for films containing BrE, being similar among all tested concentrations: after 20 min and 240 min of incubation, ca. 77% and 93% of ABTS$^{\bullet+}$ inhibition, respectively, was achieved (Figure 7b), which may be related to the melanoidins and phenolic compounds present in the BrE (Table 1).

### 3.3. *Effect of Potato Starch/BrE-Based Films on Packed Sliced Cheese Quality*

The films containing 15% BrE were chosen to pack sliced cheese due to the fact of their higher water tolerance, gas barrier, UV-protective, and antioxidant properties (Figures 6 and 7), while the thickness, mechanical properties, and wettability were similar to the other BrE dosages studied (Figures 4 and 5).

After packaging the sliced cheese with the starch/15% BrE-based films (day 0), the appearance of wrinkles in the films promoted by direct contact with the cheese was observed. This was not observed when commercial PA/PE plastic was used (Table 2). The absorption of water molecules by the starch/15% BrE-based films was corroborated by the significant cheese weight loss: cheese packed in PA/PE material resulted in only 0.12% weight loss, while a 6.52% weight loss was observed for the cheese packed in the material containing BrE after 7 days of storage (Table 2). Accordingly, an increase in the weight of 3% was observed in the starch/15% BrE-based films. No significant differences were observed from day 7 until the end of storage, suggesting that an equilibrium in the weight lost and gain was achieved. During the storage period (14 days), the starch/15% BrE-based films wrinkles were always present. In addition, the films did not lose their adhesiveness to the cheese surface. Similar behavior was observed in starch-based films with potato peel phenolic extracts when in contact with smoked fish fillets [7].

**Table 2.** Images of sliced cheese packed in potato starch/15% BrE-based films and mean values of cheese lightness ($L^*$), red–green ($a^*$), and yellow–blue ($b^*$) and total color difference ($\Delta E$) chromatic parameters over 14 days of storage (4 °C, 80% relative humidity). The PA/PE-based materials are used as a reference.

| | | Sliced Cheese Samples | | | | |
|---|---|---|---|---|---|---|
| | | Weight Loss | Chromatic Parameters | | | |
| Image | Time (Days) | (%) | $L^*$ | $a^*$ | $b^*$ | ΔE |
| | **PA/PE-based materials** | | | | | |
|  | 0 | - | 81.44 ± 0.57 $^{a}$ | −2.86 ± 0.18 $^{a}$ | 31.04 ± 0.53 $^{a}$ | - |
|  | 7 | 0.12 ± 0.04 $^{a}$ | 79.68 ± 0.50 $^{b}$ | −3.57 ± 0.04 $^{b}$ | 30.17 ± 0.31 $^{a,b}$ | 2.09 |
|  | 14 | 0.02 ± 0.03 $^{a}$ | 81.74 ± 0.91 $^{a,b}$ | −2.44 ± 0.13 $^{c}$ | 31.12 ± 0.49 $^{a}$ | 0.52 |
| | **Starch/15% BrE-based materials** | | | | | |
|  | 0 | - | 81.44 ± 0.57 $^{a}$ | −2.86 ± 0.18 $^{a}$ | 31.04 ± 0.53 $^{a}$ | - |
|  | 7 | 6.52 ± 2.16 $^{b}$ | 76.10 ± 2.01 $^{c}$ | −2.96 ± 0.21 $^{a}$ | 30.84 ± 0.49 $^{a}$ | 5.34 |
|  | 14 | 6.54 ± 1.58 $^{b}$ | 77.12 ± 1.99 $^{b,c}$ | −1.81 ± 0.19 $^{d}$ | 34.17 ± 1.59 $^{c}$ | 5.44 |

In each column, the different lowercase letters represent values that are significantly different ($p < 0.05$).

Regarding the chromatic parameters (Table 2), on the day of packaging (day 0), the cheese samples had $L^*$, $a^*$, and $b^*$ values of 81.44, −2.86, and 31.04, respectively. Over time, significant changes were observed in the cheese packed in PA/PE, mainly regarding the lightness ($L^*$) and redness ($a^*$) coordinates (79.68 and −3.57 (day 7) and 81.74 and −2.44 (day 14), respectively), which was related to the cheese darkening. When compared to the cheese packed in the PA/PE plastics, a significant decrease in the $L^*$ values of the cheese packed with starch/15% BrE-based films was observed over the 14 days of storage (76.10 and 77.12 at day 7 and 14, respectively), indicating that the lightness of the cheese tended to decrease. In addition, a significant increase in the $a^*$ and $b^*$ values of the cheese packed with starch/15% BrE at day 14 of storage (−1.81 and 34.17, respectively) was observed, reflecting a significant decrease in the green color and an increase in the cheese yellow color. The total color difference ($\Delta E$) values confirm the chromatic changes observed, varying from 0.52 and 2.09 in cheese packed in PA/PE and from 5.34 and 5.44 in cheese in starch/15% BrE-based films. All these chromatic changes resulted in the yellow-colored darkening of the cheese (Figure S4). The removal of water from the cheese's surface decreases the water activity, which is a factor reported to promote Maillard reactions between lactose and lysine residues of protein when their concentration increases [28]. Nevertheless, the migration of the brown-colored compounds from the starch/BrE packaging to the cheese surface cannot be excluded, possibly due to the fact of the hydrophobic interactions with the cheese fatty acids. However, at day 7, the $a^*$ and $b^*$ coordinates of the cheese packed with the materials with BrE did not differ from the cheese at day 0; thus, the original color of the cheese was maintained over this period, evidence not observed for the cheese packed in PA/PE (Table 2).

The cheese texture properties are expressed in terms of the maximum in puncture (N) (i.e., the maximum force required to produce deformation) and puncture work (N.mm) (i.e., the mechanical work needed to reach the rupture point) (Figure 8). These parameters allowed to evaluate the surface hardness (cheese surface in direct contact with the packaging) and the inner hardness (cheese internal layers/paste not in contact with the packaging material) of the cheese samples, respectively [20]. Compared to the cheese before packaging (day 0) or even to cheese packed in the PA/PE materials, no significant changes were observed in terms of the maximum puncture and puncture work values for the cheese packed with starch/15% BrE-based films (Figure 8). These results revealed that both the surface and inner hardness of the cheese slices were preserved over the 14 days of storage.

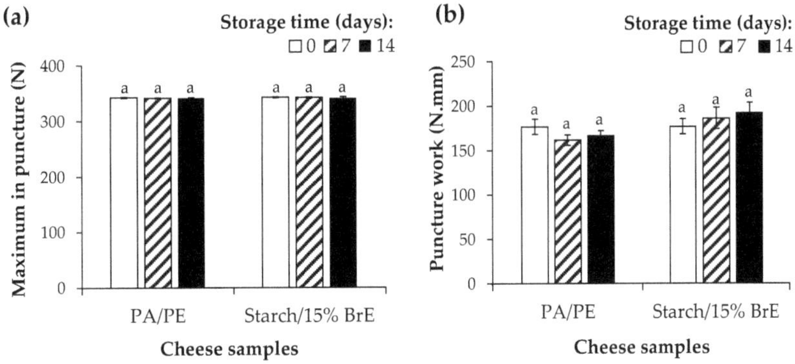


**Figure 8.** Maximum in puncture (**a**) and puncture work (**b**) of the top slices of cheese packed with starch/15% BrE-based films over 14 days of storage (4 °C, 80% relative humidity). The PA/PE-based materials are used as a reference. The values that are not significantly different ($p > 0.05$) are represented by the same lowercase letter.

Considering the volatile compounds present in the cheese samples packed with PA/PE and starch/15% BrE materials, a total of 31 compounds were determined during the 14 days of storage, belonging to the chemical families of aldehydes (7), ketones (10), alcohols (12), and acids (2) (Table S3). These chemical families were already referred in semi-hard cheese packed in poly(lactic acid) material (PLA) [29].

At day 0, a total of 46.72, 27.26, and 7.49 μg ethyl heptanoate eq./g of cheese of ketones, alcohols, and aldehydes, respectively, was determined in the cheese. During storage, a significant increase in the content of aldehydes was observed in cheese packed in PA/PE, mainly due to the higher increase in 3-methyl-butanal, nonanal, benzaldehyde, and benzeneacetaldehyde. A similar trend was observed for ketones, namely, 2-butanone, 2,3-butanedione, and 3-hydroxybutanone. These compounds, if present in amounts higher than their odor threshold (i.e., a lower concentration of a compound in the vapory phase that can be detected by smell [30]), can contribute with acrid, waxy, almond-like, burnt sugars, and grassy notes in the case of aldehydes and with acetone-like, butter, and woody aromas in the case of ketones [31]. Furthermore, the presence of acetic and butanoic acids was determined, which can contribute with vinegar and rancid odors [30,31]. On the other hand, a significant decrease was observed, mainly on day 7 of storage, in the amount and number of alcohols. This can be related to their oxidation, forming aldehydes, ketones, and acids, contributing to their increase in the cheese during storage (Figure 9, Table S3).

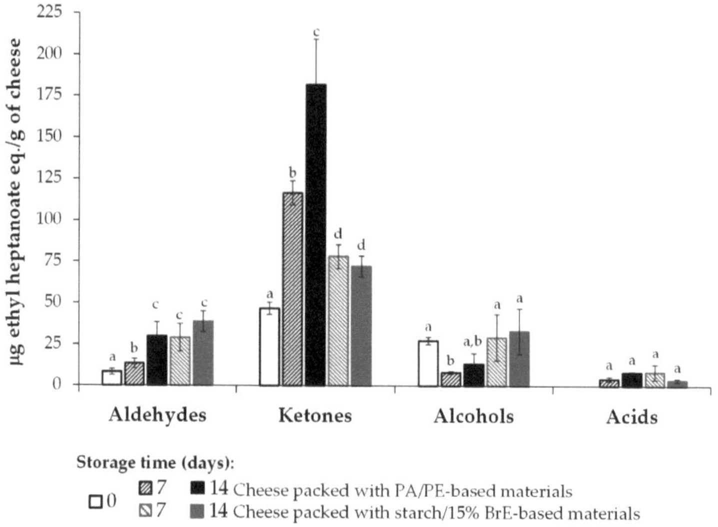


**Figure 9.** Volatile profile of the top slices of semi-hard cheese packed with starch/15% BrE-based films, stored for 14 days (4 °C, 80% relative humidity). The PA/PE-based materials are used as a reference. In each chemical family, the values that are significantly different ($p < 0.05$) are represented by different lowercase letters.

In the cheese packed in starch/15% BrE-based films, when compared to PA/PE, no significant differences were observed in the content of aldehydes and acids. However, a significantly lower concentration of ketones and a higher concentration of alcohols in the cheese were observed when stored in starch/15% BrE-based films. When compared to day 0, an increase in the aldehydes, ketones, and acids was observed, while the content of the alcohols remained similar (Figure 9, Table S3). These results indicate that BrE films can act as antioxidants, mitigating the oxidation of cheese components. In addition, over the entire storage time, no visual growth in the molds and yeasts was observed in the cheese samples, neither when the bioplastics nor when the plastics were used (images in Table 2 and Figure S4), accordingly with the absence of volatiles emitted by microbial growth.

## 4. Conclusions

The starch/15% BrE-based films revealed to be suitable for packaging semi-hard sliced cheese. They were able to preserve both the inner and surface cheese hardness over the 14 days of storage at 4 °C and 80% of relative humidity, similar to the cheese packed in PA/PE plastic. Nevertheless, a cheese weight loss and a yellow-colored darkening were observed, related to the migration of water from the cheese to the film. A similar volatile profile was also determined in the cheese packed in both materials, although the BrE constituents allowed for decreasing the content of ketones, a chemical family of compounds responsible for oxidation aromas.

The BrE, mainly constituted by carbohydrates, protein, and small amounts of phenolics, esterified fatty acids, and melanoidins, enhanced the traction resistance and promoted a plasticizing effect on the starch-based films, allowing for the development of flexible films. Additionally, the BrE decreased the films' surface hydrophilicity and water solubility, thus enhancing their tolerance to water conditions. Moreover, the potato starch/BrE-based films possessed UV-protective and antioxidant properties, which were even more relevant for the higher BrE amount (15%, *w*/*w* in relation to starch dry weight). Therefore, the BrE was revealed to contain molecules, including melanoidins, with interest to tune the mechanical and physicochemical performances of the potato starch-based films, allowing for the development of active bio-based materials suitable for use in active food packaging applications, including cheese coating.

**Supplementary Materials:** The following supporting information can be downloaded at: https://www.mdpi.com/article/10.3390/foods12061167/s1, Figure S1. Scheme of the experimental process to recover the brownish water-soluble extract (BrE) from potato chip industry frying residues; Figure S2. Graphic representation of the absorbance at 420, 405, 325, and 280 nm for the 5 ultrafiltration retentates (a) and for the 100 kDa (b); 50 kDa (c); 30 kDa (d); 10 kDa (e); 5 kDa (f) permeates. A total of 3 cycles were performed for each cut-off membrane (each cycle corresponding to the reduction until 50 mL of the 500 mL of sample introduced into the UF reservoir); Figure S3. Yield (in percentage) of each BrE ultrafiltration (UF) fraction. The different colors represent the chromatic tone of the fraction; Figure S4. Images of sliced cheese packed with potato starch/15% BrE-based films, highlighting the browning of the cheese surface after 14 days of storage; Table S1. Images and mean values of lightness ($L^*$), red–green ($a^*$), yellow–blue ($b^*$), and total color difference ($\Delta E$) of potato starch-based films containing different amounts of BrE (0%, 5%, 10%, and 15% *w*/*w* of dry starch weight); Table S2. UV–Vis absorbance values at 260 nm of the aqueous medium of potato starch-based films containing different amounts of BrE (0%, 5%, 10%, and 15% *w*/*w* of dry starch weight), after 14 days in contact with water; Table S3. Volatile compounds, organized by chemical family, determined in the top slices of the cheese packed with potato starch/15% BrE-based materials, stored for 14 days (4 °C, 80% relative humidity). The PA/PE-based materials are used as a reference.

**Author Contributions:** Conceptualization, S.P.; methodology, S.P., A.M.P., J.L., M.S.G. and E.M.W.; validation, S.P. and M.K.P.; formal analysis, S.P., A.M.P., M.S.G. and E.M.W.; investigation, S.P.; resources, M.A.C., M.K.P. and I.G.; data curation, S.P. and I.G.; writing—original draft preparation, S.P. and A.M.P.; writing—review and editing, S.P., M.R.D., F.M.N., M.K.P., M.S.G., I.G. and M.A.C.; visualization, S.P. and M.K.P.; supervision, S.P., I.G., M.K.P. and M.A.C. All authors have read and agreed to the published version of the manuscript.

**Funding:** This work received financial support from PT national funds (FCT/MCTES) for the financial support of LAQV-REQUIMTE (UIDB/50006/2020 + UIDP/50006/2020), CICECO-Aveiro Institute of Materials (UIDB/50011/2020 + UIDP/50011/2020 + LA/P/0006/2020), and CESAM (UIDB/50017/2020 + UIDP/50017/2020 + LA/P/0094/2020) research units and CQ-VR at UTAD Vila Real (UIDP/00616/2020) through PT national funds and, where applicable, cofinanced by the FEDER, within the PT2020 Partnership Agreement, Compete 2020. FCT is also thanked for the postdoc grant SFRH/BPD/117213/2016 (SP) and for the Individual Call to Scientific Employment Stimulus (IG, CEECIND/00430/2017). Thanks are also given to the financial support of Norwegian Levy on Agricultural Products through the Strategic Program FutureFoodControl (Project no 314743).

**Data Availability Statement:** The data are contained within the article or the Supplementary Materials.

**Acknowledgments:** The authors thank POTATOPLASTIC project (POCI-01-0247-FEDER-017938), financed by FEDER through POCI, to "Isolago—Indústria de Plásticos, S.A.", the project leader, and to "A Saloinha, Lda." for providing the potato byproducts. CIRCUL-A-BILITY cost action (CA19124) is thanked for funding the Short-Term Scientific Mission (STSM) of J.L., S.P. and I.G., S.P. also thanks the financial support of Norte Portugal Regional Operational Programme (NORTE 2020), under the PT 2020 Partnership Agreement, through the European Regional Development Fund (ERDF) and FSE.

**Conflicts of Interest:** The authors declare no conflict of interest.

## References

1. FAOSTAT Statistics. Potatoes Production in the World. 2020. Available online: http://faostat.fao.org/site/567/DesktopDefault.aspx?PageID=567#ancor (accessed on 19 November 2022).
2. Pedreschi, F. Frying of Potatoes: Physical, Chemical, and Microstructural Changes. *Dry. Technol.* **2012**, *30*, 707–725. [CrossRef]
3. Basuny, A.M.M.; Mostafa, D.M.M.; Shaker, A.M. Relationship Between Chemical Composition and Sensory Evaluation of Potato Chips Made from Six Potato Varieties with Emphasis on the Quality of Fried Sunflower Oil. *World J. Dairy Food Sci.* **2009**, *4*, 193–200.
4. Freitas, L.; Barbosa, J.; da Costa, A.; Bezerra, F.; Pinto, R.; de Carvalho, R. From waste to sustainable industry: How can agro-industrial wastes help in the development of new products? *Resour. Conserv. Recy.* **2021**, *169*, 105466. [CrossRef]
5. Goncalves, I.; Lopes, J.; Barra, A.; Hernandez, D.; Nunes, C.; Kapusniak, K.; Kapusniak, J.; Evtyugin, D.; da Silva, J.; Ferreira, P.; et al. Tailoring the surface properties and flexibility of starch-based films using oil and waxes recovered from potato chips byproducts. *Int. J. Biol. Macromol.* **2020**, *163*, 251–259. [CrossRef]
6. Petronilho, S.; Oliveira, A.; Domingues, M.R.; Nunes, F.M.; Coimbra, M.A.; Gonçalves, I. Hydrophobic Starch-Based Films Using Potato Washing Slurries and Spent Frying Oil. *Foods* **2021**, *10*, 2897. [CrossRef]
7. Lopes, J.; Gonçalves, I.; Nunes, C.; Teixeira, B.; Mendes, R.; Ferreira, P.; Coimbra, M.A. Potato peel phenolics as additives for developing active starch-based films with potential to pack smoked fish fillets. *Food Pack. Shelf Life* **2021**, *28*, 100644. [CrossRef]
8. Hosseinzadeh-Bandbafha, H.; Nizami, A.; Kalogirou, S.; Gupta, V.; Park, Y.; Fallahi, A.; Sulaiman, A.; Ranjbari, M.; Rahnama, H.; Aghbashlo, M.; et al. Environmental life cycle assessment of biodiesel production from waste cooking oil: A systematic review. *Renew. Sust. Energy Rev.* **2022**, *161*, 112411. [CrossRef]
9. Mesias, M.; Delgado-Andrade, C. Melanoidins as a potential functional food ingredient. *Curr. Opin. Food Sci.* **2017**, *14*, 37–42. [CrossRef]
10. Moreira, A.; Coimbra, M.; Nunes, F.; Passos, C.; Santos, S.; Silvestre, A.; Silva, A.; Rangel, M.; Domingues, M. Chlorogenic acid-arabinose hybrid domains in coffee melanoidins: Evidences from a model system. *Food Chem.* **2015**, *185*, 135–144. [CrossRef]
11. Moreira, A.; Nunes, F.; Domingues, M.; Coimbra, M. Coffee melanoidins: Structures, mechanisms of formation and potential health impacts. *Food Funct.* **2012**, *3*, 903–915. [CrossRef]
12. Petronilho, S.; Navega, J.; Pereira, C.; Almeida, A.; Siopa, J.; Nunes, F.M.; Coimbra, M.A.; Passos, C.P. Bioactive Properties of Instant Chicory Melanoidins and Their Relevance as Health Promoting Food Ingredients. *Foods* **2023**, *12*, 134. [CrossRef] [PubMed]
13. Passos, C.; Costa, R.; Ferreira, S.; Lopes, G.; Cruz, M.; Coimbra, M. Role of Coffee Caffeine and Chlorogenic Acids Adsorption to Polysaccharides with Impact on Brew Immunomodulation Effects. *Foods* **2021**, *10*, 378. [CrossRef] [PubMed]
14. Lopes, G.R.; Passos, C.P.; Petronilho, S.; Rodrigues, C.; Teixeira, J.A.; Coimbra, M.A. Carbohydrates as targeting compounds to produce infusions resembling espresso coffee brews using quality by design approach. *Food Chem.* **2021**, *344*, 128613. [CrossRef] [PubMed]
15. Vangelder, W. Conversion factor from nitrogen to protein for potato-tuber protein. *Potato Res.* **1981**, *24*, 423–425. [CrossRef]
16. Bekedam, E.; Schols, H.; Van Boekel, M.; Smit, G. High molecular weight melanoidins from coffee brew. *J. Agric. Food Chem.* **2006**, *54*, 7658–7666. [CrossRef]
17. Passos, C.; Cepeda, M.; Ferreira, S.; Nunes, F.; Evtuguin, D.; Madureira, P.; Vilanova, M.; Coimbra, M. Influence of molecular weight on in vitro immunostimulatory properties of instant coffee. *Food Chem.* **2014**, *161*, 60–66. [CrossRef]
18. Re, R.; Pellegrini, N.; Proteggente, A.; Pannala, A.; Yang, M.; Rice-Evans, C. Antioxidant activity applying an improved ABTS radical cation decolorization assay. *Free Radic. Biol. Med.* **1999**, *26*, 1231–1237. [CrossRef]
19. Oliveira, G.; Goncalves, I.; Barra, A.; Nunes, C.; Ferreira, P.; Coimbra, M. Coffee silverskin and starch-rich potato washing slurries as raw materials for elastic, antioxidant, and UV-protective biobased films. *Food Res. Int.* **2020**, *138*, 109733. [CrossRef]
20. Benedito, J.; Simal, S.; Clemente, G.; Mulet, A. Manchego cheese texture evaluation by ultrasonics and surface probes. *Int. Dairy J.* **2006**, *16*, 431–438. [CrossRef]
21. Sarfraz, J.; Hansen, A.; Haugen, J.; Le, T.; Nilsen, J.; Skaret, J.; Huynh, T.; Pettersen, M. Biodegradable Active Packaging as an Alternative to Conventional Packaging: A Case Study with Chicken Fillets. *Foods* **2021**, *10*, 1126. [CrossRef]
22. Coimbra, M.; Waldron, K.; Delgadillo, I.; Selvendran, R. Effect of processing on cell wall polysaccharides of green table olives. *J. Agric. Food Chem.* **1996**, *44*, 2394–2401. [CrossRef]
23. Subramanian, N.; White, P.; Broadley, M.; Ramsay, G. The three-dimensional distribution of minerals in potato tubers. *Ann. Bot.* **2011**, *107*, 681–691. [CrossRef]

24. Cruz-Galvez, A.; Castro-Rosas, J.; Rodriguez-Marin, M.; Cadena-Ramirez, A.; Tellez-Jurado, A.; Tovar-Jimenez, X.; Chavez-Urbiola, E.; Abreu-Corona, A.; Gomez-Aldapa, C. Antimicrobial activity and physicochemical characterization of a potato starch-based film containing acetonic and methanolic extracts of Hibiscus sabdariffa for use in sausage. *LWT-Food Sci. Technol.* **2018**, *93*, 300–305. [CrossRef]
25. Goncalves, I.; Hernandez, D.; Cruz, C.; Lopes, J.; Barra, A.; Nunes, C.; da Silva, J.; Ferreira, P.; Coimbra, M. Relevance of genipin networking on rheological, physical, and mechanical properties of starch-based formulations. *Carbohydr. Polym.* **2021**, *254*, 117236. [CrossRef]
26. Luchese, C.; Garrido, T.; Spada, J.; Tessaro, I.; de la Caba, K. Development and characterization of cassava starch films incorporated with blueberry pomace. *Int. J. Biol. Macromol.* **2018**, *106*, 834–839. [CrossRef] [PubMed]
27. Albishi, T.; John, J.; Al-Khalifa, A.; Shahidi, F. Phenolic content and antioxidant activities of selected potato varieties and their processing by-products. *J. Funct. Foods* **2013**, *5*, 590–600. [CrossRef]
28. Belitz, H.D.; Grosch, W.; Schieberle, P. *Food Chemistry*, 4th ed.; Springer: Berlin/Heidelberg, Germany, 2009.
29. Holm, V.; Mortensen, G.; Risbo, J. Quality changes in semi-hard cheese packaged in a poly(lactic acid) material. *Food Chem.* **2006**, *97*, 401–410. [CrossRef]
30. Petronilho, S.; Lopez, R.; Ferreira, V.; Coimbra, M.A.; Rocha, S.M. Revealing the Usefulness of Aroma Networks to Explain Wine Aroma Properties: A Case Study of Portuguese Wines. *Molecules* **2020**, *25*, 272–289. [CrossRef]
31. Bagler, G. FlavorDB Search: A Resource for Exploring Flavor Molecules. Center for Computational Biology, Indraprastha Institute of Information Technology Delhi (IIIT Delhi), Academic Block, 2017. Available online: http://cosylab.iiitd.edu.in/flavordb (accessed on 6 December 2022).

*foods*

MDPI

*Article*

# Preparation and Characterization of Sodium Caseinate-Coated Papers Based on Glycerol and Sorbitol Contents for Packaging Application

**Dowan Kim * and Jihyeon Hwang**

Department of Food Processing and Distribution, College of Life Science, Gangneung-Wonju National University, 7 Jukheon-gil, Gangneung 25457, Gangwon-do, Republic of Korea
* Correspondence: dowankim@gwnu.ac.kr; Tel.: +82-33-640-2337

**Abstract:** Bio-based packaging materials are promising alternatives to petroleum-based plastics. Paper-based packaging materials are candidates for improving food sustainability; however, paper has poor gas and water vapor barrier properties. In this study, entirely bio-based sodium caseinate (CasNa)-coated papers with two plasticizers, glycerol (GY) and sorbitol (SO), were prepared. The morphological and chemical structure, burst strength, tensile strength, elongation at break, air permeability, surface properties, and thermal stability of the pristine CasNa-, CasNa/GY-, and CasNa/SO-coated papers were evaluated. The use of GY and SO strongly affected the tensile strength, elongation at break, and air barrier of the CasNa/GY- and CasNa/SO-coated paper. The air barrier and flexibility of the CasNa/GY-coated papers were higher than those of the CasNa/SO-coated papers. Compared to SO, GY better coated and penetrated the CasNa matrix, which positively affected the chemical and morphological structure of the coating layer and the interaction between the coating layer and paper. Overall, CasNa/GY was superior to the CasNa/SO coating. CasNa/GY-coated papers may be a good alternative for packaging materials in the food, medical, and electronic sectors, which would promote sustainability.

**Keywords:** paper-based packaging; sodium caseinate; glycerol; sorbitol; food packaging; sustainability

**Citation:** Kim, D.; Hwang, J. Preparation and Characterization of Sodium Caseinate-Coated Papers Based on Glycerol and Sorbitol Contents for Packaging Application. *Foods* **2023**, *12*, 940. https://doi.org/10.3390/foods12050940

Academic Editors: Sílvia Petronilho and Mario M. Martinez

Received: 12 January 2023
Revised: 20 February 2023
Accepted: 21 February 2023
Published: 22 February 2023



## 1. Introduction

Petroleum-based plastics such as polyethylene (PE), polypropylene (PP), and polyethylene terephthalate (PET) are made by linking hundreds of fossil oil-based monomers by chemical reactions [1]. They have excellent chemical stability, mechanical strength, transparency, and gas barrier properties, making petroleum-based plastics suitable for various packaging applications, such as food, medical devices, and electronic devices [1,2]. However, the global increase in the use of non-renewable petroleum plastics has negatively affected the environment. In response to environment issues, the European Union has published guidelines to ban single-use plastic products such as straws, food, and beverage containers [3].

Bio-based materials are promising alternatives to reduce the use of petroleum-based plastics for packaging [2,4–7]. Paper-based packaging materials are widely used in various foods, such as snacks, powdered foods, beverages, and frozen foods. Paper-based packaging materials have various advantages, such as ease of recyclability, degradability after use, and good mechanical properties [5,7]. However, paper has poor gas and water vapor barriers because of its porous and hydrophilic structure and lack of heat sealability [5]. Therefore, paper is usually coated and laminated with various plastics, such as PE and PP, to improve its properties. Unfortunately, plastic-coated and laminated paper lose their advantages as a sustainable material because of the difficulties in separating plastic from paper and their non-biodegradability [5,8].

Extensive research has been conducted on coated paper prepared using bio-based and renewable materials, such as zein, proteins, starch, nanocellulose, and polyvinyl alcohol, instead of plastics [5–9]. Coating is a process in which non-porous paper can be obtained by covering the porous structure and surface of cellulose fiber, improving its gas and water vapor barrier properties [8,9]. Sodium caseinate (CasNa) is a natural water-soluble polymer originating from the acid precipitation and neutralization of casein from milk and cheese production [2,4]. CasNa has relatively good film forming and coating abilities, making it an attractive candidate for preparing edible and biodegradable films. Furthermore, CasNa-based films exhibit good oxygen and aroma barrier properties due to intermolecular interactions, such as hydrogen bonding, which increase the intermolecular cohesion to form films [2,4–6]. However, these chemical and physical interactions cause CasNa films to shrink during the drying process at high temperatures and become brittle [6,10]. This hydrophilic structure negatively affects the physical properties of this film, making it a poor water barrier that restricts its various applications in food packaging [2,4–6].

To solve this problem, several researchers have reported the plasticizer effect using polyols such as GY, SO, polyethylene glycol (PEG), and stearic acid (SA). Plasticizers interrupt the intermolecular interactions of protein chains and induce the hydrogen bonds between protein chains and plasticizer, which improve the workability, flexibility, and brittleness of CasNa [5,11–13]. Colak et al., reported that GY-plasticized CasNa with non-brittleness was successfully prepared via solution casting and a brown extrusion process [11]. Rezvani et al., showed that hydrophobic SA strongly affects the drying process, mechanical strength, and water vapor permeability of CasNa films [12]. Siew et al., prepared CasNa/GY and CasNa/PEG edible films and concluded that the interaction and compatibility between the plasticizer and CasNa strongly influenced the viscosity of the coating solution and the physical properties of the film [13]. Khwaldia reported that the water vapor permeability of CasNa/paraffin wax-coated papers was less permeable to water vapor and weaker than that of only CasNa-coated paper [5]. Despite considerable research on the processing and physical properties of edible CasNa films depending on a plasticizer, only a few studies have reported the properties of CasNa-coated paper.

Therefore, the present study aimed to prepare CasNa/GY-coated and CasNa/SO-coated papers using the solution blending method. In addition, their chemical and morphological structures were evaluated and compared depending on the plasticizer content to determine their suitability for various packaging applications.

## 2. Materials and Methods

### 2.1. Materials

The base paper (bleached kraft pulp, grammage: 80 g/m$^2$, brightness 84%, and thickness: 95 μm) was purchased from OPI Co., Ltd. (Anseong, Republic of Korea). The casein sodium salt from bovine milk was purchased from Merck (Seoul, Republic of Korea). GY and SO were purchased from Duksan Pure Chemical Co., Ltd. (Ansan, Republic of Korea). Deionized water was used in all experiments.

### 2.2. Preparation of Pristine CasNa, CasNa/GY, and CasNa/SO Coating Solution and Their Coated Papers

Pristine CasNa-, CasNa/GY-, and CasNa/SO-coated papers were prepared according to the compositions listed in Table 1. First, 10 g of CasNa was dissolved in 90 mL of deionized water by stirring at 600 rpm at 60 °C for 1 h. GY and SO were added to CasNa with stirring for 30 min. The prepared CasNa-based coating solutions were stored in a refrigerator at 3 °C for 24 h. The bar coating and double-layer techniques were employed to coat the CasNa-based solutions on the base paper. To create a smooth surface on the paper with an irregular surface, each coating solution was first coated onto base paper using a mayer bar (bar number: 30), and the coated papers were dried using a blow dryer at 110 °C for 1 min. Finally, double-layered pristine CasNa-, CasNa/GY-, and CasNa/SO-coated papers were prepared by coating the single-layered paper, and all of the coated papers were stored at 40 °C for 24 h.

**Table 1.** Compositions, pH, viscosity, and grammage of pristine paper, pristine CasNa–, CasNa/GY– and CasNa/SO–coated papers. Lowercase ($^{a–c}$) and uppercase letters ($^{A–D}$) indicate statistically significant differences of CasNa/GY–based coatings and CasNa/SO–based coatings, respectively ($p < 0.05$).

| Sample Code | Compositions (g) | | | | pH | Viscosity (mPa·s) | Grammage (g/m$^2$) |
|---|---|---|---|---|---|---|---|
| | CasNa | D.I. Water | GY | SO | | | |
| Base Paper | - | - | - | - | - | - | 80.0 |
| Pristine CasNa | 10.0 | 90.0 | 0.0 | - | 6.6 | 58.1 $^{a,A}$ | 90.1 $^{a,A}$ |
| CasNa/GY (1.0:0.3) | 10.0 | 90.0 | 3.0 | - | 6.7 | 50.8 $^{b}$ | 90.9 $^{ab}$ |
| CasNa/GY (1.0:0.5) | 10.0 | 90.0 | 5.0 | - | 6.7 | 45.7 $^{c}$ | 91.3 $^{ab}$ |
| CasNa/GY (1.0:0.8) | 10.0 | 90.0 | 8.0 | - | 6.9 | 45.7 $^{c}$ | 93.1 $^{b}$ |
| CasNa/SO (1.0:0.3) | 10.0 | 90.0 | - | 3.0 | 6.8 | 57.5 $^{B}$ | 91.1 $^{AB}$ |
| CasNa/SO (1.0:0.5) | 10.0 | 90.0 | - | 5.0 | 6.9 | 56.1 $^{C}$ | 92.2 $^{B}$ |
| CasNa/SO (1.0:0.8) | 10.0 | 90.0 | - | 8.0 | 6.9 | 51.7 $^{D}$ | 94.3 $^{C}$ |

### 2.3. Chemical and Morphological Characterization

#### 2.3.1. pH, Viscosity, and Grammage

The pH values of the pristine CasNa, CasNa/GY, and CasNa/SO coating solutions were determined using a pH meter (F-17 LAQUA, Horiba Scientific Ltd., Kyoto, Japan). The viscosities of the solutions were determined using a Brookfield DVE viscometer (LV, AMETEK Inc., Berwyn, IL, USA) and a spindle number of 61. The spindle was rotated at 100 rpm. The grammage of the pristine CasNa-, CasNa/GY-, and CasNa/SO-coated papers was determined gravimetrically using a microbalance (Sartorius GmbH, Goettingen, Germany).

#### 2.3.2. Morphological Structure

The top surfaces of the base paper, pristine CasNa-, CasNa/GY-, and CasNa/SO-coated papers were observed using an Inspect F scanning electron microscope (SEM; FEI Co., Ltd., Hillsboro, OR, USA).

#### 2.3.3. Chemical Structure

The chemical structure of the GY, SO, pristine CasNa-, CasNa/GY-, and CasNa/SO-coated papers were recorded on a Spectrum 65 FTIR spectrometer (Perkin Elmer Co., Ltd., Waltham, MA, USA), with a spectral frequency range of 4000–500 cm$^{-1}$ and 16 scans using the attenuated total reflection (ATR) mode.

### 2.4. Mechanical Properties

The burst strengths of the base paper, pristine CasNa-, CasNa/GY-, and CasNa/SO-coated papers were evaluated using the L&W SE 180 burst test machine (Lorentzen and Wettre Ltd., Kista, Sweden), according to KS M ISO 2785 [14]. The tensile strength and elongation at the breaks of the base paper, pristine CasNa-, CasNa/GY-, and CasNa/SO-coated papers were measured using an Instron 3367 universal test machine (UTM, Norwood, MA, USA), based on the KS M ISO 1924-2 standard protocols [15]. The load cell was 250 N, and the tensile velocity was set at 20 mm/min. All of the specimens were prepared in a rectangular shape, with 0.015 m width and 0.18 m length.

### 2.5. Air Permeability

The air permeability of the base paper, pristine CasNa-, CasNa/GY-, and CasNa/SO-coated papers was measured using the Gurley air permeability tester (ABB Lorentzen & Wettre, Kista, Stockholm, Sweden) in accordance with KS M ISO 5636-3:2013. The permeation time was measured for 100 cc of air to permeate into the coated papers [16].

### 2.6. Surface Properties

The surface properties of the base paper, pristine CasNa-, CasNa/GY-, and CasNa/SO-coated papers were evaluated using Phoenix-MT contact angle measurements (SEO Co., Ltd., Suwon, Gyeonggi-do, Republic of Korea). The surface free energies ($\gamma^s$) of the base paper and coated papers were calculated via the Owens–Wendt model, following the adhesion theory between solid and liquid [17]. Water with polar ($\gamma^s = 72.8$ mJ/m$^2$, $\gamma^P = 50.3$ mJ/m$^2$, $\gamma^D = 22.5$ mJ/m$^2$) and diiodomethane with nonpolar ($\gamma^s = 50.8$ mJ/m$^2$, $\gamma^P = 50.4$ mJ/m$^2$, $\gamma^D = 0.4$ mJ/m$^2$) characteristics were used as liquids for the analysis.

### 2.7. Thermal Properties

To determine the thermal stability of the base paper, GY-, SO-, CasNa/GY-, and CasNa/SO-coated papers, thermogravimetric analysis (TGA) was performed with a 4000 TGA analyzer (PerkinElmer Co., Ltd., Waltham, MA, USA) at a heating rate of 10 °C/min and between 50 to 700 °C under $N_2$ atmosphere, with a flow rate of 90 mL/min.

### 2.8. Statistical Analysis

Analysis of variance (ANOVA) was performed using IBM SPSS version 25 (SPSS Inc., Chicago, IL, USA). Duncan's multiple range test was used to compare the differences between means. The significance level was set at $p < 0.05$.

## 3. Results and Discussion

### 3.1. CasNa Coating Solutions pH and Viscosity, and Coated Papers Grammage

To investigate the effects of GY and SO on the pH and viscosity of the CasNa coating solutions, pH and viscosity measurements were conducted, as described in Table 1. The pH of the pristine CasNa solution was 6.6. The CasNa/GY and Cas/SO coating solutions did not significantly change the pH depending on the GY and SO content, respectively. It is reported that CasNa easily formed a gel near the isoelectric point [18]. To maintain the flowability of the CasNa coating solutions in commercial coating processes, such as gravure and comma, their pH should be maintained above the isoelectric point. The viscosity of the pristine CasNa solution was 58.1 mPa·s. The viscosities of the CasNa/GY and CasNa/SO solutions decreased as the GY and SO contents increased. Siew et al. and Barreto et al., also reported a decrease in the viscosity of CasNa solution with increasing amounts of GY and SO [13,19]. This is explained by the decrease in the hydrodynamic volume of the components of CasNa solutions and the disruption of the protein-protein (e.g., hydrogen bond, ionic bond, and disulfide bond) and protein-solvent interactions by forming hydrogen bonds between CasNa and the plasticizer [13,19].

The control of the grammage in coated paper is crucial because it affects the physical properties, such as the gas barrier, and the mechanical properties. Therefore, grammages of pristine CasNa-, CasNa/GY-, and CasNa/SO-coated papers were prepared within the range of 90.1 to 94.3 g/m$^2$.

### 3.2. Coated Papers Morphologies

In the case of coated papers, a coating solution exists inside and on the top porous surface of the paper, which forms chemical and physical interactions between the coating solution and paper. The morphological structure of the coated papers affects their thermal and mechanical properties and air permeability. In this study, the effects of GY and SO on the morphological structure of CasNa/GY- and CasNa/SO-coated papers were observed using SEM, as shown in Figure 1. The base paper not coated with CasNa showed a porous structure composed of interwoven cellulose fiber (Figure 1a). The pores on the top surface of the pristine CasNa-coated paper were well covered by the pristine CasNa coating solution, although a portion of the fibers was still identified (Figure 1b). In addition, all CasNa/GY and CasNa/SO coating solutions penetrated the fibers well and covered the top surface of the paper without cracks (Figure 1c–h). This is related to the high compatibility and wettability between each coating solutions, with or without GY and SO, and the cellulose fiber, which may result in the penetration and adherence of all coating solutions to the base paper cellulose fibers [5,20]. GY and SO with low molecular weight diffuse into the CasNa matrix and facilitate mobility, in turn reducing the internal resistance of protein matrix. In addition, it is assumed that plasticized CasNa/GY and CasNa/SO coating layers are formed in the protein-plasticizer interactions with secondary force by disrupting the protein-protein interactions. These enhance the film forming process and flexibility by increasing the free volume and motion of the protein chains [21,22].

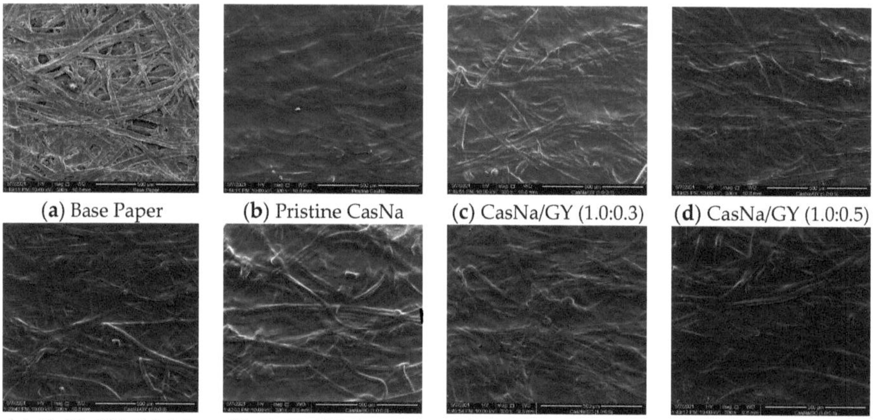

(**a**) Base Paper (**b**) Pristine CasNa (**c**) CasNa/GY (1.0:0.3) (**d**) CasNa/GY (1.0:0.5)
(**e**) CasNa/GY (1.0:0.8) (**f**) CasNa/SO (1.0:0.3) (**g**) CasNa/SO (1.0:0.5) (**h**) CasNa/SO (1.0:0.8)

**Figure 1.** SEM images of (**a**) base paper, (**b**) pristine CasNa-coated paper, (**c**–**e**) CasNa/GY–coated papers and (**f**–**h**) CasNa/SO–coated papers (×300 magnification).

As a result, all of the coated papers tested here would help improve the mechanical properties and air permeability owing to morphological changes. This result is in agreement with those reported by other researchers [5,20].

### *3.3. Coated Papers Chemical Properties*

The chemical interactions between the coating layer and cellulose are important to understand the changes in the chemical properties of coated papers depending on the plasticizer. The FTIR spectra of the CasNa/GY- and CasNa/SO-coated papers are shown in Figure 2. The CasNa/GY paper exhibited several characteristic peaks: stretching of -OH at 3293 $cm^{-1}$, stretching of CH at 2931 and 2831 $cm^{-1}$, bending of C-OH at 1415 $cm^{-1}$, stretching of CO at 1033 $cm^{-1}$, and vibration of C-C and C-O between 800–1150 $cm^{-1}$ (Figure 2a) [23,24]. According to the FTIR spectra obtained for the CasNa/GY-coated papers, characteristic peaks between 1100–918 $cm^{-1}$ appeared with the increasing GY content. In addition, the wavelengths of amide I, amide II, and amide III shifted from 1644 to 1633 $cm^{-1}$, 1515 to 1519 $cm^{-1}$, and 1236 to 1238 $cm^{-1}$, respectively. These results are consistent with previous research showing that the characteristic peak of amide II in the extruded casein sheet was shifted by the addition of GY [23,25]. The CasNa/SO paper also exhibited several characteristic peaks: stretching of -OH at 3295 $cm^{-1}$, stretching of CH at 2929 $cm^{-1}$, stretching of C-OH at 1047 and 1087 $cm^{-1}$, and bending of -OH at 1415 and 885 $cm^{-1}$ (Figure 2b) [26,27]. The characteristic peaks between 1130 and 960 $cm^{-1}$ of the CasNa/SO-coated paper gradually increased with the increase in the SO content. In addition, the wavelengths of amide I, amide II, and amide III shifted slightly from 1644 to 1633 $cm^{-1}$, 1515 to 1517, and 1236 to 1238 $cm^{-1}$, respectively. The pristine CasNa-coated paper also exhibited several characteristic peaks: stretching of -OH at 3276 $cm^{-1}$, stretching of amide I (C=O) at 1644 $cm^{-1}$, bending of amide II (NH) at 1515 $cm^{-1}$, and deformation and bending of amide III (NH and C=O) at 1236 $cm^{-1}$ (Figure 2) [10,23,28]. These results demonstrated that molecular interactions, such as hydrogen bonding between CasNa and plasticizers, and conformational rearrangement in CasNa occurred after GY or SO addition.

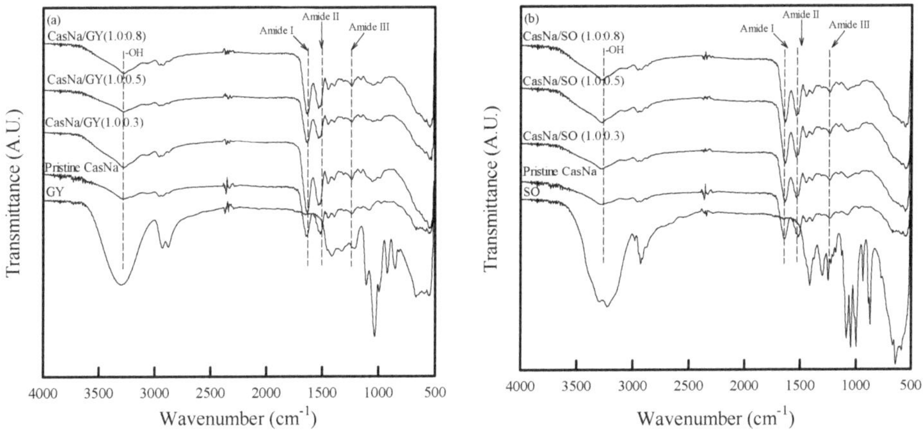

**Figure 2.** FTIR spectra of (**a**) CasNa/GY–coated papers and (**b**) CasNa/SO–coated papers.

### 3.4. Coated Papers Mechanical Properties

Packaging materials are exposed to various forces and pressures in the food and packaging processes, distribution, and storage environments. Therefore, the mechanical properties of CasNa/GY- and CasNa/SO-coated papers could correlate with the expected packaging material integrity under divergent forces and pressures in the packaging environment [10]. The mechanical properties of CasNa/GY- and CasNa/SO-coated papers were investigated as a function of the plasticizer content and compared to the base paper and pristine CasNa-coated paper, as shown in Figures 3 and 4.

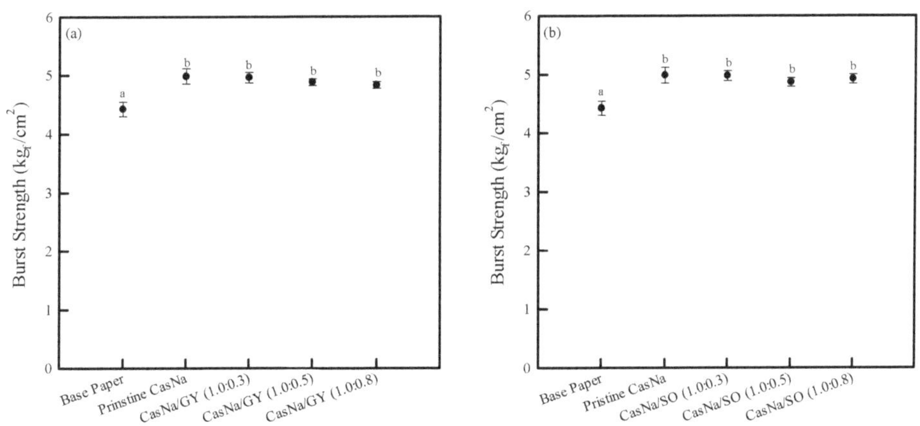

**Figure 3.** Burst strength of (**a**) CasNa/GY–coated papers and (**b**) CasNa/SO–coated papers. Lowercase letters (a,b) indicate statistically significant differences ($p < 0.05$).

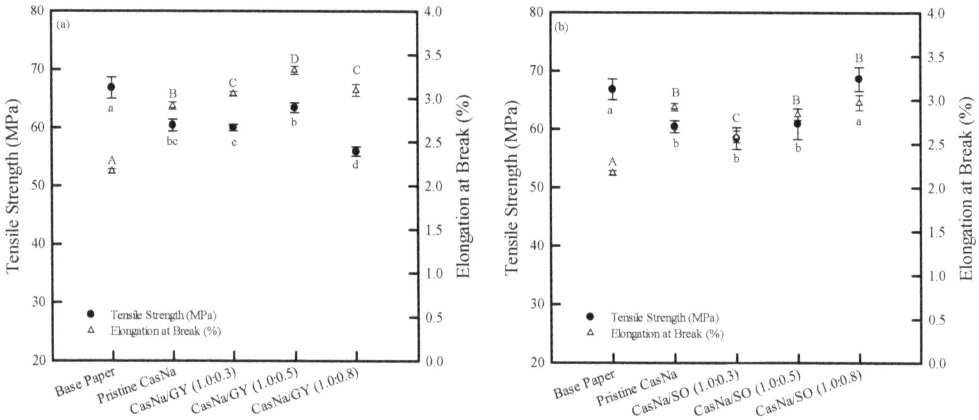


**Figure 4.** Tensile strength and elongation at break of (**a**) CasNa/GY–coated and (**b**) CasNa/SO–coated papers. Lowercase (a–d) and uppercase letters (A–D) indicate statistically significant differences of tensile strength and elongation at break, respectively ($p < 0.05$).

The base paper exhibited a burst strength of 4.4 $kg_f/cm^2$ (Figure 3). The pristine CasNa coating considerably improved the burst strength of the paper (5.0 $kg_f/cm^2$; Figure 3). This result is due to the penetration of CasNa into the pores and surface of the paper, which formed a smooth and continuous coating layer. However, the burst strength did not change significantly with the increase in the plasticizers GY and SO (Figure 3a,b).

The base paper showed a tensile strength of 67.2 $\pm$ 2.3 MPa and an elongation at break of 2.4 $\pm$ 0.08% (Figure 4). The pristine CasNa-coated paper showed a tensile strength of 66.8 $\pm$ 1.8 MPa and an elongation at break of 2.2 $\pm$ 0.02%. This was because the presence of CasNa on the surface and pores between the cellulose fibers of the base paper negatively affected the hydrogen bonding between the cellulose fibers. The increase in the GY content in CasNa decreased the tensile strength, while it increased the elongation at break (Figure 4a). This result is consistent with previous studies by Chevalier et al. and Belyamani et al. [23,29]. CasNa/GY-coated paper can be interpreted as comprising CasNa, GY, and cellulose. If the interfacial interaction between CasNa/GY and cellulose is stronger than that between the cellulose fibers, then the tensile strength and elongation at break should increase. However, the interfacial interaction between CasNa/GY and cellulose appeared to be weaker than that between the cellulose fibers. In addition, GY was previously shown to reduce the intermolecular interactions between the CasNa molecules and increase the free volume, which facilitates the mobility of CasNa [29,30]. Similarly, the elongation at break of the CasNa/SO-coated papers increased with the increasing SO content (Figure 4b). However, the tensile strength was increased after 50% SO was added to the CasNa. The slight difference between the CasNa/GY- and CasNa/SO-coated papers might be due to differences in their properties, such as their molecular weight and the state of the plasticizers. GY (92 g/mol) is a viscous liquid, and SO (182 g/mol) is a solid powder. Thus, it is difficult for SO to penetrate the CasNa matrix compared to GY. Therefore, one could assume that some SO penetrated CasNa, acting as a plasticizer, while the other SO portion formed the crystalline region in the CasNa matrix. Consequently, CasNa/SO (1.0:0.5)- and CasNa/SO (1.0:0.8)-coated papers exhibited more rigidity than the pristine CasNa- and CasNa/SO (1.0:0.3)-coated papers [30].

### 3.5. Coated Papers Air Permeability

Gas barrier properties are important for extending the shelf life and preserving the quality of various foods. In general, packaging materials must minimize the gases, including oxygen and water vapor, permeating the packaging [8,31]. Unfortunately, the paper's porous structure renders it a poor gas barrier, which is a drawback that limits its application. In this study, air barriers were evaluated to investigate the effect of pristine CasNa, CasNa/GY, and CasNa/SO coatings on paper (Figure 5). The air permeability of the base paper was 17.0 s/100 cc due to its porous structure. After coating the paper with pristine CasNa, it reached 869.7 s/100 cc because the porous structure of the paper was filled and covered with a pristine coating solution (Figure 5). Furthermore, by raising the plasticizer concentration in the CasNa solution, the air permeability of the CasNa/GY- and CasNa/SO-coated papers also increased from 869 to 40,666 s/100 cc and 869 to 33,103 s/100 cc, respectively (Figure 5a,b). GY is a relatively hydrophilic structure compared to SO. Therefore, it seemed that hydrogen bonds were strongly formed between CasNa and GY compared to between CasNa and SO, which made air permeation difficult in the CasNa/GY-coated papers. As shown in the FTIR analysis (Section 3.3), GY and SO may reduce the strong interfacial interaction between CasNa molecules, positively affecting the formation of smooth films. In addition, as seen in the SEM analysis (Section 3.2), the paper pores were filled with CasNa/GY and CasNa/SO coating solutions without brittleness, which improved the air barrier of the paper. Faust et al., reported that the oxygen barrier of glycerol-containing pea-protein isolated film was improved with an increase in the glycerol concentration [22]. They reported that adding glycerol improves the flexibility and workability of pea-protein isolated film, which positively affects the oxygen barrier property.

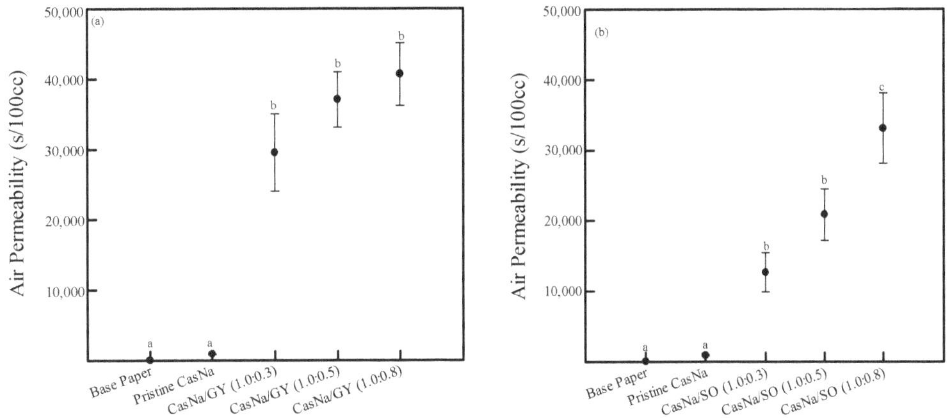


**Figure 5.** Air permeability of (**a**) CasNa/GY–coated papers and (**b**) CasNa/SO–coated papers. Lowercase letters (a–c) indicate statistically significant differences ($p < 0.05$).

### 3.6. Coated Papers Surface Properties

To investigate the surface properties of the base paper and the pristine CasNa-, CasNa/GY-, and CasNa/SO-coated papers, their static contact angles were measured using water and diiodomethane. The static contact angles of the liquids strongly depend on the changes in the chemical and morphological structures of the CasNa-based coating layer, which depends on the plasticizer. The water and diiodomethane static contact angles for the base paper were $73.9° \pm 0.62°$ and $31.0° \pm 0.35°$, respectively (Table 2). Notably, the pristine CasNa-coated paper showed decreases in the water and diiodomethane contact angles ($61.2 \pm 0.37°$ and $30.0 \pm 0.87°$; Table 2), suggesting that the hydrophilicity of the pristine CasNa-coated paper was enhanced. With the increasing GY or SO content in the CasNa, the polar surface energy changed from 11.2 to 12.5 mN and from 11.2 to 9.9 mN. The dispersive surface energy is associated with non-polarity and surface roughness. The dispersive surface energy increased slightly from 40.5 to 42.8 and from 40.5 to 43.8 mN with GY and SO, respectively (Table 2). Finally, the total surface free energies of the CasNa/GY- and CasNa/SO-coated papers were slightly higher than that of the CasNa-coated paper. Faust et al., reported that the addition of SO increased the surface energy of the polar parts in pea protein-isolated films [22]. However, in this study, the surface energy of the dispersive parts of the CasNa/GY- and CasNa/SO-coated papers increased more than the surface energy of the polar parts. This increase appears to be related to the morphology of the coated layers, as shown in Figure 1. Although the coating solutions covered the surface and pores of the fibers, the thickness of the coating layer remained uneven, owing to the morphological structure of the top surface of the base paper. To improve the physical properties of the coated paper, the coating solution must sufficiently penetrate the pores and completely spread on the paper surface [32]. In addition, the surface energy of the base paper was lower than those of the pristine CasNa-, CasNa/GY-, and CasNa/SO-coated papers (Table 2). This surface energy difference is related to the chemical structures of pristine CasNa, CasNa/GY, and CasNa/SO coating solutions, which may also affect the physical properties of the coated paper. Therefore, further studies are required on the wettability of the coated papers.

**Table 2.** Contact angles and surface energy of ([a]) CasNa/GY–coated papers and ([b]) CasNa/SO–coated papers. Lowercase ([a–c]) and uppercase letters ([A–D]) indicate statistically significant differences of CasNa/GY–coated papers and CasNa/SO–coated papers, respectively ($p < 0.05$).

| Sample Code | Static Contact Angle (°) | | Surface Energy (mN/m$^2$) | | |
|---|---|---|---|---|---|
| | Water | Diiodomethane | Polar | Dispersive | Total |
| Base Paper | $73.9 \pm 0.62$ [aA] | $31.0 \pm 0.35$ [a] | 5.1 | 41.3 | 46.4 |
| Pristine CasNa | $61.2 \pm 0.37$ [bB] | $30.0 \pm 0.87$ [bBC] | 11.2 | 40.5 | 51.7 |
| CasNa/GY (1.0:0.3) | $61.2 \pm 0.47$ [b] | $23.5 \pm 0.59$ [b] | 10.4 | 42.9 | 53.3 |
| CasNa/GY (1.0:0.5) | $60.5 \pm 0.30$ [b] | $23.5 \pm 0.47$ [b] | 10.7 | 42.9 | 53.6 |
| CasNa/GY (1.0:0.8) | $57.3 \pm 0.04$ [c] | $22.9 \pm 0.36$ [b] | 12.5 | 42.8 | 55.3 |
| CasNa/SO (1.0:0.3) | $61.9 \pm 0.22$ [BC] | $25.6 \pm 0.35$ [C] | 10.3 | 42.3 | 52.5 |
| CasNa/SO (1.0:0.5) | $62.5 \pm 0.09$ [C] | $22.9 \pm 0.54$ [B] | 9.6 | 43.3 | 52.9 |
| CasNa/SO (1.0:0.8) | $61.5 \pm 0.22$ [B] | $20.9 \pm 0.16$ [D] | 9.9 | 43.8 | 53.7 |

### 3.7. Coated Papers Thermal Stability

The base paper, pristine CasNa, CasNa/GY, and CasNa/SO-coated papers exhibited a two-step degradation process (Figure 6). The first degradation stage, at approximately 80 °C and 130 °C, was due to water evaporation. The second degradation stage, at around 230 °C and 260 °C, was due to the decomposition of the CasNa, plasticizer, and paper. The pristine CasNa-coated paper showed a 1 wt% higher decomposition temperature (70.2 °C) than that of the base paper (50.6 °C). After changing the GY content, the 1 wt% decomposition temperature increased from 70.2 °C to 77.6 °C in the CasNa/GY (1.0:0.5)-coated paper and then decreased with the rise in the GY concentration. Similarly, the pristine CasNa-coated paper also showed a higher 5 wt% decomposition temperature (278.6 °C) than the base paper (275.9 °C). With the increase in GY in CasNa, the 5 wt% decomposition temperature increased from 278.6 °C to 280.8 °C in CasNa/GY (1.0:0.3) and decreased to 266.9 °C in CasNa/GY (1.0:0.8). However, the 1 wt% and 5 wt% decomposition temperatures of the CasNa/SO-coated papers decreased with the increasing SO content.

Such a change in the thermal stability of the CasNa-, CasNa/GY-, and CasNa/SO-coated papers can be attributed to the changes originating from the interfacial interactions between CasNa, plasticizers, and paper [28,33]. Overall, the CasNa/GY- and CasNa/SO-coated papers showed lower decomposition temperatures than the pristine CasNa-coated paper. The decrease in the thermal stability of the coated papers can be explained by the effects of GY and SO on the protein-protein interaction in CasNa [28]. Tarique et al., reported that arrowroot starch films prepared using glycerol showed a three-step thermal degradation pattern. The first step in the thermal decomposition of the films was related to the loss of water from the films. The second step of thermal decomposition happened to vary in the range of approximately 125–290 °C, which was related to the loss of the glycerol compound [34]. Bareeto et al., reported that the initial decomposition temperature of milk proteins decreased, which was associated with sorbitol affecting the protein-protein interactions [33].

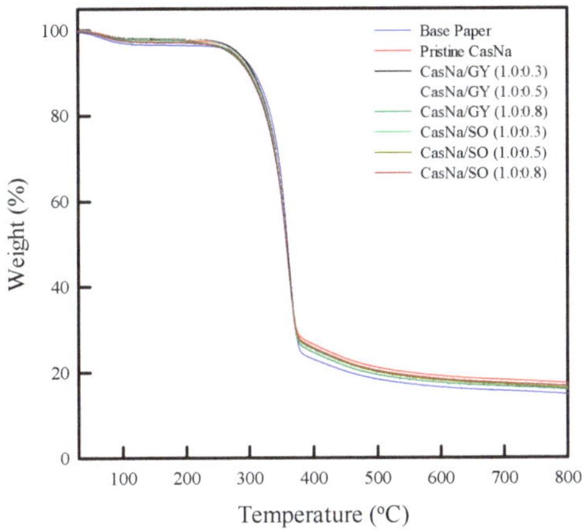


**Figure 6.** TGA curves of base paper, pristine CasNa–, CasNa/GY– and CasNa/SO–coated papers.

## 4. Conclusions

In this study, coating solutions of pristine CasNa, CasNa/GY, and CasNa/SO with different plasticizer concentrations were prepared via solution blending and coated on the surface of the base paper. SEM confirmed that the pristine CasNa-, CasNa/GY-, and CasNa/SO-coated papers were well prepared, without any cracks on the surface. Compared to the base paper and pristine CasNa-coated paper, the air barrier properties of CasNa/GY and CasNa/SO were significantly improved as the GY and SO contents increased. The CasNa/GY-coated papers exhibited higher flexibility and air barriers than CasNa/SO, although the CasNa/SO-coated papers showed lower decomposition temperatures than all the other tested papers. Furthermore, the increase in both the plasticizer contents in the CasNa solution led to a rise in the polar surface more than the dispersive surface energy. The slight difference between the CasNa/GY- and CasNa/SO-coated papers might be related to changes in the chemical and morphological structures originating from the different molecular weights and states of the plasticizers.

**Author Contributions:** Conceptualization, D.K.; Formal analysis, D.K. and J.H.; Funding acquisition, D.K.; Investigation, J.H.; Methodology, D.K. and J.H.; Project administration, D.K.; Software, J.H.; Supervision, D.K.; Validation, D.K.; Writing—original draft, D.K.; Writing—review and editing, D.K. and J.H. All authors have read and agreed to the published version of the manuscript.

**Funding:** This work was supported by the National Research Foundation of Korea (NRF) grant funded by the Korea government (MSIT) (No. NRF-2020R1G1A1101282). This work was also supported by Grant No.21162MFDS014 from the Ministry of Food and Drug Safety of Korea in 2021.

**Data Availability Statement:** The data presented in this study are available on request from the corresponding author.

**Acknowledgments:** This work was supported by the National Research Foundation of Korea (NRF) grant funded by the Korea government (MSIT). This work was also supported from the Ministry of Food and Drug Safety of Korea in 2021.

**Conflicts of Interest:** The authors declare no conflict of interest.

## References

1. Chamas, A.; Moon, H.; Zheng, J.; Qiu, Y.; Tabassum, T.; Jang, J.H.; Abu-Omar, M.; Scott, S.L.; Suh, S. Degradation rate of plastics in the environment. ACS Sustain. *Chem. Eng.* **2020**, *8*, 3494–3511. [CrossRef]

2. Qiu, Y.-T.; Wang, B.-J.; Weng, Y.-M. Preparation and characterization of genipin crosslinked and lysozyme incorporated antimicrobial sodium caseinate edible films. Food Packag. *Shelf Life* **2020**, *242*, 100601. [CrossRef]
3. Publications Office of the European Union. *Directive (EU) 2019/904 of the European Parliament and of the Council of 5 June 2019 on the Reduction of the Impact of Certain Plastic Products on the Environment (Text with EEA Relevance)*; Publications Office of the European Union: Luxembourg, 2019.
4. Valentino, M.; Volpe, S.; Giuseppe, F.A.D.; Cavella, S.; Torrieri, E. Active biopolymer coating based on sodium caseinate: Physical characterization and antioxidant activity. *Coatings* **2020**, *10*, 706. [CrossRef]
5. Khaldia, K. Water vapor barrier and mechanical properties of paper-sodium caseinate and paper-sodium caseinate-paraffin wax films. *J. Food Biochem.* **2010**, *34*, 998–1013. [CrossRef]
6. Belyamani, I.; Prochazka, F.; Assezat, G.; Debeaufort, F. Mechanical and barrier properties of extruded film made from sodium and calcium caseinates. Food Packag. *Shelf Life* **2014**, *2*, 65–72. [CrossRef]
7. Christophliemk, H.; Johansson, C.; Ullsten, H.; Järnström, L. Oxygen and water vapor transmission rates of starch-poly(vinyl alcohol) barrier coatings for flexible packaging paper. *Prog. Org. Coat.* **2017**, *113*, 218–224. [CrossRef]
8. Hamdani, S.S.; Li, Z.; Sirinakbumrung, N.; Rabnawaz, M. Zein and PVOH-based bilayer approach for plastic-free, repulpable and biodegradable oil- and water-resistant paper as a replacement for single-use plastics. *Ind. Eng. Chem. Res.* **2020**, *59*, 17856. [CrossRef]
9. Sharma, M.; Aguado, R.; Murtinho, D.; Valente, A.J.M.; Sousa, A.P.M.D.; Ferreira, J.T. A review on cationic starch and nanocellulose as paper coating components. *Int. J. Biol. Macromol.* **2020**, *162*, 578–598. [CrossRef]
10. Picchio, M.L.; Linck, G.L.; Monti, G.A.; Gugliotta, L.M.; Minari, R.J.; Igarzabal, C.I.A. Casein films crosslinked by tannic acid for food packaging applications. *Food Hydrocoll.* **2018**, *84*, 424–434. [CrossRef]
11. Colak, B.Y.; Gouance, F.; Degraeve, P.; Espuche, E.; Prochazka, F. Study of the influences of film processing conditions and glycerol amount on the water sorption and gas barrier properties of novel sodium caseinate films. *J. Membr. Sci.* **2015**, *478*, 1–11. [CrossRef]
12. Rezvani, E.; Schleining, G.; Sümen, G.; Taherian, A.R. Assessment of physical and mechanical properties of sodium caseinate and stearic acid based film-forming emulsions and edible films. *J. Food Eng.* **2013**, *116*, 598–605. [CrossRef]
13. Siew, D.C.W.; Heilmann, C.; Easteal, A.J.; Cooney, R.P. Solution and film properties of sodium caseinate/glycerol and sodium caseinate/polyethylene glycol edible coating systems. *J. Agr. Food Chem.* **1999**, *47*, 3432–3440. [CrossRef]
14. *KS M ISO 2758:2014*; Paper-Determination of Bursting Strength. Korea Agency for Technology and Standards: Maengdong-myeon, Republic of Korea, 2021.
15. *KS M ISO 1924-2:2008*; Paper and Board—Determination of Tensile Properties-Part 2: Constant Rate of Elongation Method (20 mm/min). Korea Agency for Technology and Standards: Maengdong-myeon, Republic of Korea, 2020.
16. *KS M ISO 5636-5:2013*; Paper and Board-Determination of Air Permeance (Medium Range)—Part 5: Gurley Method. Korea Agency for Technology and Standards: Maengdong-myeon, Republic of Korea, 2022.
17. Kim, D.; Lim, M.; Seo, J. Preparation of polypropylene/octadecane composite films and their use in the packaging of cherry tomatoes. *J. Appl. Polym. Sci.* **2016**, *133*, 44087. [CrossRef]
18. Kobori, T.; Matsumoto, A.; Sugiyama, S. pH-Dependent interaction between sodium caseinate and xanthan gum. *Carbohydr. Polym.* **2009**, *75*, 719–723. [CrossRef]
19. Barreto, P.L.M.; Roeder, J.; Crespo, J.S.; Maciel, G.R.; Terenzi, H.; Pires, A.T.N.; Soldi, V. Effect of concentration, temperature and plasticizer content on rheological properties of sodium caseinate and sodium caseinate/sorbitol solutions and glass transition of their films. *Food Chem.* **2003**, *82*, 425–431. [CrossRef]
20. Fabra, M.J.; Talens, P.; Chiralt, A. Microstructure and optical properties of sodium caseinate films containing oleic acid-beewax mixtures. *Food Hydrocoll.* **2009**, *23*, 676–683. [CrossRef]
21. Bocqué, M.; Voirin, C.; Lapinte, V.; Caillol, S.; Robin, J.-J. Petro-based and bio-based plasticizers: Chemical structures to plasticizing properties. *J. Polym Sci. Part A Polym. Chem.* **2016**, *54*, 11–33. [CrossRef]
22. Faust, S.; Foerster, J.; Lindner, M.; Schimid, M. Effect of glycerol and sorbitol on the mechanical and barrier properties of films based on pea protein isolate produced by high-moisture extrusion processing. *Polym. Eng. Sci.* **2022**, *62*, 95–102. [CrossRef]
23. Chevalier, E.; Assezat, G.; Prochazka, F.; Oulahal, N. Development and characterization of a novel edible extruded sheet based on different casein sources and influence of the glycerol concentration. *Food Hydrocoll.* **2018**, *75*, 182–191. [CrossRef]
24. Danish, M.; Mumtaz, M.W.; Fakhar, M.; Rashid, U. Response surface methodology based optimized purification of the residual glycerol from biodiesel production process. *Chiang Mai J. Sci.* **2017**, *44*, 1570–1582.
25. Arrieta, M.P.; Peltzer, M.A.; Garrigós, M.D.C.; Jiménez, A. Structure and mechanical properties of sodium and calcium caseinate edible active films with carvacrol. *J. Food Eng.* **2013**, *114*, 486–494. [CrossRef]
26. Castro, E.D.S.G.D.; Cassella, R.J. Direct determination of sorbitol and sodium glutamate by attenuated total reflectance Fourier transform infrared spectroscopy (ATR-FTIR) in the thermostabilizer employed in the production of yellow-fever vaccine. *Talanta* **2016**, *152*, 33–38. [CrossRef]
27. Pourfarzad, A.; Ahmadian, Z.; Habibi-Najafi, H. Interactions between polyols and wheat biopolymers in a bread model system fortified with inulin: A Fourier transform infrared study. *Heliyon* **2018**, *4*, e01017. [CrossRef]
28. Lin, H.-C.; Wang, B.-J.; Weng, Y.-M. Development and characterization of sodium caseinate edible films cross-linked with genipin. *LWT—Food Sci. Technol.* **2020**, *118*, 108813. [CrossRef]

29. Belyamani, I.; Prochazka, F.; Assezat, G. Production and characterization of sodium caseinate edible films made by brown-film extrusion. *J. Food Eng.* **2014**, *121*, 39–47. [CrossRef]
30. Kowalczyk, D.; Gustaw, W.; Świeca, M.; Baraniak, B. A study on the mechanical properties of pea protein isolate films. *J. Food Process. Preserv.* **2014**, *38*, 1726–1736. [CrossRef]
31. Kim, D.; Lee, Y.; Seo, J.; Han, H.; Khan, S.B. Preparation and properties of poly(urethane acrylate) (PUA) and tetrapod ZnO whisker (TZnO-W) composite films. *Polym. Int.* **2012**, *62*, 257–265. [CrossRef]
32. Gómez, N.; Quintana, E.; Villar, J.C. Effect of paper surface properties on coated paper wettability with different fountain solutions. *Bioresources* **2014**, *9*, 4226–4241. [CrossRef]
33. Barreto, P.L.M.; Pires, A.T.N.; Soldi, V. Thermal degradation of edible films based on milk proteins and gelatin in inert atmosphere. *Polym. Degrad. Stab.* **2003**, *79*, 147–152. [CrossRef]
34. Tarique, J.; Sapuan, S.M.; Khalina, A. Effect of glycerol plasticizer loading on the physical, mechanical, thermal, and barrier properties of arrowroot (*Maranta arundinacea*) starch biopolymers. *Sci. Rep.* **2021**, *11*, 13900. [CrossRef]

foods

MDPI

Article

# Edible Films Made of Dried Olive Leaf Extract and Chitosan: Characterization and Applications

**Michela Famiglietti [1], Alessandro Savastano [1], Rosa Gaglione [1], Angela Arciello [1], Daniele Naviglio [1] and Loredana Mariniello [1,2,*]**

[1] Department of Chemical Sciences, University of Naples "Federico II", 80126 Naples, Italy; michela.famiglietti@unina.it (M.F.); alessandro.savastano@icloud.com (A.S.); rosa.gaglione@unina.it (R.G.); angela.arciello@unina.it (A.A.); daniele.naviglio@unina.it (D.N.)

[2] Center for Studies on Bioinspired Agro-Environmental Technology (BAT), University of Naples "Federico II", 80126 Naples, Italy

* Correspondence: loredana.mariniello@unina.it

**Abstract:** Nowadays a possible strategy in food preservation consists of the use of active and functional packaging to improve safety and ensure a longer shelf life of food products. Many studies refer to chitosan-based films because of the already-known chitosan (CH) antibacterial and antifungal activity. In this work, we developed CH-based films containing Dried Olive Leaf Extract (DOLE) obtained by Naviglio extractor, with the aim to investigate the polyphenols yield and the antioxidant activity of this extract entrapped in CH-based-edible films. Olive tree cultivation produces a huge amount of byproducts that are usually simply burned. Phenolic compounds are already studied for their beneficial effects on human health. Some studies reported that phenols isolated from olive leaves have been shown to inhibit the growth of different strains of microorganisms. Thus, the antimicrobial effect of DOLE-containing films against bacterial strains (*Salmonella enterica* subsp. *enterica serovar Typhimurium* ATCC® 14028, *Salmonella enteritidis* RIVM 706, and *Enterococcus faecalis* ATCC® 29212) was tested in vitro. The DOLE component of the films is effective in inhibiting all the bacteria tested in a dose-dependent manner. Thus, it was demonstrated that these edible films can act as active bioplastics when used to wrap hamburgers in substitution for baking paper, which is normally used.

**Keywords:** olive byproducts; chitosan; active packaging; food shelf life; antioxidants; circular economy

**Citation:** Famiglietti, M.; Savastano, A.; Gaglione, R.; Arciello, A.; Naviglio, D.; Mariniello, L. Edible Films Made of Dried Olive Leaf Extract and Chitosan: Characterization and Applications. *Foods* **2022**, *11*, 2078. https://doi.org/10.3390/foods11142078

Academic Editors: Sílvia Petronilho and Mario M. Martinez

Received: 10 June 2022
Accepted: 11 July 2022
Published: 13 July 2022



## 1. Introduction

Chitosan (CH) is a natural linear copolymer derived from N-deacetylation of chitin and made up of units of N-acetyl-D-glucosamine and D-glucosamine (the latter exceeding 60%) [1], linked by beta-1,4 glycosidic linkage. Chitin is the second most abundant polymer on the earth, obtained from renewable sources, including exoskeletons of insects, arthropods such as shrimps, prawns, crabs, and cell walls of fungi [2].

Because of its biodegradability, biocompatibility, good film-forming capacity [3], and again because of its antibacterial and antifungal activity [4], CH is a very interesting polymer for applications in several industrial sectors.

The physical and chemical properties of CH depend on its molecular weight and degree of deacetylation. It presents one -$NH_2$ group and two -OH groups on each repeating unit and shows a rigid crystalline structure due to inter and intra-molecular hydrogen bonding [2]. These groups have a fundamental role in some important physicochemical characteristics including solubility and mechanical properties [3].

CH is a weak base insoluble in water and organic solvents, but amino groups make it soluble under acidic conditions below pH 6.0 [5]. It was reported that CH is effectively soluble in several organic and inorganic acids such as acetic, citric, formic, lactic, and hydrochloric [6]. The pKa value of CH strictly depends on deacetylation degree, ionic strength and the charge neutralization of -$NH_2$ groups [7], and the insoluble-soluble transition occurs usually between pH 6.5 and 6.0. Below this value, the amino groups get protonated, and CH becomes a soluble polymer [5].

The known antibacterial activity of CH can be influenced by different factors such as its average molecular weight [8] and the nature of organic acid [9] used for its dissolution. Besides the different structures of the cell wall of Gram-negative and Gram-positive bacteria, these are the reasons that can explain the diverse antimicrobial behavior of CH against both bacteria, as reported in some studies. Regardless, CH has been shown to act by different mechanisms: it can behave as a barrier that changes the cell permeability, preventing oxygen transference and thereby inhibiting the respiratory activity of bacteria [8]; it can absorb on the surface of the microbe by means of electrostatic interactions and create a thick polymeric membrane that prevents the entrance of necessary nutrients in the cells influencing physiological activities that are important for the growth of microbes [10].

Inside the cells, $NH_3^+$ groups of CH can interact with the negatively charged phosphoryl groups on the cell membrane, causing its deformation and distortion [11]; it can also diffuse until completely disrupting the cytoplasmatic membrane and provoking the leakage of electrolytes and the cell death [12]. Finally, CH can penetrate nuclei and bind DNA, inhibiting its replication ability, and can induce the synthesis of chitinase in fruit by increasing the gene expression of this enzyme that degrades the cell walls of microbes [13].

Since CH food safety (Generally Recognized as Safe (GRAS)) was improved by the US Food and Drug Administration in 2011 [14], we investigated the solubility of CH in citric acid (CA) to use this polymer as a matrix for delivering antioxidant compounds present in DOLE. Olive tree cultivation and the olive processing industry produce every year a huge amount of byproducts and residues [15]. Olive (*Olea europaea* L.) is one of the most cultivated ancient plants in the world, with about 9 million hectares occupied worldwide [16]. Typically, it grows in the tropical and temperate regions, and it represents a fundamental cultivar of the Mediterranean region, but it is also distributed in Western Asia, Arabian Peninsula, India, Northern Africa, and Iran [17]. The byproducts from the olive processing industry are mainly made up of leaves and branches obtained from the pruning of olive trees and harvesting and cleaning of olives [18]. The amount of leaves is about 25%, while branches represent around 75% of the dried weight from the total pruning residues, and they are generally used as animal feed or simply burned [19]. The chemical composition includes a lignocellulosic fraction, proteins, and extractives, with the latter reaching up to 45% [19]. Thus, various efforts toward better olive leaf waste management were carried out to extract energy or molecules. Among them, thermochemical treatments (combustion, gasification pyrolysis), biochemical treatments (anaerobic digestion and bioethanol production), drying methods (solar drying, microwave drying, freeze-drying), extraction methods, and condensation of active components are used to produce natural additives in different applicative fields [16]. Phenolic compounds, the main high-added value fraction of extractives present in olives leaves, are already studied for their several beneficial effects on human health. It was confirmed by different researchers the high antioxidant activity of these polyphenols and their effects as antihypertensive, cholesterol-lowering, cardioprotective, anti-inflammatory, antibacterial molecules and as coadjuvants in obesity treatment [20–23]. Origin, climatic conditions, moisture content, and proportions of the branches on the tree affect the chemical composition and the amount of polyphenolic compounds in olive leaves [16]. Polyphenols mainly present in olive leaves belong to five groups: oleuropeosides (oleuropein and verbascoside), flavones (luteolin-7-glucoside, apigenin-7-glucoside, diosmetin-7-glucoside, luteolin and diosmetin), flavonols (rutin), flavan-3-ols (catechin), and substituted phenols (tyrosol, hydroxytyrosol, vanillin, vanillic acid and caffeic acid) [24]. Among these the most abundant compound in olive leaves is oleuropein, which follows hydroxytyrosol, luteolin, apigenin, and verbascoside [25]. Hydroxytyrosol is more present in processed olive fruit and olive oil, while oleuropein is predominant in unprocessed olive fruit and leaves. This fact was explained due to the enzymatic reactions that occur during the maturation and processing of olives [24]. Oleuropein can prevent the formation of free radicals thanks to its ability to chelate metal ions such as $Cu^{2+}$ and $Fe^{3+}$ [26]. Moreover, both oleuropein and hydroxytyrosol have been shown to be scavengers of superoxide anions as well as of hypochlorous acid, a potent oxidant produced in vivo at the site of inflammation and a major component of chlorine-based bleaches that can often come into contact with food during manufacturing [27]. To date, different extraction methods were used to obtain high-added value from olive leaves, the traditional and most used is the solid-liquid extraction by maceration of the biomass in a solvent [28]. In the last years, new extraction methods are developed to reduce the amounts of solvent, time extraction, and sample preparation cost. Among them, supercritical fluid extraction and ultrasound-assisted extraction are more commonly used, followed by superheated liquid extraction, pressurized liquid extraction, and fractionation by solid-phase extraction. The selected extraction method, the nature of solvent, temperature, pH, and time of extraction are fundamental parameters that influence the yield and the content of polyphenolic compounds derived from olive leaves [29,30]. Obviously, the chemical characteristics of compounds make them less or more extractable by different methods: it was observed that compounds with less polarity as apigenin and luteolin are more easily obtainable by supercritical fluid extraction or pressurized liquid extraction, while maceration is more suitable to extract polar compounds such as oleuropein derivatives [31]. This work aimed to investigate the polyphenols yield of DOLE obtained by the Naviglio extractor and the antioxidant activity of this extract in chitosan-based-edible films. Our studies have also demonstrated the antioxidant activity of DOLE-containing films after in vitro oral digestion, which suggests the use of our films as a novel supplement for a diet rich in polyphenols. Moreover, we tested in vitro the antimicrobial effect of DOLE containing films against bacterial strains (*Salmonella typhimurium* ATCC® 14028, *Salmonella enteriditis* RIVM 706, and *Enterococcus faecalis* ATCC® 29212). In turn, the antimicrobial inhibitory effect of DOLE in meat samples over a time up to 20 days, in comparison to baking (parchment) paper normally used to store hamburgers, was demonstrated. We have also studied the mechanical properties of the films suggesting for DOLE a role as plasticizer. Thus, it can be assessed that multiple roles can be attributed to DOLE, nowadays a waste from olive trees pruning that can become a high-added-value product according to the principles of the circular economy.

## 2. Materials and Methods

### 2.1. Materials

CH (from shrimp shell, low molecular weight, Degree of Deacetylation $\geq$ 75%), CA (99%), urea, DPPH, NaCl were supplied by Sigma-Aldrich Company (St. Louis, MO, USA). Chemical reagents used for electrophoresis were from Bio-Rad (Segrate, Milano, Italy). Sodium hydroxide and hydrochloric acid were purchased from Sigma Aldrich Companyv (St.Louis, MO, USA). Glycerol (GLY), methanol and magnesium nitrate were from Carlo Erba S.p.A. (Milan, Italy) mTGase (Activa, WM, containing 1% of mTGase and 99% of maltodextrins) obtained from the culture of *Streptoverticillium* sp., was supplied by Prodotti Gianni (Milano, Italy). Folin–Ciocâlteu reagent was purchased by Fisher Bioreagents (Pittsburgh, PA, USA) Contact slide kit for controlling bacterial development (*Enterobacteriaceae* and *Streptococcus Faecalis*) in meat samples was purchased by Chimica Centro according to ISO 18593. Meat samples (hamburghers) were purchased from a local market. Olive leaf samples (cv "pisciottano") were collected when the pruning is traditionally done from a farm in Cilento, a seaside land of the Campania Region, Southern Italy. Olive leaves were kept in sealed bags and stored at room temperature until use.

### 2.2. DOLE Preparation by Naviglio's Principle

Before being submitted to the extraction process, olive leaves were dried in a static oven at 50 °C for two days. After that, 50 g of the samples were put in the filter bag (100 µm porosity), which was then inserted in the extraction chamber of the Naviglio extractor® (Lab Model 500 $cm^3$ capacity). Extraction was performed using 500 mL of $H_2O$ (containing 0.15% (*w*/*v*) CA, and 0.20% (*w*/*v*) potassium sorbate, pH 4.1) at room temperature and applying a 10-bar pressure for 2 min (static phase). After that, a remixing of liquid was applied for 2 min (dynamic phase). The pressure gradient is responsible for the extraction from the solid material (suction effect) as reported in Naviglio [32]. Liquid samples (10 mL), named DOLE, were collected after 2, 4, 6, and 24 h. DOLE samples were kept at 4 °C until they were used.

### 2.3. Total Polyphenols Content (TPC) Evaluation

DOLE TPC in samples obtained at different times of extraction, was determined by colorimetric in vitro assay using Folin–Ciocâlteau [33]. The calibration curve was set up with different concentrations of gallic acid (0.1–1 g/L). In particular, 0.1 mL of gallic acid solutions were mixed with 0.5 mL of 2 N Folin–Ciocâlteau reagent and 1.5 mL of freshly prepared 20% (*v*/*v*) $Na_2CO_3$, then the volume was adjusted to 10 mL with distilled water. Conversely, 0.1 mL of DOLE at different times of extraction (2-4-6-24 h) was mixed with the same reagents described. All the samples were incubated at room temperature in the dark for 1 h, then absorbance of the samples was measured at 765 nm using UV/visible Spectrophotometer (SmartSpec 3000 Bio-Rad, Segrate, Milan, Italy). The results were expressed as gallic acid equivalents (GAE). All determinations were carried out in triplicates.

### 2.4. Solubility of CH in CA

To investigate the solubility of CH in CA, different aqueous solutions of such acid were prepared (1.5–3.0% (*w*/*v*)). In each solution (50 mL) 1.5 g of CH were added under stirring. After 2 h, each solution was let stand for 1 h (without any stirring) to check the formation of precipitates. To quantify the CH dissolved, the supernatants were left to dry to obtain the dried matter. Solubility was expressed as percentage of the 1.5 g of CH added in each solution.

### 2.5. Preparation of Film Forming Solutions (FFSs) Containing DOLE

FFSs were prepared by using CA 3% (*w*/*v*) and chitosan 3% (*w*/*v*) at pH 2.5. DOLE was added in different volumes (10-15-20-40% (*v*/*v*)) after that its pH value was adjusted to 2.8 using CA 1 M. Finally, glycerol, used as a plasticizer, was mixed at 10% (*w*/*w*) (in respect to the weight of CH) in each type of FFS.

### 2.6. Zeta Potential and Particle Size Measurements

Zeta potential, average particle size, and polydispersity index of the FFSs were analyzed using the Zetasizer Nano-ZSP (Malvern®, Worcestershire, UK).

Three independent measurements were carried out on each sample of FFSs (1 mL) diluted to have a CH final concentration of 1 mg/mL.

Zetasizer Nano combines different techniques of Light Scattering to obtain a complete characterization of a colloidal system. Operating with a helium-neon laser at a fixed wavelength of 633 nm, it calculates the Zeta potential by using the Electrophoretic Light Scattering (ELS), the mean diameter of particles by using Dynamic Light Scattering (DLS) and the polydispersity index (PDI), which represents a value relative to the variance in the particle size distribution.

Three independent measurements were carried out on each sample of FFSs (1 mL) diluted to have a final concentration of 1 mg/mL.

### 2.7. Films Preparation

FFSs were poured in polypropylene Petri dishes (9 cm in diameter) and dried in a climatic chamber at 25 °C and 45% RH for 2 days. After that, the films were stored in a desiccator to balance the moisture content at 60% RH and room temperature during the subsequent analyses.

### 2.8. Antioxidant Activity

The activity to scavenge DPPH, 2,2-diphenyl-1-picrylhydrazyl, was measured for the solubilized films with all the different volumes of DOLE, for FFS 20% (*v*/*v*) and for DOLE solution (20% *v*/*v*) for one month, every 5 days.

Films were solubilized in CA 3% (*w*/*v*): 20 mg of each sample were dissolved in 500 µL of CA solution and subjected to sonication for 3 min.

For the *DPPH* assay, each sample was prepared adding to 900 µL of DPPH methanolic solution (0.005% (*w*/*v*)) 100 µL of different samples. The latter were incubated in dark for 30 min at room temperature. After that, absorbance was measured at 517 nm using a UV/visible Spectrophotometer (SmartSpec 3000 Bio-Rad, Segrate, Milan, Italy), considering methanol as blank and a sample with methanol added to DPPH solution as a control. The *DPPH scavenging activity* was calculated for each sample according to this equation:

$$\%DPPH\ scavenging\ activity = (A_{control} - A_{sample}/A_{control}) \times 100.$$

### 2.9. In Vitro Oral Digestion

After one month, the film containing 20% (*v*/*v*) DOLE was subjected to in vitro oral digestion, i.e., under simulated oral conditions according to Giosafatto et al. [34] with some modifications, and then the DPPH scavenging activity of this solution was measured to study the release of polyphenols entrapped in chitosan films.

For this analysis, 20 mg of each type of film was incubated in 500 µL of Simulated Salivary Fluid (SSF, (150 mM of NaCl, 3 mM of urea, pH 6.9) for 7 min at 170 rpm, according to Giosafatto et al. [34].

### 2.10. Film Mechanical Properties and Thickness

An Instron universal testing instrument (model no. 5543A, Instron Engineering Corp., Norwood, MA, USA) was used for mechanical properties characterization. Tensile Strength (TS), Young's Modulus (YM), and Elongation at Break (EB) were determined according to ASTM D882-18 (1997). Each film was cut into strips with a length of 40 mm and a width of 10 mm, and they were tested by using a 1 kN load cell and with a rate of grip separation of 20 mm/min. Film thickness was determined in five spots on each strip by using a digital micrometer (IP65 Alpa exacto, Alpa metrology Co., Pontoglio (BS), Italy, sensitivity 0.001 mm).

### 2.11. Bacterial Strains, Growth Conditions, and Antimicrobial Activity of DOLE-Based Films

Bacterial strains *Salmonella enteriditis* 706 RIVM, *Salmonella enterica* subsp. *Enterica serovar Typhimurium* (ATCC® 14028) and *Enterococcus faecalis* ATCC® 29212 were grown in Muller Hinton Broth (MHB, Becton Dickinson Difco, Franklin Lakes, NJ, USA) and on Tryptic Soy Agar (TSA; Oxoid Ltd., Hampshire, UK) as previously described [35]. In all the experiments, bacteria were inoculated and grown overnight in MHB at 37 °C.

In all the experiments to assess the antimicrobial activity of DOLE-based films, bacterial cells were inoculated and grown overnight in MHB at 37 °C. The next day, bacteria were transferred to a fresh Tryptic Soy Broth (TSB) tube and grown to mid- logarithmic phase. The antimicrobial activity of the films, containing or not containing DOLE, was tested by using a previously described experimental procedure [36]. Briefly, bacterial cells were diluted into TSB to approximately $2 \times 10^7$ CFU/mL and inoculated into TSA plates. A 1.5-cm$^2$ square of the edible film was placed into the center of the inoculated plate and pressed to ensure full contact with the agar surface. Plates were then incubated at 37 °C for 24 h and the bacterial growth underneath the film was evaluated.

### 2.12. DOLE Containing Film Capability to Contrast Bacterial Growth in Meat Samples

Hamburgers were prepared and both surfaces were covered with edible films containing DOLE at the following concentrations: 0%, 10%, 15%, and 20% (*v*/*v*). Hamburgers coated with baking paper were used as control. All samples were kept at 4 °C, but analyzed at different times (0 days, 5 days, 10 days, and 20 days) as follows: 10 g of each hamburger (from which the edible films or baking paper were removed) were homogenized in sterile conditions in the presence of 100 mL of $H_2O$. Contact slides (designed for identifying *Enterobacteriaceae* and *Streptococcus Faecalis*) were dipped in these solutions for 30 s and used according to the manufacturer's instructions of Liofilchem®-Contact slide 5 kit following ISO 18593.

### 2.13. Statistical Analysis

SPSS19 (Version 19, SPSS Inc., Chicago, IL, USA) software was used for all statistical analyses. One-way analysis of variance (ANOVA) and Duncan's multiple range tests ($p < 0.05$) were used to determine the significant difference among the samples.

## 3. Results

### 3.1. Olive Leaf Extract Preparation and Total Polyphenol Content (TPC)

During pruning of olive trees, a huge amount of pruning wastes is produced in all countries where the olive tree cultivation is part of the local economy. It is well known that olive tree drupes can produce one of the best oils from a nutritional point of view. Olive oil is, in fact, rich with polyphenols and vitamin E, depending on the method used [36–38]. Also, the pruning waste is rich in valuable compounds, such as polyphenols. It is described that olive leaves are particularly plentiful in highly healthy molecules, such as hydroxytyrosol and oleuropein [24]. For this reason, many items can be found, especially in the online market, that are suggested as supplements and/or cosmetic purposes. In this paper we wanted to propose the use of polyphenols extracted by olive pruning wastes as antioxidants and antimicrobial components of CH-based films to be used as both supplements for an easy administration and as films to wrap hamburgers. To reach this aim we have obtained dried olive extracts (DOLE) by referring to the Naviglio methodology [32] Naviglio extractor is an innovative technology that allows to quickly extract, from solid materials, the extractable compounds in organic and inorganic solvents compared to conventional techniques. It is based on a suction effect theorized in "Naviglio's Principle" [32]. Compression of liquid in which the solid matrix occurs at a pressure of around 8–10 bar by extracting solvent for a certain time is followed by an immediate decompression at atmospheric pressure. Thereby, a rapid release of extracting liquid, caused by a pressure gradient, is transported mechanically outside from the solid matrix the extractable compounds. Each extractive cycle consists of two phases: in the first one, also called the static phase, the system is under pressure and the liquid can penetrate the solid matrix; at the beginning of the second one, a negative pressure gradient is generated between the inside and the outside of the solid matrix to provoke the suction effect necessary to the extraction [32]. The extractions were performed at different times to assess the best protocol to obtain the highest quantity of polyphenolic molecules. As shown in Table 1, the TPC increases over time. Moreover, the amount of TPC results are notable, as expected from the literature [24]. It can be assessed that Naviglio's method is very effective, as reported in Mirpoor et al. [39] In fact, these authors compared Naviglio's method with maceration-based technology for the extraction of TPC from cardon leaves. Extraction in water was assessed to be more efficient than extraction with organic solvents, also by Sabry et al. [24]. Moreover, olive leaves derived from pruning contain, after 24 h of time extraction, a higher amount of TPC ($6.6 \pm 0.1$ g GAE/L) compared to cardon ($0.15 \pm 0.1$ g GAE/L), suggesting the importance of recovering olive leaves wastes. It is worth to note that TPC is equal to $4.4 \pm 0.2$ g GAE/L after only 2 h of time extraction.

**Table 1.** Total phenolic content in DOLE at different times of extraction.

| Hours | Polyphenols (g GAE/L) |
|---|---|
| 2 | 4.4 ± 0.2 [a] |
| 4 | 4.6 ± 0.1 [a] |
| 6 | 5.5 ± 0.2 [b] |
| 24 | 6.6 ± 0.4 [c] |

Different small letters (a–c) indicate significant differences among the values reported in each column ($p < 0.05$).

### 3.2. Chitosan Solubility in Citric Acid

Because of amino groups, CH is soluble in acidic solutions among which there is CA that is food grade. However, the solubilization depends on the degree of deacetylation (DD). Thus, we have performed experiments to assess the CA concentration to achieve the maximum solubility of CH. CA was used in substitution of acetic acid normally used in our previous papers, where chitosan-based bioplastics were prepared to be proposed to wrap foods [40]. In this work we decided to recur to CA since our main aim was to use chitosan-base edible films to convey phenolic compounds present in DOLE as supplements with a pleasant taste for consumers (which is not insured using acetic acid as the solvent). The maximum solubility was achieved using 3% (*w*/*v*) CA, as shown in Figure 1. Thus, further analyses were performed using 3% (*w*/*v*) CH dissolved in 3% (*w*/*v*) CA.

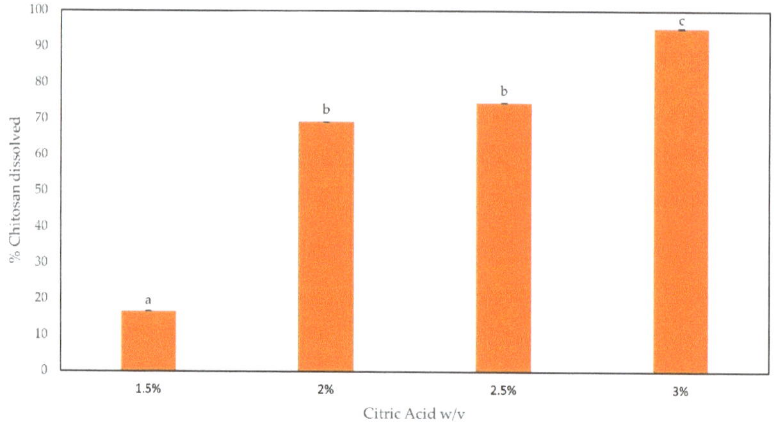


**Figure 1.** CH solubility, expressed in percentage, when dissolved in CA. Different small letters (a–c) indicate significant differences among the values reported in each bar ($p < 0.05$).

### 3.3. FFSs Characterization and Preparation of Films by Casting

FFSs prepared in absence and in presence of increasing concentrations of DOLE were analyzed via Zeta sizer to determine the Zeta potential, particle size, and PDI. Table 2 shows that all the solutions were highly stable. The addition of DOLE provoked a lowering in the Zeta potential value in respect to the control (CH-based solution in absence of DOLE), which turned out higher by increasing the amount of DOLE. The mean particle size of DOLE containing samples were lower than the one exhibited by the control and decreased by increasing the DOLE concentration. The same observation can be done regarding PDI values. The decreasing in mean particle size could be due to aggregation of CH particles provoked by polyphenolic molecules present in DOLE. These solutions were used to cast films. Obtained films are shown in Figure 2. Films appear transparent but yellowish depending on the DOLE content.

**Table 2.** Zeta potential, particle size, and PDI of FFSs containing DOLE (10–40% (*v*/*v*)). The control was set up in the absence of DOLE.

| DOLE in FFSs % (*v*/*v*) | Z Potential (mV) | Mean Particle Size (nm) | PDI |
|---|---|---|---|
| 0 | 55.71 ± 4.61 [a] | 567.00 ± 19.33 [a] | 0.67 ± 0.02 [a] |
| 10 | 40.08 ± 3.20 [c] | 568.96 ± 31.40 [a] | 0.48 ± 0.01 [b] |
| 15 | 43.59 ± 2.51 [c] | 500.68 ± 28.30 [b] | 0.50 ± 0.03 [b] |
| 20 | 49.37 ± 5.42 [b] | 426.22 ± 49.05 [c] | 0.51 ± 0.05 [b] |
| 40 | 53.55 ± 5.61 [a,b] | 419.32 ± 73.02 [c] | 0.51 ± 0.05 [b] |

Different small letters (a–c) indicate significant differences among the values reported in each column ($p < 0.05$).

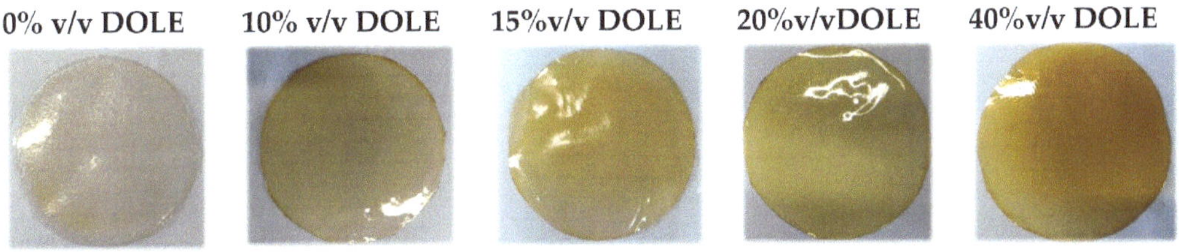


**Figure 2.** Chitosan-based films containing different volumes of DOLE (0–40% (*v*/*v*)).

### 3.4. Antioxidant Activity (AA) of DOLE-Containing Films over the Time

Figure 3 shows the AA of solubilized DOLE containing films detected from 0 time and every 5 days up to 30. Lower AA was observed in films containing 10 and 15% (*v*/*v*) DOLE even after few days of storage. Instead, the AA remained high in both kinds of films containing 20% and 40% (*v*/*v*) DOLE, respectively. The AA was also detected over the time in 20% (*v*/*v*) DOLE solutions and in 20% (*v*/*v*) DOLE FFS. It can be noticed that the AA values in solubilized films were lower than DOLE-containing solutions and FFS. In Figure 3, we report only the results obtained with 20% (*v*/*v*) DOLE containing samples, but similar results were obtained with 40% (*v*/*v*) DOLE samples. A possible explanation could be that the phenolic compounds present in DOLE exert a lower activity when entrapped in the chitosan film matrix. A reduction of AA due to cardoon leaf extracts (CLE) entrapped in cardoon protein-based films was also reported in Mirpoor et al. [39], even though for a longer time (70 days). It must be noticed that DOLE-containing films showed a high AA even after 30 days (70%) compared to CLE-containing films at time 0 (60%) [39].

### 3.5. In Vitro Oral Digestion

In order to ensure the release of polyphenols entrapped in CH films, samples (20% *v/v* DOLE) stored for 30 days were subjected to in vitro oral digestion, under simulated oral conditions, and then the DPPH scavenging activity of these solutions was measured. The results in Figure 4 showed that the release of polyphenols from 30 days storage films occurred properly after in vitro oral digestion. The AA results of such films were compared to the ones obtained from the controls represented by solubilized films stored for the same time (30 days). The AA of both groups of different samples are perfectly comparable, demonstrating that polyphenols entrapped in the films are still active after in vitro simulated oral digestion. Thus, we can affirm that DOLE containing films could be used to delivery polyphenols in a novel supplement for an effective application into the nutraceutical industry.

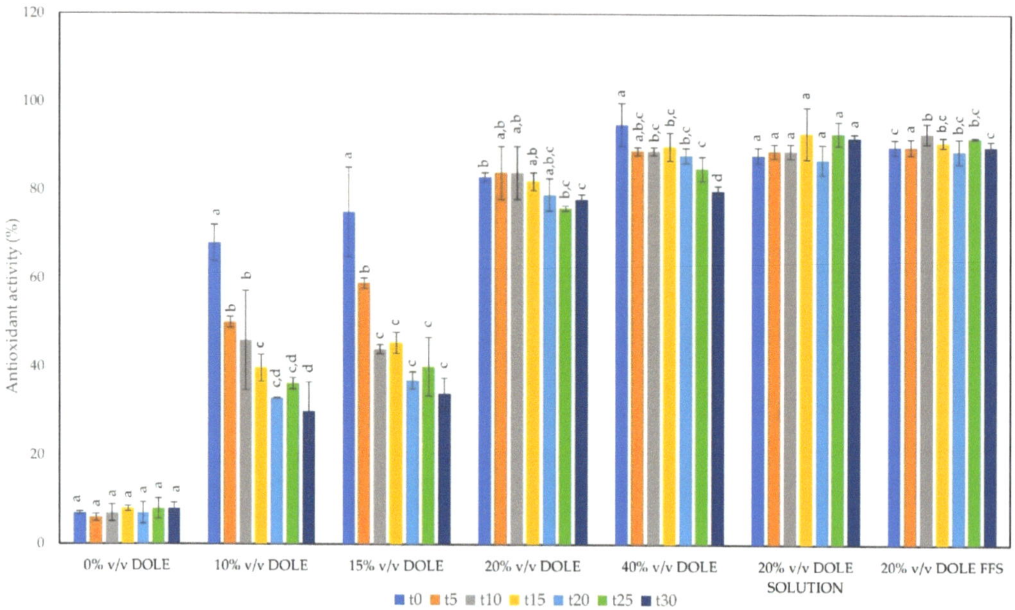


**Figure 3.** Antioxidant activity by DPPH assay of solubilized films (DOLE 0-10-15-20-40% ($v/v$)), 20% ($v/v$) DOLE solutions, and 20% ($v/v$) DOLE Film Forming Solutions (FFS) at different days of storage. Different small letters (a–d) indicate significant differences among the values reported in each bar ($p < 0.05$).

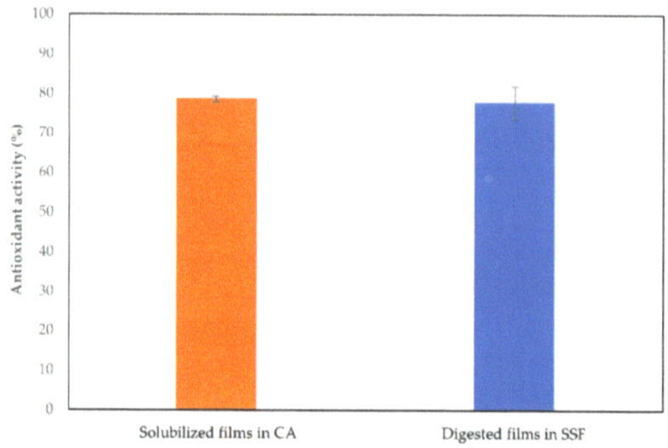


**Figure 4.** Antioxidant activity by DPPH assay of 20% ($v/v$) DOLE films subjected to in vitro oral digestion after 30 days compared to the AA of solubilized films after 30 days.

### *3.6. Mechanical Properties*

Since DOLE visually seemed to have a plasticizer effect on chitosan-based films, we studied the mechanical properties of these films to verify the possible influence of DOLE on Tensile Strength (TS), Elongation at Break (EB), and Young's Modulus (YM). Samples thickness was also measured.

As shown in Figure 5, the medium thickness of all samples does not differ significatively from the control. Regarding mechanical properties, we note that they are influenced by the presence of DOLE. We can observe, proportionally with the increment of DOLE concentration, an increase of EB, and a decrease of TS and YM, also with respect to the control without DOLE. These results suggest a role as plasticizer for DOLE in addition to the one exerted by glycerol. Similar results were obtained by Mariniello et al., using grape juice as a plasticizer in *N. sativa* protein-based films [41].

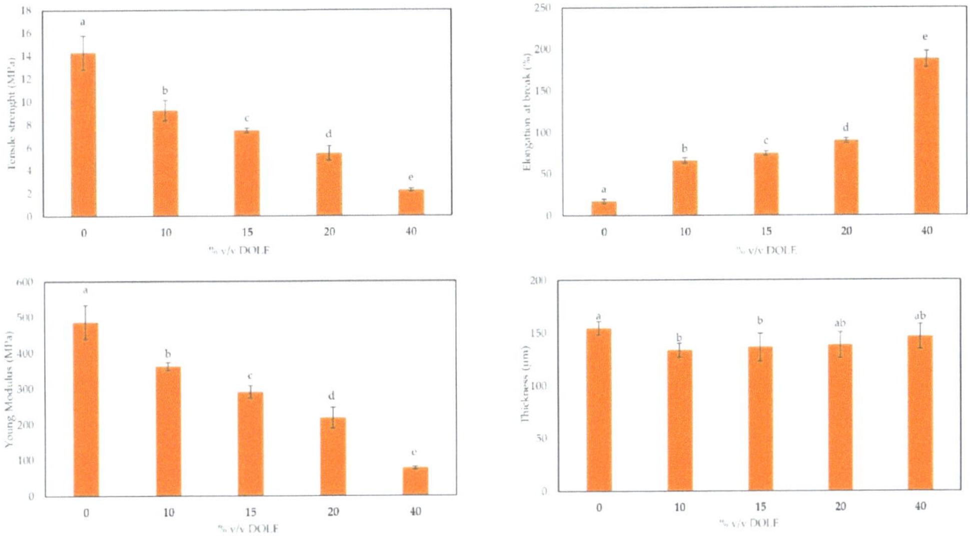


**Figure 5.** Properties of chitosan-based films with different % ($v/v$) of DOLE (0–40%). On the top: from left to right Tensile Strength and Elongation at Break. On the bottom: from left to right Young's Modulus and Thickness. Different small letters (a–e) indicate significant differences among the values reported in each bar ($p < 0.05$).

### *3.7. Antimicrobial Activity of Films Containing DOLE*

Films prepared in the absence or in the presence of DOLE at different concentrations (10%, 20%, or 40% ($v/v$)) were analyzed for their antimicrobial activity towards Gram-negative bacteria (*S. enteriditis* 706 RIVM, *S. enterica* subsp. *enterica serovar* Typhimurium ATCC® 14028) and a Gram-positive bacterial strain *E. faecalis* ATCC® 29212. Usually, these strains are involved in food contamination and spoilage. In particular, the presence of *E. faecalis* suggests that a contamination by the two other pathogen strains may occur. Thus, small squares (1.5 cm × 1.5 cm) of each film were placed into the inoculated plate in full contact with the agar surface. After 24 h at 37 °C, only the films containing DOLE were found to be able to significantly inhibit bacterial growth with the strongest effects observed at the highest content of DOLE (Figure 6a,b). The DOLE component of the films is effective in inhibiting the growth of all the bacteria tested in a dose dependent manner. Other studies suggested a broad antimicrobial activity of olive leaves extract in a concentration dependent-manner [42]. Taking advantage of these results, we have tested the capability of DOLE-containing films as active bioplastics, able to control the bacterial spoilage of meat hamburgers over time. To this aim, we have used the Liofilchem® contact slide 5 kit, which allows selective detection of *Enterobacteriaceae* and enterococci. In particular, employed slides contain media to selectively identify *E. coli* ATCC® 25922 and *Enterococcus faecalis* ATCC® 19433. As samples, we have chosen meat hamburgers and covered each side with our edible films containing increasing amounts of DOLE. As controls we used hamburgers coated with baking paper, as is commonly done. We kept all samples stored at 4 °C for different times (0, 5, 10, and 20 days). Afterwards, meat samples were treated as described in materials and method in Section 2.12. As shown in Figure 7, the most effective edible films are the ones containing 15% and 20% ($v/v$) DOLE after 20 days storage, demonstrating that the antimicrobial activity of phenolic compounds is still valuable over time, especially against *E. coli*. Samples were also tested for the presence of *E. faecalis* bacteria, that were totally absent even in the control at time 0 (data not shown). Thus, we can assess that DOLE-containing films exhibit antimicrobial activity against all the bacteria tested both in vitro (Figure 6) and as edible films used to protect hamburgers over time (Figure 7).

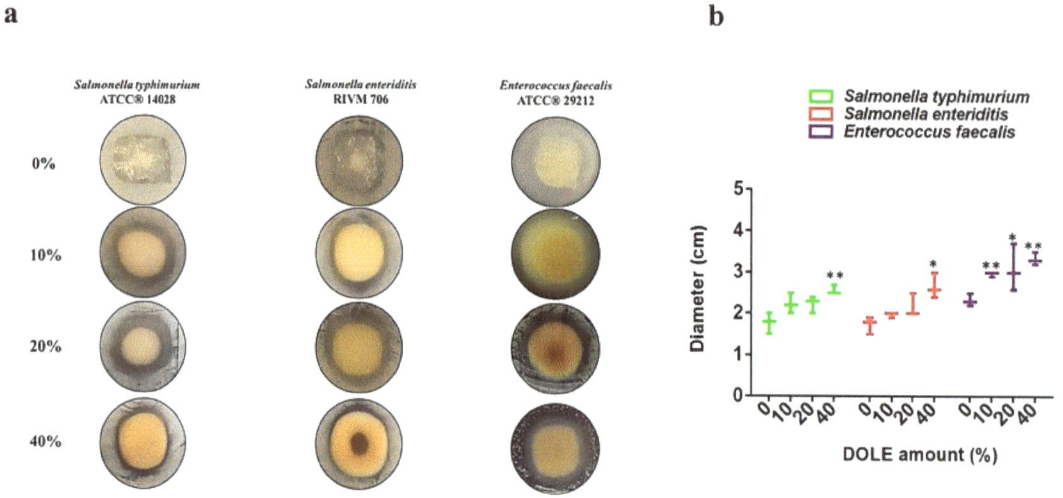


**Figure 6.** Antimicrobial activity of DOLE-based films. Antimicrobial effects were evaluated by analyzing the growth of the bacterial cells held in direct contact with the films (**a**). The diameter of the inhibition circle was determined for each bacterial strain in contact with the different films and shown in the graph (**b**). The experiments were carried out in triplicate. Statistical analyses were performed by using Student's *t*-test. Significant differences were indicated as * $p < 0.05$ or ** $p < 0.01$.

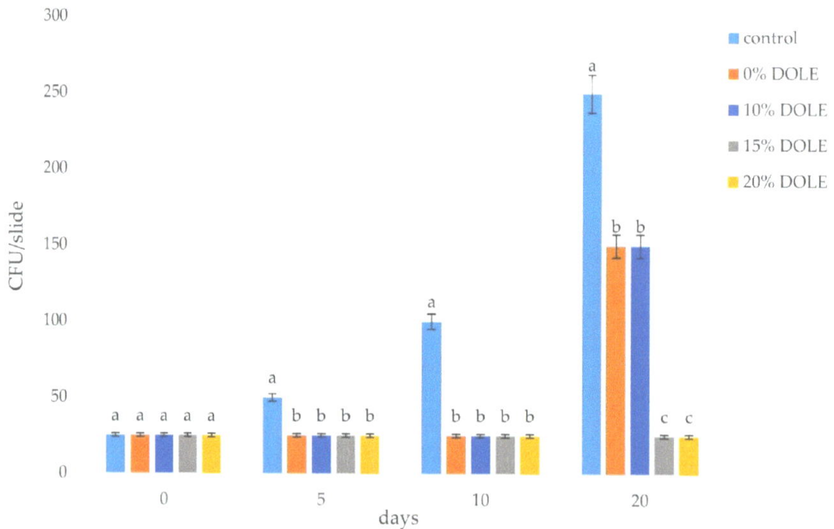


**Figure 7.** CFU (*Enterobacteriaceae*) in meat samples covered with DOLE-containing films. Controls represent samples covered with baking paper. Different small letters (a–c) indicate significant differences among the values reported in each bar ($p < 0.05$).

## 4. Conclusions

In the present paper it was demonstrated that olive byproducts can be recovered to obtain a high added-value product possibly used in the food industry. It showed the effectiveness of polyphenols contained in DOLE and entrapped in a chitosan-based matrix in acting as antimicrobials and antioxidants. Obtained films can be classified as edible and, thus, could be used as supplements to reinforce the human diet with polyphenols. Moreover, such films can be used to wrap meat hamburgers to delay spoilage during storage due to common microbial contaminants.

**Author Contributions:** M.F., A.S. and L.M. planned and designed the work. D.N. and A.S. extracted polyphenol and antioxidant properties from olive leaves. R.G. and A.A. performed antimicrobial activity in vitro. All other experimental results were performed by M.F. and A.S. M.F. and L.M. wrote the article. All authors have read and agreed to the published version of the manuscript.

**Funding:** This research was funded by the grant PRIN: Progetti di Ricerca di Interesse Nazionale—Bando 2017—"CARDoon valorisation by InteGrAted biorefiNery (CARDIGAN)" Ministero Italiano dell'Università e della Ricerca (MUR) (COD. 2017KBTK93).

**Institutional Review Board Statement:** Not applicable.

**Data Availability Statement:** The data presented in this study are available on request from the corresponding author.

**Acknowledgments:** The authors would like to thank Maria Fenderico for her technical support.

**Conflicts of Interest:** The authors declare no conflict of interest.

## References

1. Elsabee, M.Z.; Abdou, E.S. Chitosan Based Edible Films and Coatings: A Review. *Mater. Sci. Eng. C* **2013**, *33*, 1819–1841. [CrossRef]
2. Shukla, S.K.; Mishra, A.K.; Arotiba, O.A.; Mamba, B.B. Chitosan-Based Nanomaterials: A State-of-the-Art Review. *Int. J. Biol. Macromol.* **2013**, *59*, 46–58. [CrossRef]
3. Mujtaba, M.; Morsi, R.E.; Kerch, G.; Elsabee, M.Z.; Kaya, M.; Labidi, J.; Khawar, K.M. Current Advancements in Chitosan-Based Film Production for Food Technology; A Review. *Int. J. Biol. Macromol.* **2019**, *121*, 889–904. [CrossRef]
4. No, H.K.; Young Park, N.; Ho Lee, S.; Meyers, S.P. Antibacterial Activity of Chitosans and Chitosan Oligomers with Different Molecular Weights. *Int. J. Food Microbiol.* **2002**, *74*, 65–72. [CrossRef]
5. Melro, E.; Antunes, F.E.; da Silva, G.J.; Cruz, I.; Ramos, P.E.; Carvalho, F.; Alves, L. Chitosan Films in Food Applications. Tuning Film Properties by Changing Acidic Dissolution Conditions. *Polymers* **2021**, *13*, 1. [CrossRef]
6. El Knidri, H.; Belabeed, R.; Addaou, A.; Laajeb, A.; Lahsini, A. Extraction, Chemical Modification and Characterization of Chitin and Chitosan. *Int. J. Biol. Macromol.* **2018**, *120*, 1181–1189. [CrossRef]
7. Sorlier, P.; Denuzière, A.; Viton, C.; Domard, A. Relation between the Degree of Acetylation and the Electrostatic Properties of Chitin and Chitosan. *Biomacromolecules* **2001**, *2*, 765–772. [CrossRef]
8. Dotto, G.L.; Vieira, M.L.G.; Pinto, L.A.A. Use of Chitosan Solutions for the Microbiological Shelf Life Extension of Papaya Fruits during Storage at Room Temperature. *LWT Food Sci. Technol.* **2015**, *64*, 126–130. [CrossRef]
9. Vimaladevi, S.; Panda, S.K.; Xavier, K.A.M.; Bindu, J. Packaging Performance of Organic Acid Incorporated Chitosan Films on Dried Anchovy (*Stolephorus Indicus*). *Carbohydr. Polym.* **2015**, *127*, 189–194. [CrossRef]
10. Al-Naamani, L.; Dobretsov, S.; Dutta, J. Chitosan-Zinc Oxide Nanoparticle Composite Coating for Active Food Packaging Applications. *Innov. Food Sci. Emerg. Technol.* **2016**, *38*, 231–237. [CrossRef]
11. Del Hoyo-Gallego, S.; Pérez-Álvarez, L.; Gómez-Galván, F.; Lizundia, E.; Kuritka, I.; Sedlarik, V.; Laza, J.M.; Vila-Vilela, J.L. Construction of Antibacterial Poly(Ethylene Terephthalate) Films via Layer by Layer Assembly of Chitosan and Hyaluronic Acid. *Carbohydr. Polym.* **2016**, *143*, 35–43. [CrossRef]
12. Youssef, A.M.; El-Sayed, S.M.; El-Sayed, H.S.; Salama, H.H.; Dufresne, A. Enhancement of Egyptian Soft White Cheese Shelf Life Using a Novel Chitosan/Carboxymethyl Cellulose/Zinc Oxide Bionanocomposite Film. *Carbohydr. Polym.* **2016**, *151*, 9–19. [CrossRef]
13. Ochoa-Velasco, C.E.; Guerrero-Beltrán, J.Á. Postharvest Quality of Peeled Prickly Pear Fruit Treated with Acetic Acid and Chitosan. *Postharvest Biol. Technol.* **2014**, *92*, 139–145. [CrossRef]
14. Abdel-rahman, A.; Anyangwe, N.; Carlacci, L.; Casper, S.; Danam, R.P.; Enongene, E.; Erives, G.; Fabricant, D.; Gudi, R.; Hilmas, C.J.; et al. The Safety and Regulation of Natural Products Used as Foods and Food Ingredients. *Toxicol. Sci.* **2011**, *123*, 333–348. [CrossRef]
15. Kiritsakis, K.; Goula, A.M.; Adamopoulos, K.G.; Gerasopoulos, D. Valorization of Olive Leaves: Spray Drying of Olive Leaf Extract. *Waste Biomass Valoriz.* **2018**, *9*, 619–633. [CrossRef]
16. Espeso, J.; Isaza, A.; Lee, J.Y.; Sörensen, P.M.; Jurado, P.; de Avena-Bustillos, R.J.; Olaizola, M.; Arboleya, J.C. Olive Leaf Waste Management. *Front. Sustain. Food Syst.* **2021**, *5*, 1–13. [CrossRef]
17. Özcan, M.M.; Matthäus, B. A Review: Benefit and Bioactive Properties of Olive (*Olea Europaea* L.) Leaves. *Eur. Food Res. Technol.* **2017**, *243*, 89–99. [CrossRef]

18. Molina-Alcaide, E.; Yáñez-Ruiz, D.R. Potential Use of Olive By-Products in Ruminant Feeding: A Review. *Anim. Feed Sci. Technol.* **2008**, *147*, 247–264. [CrossRef]
19. Manzanares, P.; Ruiz, E.; Ballesteros, M.; Negro, M.J.; Gallego, F.J.; López-Linares, J.C.; Castro, E. Residual Biomass Potential in Olive Tree Cultivation and Olive Oil Industry in Spain: Valorization Proposal in a Biorefinery Context. *Span. J. Agric. Res.* **2017**, *15*, 1–12. [CrossRef]
20. Herrero, M.; Temirzoda, T.N.; Segura-Carretero, A.; Quirantes, R.; Plaza, M.; Ibañez, E. New Possibilities for the Valorization of Olive Oil By-Products. *J. Chromatogr. A* **2011**, *1218*, 7511–7520. [CrossRef]
21. Susalit, E.; Agus, N.; Effendi, I.; Tjandrawinata, R.R.; Nofiarny, D.; Perrinjaquet-Moccetti, T.; Verbruggen, M. Olive (*Olea Europaea*) Leaf Extract Effective in Patients with Stage-1 Hypertension: Comparison with Captopril. *Phytomedicine* **2011**, *18*, 251–258. [CrossRef]
22. Jemai, H.; Feki, A.E.L.; Sayadi, S. Antidiabetic and Antioxidant Effects of Hydroxytyrosol and Oleuropein from Olive Leaves in Alloxan-Diabetic Rats. *J. Agric. Food Chem.* **2009**, *57*, 8798–8804. [CrossRef]
23. Khalatbary, A.R.; Zarrinjoei, G.R. Anti-Inflammatory Effect of Oleuropein in Experimental Rat Spinal Cord Trauma. *Iran. Red Crescent Med. J.* **2012**, *14*, 229–234.
24. Sabry, O.M.M. Review: Beneficial Health Effects of Olive Leaves Extracts. *J. Nat. Sci. Res.* **2014**, *4*, 19.
25. Benavente-García, O.; Castillo, J.; Lorente, J.; Ortuño, A.; Del Rio, J.A. Antioxidant Activity of Phenolics Extracted from *Olea Europaea* L. Leaves. *Food Chem.* **2000**, *68*, 457–462. [CrossRef]
26. Andrikopoulos, N.; Kaliora, A.; Assimopoulou, A.; Papageorgiou, V. Inhibitory Activity of Minor Polyphenolic and Nonpolyphenolic Constituents of Olive Oil Against In Vitro Low-Density Lipoprotein Oxidation. *J. Med. Food* **2002**, *5*, 1–7. [CrossRef]
27. Visioli, F.; Poli, A.; Galli, C. Antioxidant and Other Biological Activities of Phenols from Olives and Olive Oil. *Med. Res. Rev.* **2002**, *22*, 65–75. [CrossRef]
28. Kashaninejad, M.; Sanz, M.T.; Blanco, B.; Beltrán, S.; Niknam, S.M. Freeze Dried Extract from Olive Leaves: Valorisation, Extraction Kinetics and Extract Characterization. *Food Bioprod. Process.* **2020**, *124*, 196–207. [CrossRef]
29. Ghomari, O.; Sounni, F.; Massaoudi, Y.; Ghanam, J.; Drissi Kaitouni, L.B.; Merzouki, M.; Benlemlih, M. Phenolic Profile (HPLC-UV) of Olive Leaves According to Extraction Procedure and Assessment of Antibacterial Activity. *Biotechnol. Rep.* **2019**, *23*, e00347. [CrossRef]
30. Rahmanian, N.; Jafari, S.M.; Wani, T.A. Bioactive Profile, Dehydration, Extraction and Application of the Bioactive Components of Olive Leaves. *Trends Food Sci. Technol.* **2015**, *42*, 150–172. [CrossRef]
31. Taamalli, A.; Arráez-Román, D.; Barrajón-Catalán, E.; Ruiz-Torres, V.; Pérez-Sánchez, A.; Herrero, M.; Ibañez, E.; Micol, V.; Zarrouk, M.; Segura-Carretero, A.; et al. Use of Advanced Techniques for the Extraction of Phenolic Compounds from Tunisian Olive Leaves: Phenolic Composition and Cytotoxicity against Human Breast Cancer Cells. *Food Chem. Toxicol.* **2012**, *50*, 1817–1825. [CrossRef]
32. Naviglio, D. Naviglio's Principle and Presentation of an Innovative Solid-Liquid Extraction Technology: Extractor Naviglio®. *Anal. Lett.* **2003**, *36*, 1647–1659. [CrossRef]
33. Dewanto, V.; Xianzhong, W.; Adom, K.K.; Liu, R.H. Thermal Processing Enhances the Nutritional Value of Tomatoes by Increasing Total Antioxidant Activity. *J. Agric. Food Chem.* **2002**, *50*, 3010–3014. [CrossRef]
34. Giosafatto, C.V.L.; Al-Asmar, A.; D'Angelo, A.; Roviello, V.; Esposito, M.; Mariniello, L. Preparation and Characterization of Bioplastics from Grass Pea Flour Cast in the Presence of Microbial Transglutaminase. *Coatings* **2018**, *8*, 435. [CrossRef]
35. Gaglione, R.; Cesaro, A.; Olmo, E.D.; Di Girolamo, R.; Tartaglione, L.; Pizzo, E.; Arciello, A. Cryptides Identified in Human Apolipoprotein B as New Weapons to Fight Antibiotic Resistance in Cystic Fibrosis Disease. *Int. J. Mol. Sci.* **2020**, *21*, 2049. [CrossRef]
36. Abdalrazeq, M.; Jaradat, N.; Qadi, M.; Giosafatto, C.V.L.; Dell'olmo, E.; Gaglione, R.; Arciello, A.; Porta, R. Physicochemical and Antimicrobial Properties of Whey Protein-Based Films Functionalized with Palestinian Satureja Capitata Essential Oil. *Coatings* **2021**, *11*, 1364. [CrossRef]
37. Pizarro, M.L.; Becerra, M.; Sayago, A.; Beltrán, M.; Beltrán, R. Comparison of Different Extraction Methods to Determine Phenolic Compounds in Virgin Olive Oil. *Food Anal. Methods* **2013**, *6*, 123–132. [CrossRef]
38. Gimeno, E.; Castellote, A.I.; Lamuela-Raventós, R.M.; De la Torre, M.C.; López-Sabater, M.C. The Effects of Harvest and Extraction Methods on the Antioxidant Content (Phenolics, alfa -Tocopherol, and beta-Carotene) in Virgin Olive Oil. *Food Chem.* **2002**, *78*, 207–211. [CrossRef]
39. Mirpoor, S.F.; Varriale, S.; Porta, R.; Naviglio, D.; Spennato, M.; Gardossi, L.; Giosafatto, C.V.L.; Pezzella, C. A Biorefinery Approach for the Conversion of Cynara Cardunculus Biomass to Active Films. *Food Hydrocoll.* **2022**, *122*, 107099. [CrossRef]
40. Porta, R.; Mariniello, L.; di Pierro, P.; Sorrentino, A.; Giosafatto, C.V.L. Transglutaminase Crosslinked Pectinand Chitosan-Based Edible Films: A Review. *Crit. Rev. Food Sci. Nutr.* **2011**, *51*, 223–238. [CrossRef]
41. Yaseen, D.; Sabbah, M.; Al-Asmar, A.; Altamimi, M.; Famiglietti, M.; Giosafatto, C.V.L.; Mariniello, L. Functionality of Films from Nigella Sativa Defatted Seed Cake Proteins Plasticized with Grape Juice: Use in Wrapping Sweet Cherries. *Coatings* **2021**, *11*, 1383. [CrossRef]
42. Pereira, A.P.; Ferreira, I.C.F.R.; Marcelino, F.; Valentão, P.; Andrade, P.B.; Seabra, R.; Estevinho, L.; Bento, A.; Pereira, J.A. Phenolic Compounds and Antimicrobial Activity of Olive (*Olea Europaea* L. *Cv.* Cobrançosa) Leaves. *Molecules* **2007**, *12*, 1153–1162. [CrossRef]

*foods* MDPI

*Article*

# Preparation and Properties of Blended Composite Film Manufactured Using Walnut-Peptide–Chitosan–Sodium Alginate

**Wenqi Yan [1], Haochen Sun [1], Wenxin Liu [1] and Hao Chen [1,2,*]**

1 Marine College, Shandong University (Weihai), No. 180 Wenhua West Road, Gao Strict, Weihai 264209, China; 201900810230@mail.sdu.edu.cn (W.Y.); 201900810229@mail.sdu.edu.cn (H.S.); liuwenxin1999@163.com (W.L.)

2 The Key Laboratory of Synthetic and Biological Colloids, Ministry of Education, Jiangnan University, No. 1800 Lihu Road, Wuxi 214122, China

* Correspondence: chenhao_2016@sdu.edu.cn; Tel.: +86-17863067833

**Abstract:** In this study, layer-by-layer assembly was performed to prepare sodium alginate (SA) layer and walnut-peptide–chitosan (CS) bilayer composite films. Genipin was adopted to crosslink CS and walnut peptide. The properties of walnut peptide-CS-SA composite film were determined, and the influence of material ratio on the performance of composite film was explored. According to the results, the mechanical tensile property, oil absorption property, and water vapor barrier property of the composite film were improved with the presence of genipin. Moreover, the proportion of CS and walnut peptide had significant effects on color, transmittance, mechanical properties, barrier properties, and antioxidant properties of the composite films. Among them, the composite film containing 1% ($w/v$) CS, 1% ($w/v$) walnut peptide, and 0.01% ($w/v$) genipin showed the best performance, with a tensile strength of 3.65 MPa, elongation at break of 30.82%, water vapor permeability of 0.60 $g{\cdot}mm{\cdot}m^{-2}{\cdot}h^{-1}{\cdot}kPa^{-1}$, oil absorption of 0.85%, and the three-phase electrochemistry of 2,2-diphenyl-1-picrylhydrazyl (DPPH) radical scavenging rate of 25.59%. Under this condition, the tensile property, barrier property, and oxidation resistance of the composite film are good, which can provide a good preservation effect for food, and has great application potential.

**Keywords:** sodium alginate; chitosan; walnut peptide; composite film; layer by layer assembly

**Citation:** Yan, W.; Sun, H.; Liu, W.; Chen, H. Preparation and Properties of Blended Composite Film Manufactured Using Walnut-Peptide–Chitosan–Sodium Alginate. *Foods* **2022**, *11*, 1758. https://doi.org/10.3390/foods11121758

Academic Editors: Sílvia Petronilho and Mario M. Martinez

Received: 30 March 2022
Accepted: 11 June 2022
Published: 15 June 2022



## 1. Introduction

Traditional plastics and their products are not easy to be degraded in nature. After entering the environmental system, they can stay in the soil and water for a long time, forming "white pollution". Incineration may also produce toxic substances, which pose a serious threat to the ecological environment [1]. The work to solve the problem of plastic pollution mainly focuses on three aspects: first, developing substitutes for plastics and their products; second, developing degradable plastics; and third, increasing the recycling of plastics and their products [2–6]. However, over the last decade, there has been a growing interest in the development and use of bio-based packaging materials to replace the conventional synthetic packaging materials needed to preserve and protect foods. The development of innovative edible films has emerged as a new research area in food science [7]. Edible film is made of natural edible biomacromolecules (such as protein, lipid, sugar, etc.) as raw materials, and added with edible plasticizers, crosslinking agents, and other substances, which are edible through the interaction between molecules. The selection of edible materials with good biocompatibility, or the production of packaging films for one particular application is very important to improve the quality of fresh and processed food, and increase the shelf-life of the food and its packaging. Edible films can also be used as a good carrier for functional substances, such as antioxidants, antibacterial agents, flavors, probiotics, etc. It can not only prolong the shelf-life of food, but also improve its nutritional value, and has a good application prospect [8–10].

SA and CS have shown great potential for use as food coatings due to their biodegradability, biocompatibility, nontoxicity, and versatile chemical and physical properties [11]. SA is one of the most versatile biodegradable polymers, which is a linear polysaccharide extracted from brown algae. The chemical structure of alginate is composed of β-d-mannuronic acid and β-l-guluronic acid [12]. On the other hand, CS is found in the shells of crustaceans, such as shrimp and crabs. In molecular terms, CS is achieved by the deacetylation of chitin [13].

Because both CS and SA have strong hydrophilicity, the performance of single film is poor. In order to improve the single-film-forming performance of CS and SA, the electrostatic reaction between CS and SA was used to directly blend the two films. However, direct blending may result in a strong reaction between -$NH^{3+}$ on CS and -$COO^-$ on SA, resulting in a water-insoluble mixture. The viscosity of the film solution is large and uneven, resulting in poor film quality [14]. In recent years, the technology of layer-by-layer (LBL) assembly is gradually emerging. Macromolecules with opposite charges are deposited in turn, and the noncovalent interactions between polymer layers can endow bilayer films or multilayer films with better mechanical properties and barrier properties [15]. This technology can overcome the above shortcomings. Different layers in CS-SA bilayer films are not simply superimposed, but intermolecular interactions occur. Under this electrostatic interaction, molecules in the films are combined [16].

Particularly, crosslinking with genipin, which is a reagent of natural origin, improves the swelling, water resistance, and mechanical properties of the films [17]. Genipin has recently been used in biomedical applications and for controlled drug release, due to its biocompatibility and low toxicity [18]. Genipin also has excellent pharmacological value, such as protecting the liver and gallbladder, anti-inflammatory, antibacterial, anti-tumor, treating gastritis, treating diabetes, etc. [19]. Studies have shown that synthetic crosslinking reagents such as glyoxal, glutaraldehyde, and epichlorohydrin are all more or less cytotoxic, and may impair the biocompatibility of chitosan [20]. In this regard, genipin extracted from gardenia fruits is obviously advantageous. As a water-soluble bi-functional crosslinker, it reacts promptly with chitosan. The resulting crosslinked products have better biocompatibility and stability for use [20,21]. The genipin-crosslinked films' total soluble matter, water vapor permeability, tensile strength, ultraviolet and visible light barriers, and thermal properties were improved in relation to the uncrosslinked film, whereas their moisture content and microstructure were not affected, demonstrating the effectiveness of genipin on improving the properties of gelatin-based films and the promising application of films as food packaging [22,23]. However, the crosslinking agent, genipin, is seldom used in the related research of edible film at present, and the excellent crosslinking ability and biocompatibility of genipin still need to be further explored.

Compared with walnut protein, walnut polypeptide has better solubility, emulsifying property, oil absorption, and oxidation resistance, and is more easily absorbed and utilized by the human body; walnut polypeptide has good nutrition and health care functions, but poor solubility, and is very limited in application in the food field [24]. Research has shown that walnut peptide is rich in amino acid residues such as glutamic acid, arginine, and aspartic acid, related to its antioxidant property. In addition, walnut peptide also has a certain anti-cancer performance, and has great application potential in the fields of medicine and food, and further exploration is needed to expand the scope of application of walnut peptide [25]. In this paper, walnut peptide is added into the composite film as a functional active ingredient, which enhanced the preservative effect of the composite film and also provided a new idea for the application of walnut peptide in production.

In order to achieve better functional properties, the application of multilayers is an available way which can take advantage of the unique properties of several materials [26,27]. In this study, CS and SA were used as the main film-forming matrix, and walnut peptide was added to enhance the antioxidant performance of the composite film. Genipin was added to chemically crosslink CS and walnut peptide, and the walnut-peptide–CS–SA composite film was prepared by a layer-by-layer assembly method. The effects of the ratio of walnut polypeptide to CS on the performances of the composite film are explored, and the performances of the composite film were measured, including thickness, mechanical tensile property, color difference value, light transmittance, water vapor transmittance, oil absorption, and antioxidant property. Further, a novel composite edible film with excellent performance and green safety is prepared.

## 2. Materials and Methods

### 2.1. Materials

SA (chemically pure, viscosity (10 g/L, 20 °C) ≥0.02 Pa·s) and CS (≥80% deacetylated, viscosity 50–800 mPa·s) were purchased from Sinopharm Chemical Reagent Co., Ltd. (Shanghai, China). Walnut peptide powder (98% purity) was obtained from Ningshan Guosheng Biotechnology Co., Ltd. (Shanxi, China). Glycerol (analytically pure) and potassium bromide (analytically pure) were from Tianjin Beichen District Fangzheng Reagent Factory (Tianjin, China), and genipin (>95% purity) was from ChangSha ZhiXin Biotechnology Co., Ltd. (Changsha, China). Acetic acid (analytically pure) and hydrochloric acid (analytically pure) were from Yantai Shuangshuang Chemical Co., Ltd. (Shandong, China). Anhydrous (analytically pure) was from Tianjin Zhiyuan chemical reagent Co., Ltd. (Tianjin, China). Soybean oil was from Yihai Jiali Cereal&Oil Food Industry Co., Ltd. (Guangdong, China). DPPH (97% purity) was from Kaima biochemical Co., Ltd. (Tianjin, China). Absolute alcohol (analytically pure) was from Tianjin Fuyu Fine Chemical Co., Ltd. (Tianjin, China).

### 2.2. Preparation of Film Solutions and Film Formation

#### 2.2.1. Sodium Alginate Monolayer Film Production

SA was weighed at 1.5 g, and then slowly added into 100 mL distilled water while stirring. The solutions were stirred continuously for 1 h using a magnetic stirrer (RH-KT/C, IKA Works Guangzhou, Guangzhou, China) in a 40 °C water bath until the powders were completely dissolved. Then, 2 mL glycerol was added, stirred evenly, and made into a 100 mL film solution. The film solution was degassed by ultrasonic cleaner (KQ-300B, Kunshan Shumei Super Sound Instrument Co., Ltd., Kunshan, China) for 10 min. Then, 15 g of the solution was poured into a plastic petri dish with a diameter of 90 mm. The solution was evenly tiled and dried in a constant-temperature air-blowing drying oven at 30 °C for 24 h to form the SA monolayer.

#### 2.2.2. Chitosan–Walnut-Peptide Covalent Complex Production

A certain amount of CS powder was weighed according to the proportion in Table 1 and dissolved in 1% (*v*/*v*, %) acetic acid solution, and magnetically stirred until completely dissolved. According to the proportion in Table 1, a certain amount of walnut peptide powder was dissolved in distilled water and magnetically stirred at 60 °C for 30 min. The walnut peptide aqueous solution was mixed with CS acetic acid solution, and 2 mL plasticizer glycerol was added.

**Table 1.** Composition of different composite films.

| Sample Number | SA (*w*/*v*, %) | Glycerol (Lower Layer) (*v*/*v*, %) | CS (*w*/*v*, %) | Walnut Peptide (*w*/*v*, %) | Glycerol (Upper Layer) (*v*/*v*, %) | Genipin (*w*/*v*, %) |
|---|---|---|---|---|---|---|
| S-C0 | 1.5 | 2.0 | 1.0 | 1.0 | 2.0 | 0 |
| S-CG1 | 1.5 | 2.0 | 2.0 | 0 | 2.0 | 0.01 |
| S-CG2 | 1.5 | 2.0 | 1.5 | 0.5 | 2.0 | 0.01 |
| S-CG3 | 1.5 | 2.0 | 1.0 | 1.0 | 2.0 | 0.01 |
| S-CG4 | 1.5 | 2.0 | 0.5 | 1.5 | 2.0 | 0.01 |
| S-CG5 | 1.5 | 2.0 | 0 | 2.0 | 2.0 | 0.01 |

S-CG1~S-CG5 were mixed with genipin acetic acid aqueous solution, and S-C0 was the control group without genipin. HCl (1 mol/L) was used to adjust the pH of all the six kinds of mixed solution to 3.0 to prepare the 100 mL membrane solution. The solution was degassed in an ultrasonic cleaner for 10 min, and cooled to room temperature for further use.

#### 2.2.3. Walnut-Peptide–Chitosan–Sodium Alginate Composite Edible Film Production

The preparation process is shown in Figure 1. A total of 17 g CS–walnut-peptide solution was poured onto the SA monolayer that was about to be completely dried, spread evenly, and dried in a constant-temperature blast-drying oven at 30 °C for 24 h to form a film. After drying, the film was taken down and put in a dry dish (RH60%, room temperature) containing saturated KBr solution to return moisture.

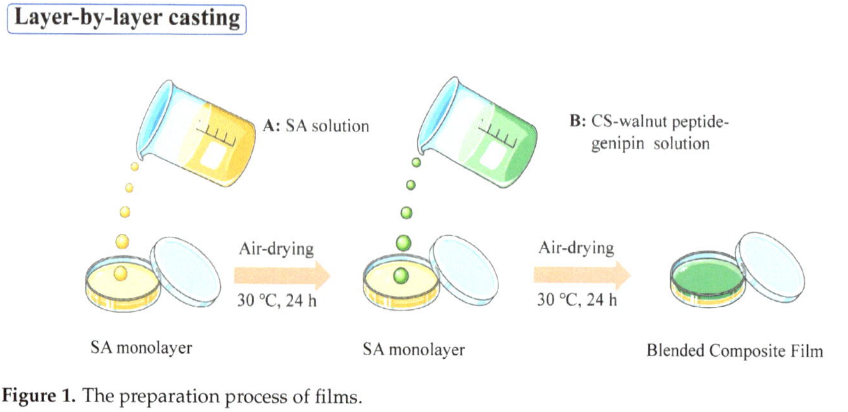


**Figure 1.** The preparation process of films.

### 2.3. *Fourier-Transform Infrared (FTIR) Spectroscopy*

The FTIR analysis was carried out using a PIKE MIRacle Universal ATR system in a Tensor 27 FTIR-ATR spectrometer (Bruker Optik GmbH, Ettlingen, Germany) to obtain the FTIR spectra of six sample films and raw material powers (SA, CS, walnut peptide, and genipin). Each sample film was cut into a small piece directedly spread on the scanning table, and each raw material powder was mixed with KBr using a hydraulic press to form a pellet. Then, the samples were kept in the sample recording chamber under reduced vacuum. The scanning and the resolution range were 500–4000 $cm^{-1}$ and 4 $cm^{-1}$, respectively.

### 2.4. *Mechanical Test*

The composite film was cut into strips of 50 mm × 10 mm, and both ends were clamped flatly on the tensile probe of the texture analyzer (Universal TA, Shanghai Tengba Instrument Technology Co., Ltd., Shanghai, China). The initial spacing was set to be 20 mm, and the tensile speed was set to be 60 mm/min. Three parallels were made for each composite film, and the maximum tensile force and elongation length of the film at break were recorded. The results were averaged. The tensile strength (TS, MPa) and elongation at break (EB, %) were calculated by following Equations (1) and (2).

$$\mathrm{TS} = \frac{F}{d \times W} \tag{1}$$

$$\mathrm{EB} = \frac{L_1 - L_0}{L_0} \times 100\% \tag{2}$$

where $F$ denotes the maximum tension when the film breaks, N; $d$ represents the thickness of the film, mm; $W$ represents the width of the film sample, mm; $L_0$ is the length of the stretching front film, mm; $L_1$ is the length of the stretched film, mm.

### 2.5. Chromatic Properties

The color parameters ($L$: lightness/brightness, $a$: redness/greenness, and $b$: yellowness/blueness) of the film were measured by a colorimeter (NR110, Shenzhen 3nh Technology Co., Ltd., Shenzhen, China), and the colorimeter was used after calibration by standard white film. Taking S-CG1 sample as reference, the color difference was calculated, and the average values of three composite film samples were taken. The color difference ($\Delta$E) was calculated by following Equation (3).

$$\Delta\mathrm{E} = \sqrt{(L - L_1)^2 + (a - a_1)^2 + (b - b_1)^2} \tag{3}$$

where $L$, $a$, and $b$ are the color parameter values of film samples; $L_1$, $a_1$, and $b_1$ are the color parameter values of the control film (S-CG1).

### 2.6. Transmittance

The light transmittance was measured by UV-visible spectrophotometer (T6, Beijing Persee General Instrument Co., Ltd., Beijing, China). The edible film was cut into 1.2 cm × 4 cm samples, which were closely attached to the inner side of the cuvette (using an empty cuvette as the control). The light transmittance of the samples was measured at the wavelength of 560 nm, and three parallel samples were measured, and the average value was taken [28]. The transmittance (T, %) was calculated by following Equation (4).

$$\mathrm{T} = 0.1^{A_{560}} \times 100\% \tag{4}$$

### 2.7. Water Vapor Permeability (WVP)

Anhydrous $CaCl_2$ (3 g) was added to a 40 mm × 25 mm weighing bottle, which was then sealed with a film, and the initial mass of the weighing bottle was recorded. The weighing bottle was then placed in a drying dish containing a saturated KBr solution at room temperature and 60% relative humidity, and the mass of the weighing bottle was measured every 24 h until the weighing bottle weight did not change, and the final mass was recorded [29]. The WVP was calculated by following Equation (5).

$$\mathrm{WVP} = \frac{\Delta m \times d}{At\Delta P} \tag{5}$$

where WVP is the water vapor permeability of the film, $g \cdot mm \cdot m^{-2} \cdot h^{-1} \cdot kPa^{-1}$; $\Delta m$ is the added mass of the weighing bottle, g; $A$ is the surface area of the composite film, $m^2$; $t$ is the interval time, h; $d$ is the film thickness, mm; $\Delta P$ is the vapor pressure difference on both sides of the composite film, kPa.

### 2.8. Oil Absorption

The film was cut into a square with a size of 25 mm × 25 mm, weighed, and laid flat on the filter paper. Then, 4 mL of soybean oil was poured onto the film and allowed to stand for 0.5 h. Oil on the surface of the dry film was absorbed by the filter paper, and the film weight was measured. The oil absorption of the film was expressed as the ratio of the increment of the film weight to the original weight of the film. The oil absorption was calculated by following Equation (6).

$$\text{Oil absorption} = \frac{m_2 - m_1}{m_1} \times 100\% \tag{6}$$

where $m_1$ is the mass before oil absorption of the film, g; $m_2$ is the mass of the film after oil absorption, g.

### 2.9. Antioxidant Activity

The film sample (0.3 g) was weighed into 30 mL of distilled water, and after 24 h, when it was dissolved completely, 30 mL of ethanol was added. The composite film solution was centrifuged at 4000 r/min for 20 min, and the supernatant was extracted as the film extract. First, 1 mL of the supernatant was mixed with 5 mL of 0.01 mmol/L DPPH ethanol solution, and the mixture was put in the dark for reaction for 30 min. The absorbance, $A_i$, of the solution at 517 nm, and the absorbance, $A_j$, of 1 mL film extract and 5 mL ethanol at the same time, as well as the absorbance, $A_o$, of sample solvent (0.5 mL distilled water + 0.5 mL ethanol) were measured after standing in the dark for 30 min with 5 mL 0.01 mmol/L DPPH ethanol solution [30]. The antioxidant activity was shown as DPPH radical clearance, and it was calculated by following Equation (7).

$$\text{DPPH free radical scavenging rate\%} = \frac{A_o - A_i - A_j}{A_o} \times 100\% \tag{7}$$

### 2.10. Statistical Analysis

In the experiment, three groups of parallel samples were set for each sample, and the data were processed by Microsoft Excel 2010. The significant difference between the different samples was evaluated by Duncan analysis in SPSS 21 ($p < 0.05$). The images were drawn by Origin 2019 and KingDraw 3.0.

## 3. Results and Discussion

### 3.1. FTIR Analysis

The FTIR-ATR technique was used to evaluate the functional groups of the materials, and to detect possible changes in the proportion of raw materials. As an excellent natural biological crosslinking agent, genipin can not only improve the swelling, water resistance, and mechanical properties of the films [17], but also chemically crosslinks CS and walnut peptide and improves the density of the membrane structure [31,32]. FTIR-ATR spectroscopy was used for the analysis of intermolecular interactions between the genipin, CS, and walnut peptide, which could be reflected in infrared spectra by the location and intensity of characteristic absorption bands, as shown in Figure 2a,b.

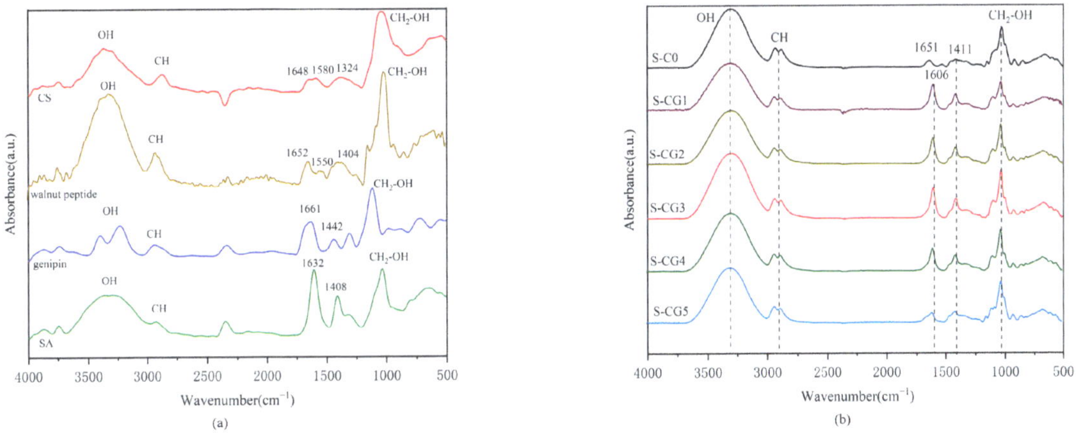


**Figure 2.** FTIR spectra of raw material (**a**) powders, including CS, walnut peptide, genipin, and SA, and six sample films described in Table 1 (**b**).

A strong and broad adsorption band at around 3500–3000 cm$^{-1}$ corresponded to -OH stretching, which was affected by the intermolecular or intramolecular hydrogen bonds [33]. The band corresponding to asymmetric stretching vibrations of the methylene group occurred at approximately 2950–2900 cm$^{-1}$ [34]. In addition, the absorption peaks at the range of 1030–1010 cm$^{-1}$ correspond to the stretching of $CH_2$-OH, respectively [35].

In Figure 2a, both CS, with a degree of deacetylation of 85%, and walnut peptide displayed two strong vibrations at 1652–1648 and 1580–1550 cm$^{-1}$. These have previously been assigned to amide I and amide II vibrations [36]. Besides, they had N-H and C-N absorption peaks at 1580 and 1324 cm$^{-1}$ [37]. The carboxylate anion –COO$^{-}$ of SA had two characteristic absorption peaks at 1632 and 1408 cm$^{-1}$, and the ester bond COO of genipin also had two characteristic absorption peaks at 1661 and 1428 cm$^{-1}$, corresponding to the stretching vibrations of C=O and C-O, respectively [38,39]. In Figure 2b, the band at 1661 cm$^{-1}$, which represents the carbonyl group of ester in genipin, shifted to a lower frequency that appeared at 1638–1642 cm$^{-1}$ in crosslinked films. This is due to the crosslinking reaction between genepin and CS, that is, the nucleophilic attack by the amino group of CS on the olefinic carbon atom of genipin, resulting in the opening of the dihydropyran ring and the formation of tertiary amine. The subsequent slower reaction is the formation of amide through the reaction of the amino group on chitosan with the ester group of genipin [35,40]. Walnut peptides also have many amino groups that can react with genipin. Furthermore, the increase observed in the peaks at around 1411 cm$^{-1}$ can be assigned to absorptions from C-N stretching vibrations, respectively, more numerous after crosslinking with genipin [41]. In addition, the absorption peak at 1028 cm$^{-1}$ could be attributed to the interaction between the -OH group of glycerol (used as a plasticizer) and genipin in all the film samples [42]. This is similar to the results of Kumar et al. [43].

### *3.2. Mechanical Properties*

The composite films were successfully prepared, and the average thickness of each sample was characterized. The results showed that with the increase of walnut peptide content, the thickness of the films gradually increased from 0.1505 mm to 0.2133 mm, and the structure of composite film became loose. In order to evaluate the mechanical properties of composite films with different raw material ratios, the TS (tensile strength) and EB (elongation at break) of composite films were measured. It can be seen from Figure 3 that the TS of the composite film first increased and then decreased with the decreasing mass ratio of CS/walnut peptide in the sample added with the crosslinking agent, genipin. The changes of TS and EB were likely due to the micro network structure and intermolecular force in the film matrix formed by the crosslinking of CS, walnut peptide, and genipin [44].

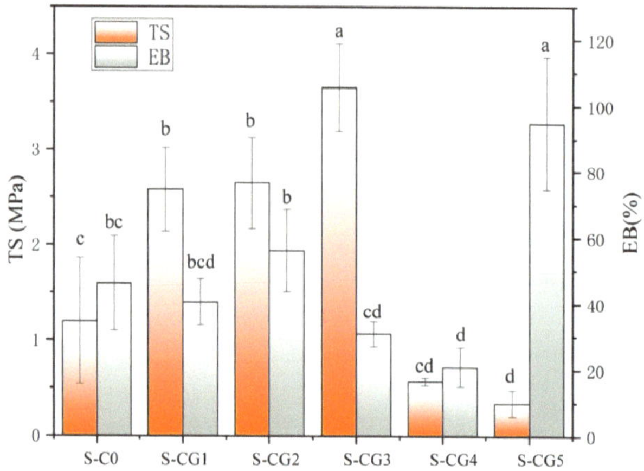


**Figure 3.** The tensile strength and elongation at break of composite films. Results are expressed as average, and different letters (a–d) indicate significant differences ($p < 0.05$).

TS is the maximum tensile strength that a film can sustain, thus it expresses the maximum stress developed in the film during the tensile testing [45]. The TS of the S-CG3 sample with a CS/walnut peptide mass ratio of 1:1 was the highest (3.65 MPa), which was significantly higher than that of S-C0 sample without genipin. This is because the crosslinking effect of genipin made the covalent crosslinking between CS and walnut peptide form a Schiff base [46], which made the film structure closer than CS-SA film and walnut peptide-SA film, and it could better resist the tearing of external forces. However, when the walnut peptide ratio was too large, the strong interaction between CS and SA was replaced by the weak interaction between peptides, and the tensile properties of the bilayer film were also affected. Moreover, too high amounts of walnut peptides possibly caused inhomogeneity of the polymer matrices, which subsequently reduced the mechanical properties [47].

EB is a measure of the film's capacity for stretching, and also refers to the maximum change in length of the test specimen before breaking [48]. Elongation of the films depended on several factors, including homogeneity of the matrices and plasticization, which increased the flexibility and deformability of the polymers [49]. The elongation at break of the composite film first increased and then decreased from S-CG1 to S-CG4. The binding sites of CS and walnut peptides were limited, so when the proportion of walnut peptides gradually increased, the binding between the two tended to be saturated [50]. The crosslinking agent, genipin, made the film dense, which made the film not easy to be stretched, so the elongation at break of the S-CG3 sample was smaller than that of the S-C0 sample without crosslinking. When the upper film did not contain CS, the polypeptide structure of walnut peptide was loose, which made the film sticky and flexible, and caused maximum elongation at break.

### 3.3. Chromatic Properties

The measurement results of the chromaticity values, $L$, $a$, and $b$, of each composite film sample and the color difference, $\Delta E$, of S-CG1 to S-CG5 samples are shown in Figure 4. The addition of plant-derived materials commonly gives color to the composite films, thereby affecting the light transmission [51]. Genipin can react with amino acid or proteins to form dark blue pigments [52]. The color of the composite film without adding genipin was pale yellow, and that of the composite film with adding genipin was blue-green, which was due to the gardenia blue pigment generated by the reaction of genipin with the amino groups of CS and walnut peptide [46].

The reaction mechanism is shown in Figure 5. The more adequate their action, the more pigment generated, and the deeper the blue color. The results showed that with the increase of the proportion of walnut peptide, the *L*, *a*, and *b* values of the composite film fluctuated, and the ΔE values of S-CG2-S-CG5 samples were higher, indicating that, compared with the S-CG1 sample without walnut peptide, it had a significant color difference. In summary, the reaction mode and reaction adequacy between walnut peptide and genipin or CS greatly affected the color of the composite film.

### 3.4. Transmittance Properties

The transmittance of composite film can reflect the compatibility of film components. If the transmittance of composite film is very low, the material compatibility is poor. As shown in Figure 6, the transmittance of the S-C0 sample without genipin crosslinking was the highest, which was 85.06%. Pigments were generated in samples S-CG1~S-CG5 due to the crosslinking reaction of genipin, resulting in a decrease in the transmittance of the film, which fluctuated between 31.86% and 42.38%. This was due to the light reflection and scattering at the interface of the continuous phase and the dispersed phase when CS and walnut peptides were combined, resulting in changes in the transmittance. Moreover, increased filler contents possibly filled up the void volume between polymer molecules, and hence, blocked the transmission path of the light [53]. The mixed systems of different layers of the composite film have heterogeneity, which is also the reason for the decrease of transmittance.

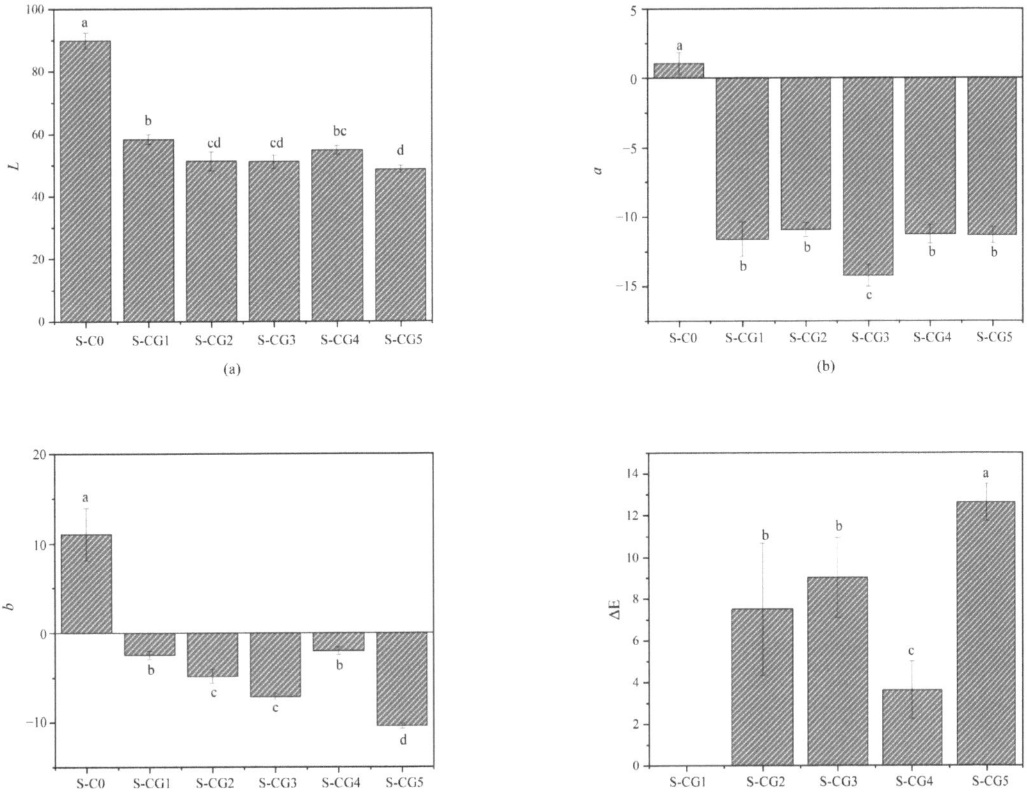


**Figure 4.** The chroma values of composite films. (**a**) *L*: lightness/brightness, (**b**) *a*: redness/greenness, (**c**) *b*: yellowness/blueness, (**d**) ΔE: the color difference. Results are expressed as average, and different letters (a–d) indicate significant differences ($p < 0.05$).

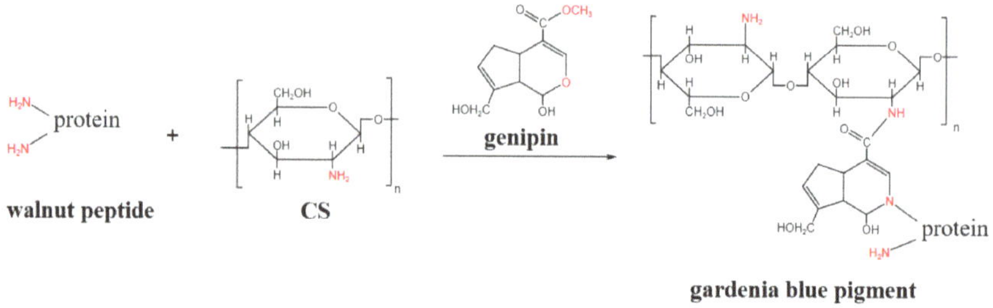


**Figure 5.** Proposed mechanism of genipin reaction with CS and walnut peptide to produce gardenia blue pigment.

### 3.5. Water Vapor Transmission Rate

Water vapor permeability is an important index to measure the barrier performance of film. Low water vapor permeability is beneficial to prevent water exchange between food and environment in wet environments. The measurement results of water vapor permeability of each composite film sample are shown in Figure 7. Among them, the water vapor permeability of S-CG1 and S-CG3 samples was the lowest, which was 0.60 $g{\cdot}mm{\cdot}m^{-2}{\cdot}h^{-1}{\cdot}kPa^{-1}$, indicating that the CS-SA bilayer structure and CS- walnut peptide crosslinking system could play a role in blocking water. S-CG1, S-CG2, and S-CG3 showed low WVP; it is possible that at lower amounts of walnut peptides, the film matrices were denser, which limited the diffusion of water vapor, giving lower WVP [54]. Walnut peptides have strong hygroscopicity. When the proportion of walnut peptide increased, the water vapor permeability increased; it is probable that the walnut peptide and CS could not be fully crosslinked, or high amounts of walnut peptides caused non-homogeneous structures and disrupted networks [55].

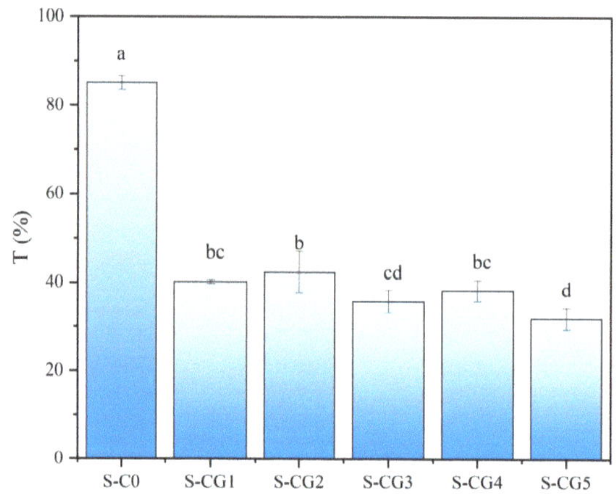


**Figure 6.** The transmittance of composite films. Results are expressed as average, and different letters (a–d) indicate significant differences ($p < 0.05$).

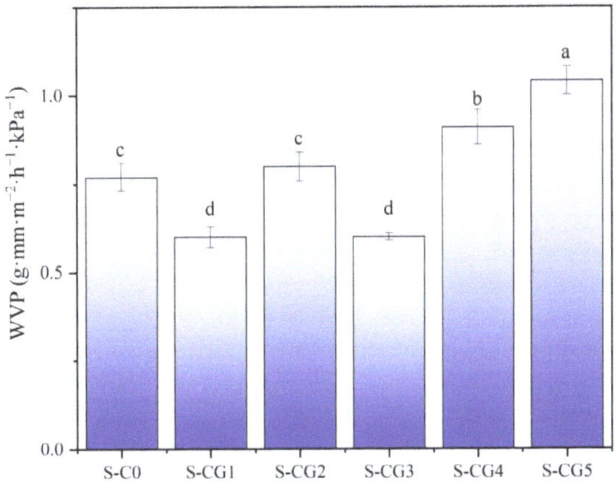


**Figure 7.** The water vapor permeability of composite films. Results are expressed as average, and different letters (a–d) indicate significant differences ($p < 0.05$).

### 3.6. Oil Absorbency Capacity

It can be seen from Figure 8 that the oil absorption of the composite film first decreased and then increased with the increase of the walnut peptide addition ratio. After the addition of genipin, the oil absorption of the film decreased, and the oil absorption of the S-CG3 sample was the weakest, which was 0.85%. This is because the CS and walnut peptide crosslinked to form a dense network structure, blocking the oil into the film, which, to a certain extent, weakened the film oil absorption. However, walnut peptide containing a high content of hydrophobic amino residues has high oil absorption, so when the proportion of walnut peptide increased to a certain range, the oil absorption of the film increased [56]. In addition, with the increase of the walnut-peptide–chitosan ratio, the non-homogeneous structures and void formation were formed, which also promoted the improvement of oil absorption.

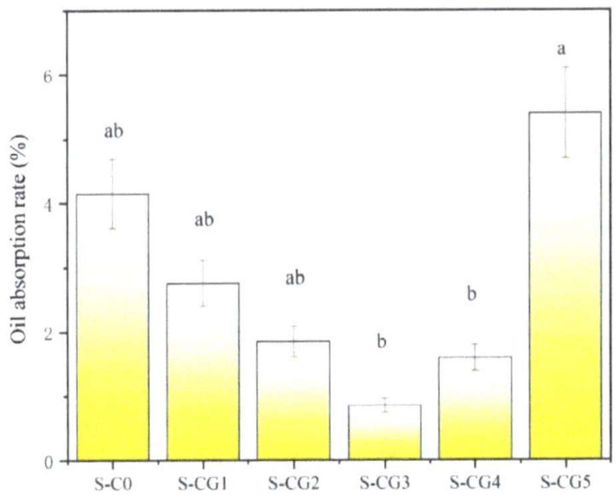


**Figure 8.** The oil absorption of composite films. Results are expressed as average, and different letters (a, b) indicate significant differences ($p < 0.05$).

### 3.7. Antioxidant Activity

The oxidation resistance of the composite film is an important standard to measure the effect of packaging preservation. Previous studies have shown that walnut polypeptides are macromolecular substances synthesized by amino acids such as glutamic acid, arginine, and aspartic acid. The molecular structure of walnut polypeptides contains plenty of amino groups, which have DPPH free radical scavenging activity and ACE inhibitory activity [57,58]. Besides, CS has active hydroxyl groups, indicating stronger DPPH free radical scavenging ability [59–61]. In addition, the layer of SA, with certain antioxidant activity, can also improve the scavenging ability of composite films [62].

In the experiment, walnut peptide was added to endow the composite film with oxidation resistance, and the DPPH radical scavenging ability of the composite film samples was measured to evaluate the oxidation resistance of the composite film. As can be seen in Figure 9, with the increase of the walnut peptide addition ratio, the DPPH radical scavenging rate of the film increased gradually, and the highest scavenging rate of the S-CG5 sample was 30.14%. However, the S-CG1 sample without walnut peptide also showed a certain antioxidant capacity, because CS also had a certain antioxidant capacity, and its antioxidant capacity was related to the degree of deacetylation [63]. In summary, walnut peptides could improve the oxidation resistance of composite films.

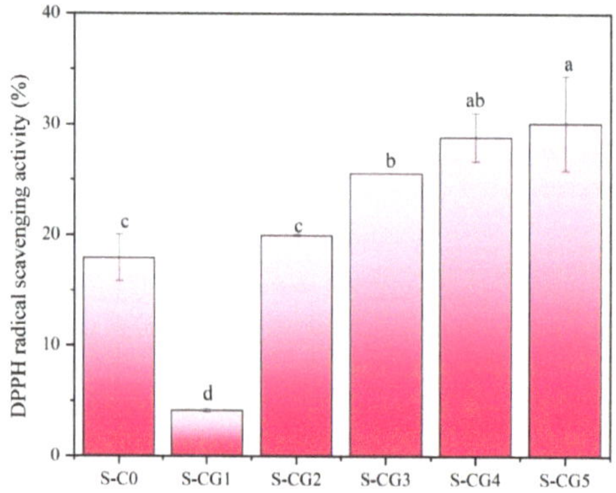


**Figure 9.** The DPPH free radical scavenging activities of composite films. Results are expressed as average, and different letters (a–d) indicate significant differences ($p < 0.05$).

## 4. Conclusions

Layer-by-layer assembly was performed to prepare SA layer and walnut-peptide–CS layer bilayer composite films. Genipin crosslinked CS and walnut peptide to form a network structure that can improve the mechanical strength of the composite film. In addition, the tensile property, barrier property, and oxidation resistance of the composite film are good. The appropriate ratio of CS and walnut peptide can create the crosslinking reaction more fully and improve the tensile strength and elongation at break of the film. The pigments generated by the crosslinking reaction of genipin could affect the color and optical properties of the film. The crosslinking of CS and walnut peptide by genipin makes the film dense, and oil and water cannot easily penetrate the composite film, which is conducive to the preservation of food. However, when the walnut peptide ratio increases, its moisture absorption and oil absorption will affect the barrier performance of the film. The addition ratio of walnut peptides was positively correlated with the DPPH free radical scavenging rate of the composite film. The walnut peptides endowed the composite film with antioxidant capacity, so that the composite film had better preservation function. Comprehensively, the composite film with a CS/walnut peptide ratio of 1:1 and genipin addition had the best performance. The tensile strength was 3.65 MPa, the elongation at break was 30.82%, the water vapor permeability was 0.60 $g \cdot mm \cdot m^{-2} \cdot h^{-1} \cdot kPa^{-1}$, the oil absorption was 0.85%, and the DPPH radical scavenging rate was 25.59%. The composite film had good mechanical properties and barrier properties, which could provide good preservation effects.

**Author Contributions:** Conceptualization, W.Y., H.S. and H.C.; Data curation, W.Y. and H.S.; Formal analysis, W.Y. and W.L.; Funding acquisition, H.C.; Investigation, W.L.; Methodology, W.Y. and H.S.; Project administration, H.C.; Software, W.Y. and H.S.; Supervision, W.Y. and H.C.; Validation, W.Y.

and H.S.; Visualization, W.Y.; Writing—original draft, W.Y. and H.S.; Writing—review and editing, H.C. All authors have read and agreed to the published version of the manuscript.

**Funding:** This research was funded by the Natural Science Foundation of Shandong Province (No. ZR2019BC053); the National Natural Science Foundation of China (No. 32102129); the Key Laboratory of Synthetic and Biological Colloids, Ministry of Education, Jiangnan University (No. 1022050205219730/008); and the Young Scholars Program of Shandong University, Weihai (No. 202208).

**Institutional Review Board Statement:** Not applicable.

**Informed Consent Statement:** Not applicable.

**Data Availability Statement:** The data are available from the corresponding author.

**Conflicts of Interest:** The authors declare no conflict of interest. The funders had no role in the design of the study; in the collection, analyses, or interpretation of data; in the writing of the manuscript, or in the decision to publish the results.

## References

1. Qin, Z.-H.; Mou, J.-H.; Chao, C.Y.H.; Chopra, S.S.; Daoud, W.; Leu, S.-y.; Ning, Z.; Tso, C.Y.; Chan, C.K.; Tang, S.; et al. Biotechnology of Plastic Waste Degradation, Recycling, and Valorization: Current Advances and Future Perspectives. *ChemSusChem* **2021**, *14*, 4103–4114. [CrossRef] [PubMed]
2. Agarwal, S. Biodegradable Polymers: Present Opportunities and Challenges in Providing a Microplastic-Free Environment. *Macromol. Chem. Phys.* **2020**, *221*, 2000017. [CrossRef]
3. Choudhary, K.; Sangwan, K.S.; Goyal, D. Environment and economic impacts assessment of PET waste recycling with conventional and renewable sources of energy. *Procedia CIRP* **2019**, *80*, 422–427. [CrossRef]
4. Fenton, R.L.; Duplan, A.; Hall, B.L. Modifying an Enzyme to Improve PET Plastic Recycling. *FASEB J.* **2020**, *34*, 1. [CrossRef]
5. Rusdi, S.; Destian, R.A.; Rahman, F.; Chafidz, A. Preparation and Characterization of Bio-Degradable Plastic from Banana Kepok Peel Waste. In Proceedings of the Materials Science Forum; Trans Tech Publications Ltd.: Freienbach, Switzerland, 2020; Volume 981, pp. 132–137. [CrossRef]
6. Thiounn, T.; Smith, R.C. Advances and approaches for chemical recycling of plastic waste. *J. Polym. Sci.* **2020**, *58*, 1347–1364. [CrossRef]
7. Milani, J.M.; Nemati, A. Lipid-Based Edible Films and Coatings: A Review of Recent Advances and Applications. *J. Packag. Technol. Res.* **2022**, *6*, 11–22. [CrossRef]
8. Brzoska, N.; Mueller, M.; Nasui, L.; Schmid, M. Effects of film constituents on packaging-relevant properties of sodium caseinate-based emulsion films. *Prog. Org. Coat.* **2018**, *114*, 250–258. [CrossRef]
9. Galus, S.; Arik Kibar, E.A.; Gniewosz, M.; Kraśniewska, K. Novel materials in the preparation of edible films and coatings—A review. *Coatings* **2020**, *10*, 674. [CrossRef]
10. Gennadios, A.; Hanna, M.A.; Kurth, L.B. Application of edible coatings on meats, poultry and seafoods: A review. *LWT-Food Sci. Technol.* **1997**, *30*, 337–350. [CrossRef]
11. Bilbao-Sainz, C.; Chiou, B.S.; Punotai, K.; Olson, D.; Williams, T.; Wood, D.; Rodov, V.; Poverenov, E.; McHugh, T. Layer-by-layer alginate and fungal chitosan based edible coatings applied to fruit bars. *J. Food Sci.* **2018**, *83*, 1880–1887. [CrossRef]
12. Rahmani, B.; Hosseini, H.; Khani, M.; Farhoodi, M.; Honarvar, Z.; Feizollahi, E.; Shokri, B.; Shojaee-Aliabadi, S. Development and characterisation of chitosan or alginate-coated low density polyethylene films containing Satureja hortensis extract. *Int. J. Biol. Macromol.* **2017**, *105*, 121–130. [CrossRef] [PubMed]
13. Campos, C.A.; Gerschenson, L.N.; Flores, S.K. Development of edible films and coatings with antimicrobial activity. *Food Bioprocess Technol.* **2011**, *4*, 849–875. [CrossRef]
14. Nair, M.S.; Tomar, M.; Punia, S.; Kukula-Koch, W.; Kumar, M. Enhancing the functionality of chitosan-and alginate-based active edible coatings/films for the preservation of fruits and vegetables: A review. *Int. J. Biol. Macromol.* **2020**, *164*, 304–320. [CrossRef] [PubMed]
15. Silva, J.M.; García, J.R.; Reis, R.L.; García, A.J.; Mano, J.F. Tuning cell adhesive properties via layer-by-layer assembly of chitosan and alginate. *Acta Biomater.* **2017**, *51*, 279–293. [CrossRef] [PubMed]
16. Nie, B.; Wang, H.; Zhang, Y.; Rao, C.; Wang, H.; Gao, X.; Li, W.; Niu, B. Effect of sodium alginate/phosphate-stabilized amorphous calcium carbonate nanoparticles on chitosan membranes. *Food Biosci.* **2022**, *46*, 101570. [CrossRef]
17. Shah, R.; Stodulka, P.; Skopalova, K.; Saha, P. Dual crosslinked collagen/chitosan film for potential biomedical applications. *Polymers* **2019**, *11*, 2094. [CrossRef]
18. Koc, F.E.; Altıncekic, T.G. Investigation of gelatin/chitosan as potential biodegradable polymer films on swelling behavior and methylene blue release kinetics. *Polym. Bull.* **2021**, *78*, 3383–3398. [CrossRef]

19. Jo, M.J.; Jeong, S.; Yun, H.K.; Kim, D.Y.; Kim, B.R.; Kim, J.L.; Na, Y.J.; Park, S.H.; Jeong, Y.A.; Kim, B.G. Genipin induces mitochondrial dysfunction and apoptosis via downregulation of Stat3/mcl-1 pathway in gastric cancer. *BMC Cancer* **2019**, *19*, 739. [CrossRef]
20. Muzzarelli, R.A.A. Genipin-crosslinked chitosan hydrogels as biomedical and pharmaceutical aids. *Carbohydr. Polym.* **2009**, *77*, 1–9. [CrossRef]
21. Nunes, C.; Maricato, É.; Cunha, Â.; Nunes, A.; Silva, J.A.L.d.; Coimbra, M.A. Chitosan–caffeic acid–genipin films presenting enhanced antioxidant activity and stability in acidic media. *Carbohydr. Polym.* **2013**, *91*, 236–243. [CrossRef]
22. Kildeeva, N.; Chalykh, A.; Belokon, M.; Petrova, T.; Matveev, V.; Svidchenko, E.; Surin, N.; Sazhnev, N. Influence of genipin crosslinking on the properties of chitosan-based films. *Polymers* **2020**, *12*, 1086. [CrossRef] [PubMed]
23. Iahnke, A.O.e.S.; Stoll, L.; Bellé, A.S.; Hertz, P.F.; Rios, A.d.O.; Rahier, H.; Flôres, S.H. Gelatin capsule residue-based films crosslinked with the natural agent genipin. *Packag. Technol. Sci.* **2020**, *33*, 15–26. [CrossRef]
24. Jahanbani, R.; Ghaffari, S.M.; Salami, M.; Vahdati, K.; Sepehri, H.; Sarvestani, N.N.; Sheibani, N.; Moosavi-Movahedi, A.A. Antioxidant and anticancer activities of walnut (*Juglans regia* L.) protein hydrolysates using different proteases. *Plant Foods Hum. Nutr.* **2016**, *71*, 402–409. [CrossRef]
25. Hardman, W.E.; Primerano, D.A.; Legenza, M.T.; Morgan, J.; Fan, J.; Denvir, J. mRNA expression data in breast cancers before and after consumption of walnut by women. *Data Brief* **2019**, *25*, 104050. [CrossRef] [PubMed]
26. Wang, H.; Gong, X.; Miao, Y.; Guo, X.; Liu, C.; Fan, Y.-Y.; Zhang, J.; Niu, B.; Li, W. Preparation and characterization of multilayer films composed of chitosan, sodium alginate and carboxymethyl chitosan-ZnO nanoparticles. *Food Chem.* **2019**, *283*, 397–403. [CrossRef] [PubMed]
27. Zhuang, C.; Jiang, Y.; Zhong, Y.; Zhao, Y.; Deng, Y.; Yue, J.; Wang, D.; Jiao, S.; Gao, H.; Chen, H. Development and characterization of nano-bilayer films composed of polyvinyl alcohol, chitosan and alginate. *Food Control* **2018**, *86*, 191–199. [CrossRef]
28. Dong, M.; Tian, L.; Li, J.; Jia, J.; Dong, Y.; Tu, Y.; Liu, X.; Tan, C.; Duan, X. Improving physicochemical properties of edible wheat gluten protein films with proteins, polysaccharides and organic acid. *LWT* **2022**, *154*, 112868. [CrossRef]
29. Zareie, Z.; Yazdi, F.T.; Mortazavi, S.A. Development and characterization of antioxidant and antimicrobial edible films based on chitosan and gamma-aminobutyric acid-rich fermented soy protein. *Carbohydr. Polym.* **2020**, *244*, 116491. [CrossRef]
30. Zhao, R.; Guan, W.; Zhou, X.; Lao, M.; Cai, L. The physiochemical and preservation properties of anthocyanidin/chitosan nanocomposite-based edible films containing cinnamon-perilla essential oil pickering nanoemulsions. *LWT* **2022**, *153*, 112506. [CrossRef]
31. Chang, Y.; Tsai, C.-C.; Liang, H.-C.; Sung, H.-W. Reconstruction of the right ventricular outflow tract with a bovine jugular vein graft fixed with a naturally occurring crosslinking agent (genipin) in a canine model. *J. Thorac. Cardiovasc. Surg.* **2001**, *122*, 1208–1218. [CrossRef]
32. Yoo, J.S.; Kim, Y.J.; Kim, S.H.; Choi, S.H. Study on genipin: A new alternative natural crosslinking agent for fixing heterograft tissue. *Korean J. Thorac. Cardiovasc. Surg.* **2011**, *44*, 197. [CrossRef] [PubMed]
33. Sheela, T.; Bhajantri, R.; Nambissan, P.; Ravindrachary, V.; Lobo, B.; Naik, J.; Rathod, S.G. Ionic conductivity and free volume related microstructural properties of LiClO4/PVA/NaAlg polymer composites: Positron annihilation spectroscopic studies. *J. Non-Cryst. Solids* **2016**, *454*, 19–30. [CrossRef]
34. Carneiro, J.; Tedim, J.; Ferreira, M. Chitosan as a smart coating for corrosion protection of aluminum alloy 2024: A review. *Prog. Org. Coat.* **2015**, *89*, 348–356. [CrossRef]
35. Hisham, S.F.; Kasim, S.H.; Abu Bakar, S.; Mohd Sabri, S.N.; Mastor, A.; Abdul Manaf, A.Y.; Abu, N.; Noorsal, K.; Abdul Rashid, A.H. Cross-linked effects by genipin on physicochemical properties of chitosan film. In Proceedings of the Advanced Materials Research; Trans Tech Publications Ltd.: Freienbach, Switzerland, 2016; Volume 1133, pp. 108–112. [CrossRef]
36. Osman, Z.; Arof, A.K. FTIR studies of chitosan acetate based polymer electrolytes. *Electrochim. Acta* **2003**, *48*, 993–999. [CrossRef]
37. Lawrie, G.; Keen, I.; Drew, B.; Chandler-Temple, A.; Rintoul, L.; Fredericks, P.; Grøndahl, L. Interactions between alginate and chitosan biopolymers characterized using FTIR and XPS. *Biomacromolecules* **2007**, *8*, 2533–2541. [CrossRef] [PubMed]
38. Li, J.; Ma, J.; Chen, S.; He, J.; Huang, Y. Characterization of calcium alginate/deacetylated konjac glucomannan blend films prepared by Ca2+ crosslinking and deacetylation. *Food Hydrocoll.* **2018**, *82*, 363–369. [CrossRef]
39. Gholizadeh, B.S.; Buazar, F.; Hosseini, S.M.; Mousavi, S.M. Enhanced antibacterial activity, mechanical and physical properties of alginate/hydroxyapatite bionanocomposite film. *Int. J. Biol. Macromol.* **2018**, *116*, 786–792. [CrossRef]
40. Aldana, A.A.; González, A.; Strumia, M.C.; Martinelli, M. Preparation and characterization of chitosan/genipin/poly (N-vinyl-2-pyrrolidone) films for controlled release drugs. *Mater. Chem. Phys.* **2012**, *134*, 317–324. [CrossRef]
41. Klein, M.P.; Hackenhaar, C.R.; Lorenzoni, A.S.; Rodrigues, R.C.; Costa, T.M.; Ninow, J.L.; Hertz, P.F. Chitosan crosslinked with genipin as support matrix for application in food process: Support characterization and β-d-galactosidase immobilization. *Carbohydr. Polym.* **2016**, *137*, 184–190. [CrossRef]
42. Liu, F.; Chiou, B.-S.; Avena-Bustillos, R.J.; Zhang, Y.; Li, Y.; McHugh, T.H.; Zhong, F. Study of combined effects of glycerol and transglutaminase on properties of gelatin films. *Food Hydrocoll.* **2017**, *65*, 1–9. [CrossRef]
43. Kumar, S.; Mudai, A.; Roy, B.; Basumatary, I.B.; Mukherjee, A.; Dutta, J. Biodegradable hybrid nanocomposite of chitosan/gelatin and green synthesized zinc oxide nanoparticles for food packaging. *Foods* **2020**, *9*, 1143. [CrossRef] [PubMed]

44. Arrieta, M.P.; Peltzer, M.A.; del Carmen Garrigós, M.; Jiménez, A. Structure and mechanical properties of sodium and calcium caseinate edible active films with carvacrol. *J. Food Eng.* **2013**, *114*, 486–494. [CrossRef]
45. Scartazzini, L.; Tosati, J.; Cortez, D.; Rossi, M.; Flôres, S.; Hubinger, M.; Di Luccio, M.; Monteiro, A. Gelatin edible coatings with mint essential oil (Mentha arvensis): Film characterization and antifungal properties. *J. Food Sci. Technol.* **2019**, *56*, 4045–4056. [CrossRef]
46. Butler, M.F.; Ng, Y.-F.; Pudney, P.D.A. Mechanism and kinetics of the crosslinking reaction between biopolymers containing primary amine groups and genipin. *J. Polym. Sci. Part A Polym. Chem.* **2003**, *41*, 3941–3953. [CrossRef]
47. Harnkarnsujarit, N.; Li, Y. Structure–property modification of microcrystalline cellulose film using agar and propylene glycol alginate. *J. Appl. Polym. Sci.* **2017**, *134*, 45533. [CrossRef]
48. Nandane, A.S.; Jain, R. Study of mechanical properties of soy protein based edible film as affected by its composition and process parameters by using RSM. *J. Food Sci. Technol.* **2015**, *52*, 3645–3650. [CrossRef]
49. Leelaphiwat, P.; Pechprankan, C.; Siripho, P.; Bumbudsanpharoke, N.; Harnkarnsujarit, N. Effects of nisin and EDTA on morphology and properties of thermoplastic starch and PBAT biodegradable films for meat packaging. *Food Chem.* **2022**, *369*, 130956. [CrossRef]
50. Kumari, N.; Bangar, S.P.; Petrů, M.; Ilyas, R.A.; Singh, A.; Kumar, P. Development and Characterization of Fenugreek Protein-Based Edible Film. *Foods* **2021**, *10*, 1976. [CrossRef]
51. Klinmalai, P.; Srisa, A.; Laorenza, Y.; Katekhong, W.; Harnkarnsujarit, N. Antifungal and plasticization effects of carvacrol in biodegradable poly (lactic acid) and poly (butylene adipate terephthalate) blend films for bakery packaging. *LWT* **2021**, *152*, 112356. [CrossRef]
52. Touyama, R.; Takeda, Y.; Inoue, K.; Kawamura, I.; Yatsuzuka, M.; Ikumoto, T.; Shingu, T.; Yokoi, T.; Inouye, H. Studies on the Blue Pigments Produced from Genipin and Methylamine. I. Structures of the Brownish-Red Pigments, Intermediates Leading to the Blue Pigments. *Chem. Pharm. Bull.* **1994**, *42*, 668–673. [CrossRef]
53. Chatkitanan, T.; Harnkarnsujarit, N. Development of nitrite compounded starch-based films to improve color and quality of vacuum-packaged pork. *Food Packag. Shelf Life* **2020**, *25*, 100521. [CrossRef]
54. Katekhong, W.; Wongphan, P.; Klinmalai, P.; Harnkarnsujarit, N. Thermoplastic starch blown films functionalized by plasticized nitrite blended with PBAT for superior oxygen barrier and active biodegradable meat packaging. *Food Chem.* **2022**, *374*, 131709. [CrossRef] [PubMed]
55. Bumbudsanpharoke, N.; Wongphan, P.; Promhuad, K.; Leelaphiwat, P.; Harnkarnsujarit, N. Morphology and permeability of bio-based poly (butylene adipate-co-terephthalate) (PBAT), poly (butylene succinate) (PBS) and linear low-density polyethylene (LLDPE) blend films control shelf-life of packaged bread. *Food Control* **2022**, *132*, 108541. [CrossRef]
56. Zhang, G.; Fang, S.; Regenstein, J.M.; Wang, F. Preparation, characterization and stability of nanoliposomes loaded with peptides from defatted walnut (*Juglans regia* L.) meal. *J. Food Sci. Technol.* **2022**. [CrossRef]
57. Matsui, R.; Honda, R.; Kanome, M.; Hagiwara, A.; Matsuda, Y.; Togitani, T.; Ikemoto, N.; Terashima, M. Designing antioxidant peptides based on the antioxidant properties of the amino acid side-chains. *Food Chem.* **2018**, *245*, 750–755. [CrossRef] [PubMed]
58. Jariteh, M.; Ebrahimzadeh, H.; Niknam, V.; Mirmasoumi, M.; Vahdati, K. Developmental changes of protein, proline and some antioxidant enzymes activities in somatic and zygotic embryos of Persian walnut (*Juglans regia* L.). *Plant Cell Tissue Organ Cult. (PCTOC)* **2015**, *122*, 101–115. [CrossRef]
59. Abd El-Hack, M.E.; El-Saadony, M.T.; Shafi, M.E.; Zabermawi, N.M.; Arif, M.; Batiha, G.E.; Khafaga, A.F.; Abd El-Hakim, Y.M.; Al-Sagheer, A.A. Antimicrobial and antioxidant properties of chitosan and its derivatives and their applications: A review. *Int. J. Biol. Macromol.* **2020**, *164*, 2726–2744. [CrossRef]
60. Xue, C.; Yu, G.; Hirata, T.; Terao, J.; Lin, H. Antioxidative activities of several marine polysaccharides evaluated in a phosphatidylcholine-liposomal suspension and organic solvents. *Biosci. Biotechnol. Biochem.* **1998**, *62*, 206–209. [CrossRef]
61. Tomida, H.; Fujii, T.; Furutani, N.; Michihara, A.; Yasufuku, T.; Akasaki, K.; Maruyama, T.; Otagiri, M.; Gebicki, J.M.; Anraku, M. Antioxidant properties of some different molecular weight chitosans. *Carbohydr. Res.* **2009**, *344*, 1690–1696. [CrossRef]
62. Kelishomi, Z.H.; Goliaei, B.; Mahdavi, H.; Nikoofar, A.; Rahimi, M.; Moosavi-Movahedi, A.A.; Mamashli, F.; Bigdeli, B. Antioxidant activity of low molecular weight alginate produced by thermal treatment. *Food Chem.* **2016**, *196*, 897–902. [CrossRef]
63. Panda, P.; JenMing, Y.; YenHsiang, C.; WeiWen, S. Modification of different molecular weights of chitosan by p-Coumaric acid: Preparation, characterization and effect of molecular weight on its water solubility and antioxidant property. *Int. J. Biol. Macromol.* **2019**, *136*, 661–667. [CrossRef] [PubMed]

*foods*

MDPI

*Article*

# Hydrophobic Starch-Based Films Using Potato Washing Slurries and Spent Frying Oil

**Sílvia Petronilho [1,2,*], André Oliveira [1], M. Rosário Domingues [1,3,4], Fernando M. Nunes [2], Manuel A. Coimbra [1] and Idalina Gonçalves [5]**

1 LAQV-REQUIMTE, Department of Chemistry, Campus Universitário de Santiago, University of Aveiro, 3810-193 Aveiro, Portugal; andreoliveira98@ua.pt (A.O.); mrd@ua.pt (M.R.D.); mac@ua.pt (M.A.C.)
2 Chemistry Research Centre-Vila Real, Department of Chemistry, University of Trás os-Montes and Alto Douro, Quinta de Prados, 5001-801 Vila Real, Portugal; fnunes@utad.pt
3 Mass Spectrometry Centre, Department of Chemistry, Campus Universitário de Santiago, University of Aveiro, 3810-193 Aveiro, Portugal
4 CESAM, Centre for Environmental and Marine Studies, Department of Chemistry, Campus Universitário de Santiago, University of Aveiro, 3810-193 Aveiro, Portugal
5 CICECO—Aveiro Institute of Materials, Department of Materials and Ceramic Engineering, Campus Universitário de Santiago, University of Aveiro, 3810-193 Aveiro, Portugal; idalina@ua.pt
* Correspondence: silviapetronilho@ua.pt

**Abstract:** Starch is a promising candidate for preparing biodegradable films with useful gas barriers and thermoplastic capabilities. However, these materials are hydrophilic and brittle, thus limiting their application range. To overcome these drawbacks, it has been hypothesized that starch can be hydrophobized and plasticized during the starch-based film production using a single-step approach and following transesterification principles. In this work, KOH powder and spent frying oil (SFO) were used as an alkaline catalyst and a source for triacylglycerides, respectively, to promote the modification of starch. Different ratios of SFO (*w*/*w* related to the dried starch weight) were tested. When compared to the neat films (without a catalyst and SFO), the incorporation of at least 15% SFO/KOH gave rise to transparent, hydrophobic (water contact angles of *ca.* 90°), stretchable (*ca.* 20×), elastic (*ca.* 5×), and water tolerant starch-based films, contrary to the films produced without the catalyst. ATR-FTIR and $^1$H NMR revealed structural differences among the produced films, suggesting that starch was modified with the SFO-derived fatty acids. Therefore, adding KOH during the potato starch/spent frying oil-based film's production was determined to be a promising in situ strategy to develop starch-based materials with improved hydrophobicity and flexibility, while valorizing the potato chip industry's byproducts.

**Keywords:** potato chip byproducts; biobased films; alkaline catalyst; transesterification; single step; hydrophobicity; stretchability; water vapor barrier

**Citation:** Petronilho, S.; Oliveira, A.; Domingues, M.R.; Nunes, F.M.; Coimbra, M.A.; Gonçalves, I. Hydrophobic Starch-Based Films Using Potato Washing Slurries and Spent Frying Oil. *Foods* **2021**, *10*, 2897. https://doi.org/10.3390/foods10122897

Academic Editor: João Borges Laurindo

Received: 12 October 2021
Accepted: 18 November 2021
Published: 23 November 2021



## 1. Introduction

Starch, a carbohydrate composed of linked D-glucose units in the form of amylose and amylopectin [1], stands out as a promising natural polymer for developing biodegradable films due to its film-forming ability with thermoplastic and interesting gas barrier properties [2–4], along with its high availableness, reduced cost, and renewability [5]. To avoid the use of edible starch in the material's development, starch obtained from agri-food industry byproducts has been considered [6,7]. For instance, starch recovered from the potato industry's washing slurries recently showed physicochemical, thermal, and rheological properties that were competitive with commercially available starch (directly extracted from the potato) and suitable for use in the development of biobased materials [8]. Nevertheless, a shortcoming of starch is its inherent hydrophilic nature [9] due to its hydroxyl-rich structure that can form hydrogen bonds with water, thus it creates materials that have a high water sensitivity and are of a hygroscopic nature [10]. This drawback has limited the application range of starch-based material and has hampering its widespread usage in the biobased and biodegradable plastics industry.

Chemical modification has been studied [5,11,12] for improving the hydrophobicity of starch, which mainly stands on the replacement of the hydroxyl functional groups of the starch's backbone with huge hydrophobic groups that have no capacity to form hydrogen bonding with water molecules [13]. The direct esterification of starch with free fatty acids [14–16] or with fatty acid derivatives, such as chlorides [10,17] and anhydrides [18–20], which are thermodynamically favorable, is one of the strategies used. Besides hydrophobization, starch esterification with fatty acids also increases its thermal stability and processability properties [5,11,12]. However, these multi-step approaches are time consuming and normally involve the use of organic solvents or mixtures of solvents to first solubilize the starch before proceeding with its chemical modification [10,17–20]. Moreover, an enzymatic pre-treatment using triacylglyceride sources (such as vegetable oils) is also normally required to release the fatty acid derivatives for further use in starch esterification [14–16]. Transesterification with highly pure soybean oil [21] and analytical grade esters, such as vinyl laurate and stearate [22], both in DMSO and when using alkaline catalysts, has also been applied to modify the hydrophilic nature of commercial starch. Similarly, hydrophobic starch can also be obtained from analytical grade fatty acid methyl esters (FAMEs) using densified carbon dioxide in the presence of potassium carbonate in a high-pressure reactor [23]. When used in paper coating, the transesterified starch was able to provide high hydrophobicity and prolonged water resistance to the cellulose-based materials [21].

To the best of our knowledge, none of the already published works deal with the modification of starch during the formation of biobased films, but have always focused on a two-step reaction procedure for the modification of granular starch followed by its processing as a biomaterial. In this work, it was hypothesized that, in the presence of potassium hydroxide (KOH), starch can be in situ hydrophobized and plasticized during the production of starch/potato chips' spent frying oil-based films, following transesterification principles. The reaction mechanism involved in the transesterification of starch in the presence of triacylglycerides is presumably similar to the alcohols' transesterification with triacylglycerides, forming fatty acid esters [24,25]. In this reaction, the alcohol is transformed into an alkoxide by an alkaline catalyst (such as KOH used in this work). After this, the triacylglycerides' carbonyl group is attacked by the alkoxide, thus forming an alkyl ester and an anion of diacylglyceride that instantaneously captures a proton from the medium. Without the presence of the catalyst, no transesterification reaction occurs between the oil-derived fatty acids and the OH groups of the starch. The conditions that were used needed to be a compromise between promoting the transesterification in the presence of KOH as catalyst and avoiding the saponification reactions that can be promoted by KOH.

Under a circular economy concept, and as a sustainable alternative of channeling foodstuffs for materials purposes, starch and esterified fatty acids were recovered from potato washing slurries and potato chip spent frying oil, respectively, two non-value byproducts of the potato chip industry.

## 2. Materials and Methods

### 2.1. Materials

Potato chip industry byproducts, namely washing slurries and frying residues, were provided by A Saloinha, Lda. (Mafra, Portugal). Starch was recovered from the lyophilized potato washing slurries, which was formed by 25 % amylose and had a gelatinization temperature and enthalpy of 59–71 °C and 12.5 J/g, respectively [8].

Potassium hydroxide (KOH, 99.9% purity) was purchased from José Manuel Gomes dos Santos (Odivelas, Portugal), glycerol was from Scharlab S.L. (Barcelona, Spain), and sodium azide, chloroform, and methanol were from Sigma-Aldrich (Lisboa, Portugal). Analytical grade reagents were used with no further purification.

### 2.2. Recovery and Characterization of Potato Chip Spent Frying Oil

Potato chip spent frying oil (SFO) was recovered from 60 g of potato chip industry frying residues using Soxhlet extraction, at 70 °C, for 5 h (*ca.* 15 extraction cycles) with a chloroform/methanol mixture (2:1 *v*/*v*). Then, the remaining water was removed using anhydrous sodium phosphate, and the solvent was evaporated in a rotary evaporator until the samples' weight was constant. The recovered SFO yield (g of lipid per 100 g of potato chip frying residues) was determined in three independent replicates. The SFO was stored in dark conditions, at room temperature, until further use.

The fatty acid methyl esters (FAMEs) were determined after the SFO fraction alkaline-catalyzed transesterification [26] using 5 mg of SFO andheptadecanoate methyl ester (1.5 mg/mL, in *n*-hexane) and KOH methanolic solution (2 M) as the internal standard and the catalyst, respectively. The FAMEs were analyzed in a GC-FID Perkin Elmer Clarus 400 equipped with a DB-FFAP column (30 m × 0.32 mm (I.D.) × 0.25 μm film thickness, J&W Scientific Inc., Folsom, CA, USA). The injection was in split mode with 20:1 (5 min) ratio. The GC injection port and the detector were programmed at 245 and 250 °C, respectively. The oven program followed 3 temperature ramps as already described [8]. The identification of the FAME compounds was based on the comparison of their retention times with the ones obtained by a commercial FAME mixture ($C_8$–$C_{24}$) injection.

The peroxide value of the recovered SFO was evaluated by titration according to the American Oil Chemists' Society (AOCS) official methods [27]. Briefly, 5 g SFO was dissolved in 30 mL acetic acid/chloroform mixture (3:2 *v*/*v*) to which saturated potassium iodide solution (0.5 mL) was then added. This mixture was then incubated in the dark (1 min) and diluted with distilled water (30 mL). The titration occurred with a sodium thiosulfate solution (0.01 N) in the presence of a starch suspension (10%, 0.5 mL) until the dark blue color disappeared. A total of 3 independent replicates were performed and the peroxide value was expressed in milliequivalents of peroxide per kg of the oil sample.

### 2.3. Starch-Based Film Production

The starch-based films were produced following a previously described solvent casting method [8]. Briefly, a suspension of potato starch (4% *w*/*v*) containing KOH (10% *w*/*w* related to the SFO weight basis), as alkaline catalyst, glycerol (30% *w*/*w* related to the starch dry weight basis), as a plasticizer, and SFO in different proportions (10%, 15%, and 20% *w*/*w* related to starch dry weight basis). Each dispersion was gelatinized at 95 °C ± 0.1, for a total of 45 min with constant stirring (*ca.* 500 rpm). The obtained gelatinized suspension was then degassed with a vacuum pump, spread onto plexiglass plates, and dried (25 °C, overnight). Films without KOH and SFO were used as controls.

### 2.4. Film Characterization

#### 2.4.1. Wettability

Water contact angles (WCAs) of the developed films' surfaces were evaluated using a tensiometer (OCA 20, Dataphysics) equipped with an automatic image capture system (Dataphysics SCA20 M4), as described [7]. Briefly, 3 μL ultrapure water were dropped in the film strips' surfaces (1 cm × 10 cm) and the WCAs were calculated following the Laplace–Young equation [28]. At least 10 droplet images were obtained along each strip length. The measurements were performed on the two surfaces of each produced starch-based film, i.e., the one exposed to air during casting (film upper surface) and the other one which was in contact with the plexiglass plates (film lower surface).

### 2.4.2. Mechanical Properties

The mechanical properties of each developed starch-based film were determined using texture analyzer apparatus (model TA. Hdi, Stable Micro Systems), as already described [7]. Briefly, each film was cut (10 strips each of size $1 \times 9$ cm) and the thickness was measured at 4 different points using a digital micrometer (*ca.* 0.001 mm accuracy; Mitutoyo Corporation, Kanagawa, Japan). A total of 6 specimens of each developed film were tested. Tensile strength, elongation at break, and Young's modulus values were calculated based on the ASTM D 882 83 standard method.

### 2.4.3. Fourier-Transform Infrared (FTIR) Spectroscopy Analysis

The FTIR analysis of the films was determined by a Golden Gate single reflection diamond ATR system in a Perkin Elmer Spectrum BX spectrometer (Perkin Elmer, Inc., Hopkinton, MA, USA). FTIR spectra were taken at a resolution of 16 $cm^{-1}$ and with double bi-directional scans, 32 co-added scans, and a wave number range between 4000 and 500 $cm^{-1}$ (mid infrared region). The FTIR analyses were performed in five replicates of the film squares (4 $cm^2$) that had been previously washed in *n*-hexane. To allow a comparison among the different film samples under study each spectrum was normalized.

### 2.4.4. Proton Nuclear Magnetic Resonance Spectroscopy ($^1H$ NMR) Analysis

The NMR measurements were conducted to obtain detailed information on the structure of the developed films. The $^1H$ NMR spectra were acquired on a Bruker Advance III TM HD 500 MHz liquid-state NMR spectrometer (narrow bore), 11.75 T magnetic field (ultrashield ascend). Prior to the analysis, each film sample was dissolved in dimethyl sulfoxide (DMSO)-d6 at room temperature. The experiments were carried out at 10 °C with a double resonance 5 mm broadband observation probe with inverse capabilities and $N_2$ cryocontrol unit for prodigy probe with z-gradients. All chemical shifts were in parts per million (ppm). The measurements were performed in triplicate, each one corresponding to an independent experiment.

### 2.4.5. Chromatic Properties

The chromatic properties of each film were evaluated using tristimulus colorimetry (CIELab). The $L^*$ (luminosity), $a^*$ (red/green color), and $b^*$ (yellow/blue color) chromatic coordinates were analyzed using a CR-400 Chroma Meter [29]. The films' total color difference (ΔE) was calculated using Equation (1):

$$\Delta E = \sqrt{(\Delta L^*)^2 + (\Delta a^*)^2 + (\Delta b^*)^2} \quad (1)$$

where $\Delta L^*$, $\Delta a^*$, and $\Delta b^*$ are the $L^*$, $a^*$, and $b^*$ differences among films with KOH and SFO and the neat starch-based films.

### 2.4.6. Water Solubility

Each film square (4 $cm^2$) was weighed, immersed in an aqueous solution (30 mL) containing 0.02% (*v*/*v*) of sodium azide, and kept with orbital agitation (80 rpm, room temperature, 7 days). The films were then dried (105 °C, overnight), cooled down to room temperature, and weighed [7]. The weight loss percentage was determined to evaluate the solubility of each film (Equation (2)):

$$\%\ \text{Water solubility} = \frac{W_f - W_i}{W_f} \times 100 \quad (2)$$

where $W_f$ and $W_i$ are the films' weight before (with no residual humidity, dried following the aforementioned conditions) and after the incubation into distilled water. All measurements were performed in three independent replicates.

#### 2.4.7. Water Vapor Transmission Rate

The water vapor transmission rate (WVTR) of each film (samples with 28 mm diameter) was assessed using test dishes and a chamber kept at 23 ± 2 °C with 53% relative humidity and an air velocity of *ca.* 160 m/min, as already described [7]. WVTR was calculated following Equation (3):

$$\text{WVTR}\left(\frac{\text{g}}{\text{m}^2} \times \text{day}\right) = \frac{24 \times \text{X}}{\text{A} \times \text{Y}} \quad (3)$$

where X refers to the weight gain expressed in grams, Y refers to the time in h for the weight gain of X, and A is the area exposed of each specimen ($m^2$). Thee independent replicates were determined for each formulation.

#### 2.4.8. Statistical Analysis

The data from wettability, mechanical, chromatic, water solubility, and water vapor transmission rate were statistically evaluated by applying the Student's *t*-test with a level of significant difference of 95% and $p < 0.05$.

## 3. Results and Discussion

Potato chip industry frying residues add a total of 42% (*w*/*w*) spent frying oil (SFO), mainly composed by oleic (50%), palmitic (33%), linoleic (12%), and stearic (3%) acid residues (Figure S1). The SFO peroxide value, indicative of oil oxidation promoted by thermal frying conditions [30], was 7.2 ± 0.5 milliequivalent of oxygen/kg of oil. Similar peroxide values have been reported for sunflower and soyabean oils (*ca.* 7.6 and 7.7 milliequivalent of oxygen/kg of oil, respectively) with 4 frying at 170–180 °C [27].

The influence of an alkaline catalyst (KOH) and SFO on the formulation of potato starch-based films was studied by measuring the films' wettability, mechanical, chromatic, water solubility, and water vapor barrier properties.

### 3.1. *Wettability of Potato Starch/SFO/KOH-Based Films*

Figure 1 reveals the similarity of the water contact angles (WCAs) on both sides of the neat potato starch-based films (without SFO and KOH), highlighting their compositional homogeneity and hydrophilic character (*ca.* 50°). The incorporation of SFO significantly decreased the films hydrophilicity at the upper side (from *ca.* 50° to *ca.* 80°), keeping the WCA at *ca.* 50° in the films' lower surface. This higher WCA variability between the starch/SFO-based films' surfaces can be explained by the phase separation concerning water and SFO during casting and/or due to the micelles' migration promoted by the different compounds' density, similarly to what was observed in starch-based films when using waxes and oils recovered from potato industry byproducts [8]. These wettability profiles were similar for all the tested SFO loadings. In the presence of KOH, both upper and lower sides of the starch/SFO-based films became hydrophobic (*ca.* 90° on both sides), particularly when KOH and 15 and 20% SFO were used. Indeed, films containing KOH and 10% SFO still showed surfaces with heterogeneous wettability (*ca.* 115° at the upper and *ca.* 80° at the lower surface), showing the migration of lipidic components onto the films upper surface. The homogeneous wettability of the films containing the highest SFO concentrations suggests the starch modification with SFO-derived fatty acids through transesterification reactions.

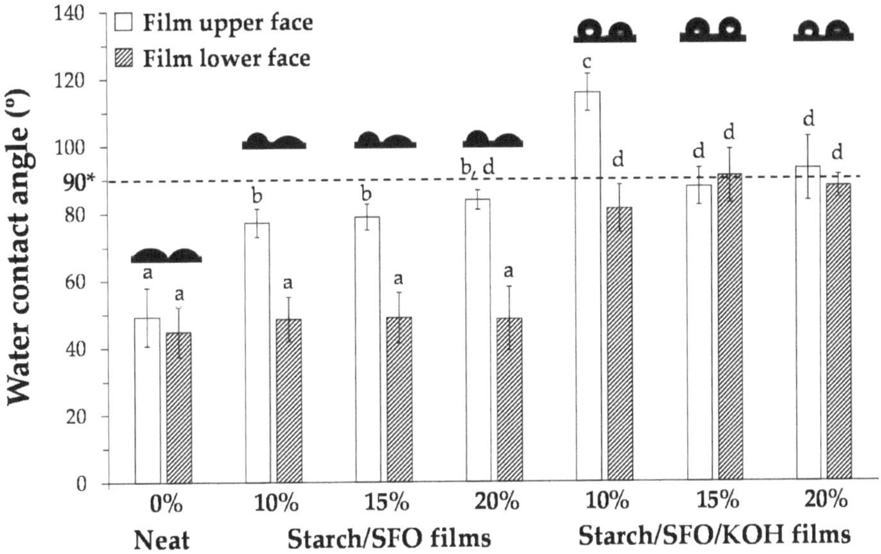


**Figure 1.** Water contact angles of "upper" and "lower" surfaces of potato starch-based films containing different SFO (spent frying oil) amounts (0, 10, 15 and 20% *w/w* related to starch dry weight) in the absence and presence of KOH (10% *w/w* related to the SFO weight). Different lowercase letters represent significantly different values ($p < 0.05$). * Hydrophobicity benchmark [31].

### 3.2. Thickness and Mechanical Properties of Potato Starch/SFO/KOH-Based Films

The neat potato starch-based films revealed a thickness of *ca.* 72 µm and a brittleness performance, reflected by their low tensile strength (*ca.* 7 MPa), elongation at break (*ca.* 1%), and high rigidity (Young's modulus of *ca.* 507 MPa), following the tendency already reported [7,32]. Compared to the neat films, SFO incorporation increased the films' elongation at break and decreased their tensile strength and Young's modulus. Except for the 15% SFO amount, SFO had no significant influence in the films thickness. This amount of SFO seemed to promote the higher dispersion of starch and oil with a consequent maximum thickness. When the amount of oil increased to 20% an oil phase seems to be formed, explaining the decrease in the thickness of the films when compared to the 15% SFO. Moreover, the mechanical behavior was similar for all the SFO loadings tested (Figure 2), being more evident for the highest SFO amount used. The SFO presence in the starch matrix might have contributed to the decrease in the hydrogen cohesion forces within the amylose network, conferring a more flexible behavior to the starch/SFO-based films, similarly to what was observed when other lipidic compounds, such as potato waxes [8] and free fatty acids (palmitic, stearic, and oleic) were used [33]. Accordingly, the use of 1–3% oil in relation to starch (*w/w*) was shown not to be enough to change the starch-based films' mechanical performance [8].

In the presence of KOH, although following the same mechanical trend as the films that only used SFO, the increase in the films' elongation at break was even more evident when 15 and 20% SFO were used, as well as the decrease in the tensile strength and Young's modulus. Therefore, for films containing 15 and 20% SFO, KOH increased by *ca.* 20× the stretchability and *ca.* 5× the elasticity of the neat materials, indicating that the starch might be modified with the SFO-derived fatty acids. The 10% SFO/KOH films presented an elongation, tensile strength, and Young's Modulus values similar to the ones observed for the neat films. This can be explained by the modification of the starch to a certain extent, with an amount of free oil that was still high but not enough to change the mechanical performance of the starch-based films, when compared to the matrix containing a high amount of oil (15 and 20% SFO) that did not easily disperse in the aqueous starch medium. Thus, contrasting to starch/10% SFO/KOH-based films, 15 and 20% SFO amounts revealed to be in sufficient quantities to hinder the SFO-derived fatty acids dispersion, thus promoting the starch matrix discontinuity and decreasing the cohesion forces of the starch structure.

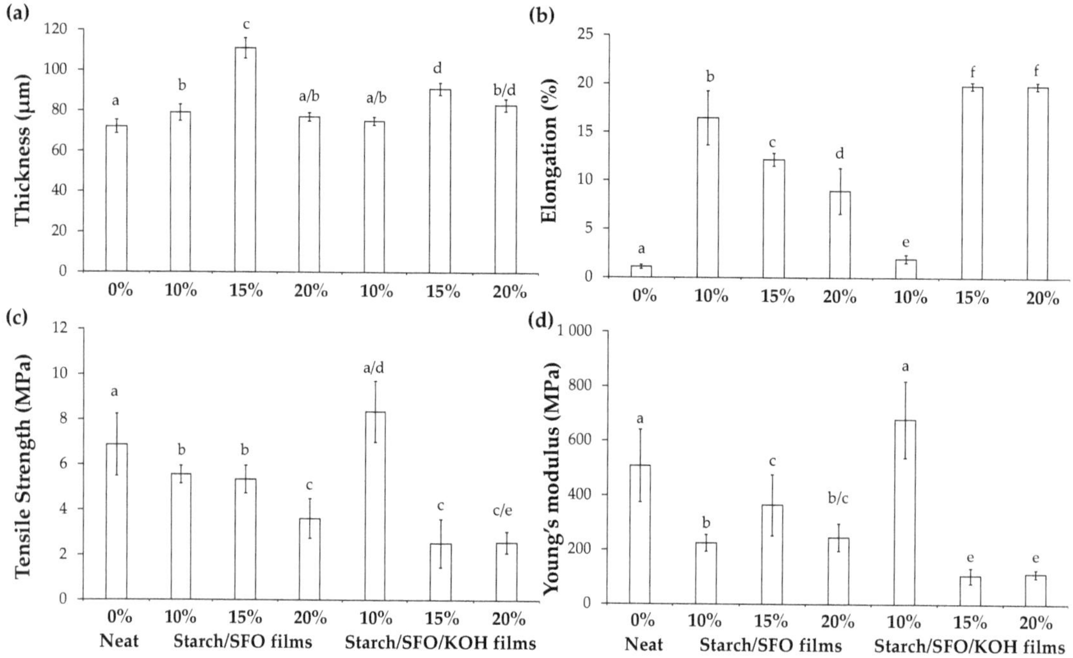


**Figure 2.** Thickness (**a**) and mechanical properties (elongation at break (**b**), tensile strength (**c**), and Young's modulus (**d**) of potato starch-based films different SFO amounts (0, 10, 15 and 20% *w*/*w* related to starch dry weight) in the absence and presence of KOH (10% *w*/*w* related to the SFO weight). Different lowercase letters represent significantly different values ($p < 0.05$).

Based on wettability (Figure 1) and mechanical property results (Figure 2), the most hydrophobic, stretchable, and elastic properties were obtained when KOH and 15 and 20% SFO were incorporated into the starch-based formulations and no significant differences were observed among these two SFO amounts. With the aim of validating whether the starch network chemical modification was responsible for these changes, the structural (FTIR and $^1$H RMN assays), chromatic, water solubility, and water vapor barrier properties of starch-, starch/15% SFO-, and starch/15% SFO/KOH-based films were studied.

### *3.3. Structural Characterization of Potato Starch/15% SFO/KOH-Based Films*

#### 3.3.1. FTIR Analysis

The FTIR profiles of the three developed formulations (Figure 3) present peaks at 1157 cm$^{-1}$, 1080 cm$^{-1}$, and 998 cm$^{-1}$, which were characteristic of starch [34]. Besides, another peak at 1634 cm$^{-1}$ appears, corresponding to the water molecules absorbed in the starch-based films. A strong and broad adsorption band at around 3000–3500 cm$^{-1}$, attending its maximum at 3300 cm$^{-1}$, was also observed, which is related to the hydroxyl groups of the potato starch. The FTIR spectra of both starch-based films with 15% SFO and 15% SFO/KOH showed a new band at 1740 cm$^{-1}$ corresponding to the carbonyl group (C=O) that can be related, in the first case, to the fatty esters from the triacylglycerides of the SFO, or even, in the second case, from starch esters that can be formed in the presence of KOH [22]. Besides, the aliphatic region from the long SFO fatty acid chains is also clearly present at 2916 cm$^{-1}$ and 2851 cm$^{-1}$ in films containing 15% SFO and 15% SFO/KOH.

Furthermore, only in the presence of KOH, a small peak at 1560 $cm^{-1}$ was observed, corresponding to the carboxylate anion group ($COO^-$) generated during the alkaline saponification of the SFO-derived fatty acids [35], suggesting that they can also be involved in the modification of the starch-based film properties.

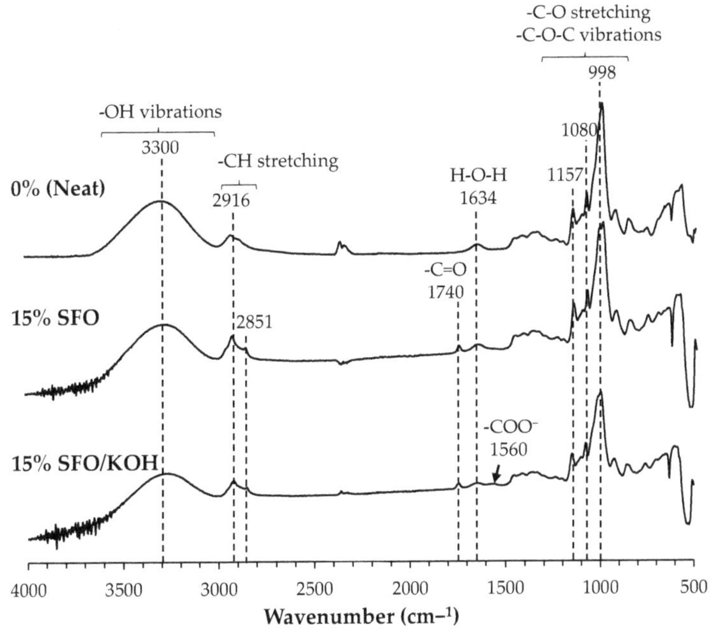


**Figure 3.** FTIR spectra of potato starch-based films containing 0% (neat), 15% SFO, and 15% SFO/KOH.

3.3.2. $^1$H NMR Analysis

For the three developed formulations, the strongest signals were assigned as water (3.34 ppm) and DMSO-d6 (2.50 ppm) (Figure 4) [10]. In the $^1$H NMR spectrum of the neat film the starch characteristic proton resonance signals in the range of 3.50 ppm to 5.65 ppm were visible, which corresponded to the glucose ring protons. Particularly, the resonance signals at 5.40, 5.50 and 4.58 ppm were assigned to the 3 hydroxyl groups on C-2, C-3, and C-6 (OH-2, 3, and 6), respectively [15]. Moreover, at 5.10 ppm and 4.90 ppm it was possible to observe the signals of the anomeric hydrogen atom, denoted as H-1 and H-1′, corresponding to the starch $\alpha$1,4 and $\alpha$1,6 linkages signals, respectively [36].

For the $^1$H NMR spectrum of starch-based films with 15% SFO, the characteristic proton resonance bands of glucose units were still present, but new signals appeared at 0.87 ppm associated to the protons of the terminal methyl group ($CH_3$), at 2.22 ppm associated to the methylene group ($CH_2$) that occurs immediately before the carbonyl group of the ester, denoted as $CH_2$(CO), which was linked to a second oxygen atom bonded to the carbon atom in the carbonyl group by a single bond. Other proton peaks at 1.24 ppm, 1.44 ppm, and 1.96 ppm, associated to other protons of different methylene groups of the SFO fatty acid chain [15], were also observed in films containing 15% SFO. However, in the presence of KOH, it was observed that the appearance of new spectral signals in the region of the aliphatic hydrogen atoms ($CH_2$ groups) of the fatty acid chain (1.20 ppm, 1.67 ppm, 1.91 ppm, 2.12 ppm), as well as on the region of the terminal methyl groups at 0.76 ppm, denoted as $CH_3$, were possible due to the chemical environment changes promoted by the starch modification in the presence of KOH. Furthermore, changes in the proton resonance of the OH-2 and OH-3 glucose units were observed (Figure 4), suggesting that the modification of the starch that occurred at the C-2 or C-3 positions, could have resulted from changed chemical environments due to the modification of the starch by the SFO-derived fatty acids in the presence of KOH. These changes may be due to the substitution of hydrogen atoms in the starch hydroxyl groups by the SFO-derived fatty acids.

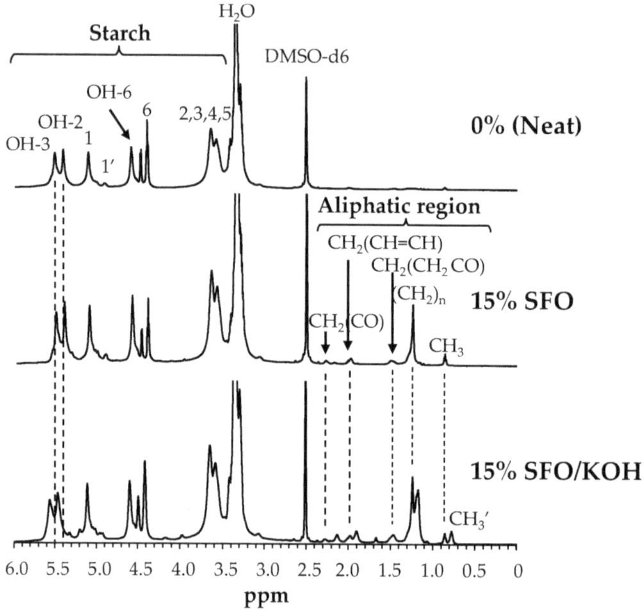


**Figure 4.** $^1$H NMR spectra of potato starch-based films containing 0% (neat), 15% SFO, and 15% SFO/KOH, in DMSO-d6.

### 3.4. *Chromatic Properties of Potato Starch/15% SFO/KOH-Based Films*

The $L^*$ values determined among the three tested film conditions were statistically similar (Table 1), revealing that, even in the presence of KOH, the incorporation of 15% SFO maintained the luminosity of the neat materials. However, both a* and b* values significantly changed. For starch-based films with 15% SFO and 15% SFO/KOH, the a* values decreased from 1.54 (neat film) to 1.27 and 0.89, respectively, while the b* values increased from −1.64 (neat film) to −0.99 and 0.23, respectively. Therefore, both SFO and SFO/KOH conferred a light yellowish coloration to the native films, this effect being more pronounced when KOH was included in the system. Moreover, the total color difference ($\Delta$E) values varied from 0.75 to 1.98 for films with 15% SFO and 15% SFO/KOH, respectively. These differences may be related to the yellow-colored oxidized compounds that were present in the recovered SFO. These chromatic variations are in accordance to the bibliography, where the $\Delta$E value was impacted by the incorporation of natural potato peel waxes [8] and beeswax, candelilla stalks, and carnauba leaves [37]. However, as can be observed in real images of the films in Table 1, the intrinsic films' transparency was un-compromised by the usage of SFO and SFO/KOH.

### 3.5. *Solubility and Water Vapor Barrier Properties of Potato Starch/15% SFO/KOH-Based Films*

When in contact with aqueous environments, the neat films (0%) lost *ca.* 22% of their initial weight. A similar effect was observed for the starch/15% SFO-based films (*ca.* 23%). However, the presence of KOH minimized the weight loss (*ca.* 18%) of starch/15% SFO-based films, thus increasing their water tolerance (Figure 5a). This phenomenon supports the starch transesterification with SFO-derived fatty acids, decreasing the hydrogen bonding among the starch hydroxyl groups and the surrounding water molecules. As all starch-based films contained glycerol as plasticizing agent, the high-water solubility of this very hydrophilic molecule can explain the verified films' weight loss after immersion into water [7].

**Table 1.** Real images and mean values of lightness ($L^*$), red-green ($a^*$), yellow-blue ($b^*$), and total color difference ($\Delta E$) of potato starch-based films containing 0% (neat), 15% SFO (spent frying oil), and 15% SFO/KOH. Different lowercase letters represent significantly different values ($p < 0.05$).

| Starch-Based Films | Real Image | $L^*$ | $a^*$ | $b^*$ | $\Delta E$ |
|---|---|---|---|---|---|
| 0% (neat) |  | $86.76 \pm 0.42$ [a] | $1.54 \pm 0.04$ [a] | $-1.64 \pm 0.16$ [a] | - |
| 15% SFO |  | $87.06 \pm 0.32$ [a] | $1.27 \pm 0.07$ [b] | $-0.99 \pm 0.29$ [b] | 0.75 |
| 15% SFO/KOH |  | $86.66 \pm 0.35$ [a] | $0.89 \pm 0.07$ [c] | $0.23 \pm 0.31$ [c] | 1.98 |

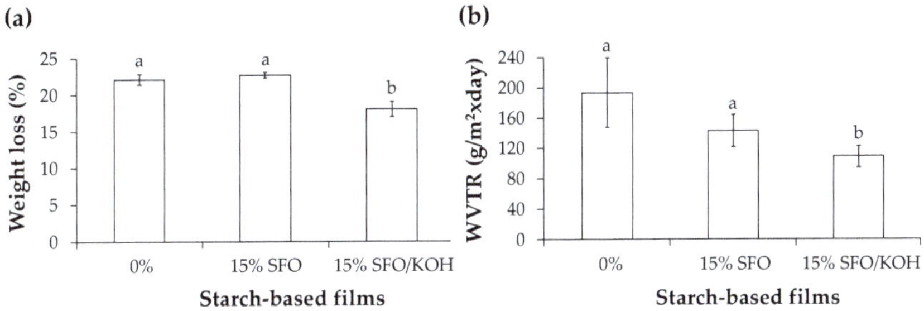


**Figure 5.** Weight loss (**a**) and water vapor transmission rate (WVTR) (**b**) of potato starch-based films containing 0%, 15% SFO, and 15% SFO/KOH. Different lowercase letters represent significantly different values ($p < 0.05$).

The presence of KOH allowed the significant decrease of the WVTR of the starch-based films (Figure 5b). The incorporation of only 15% SFO during film production decreased the films' WVTR, although did not show statistic difference when compared to the neat materials. Since the water vapor permeability occurred through the hydrophilic molecules that constituted the films [38], including amylose and amylopectin from potato starch, as well as glycerol, the highest WVTR decrease observed in the starch/15% SFO/KOH-based films reinforced the starch hydrophobization hypothesis by transesterification with the SFO fatty acids.

## 4. Conclusions

The incorporation of KOH during the starch/SFO-based films production revealed to be a promissory strategy for developing water tolerant biobased materials with enhanced stretchability and elasticity without changing their transparency, opening the possibility to extend the use of starch-based bioplastics to fields that include packaging applications. FTIR and $^1$H NMR suggested the possible modification of the starch structure by the presence of chemically bounded SFO-derived fatty acids. Therefore, this work establishes an in situ single-step approach for the simultaneous hydrophobization and plasticization of potato starch-based films, offering a new valorization strategy for the potato chip industry byproducts. The results obtained with the casting process used paves the way for the application of starch/SFO/KOH formulations by the plastic industries in their extrusion processes.

**Supplementary Materials:** The following are available online at https://www.mdpi.com/article/10.3390/foods10122897/s1, Figure S1: Fatty acids composition (determined as FAME) of the spent frying oil recovered from potato chip industry frying residues.

**Author Contributions:** Conceptualization, S.P.; methodology, S.P., A.O. and I.G.; software, A.O.; validation, S.P., I.G. and M.A.C.; formal analysis, S.P. and A.O.; investigation, S.P.; resources, M.A.C.; data curation, A.O.; writing—original draft preparation, S.P. and A.O.; writing—review and editing, S.P., M.R.D., F.M.N., I.G. and M.A.C.; visualization, S.P.; supervision, S.P., I.G. and M.A.C. All authors have read and agreed to the published version of the manuscript.

**Funding:** This work received financial support from PT national funds (FCT/MCTES) for the financial support to LAQV-REQUIMTE (FCT UIDB/50006/2020), CICECO-Aveiro Institute of Materials (FCT UIDB/50011/2020 and UIDP/50011/2020), and CESAM (FCT UIDB/50017/2020; UIDP/50017/2020) research units, and CQ-VR at UTAD Vila Real (UIDP/00616/2020), and, when applicable, co-financed by the FEDER, within the PT2020 Partnership Agreement and Compete 2020, and the Portuguese NMR Network. FCT is thanked for the post-doc grant (SP, SFRH/BPD/117213/2016) and for the Individual Call to Scientific Employment Stimulus (IG, CEECIND/00430/2017).

**Acknowledgments:** SP also gives thanks for the financial support of Norte Portugal Regional Operational Programme (NORTE 2020), under the PT 2020 Partnership Agreement, through the European Regional Development Fund (ERDF) and FSE. The authors would also like to thank the POTATOPLASTIC project (POCI-01-0247-FEDER-017938), financed by FEDER through POCI, to "Isolago—Indústria de Plásticos, S.A.", the project leader, and to "A Saloinha, Lda." for providing the potato byproducts. The authors also thank the NMR spectrometers that are part of the National NMR Network (PTNMR) and are partially supported by Infrastructure Project N° 022161 (co-financed by FEDER through COMPETE 2020, POCI, and PORL, and FCT through PIDDAC).

**Conflicts of Interest:** The authors declare no conflict of interest. The funders had no role in the design of the study; in the collection, analyses, or interpretation of data; in the writing of the manuscript; or in the decision to publish the results.

## References

1. Copeland, L.; Blazek, J.; Salman, H.; Tang, M. Form and functionality of starch. *Food Hydrocoll.* **2009**, *23*, 1527–1534. [CrossRef]
2. Tsang, Y.; Kumar, V.; Samadar, P.; Yang, Y.; Lee, J.; Ok, Y.; Song, H.; Kim, K.; Kwon, E.; Jeon, Y. Production of bioplastic through food waste valorization. *Environ. Int.* **2019**, *127*, 625–644. [CrossRef]
3. Jogi, K.; Bhat, R. Valorization of food processing wastes and by-products for bioplastic production. *Sustain. Chem. Pharm.* **2020**, *18*, 100326. [CrossRef]
4. Cazon, P.; Velazquez, G.; Ramirez, J.; Vazquez, M. Polysaccharide-based films and coatings for food packaging: A review. *Food Hydrocoll.* **2017**, *68*, 136–148. [CrossRef]
5. Ashogbon, A.; Akintayo, E. Recent trend in the physical and chemical modification of starches from different botanical sources: A review. *Starch-Starke* **2014**, *66*, 41–57. [CrossRef]
6. Lopes, J.; Gonçalves, I.; Nunes, C.; Teixeira, B.; Mendes, R.; Ferreira, P.; Coimbra, M.A. Potato peel phenolics as additives for developing active starch-based films with potential to pack smoked fish fillets. *Food Packag. Shelf Life* **2021**, *28*, 100644. [CrossRef]
7. Oliveira, G.; Gonçalves, I.; Barra, A.; Nunes, C.; Ferreira, P.; Coimbra, M. Coffee silverskin and starch-rich potato washing slurries as raw materials for elastic, antioxidant, and UV-protective biobased films. *Food Res. Int.* **2020**, *138*, 109733. [CrossRef]
8. Gonçalves, I.; Lopes, J.; Barra, A.; Hernandez, D.; Nunes, C.; Kapusniak, K.; Kapusniak, J.; Evtyugin, D.; da Silva, J.; Ferreira, P.; et al. Tailoring the surface properties and flexibility of starch-based films using oil and waxes recovered from potato chips byproducts. *Int. J. Biol. Macromol.* **2020**, *163*, 251–259. [CrossRef]
9. Jimenez, A.; Fabra, M.; Talens, P.; Chiralt, A. Edible and Biodegradable Starch Films: A Review. *Food Bioprocess Technol.* **2012**, *5*, 2058–2076. [CrossRef]
10. Zhang, Z.; Macquarrie, D.; Clark, J.; Matharu, A. Chemical modification of starch and the application of expanded starch and its esters in hot melt adhesive. *RSC Adv.* **2014**, *4*, 41947–41955. [CrossRef]
11. Haroon, M.; Wang, L.; Yu, H.; Abbasi, N.M.; Zain-ul-Abdin; Saleem, M.; Khan, R.; Ullah, R.; Chen, Q.; Wu, J. Chemical modification of starch and its application as an adsorbent material. *RSC Adv.* **2016**, *6*, 78264–78285. [CrossRef]
12. Cunha, A.; Gandini, A. Turning polysaccharides into hydrophobic materials: A critical review. Part 2. Hemicelluloses, chitin/chitosan, starch, pectin and alginates. *Cellulose* **2010**, *17*, 1045–1065. [CrossRef]

13. Masina, N.; Choonara, Y.; Kumar, P.; du Toit, L.; Govender, M.; Indermun, S.; Pillay, V. A review of the chemical modification techniques of starch. *Carbohydr. Polym.* **2017**, *157*, 1226–1236. [CrossRef]
14. Zarski, A.; Ptak, S.; Siemion, P.; Kapusniak, J. Esterification of potato starch by a biocatalysed reaction in an ionic liquid. *Carbohydr. Polym.* **2016**, *137*, 657–663. [CrossRef]
15. Zarski, A.; Bajer, K.; Zarska, S.; Kapusniak, J. From high oleic vegetable oils to hydrophobic starch derivatives: I. Development and structural studies. *Carbohydr. Polym.* **2019**, *214*, 124–130. [CrossRef] [PubMed]
16. Zarski, A.; Bajer, K.; Raszkowska-Kaczor, A.; Rogacz, D.; Zarska, S.; Kapusniak, J. From high oleic vegetable oils to hydrophobic starch derivatives: II. Physicochemical, processing and environmental properties. *Carbohydr. Polym.* **2020**, *243*, 116499. [CrossRef]
17. Fang, J.; Fowler, P.; Sayers, C.; Williams, P. The chemical modification of a range of starches under aqueous reaction conditions. *Carbohydr. Polym.* **2004**, *55*, 283–289. [CrossRef]
18. Chi, H.; Xu, K.; Wu, X.; Chen, Q.; Xue, D.; Song, C.; Zhang, W.; Wang, P. Effect of acetylation on the properties of corn starch. *Food Chem.* **2008**, *106*, 923–928. [CrossRef]
19. Pietrzyk, S.; Juszczak, L.; Fortuna, T.; Ciemniewska, A. Effect of the oxidation level of corn starch on its acetylation and physicochemical and rheological properties. *J. Food Eng.* **2014**, *120*, 50–56. [CrossRef]
20. Diop, C.; Li, H.; Xie, B.; Shi, J. Effects of acetic acid/acetic anhydride ratios on the properties of corn starch acetates. *Food Chem.* **2011**, *126*, 1662–1669. [CrossRef] [PubMed]
21. Le, P.T.; Nguyen, K.T. Hydrophobizing cellulose surfaces via catalyzed transesterification reaction using soybean oil and starch. *Heliyon* **2020**, *6*, e05559. [CrossRef]
22. Junistia, L.; Sugih, A.; Manurung, R.; Picchioni, F.; Janssen, L.; Heeres, H. Synthesis of Higher Fatty Acid Starch Esters using Vinyl Laurate and Stearate as Reactants. *Starch-Starke* **2008**, *60*, 667–675. [CrossRef]
23. Hermawan, E.; Rosyanti, L.; Megasari, L.; Sugih, A.K.; Muljana, H. Transesterification of Sago Starch Using Various Fatty Acid Methyl Esters in Densified $CO_2$. *Int. J. Chem. Eng. Appl.* **2015**, *6*, 152–155. [CrossRef]
24. Guthrie, J. Concerted mechanism for alcoholysis of esters: An examination of the requirements. *J. Am. Chem. Soc.* **1991**, *113*, 3941–3949. [CrossRef]
25. Schuchardt, U.; Sercheli, R.; Vargas, R. Transesterification of vegetable oils: A review. *J. Braz. Chem. Soc.* **1998**, *9*, 199–210. [CrossRef]
26. Cruz, M.; Bastos, R.; Pinto, M.; Ferreira, J.; Santos, J.; Wessel, D.; Coelho, E.; Coimbra, M. Waste mitigation: From an effluent of apple juice concentrate industry to a valuable ingredient for food and feed applications. *J. Clean. Prod.* **2018**, *193*, 652–660. [CrossRef]
27. Goswami, G.; Bora, R.; Rathore, M.S. Oxidation of cooking oils due to repeated frying and human health. *Int. J. Sci. Technol. Manag.* **2015**, *4*, 495–501.
28. Degennes, P. Weitting—Statics and dynamics. *Rev. Mod. Phys.* **1985**, *57*, 827–863. [CrossRef]
29. Luchese, C.; Garrido, T.; Spada, J.; Tessaro, I.; de la Caba, K. Development and characterization of cassava starch films incorporated with blueberry pomace. *Int. J. Biol. Macromol.* **2018**, *106*, 834–839. [CrossRef]
30. Petronilho, S.; Neves, B.; Melo, T.; Oliveira, S.; Alves, E.; Barros, C.; Nunes, F.; Coimbra, M.; Domingues, M. Characterization of Non-Volatile Oxidation Products Formed from Triolein in a Model Study at Frying Temperature. *J. Agric. Food Chem.* **2021**, *69*, 3466–3478. [CrossRef] [PubMed]
31. Law, K. Definitions for Hydrophilicity, Hydrophobicity, and Superhydrophobicity: Getting the Basics Right. *J. Phys. Chem. Lett.* **2014**, *5*, 686–688. [CrossRef] [PubMed]
32. Gonçalves, I.; Hernandez, D.; Cruz, C.; Lopes, J.; Barra, A.; Nunes, C.; da Silva, J.; Ferreira, P.; Coimbra, M. Relevance of genipin networking on rheological, physical, and mechanical properties of starch-based formulations. *Carbohydr. Polym.* **2021**, *254*, 117236. [CrossRef] [PubMed]
33. Jimenez, A.; Fabra, M.; Talens, P.; Chiralt, A. Phase transitions in starch based films containing fatty acids. Effect on water sorption and mechanical behaviour. *Food Hydrocoll.* **2013**, *30*, 408–418. [CrossRef]
34. Cerna, M.; Barros, A.; Nunes, A.; Rocha, S.; Delgadillo, I.; Copikova, J.; Coimbra, M. Use of FT-IR spectroscopy as a tool for the analysis of polysaccharide food additives. *Carbohydr. Polym.* **2003**, *51*, 383–389. [CrossRef]
35. Gomez, N.; Abonia, R.; Cadavid, H.; Vargas, I. Chemical and Spectroscopic Characterization of a Vegetable Oil used as Dielectric Coolant in Distribution Transformers. *J. Braz. Chem. Soc.* **2011**, *22*, 2292–2303. [CrossRef]
36. Siyamak, S.; Laycock, B.; Luckman, P. Synthesis of starch graft-copolymers via reactive extrusion: Process development and structural analysis. *Carbohydr. Polym.* **2020**, *227*, 115066. [CrossRef]
37. Muscat, D.; Adhikari, R.; McKnight, S.; Guo, Q.; Adhikari, B. The physicochemical characteristics and hydrophobicity of high amylose starch-glycerol films in the presence of three natural waxes. *J. Food Eng.* **2013**, *119*, 205–219. [CrossRef]
38. Ballesteros-Mártinez, L.; Pérez-Cervera, C.; Andrade-Pizarro, R. Effect of glycerol and sorbitol concentrations on mechanical, optical, and barrier properties of sweet potato starch film. *NFS J.* **2020**, *20*, 1–9. [CrossRef]

foods

MDPI

*Review*

# Sustainable and Bio-Based Food Packaging: A Review on Past and Current Design Innovations

**Florencia Versino [1,2,*], Florencia Ortega [1,3], Yuliana Monroy [1], Sandra Rivero [1,3], Olivia Valeria López [4] and María Alejandra García [1,3]**

1. Centro de Investigación y Desarrollo en Criotecnología de Alimentos (CIDCA), UNLP-CONICET-CICPBA, 47 y 116, La Plata 1900, Argentina
2. Facultad de Ingeniería, Universidad Nacional de La Plata (UNLP), 47 y 115, La Plata 1900, Argentina
3. Facultad de Ciencias Exactas, Universidad Nacional de La Plata (UNLP), 47 y 115, La Plata 1900, Argentina
4. Planta Piloto de Ingeniería Química (PLAPIQUI), UNS-CONICET, Camino La Carrindanga km.7, Bahía Blanca 8000, Argentina

* Correspondence: florencia.versino@ing.unlp.edu.ar

**Abstract:** Food loss and waste occur for many reasons, from crop processing to household leftovers. Even though some waste generation is unavoidable, a considerable amount is due to supply chain inefficiencies and damage during transport and handling. Packaging design and materials innovations represent real opportunities to reduce food waste within the supply chain. Besides, changes in people's lifestyles have increased the demand for high-quality, fresh, minimally processed, and ready-to-eat food products with extended shelf-life, that need to meet strict and constantly renewed food safety regulations. In this regard, accurate monitoring of food quality and spoilage is necessary to diminish both health hazards and food waste. Thus, this work provides an overview of the most recent advances in the investigation and development of food packaging materials and design with the aim to improve food chain sustainability. Enhanced barrier and surface properties as well as active materials for food conservation are reviewed. Likewise, the function, importance, current availability, and future trends of intelligent and smart packaging systems are presented, especially considering biobased sensor development by 3D printing technology. In addition, driving factors affecting fully biobased packaging design and materials development and production are discussed, considering byproducts and waste minimization and revalorization, recyclability, biodegradability, and other possible ends-of-life and their impact on product/package system sustainability.

**Keywords:** barrier properties; active material; intelligent packaging; circular economy; biobased inks and dyes; bioadhesives; bioplastic; agri-food by-products

**Citation:** Versino, F.; Ortega, F.; Monroy, Y.; Rivero, S.; López, O.V.; García, M.A. Sustainable and Bio-Based Food Packaging: A Review on Past and Current Design Innovations. *Foods* **2023**, *12*, 1057. https://doi.org/10.3390/foods12051057

Academic Editors: Sílvia Petronilho and Mario M. Martinez

Received: 31 December 2022
Revised: 14 February 2023
Accepted: 21 February 2023
Published: 2 March 2023



## 1. Introduction

Food packaging comprises items, such as containers, cups, tableware, straws, bags, wraps, and boxes, that protect or contain food. Within a growing urbanized population food packaging is needed for the transport, storage, and consumption of food products. By 2050 the world's population is estimated to reach 9.7 billion, two-thirds of which will be living in urban areas with the consequent increase in food requirements and changes in food-consumption patterns [1,2]. People living in cities tend to lead more hectic lifestyles, spending little time buying or preparing food, preferring processed food to fresh foods, and purchasing mainly in supermarkets or convenience stores to save time. Such patterns are clearly evidenced nowadays in high-income countries, though small shops and traditional grocery stores are going out of business and being replaced by supermarkets even in developing countries, thus leading to lager supply chains strictly dependent on packaged food [2,3].

A large portion of food is lost along the supply chain due to damage caused by different environmental factors, such as moisture, oxidation, thermal degradation, or microbial contamination [4]. On average, 30% of the edible part of the global food production is lost or wasted in the supply chain [2]. Thus, several research and policies have focused on reducing such losses [5–7]. Bishop et al. reported that, only considering UK households, over 2 million tons of fruit and vegetables waste are expected to be generated every year by 2030, along with 105.7 kilo tons of plastic waste from the packaging in which is purchased [8]. Therefore, not only does an increase in food demands entail greater land-use requirements, soil degradation, biodiversity loss, eutrophication, freshwater scarcity, energy resources depletion, and overall climate change by the agri-food industry, but also more plastic-waste pollution from agriculture and food packaging. Furthermore, a large portion of food packaging is discarded with food waste, making it difficult and expensive to separate in waste management systems, ending up in less preferred waste treatments and incurring an environmental burden across both food and plastic value chains [8,9].

Most food packaging is directly disposed of after use (95%) and more than a third do not enter the recollection systems [10]. In May 2018, the European Commission approved a directive by which single-use plastics were prohibited or at least restricted, intending to reduce plastic waste pollution of seas and oceans [11]. The set of approved standards comprises requirements for product labeling, waste collection, and public awareness regarding the environmental problem and responsible consumption. Since then, a slight reduction in fossil-based plastic and a mild increase in biobased plastic and post-consumer recycled plastic production has been reported in Europe, though the overall global plastic production has continued to grow (reaching 390.7 million tons in 2021) [12]. Aiming to develop more sustainable food packaging, some significant efforts have been focused on biobased and/or biodegradable materials, especially bioplastics, paper, and cardboard. Alternative cellulose sources to wood have been studied aiming at tree and biodiversity conservation in forests and rainforest ecosystems [13–16]. Plant-based bioplastics, both biodegradable (such as poly-lactic acid (PLA), polyhydroxybutyrate (PHB), or biopolymer-based plastics) and non-degradable (biobased polyolefins: e.g., bio-PE, bio-PP, and bio-PET), are currently being commercialized and agri-food industry by-products and waste are being studied as raw materials to minimize agriculture environmental impact and land-use [15,17–24]. However, enhanced recyclability capacity, retrieval quantity, and good separation through waste management systems are crucial for non-biodegradable bioplastics to be effectively recycled through conventional recycling infrastructure and technology [25]. Compostable bioplastics may be preferred to non-degradable plastics for single-use applications, such as food packaging and foodservice ware, when reutilization or reduction is not possible and recyclability is limited [26,27]. Nonetheless, even for this type of material, adequate waste stream management and industrial composting facilities are needed, since compostability will only occur under specific conditions [27–29]. Several studies have been conducted on the environmental impact of biobased materials over fossil conventional plastic packaging [8,28,30–42], yet no general result has been reported on which is the more sustainable alternative. The overall environmental impact of the package depends on the packed food product, distribution logistic and distances, clean energy resources, available waste management systems, re-use frequency, recyclability and/or biodegradability, end-of-life scenario selected for the packaging material, and overall cost of production, use, and disposal [27,43–45].

Regarding the further reduction of food waste and losses, two main strategies are followed. On the one hand, there is agri-food industry by-products reduction and revalorization as feedstock for biomaterials and active compounds recovery, aiming to reduce by-products treatment costs and overall environmental impact [46,47]. On the other hand, the development of active and intelligent packaging to prevent food spoilage and increase food safety is being performed [4,40]. Extended shelf-life increases the chances of the food to be eaten with preserved nutritional value and reduced toxicity risk. Besides, the use of sensors and new identification technologies may provide better detection of food degradation markers to optimize food distribution and conservation and thus reduce food loss [48].

The food packaging industry is a thriving market that plays a key role in the modern economy. Food packaging design is crucial to guarantee food safety throughout the supply chain, optimize storage and transportation, and facilitate the communication of product information to retailers and customers [49]. Approximately 9% of the vast amount of research reported on food packaging (41,602 publications from 1960–2022) focuses on packaging design. More sustainable food packaging design aims for minimum environmental burden at the end-of-life of the food product and package. Consequently, research and innovation in food packaging are growing fast. The literature review within the last decade indicates an accelerated increase in annual publications involving sustainable food packaging (19–35% annual increase) together with active and intelligent food packaging (13–33% annual increase). Among food, packaging paper is the second (34%) most used and studied material after plastic (37% both rigid and flexible) [50]. A sustained increase in work related to innovations and innovative food packaging is observed, with an incipient but fast increase in studies on biobased, especially bioplastic food packaging, and active and intelligent packaging for food, both showing a remarkable growth in the last couple of years (Figure 1). Such tendencies evidence the work towards developing environmentally friendly and sustainable food packaging material focusing on: (i) enhancing biopolymers performance through physical, chemical, and enzymatic modification and composite formulation, (ii) developing natural-based nanocomposites for active and intelligent sustainable materials, (iii) synthesizing new biobased and biodegradable materials, (iv) scaling up bioplastic processes, (v) searching for new renewable sources, especially undervalued raw materials, and (vi) improving materials reutilization and recycling capacity [17,23,24,47,51–61].

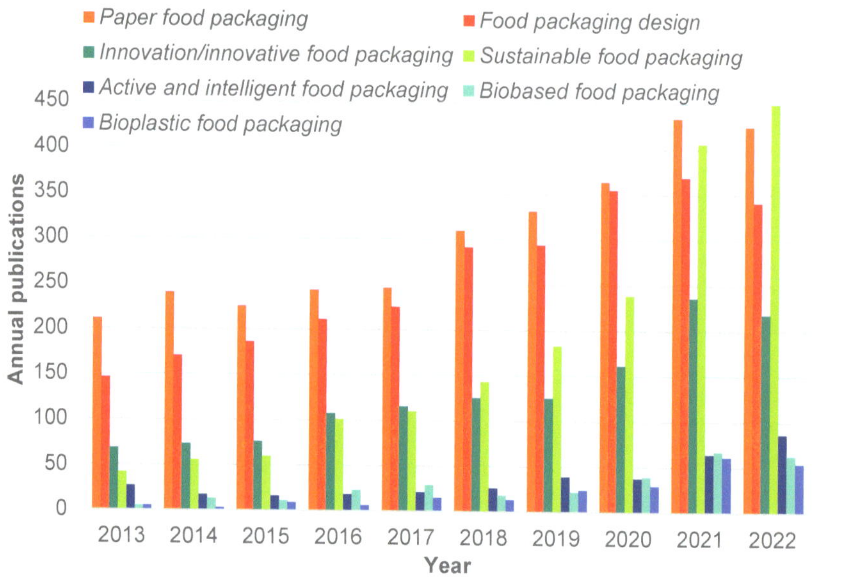


**Figure 1.** Total number of publications reported in literature in the last ten years on food packaging for specific keywords. Literature review carried out on Scopus search engine of bibliographic databases.

Novel packaging systems are aimed not only at ensuring food safety and traceability, but have also gained great importance for building more sustainable food chains, reducing food losses and waste, as well as the overall packaging environmental impact. In previous works, a recompilation of the latest research and development on biobased packaging materials and composites form agri-food waste and byproduct was conducted, with a special focus on materials properties and processing technologies for greener production [55,61,62]. Thus, this review gathers the advances reported in the last decade in the research and development on packaging materials and design aiming to improve food chain sustainability within a circular economy paradigm. Biobased materials with enhanced barrier and surface properties, and specific activity for food conservation, such as antimicrobial or antifog properties, were reviewed. An overview on recent advances in intelligent packaging development, especially biobased sensor development by 3D printing technology, is provided. A further assessment of fully biobased packaging design and materials, considering waste minimization and revalorization, recyclability, biodegradability, and compatibility of the packaged food products with current waste management technologies is given.

## 2. Sustainable Food Packaging Design

Considering that over 80% of the environmental impact of a product is determined at the design stage, design plays a key role in its sustainability [63]. In food packaging, an efficient design can increase the sustainability of the packaging and the food product it contains by minimizing both food and packaging waste at their end-of-life. However, ensuring food preservation accounts for a larger portion of the overall life-cycle environmental impact of the packed food product. On average, the packaging is estimated to comprise only 10% of the energy input required for a single person's weekly food consumption and can ensure that the residual 90% is not wasted throughout the supply chain [64]. The relative environmental impact that packaging has on a food product depends on the food type, mainly on its perishability, cost, and overall emissions and resources depletion in production. For example, greenhouse gasses (GHGs) emission from dairy products and meats is greater than those from fruits and vegetables by 13–18%, largely exceeding GHGs in packaging manufacturing and end-of-life [65]. Therefore, focus on minimizing food waste for animal food products would yield greater benefits to the system, while emphasis should be placed on reducing the impact of the packaging of fruit and vegetables [66]. Life cycle assessment (LCA) of both the food product and the preferred packaging system should be evaluated in each case for better decision-making, considering real available waste management systems and possible optimized alternatives. Consequently, apart from the protection of the product, designing the most effective and sustainable packaging is a complex process involving numerous sectors implicated in the whole supply chain, including the intended market [28,64].

To minimize the environmental impact of the packaging system, it is important to consider its capacity to contain, protect, and preserve the product as to extend its shelf-life and guarantee food safety, but it should also be adequately sized, easy-to-open and easy-to-empty, and with clearly accessible information to prevent food from being wasted. In addition, the packaging materials should meet the desired mechanical and barrier properties remaining as light-weighted as possible, food-safe, ideally reusable or recyclable, and disposed of with minimum to zero pollution. Several research efforts have focused on enhancing materials barrier properties for sustainable food packaging (See Section 3).

Most of the design decisions are aimed to reduce packaging and food waste, and all of them indirectly have an economic impact. A large portion of domestically wasted food could be because the packaging does not meet consumer needs, such as packages that are too large or difficult to empty [65,66]. For instance, easily emptied packaging systems reduce both product waste and cost per unit and make cleansing easier for better recyclability of the packaging material, therefore lowering water use for this matter. There are various research and novel start-ups commercializing easy emptying packaging coating, especially biobased and food-safe materials [67–69]. In this regard, LiquidGlide®, a spin-off from the Varanasi Lab at MIT (Massachusetts Institute of Technology), has launched a novel, biobased, durable, superhydrophobic material for packaging applications.

Additional packaging attributes that can influence food waste comprise resealability, easy-to-open, grip, portioning or dosage, and communication of food safety/freshness information [70,71]. Innovations in intelligent and smart food packaging (See Sections 4.2 and 5.2) aim to facilitate detection, recording, and communication of food quality and safety to overcome problems in expiration date labeling lack of uniformity within the global market that can lead to confusion in the actual state of the product [66]. Unclear or wrong date labeling may therefore cause potential food intoxication in consumers or unnecessary disposal of still fresh products, increasing human health risks and food waste. Nonetheless, further emphasis on the LCA need of these packaging/product systems is required to assure net environmental benefits, focusing on intelligent and smart systems cost, reusability, and recyclability [66,72].

It is important to identify the packaging functions that influence food waste to achieve more efficient packaging systems. Good mechanical resistance helps protect the product from damage due to vibrations during transport on rough roads and compression during storage as provides protection from contamination to guarantee food safety [70]. Besides, damaged products may also be rejected by customers. A real understanding of the natural characteristics and shelf-life of the food is needed to select the best packaging system. Verghese et al. reported significant annual losses of bananas in Australia partially because of improper packaging, which could be significantly overcome by replacing reusable plastic crates with more stable corrugated carton crates and clear bags for fruit neck protection and further in situ retailer owned ripening rooms [64]. Similarly, a novel non-toxic biobased *aloe vera* coating on papaya fruit has proven to delay ripening and significantly extend their shelf-life [73].

Materials with barrier properties have been widely studied to reduce food perishability (See Section 3). Depending on the food matrix characteristics, the diffusion or transmission of different substances, e.g., oxygen, carbon dioxide, ethylene, nitrogen, water vapor, and volatile molecules, microorganisms, light, or heat will need to be controlled to avoid undesired chemical and physical reactions in the food product that could lead to unpleasant odors or flavors, or even spoilage. Multi-layered packaging systems have the potential to provide barrier properties to multiple undesired or hazardous compounds simultaneously or provide the barrier materials with the mechanical or thermal resistance needed, and reversely. Currently, multi-layered packaging is widely used in the food and beverage industry, such as juice and milk cartons, usually consisting of 75% paperboard, 20% of plastic (mostly PE), and 5% of aluminum foil [74]. Despite their potential for increasing food shelf-life and mostly entailing less material to achieve the same functionality as mono-materials, their recyclability is rather difficult [75,76]. Palombini et al. studied the sustainability of packaging used for organic food in Brazil, showing that the dominant features in poor or non-recyclable polymer packaging were metallization and the use of opaque materials, polymer blends, the presence of adhesives for labeling or multi-layered packaging, and printed labeling over the packaging [76]. Better sorting of disposed multi-layer packaging would be required for high-quality recycled materials, though these are currently found in different recycling pathways, hampering their recyclability. Kiaser et al. clearly overview the two main available technologies for multi-layered packaging recycling: compatibilization methods (for polymer-based packaging) and component separation by dissolution–reprecipitation technique or delamination for independent recycling of each component when possible [77]. The major limitations of these methods are that: (i) common compatibilization does not seem to be feasible for post-consumer packaging due to fluctuation in composition and the need for more rigorous labeling and sorting of packaging waste stream; (ii) even though the dissolution–reprecipitation method could be available for the recycling of existing materials in the near future, it is a very high energy demand process; (iii) systematic delamination could be an ecologically and economically more sustainable solution if separation and adequate sorting of the delaminated layers are possible [77]. In this regard, further investigation of the optimization of multi-layered packaging recycling routes is needed. Alternatively, fully biodegradable multi-layer packaging systems designed to be composted at their end-of-life could probably reduce food waste and at the same time both avoid problems in recyclability and reduce the environmental impact of this type of food packaging. Yet, more research is needed considering LCA of the proposed materials, contemplating that a large proportion of disposed packaging labeled as compostable or biodegradable is not correctly treated. Misdirection occurs either because these materials are thrown away without proper classification or do not even enter the urban waste treatment streams, e.g., non-biodegradable plastic packaging, or are rejected from recycling plant due to a lack of available processing technologies or a shortage in volume needed for cost effective recycling, or are separated in composting facilities by mistake to avoid compost contamination or because there is no adequate separation technology to tell one type of plastic from another when the material is mixed with other biowaste [27,78,79].

Furthermore, novel composite and nanocomposite materials have also been studied to enhance materials barrier properties for food packaging. A focus on more sustainable materials fully biobased and biodegradable especially aiming to use by-products and waste from the agri-food industry is extensively reported in the literature review [14,17,47,55–57,59,60,62,80,81] (See Sections 3 and 5.1). For instance, Báscón-Villegas et al. recently published a work on composite PLA and polybutylene adipate-co-terephthalate (PBAT) with lignocellulose nanofibers obtained from wheat straw [82]. Similarly, Silva dos Santos et al. formulated a coating paperboard packaging based on chitosan, palmitic acid, and active carbon that provides increased fat and moisture barrier for paper packaging preserving biodegradability and recyclability [83]. Besides, many new active properties can be imparted to packaging materials to extend food shelf-life [4,84,85]. For example, Manfredi et al. highlighted that milk waste savings within the fresh milk supply chain by the application of an antimicrobial coating on regular multi-layer paperboard packaging is greater than the impact generated by the coating production [86]. Active packaging contains compounds are able to preserve the organoleptic or sensory characteristics to ensure the food product's quality. Of particular interest are active packaging materials containing natural antioxidants and antimicrobials, which not only extend the shelf-life of packaged products by preventing rancidity reactions, but also prevent the growth of foodborne pathogens [55,87,88]. Biobased nanocomposites have been studied for innovation in sensors for the better detection of food degradation markers and thus a reduction in food loss. In this regard, active and intelligent packaging is currently receiving great interest, although they are still being researched and are not widely commercialized. A thorough revision of recent research on these materials has been conducted in the following Sections 4 and 5.2.

The development of advanced methods for recycling and remanufacturing should be considered in designing new circular and sustainable systems. Some key aspects related include recyclable and non-hazardous materials; efficient logistic and transportation systems; energy and resource efficient technologies; and product design aimed at recyclability, remanufacturability, and reusability [89]. Even though the goal of reuse and recycling is entailed within the circular economy model, this is poorly achieved for food packaging since it is difficult to guarantee the quality and safety of materials for food contact from the available waste management systems. Currently, reuse is only viable for refillable and cleanable containers (e.g., glass bottles, stainless-steel containers) [75]. In the long term, chemical safety of recycled food packaging could be achieved by replacing unsafe substances from materials design, avoiding their entrance to the recycling stream. However, a period of 10–30 years is estimated for a significant reduction in contaminants levels [75]. In the short term, clear labeling of packaging materials, consumers' education to identify more sustainable packaging alternatives and more efficient safety and sorting systems within the waste management and recycling systems could contribute to a more sustainable use of food packaging [71,75]. Besides, sustainable recycling systems need to be cost effective in terms of the selected recycling technology, material quality of recycled products (depending on the type and conditions of secondary raw materials recollected from waste stream), availability (in terms of economy of scale), and recollection and sorting cost [28,90,91]. Facilitating recyclability has not only environmental and economic benefits, but also a social contribution, considering that dry waste sorting and selling constitute the income source for low-income workers in some countries [76]. In addition, packaging design can considerably affect the logistics in transportation, handling, and storage throughout the supply chain [92,93]. Such decisions have direct implications in distribution times and product preservation. However, the best packaging systems depends on the frequency of reuse, transportation routes and distances, and end-of-life treatment [41,64,93–96]. As regards resources reduction, water and energy saving technologies, by-products and biowaste revalorization are being achieved by the development of new biobased packaging materials and novel heat insulating materials that can help reduce energy in preserving thermosensitive foods as well as reducing their spoilage rate [17,55,56,62,97–100].

To reduce the environmental impact of biodegradable and compostable bioplastics, the preferred route of disposal and treatment is composting. However, most packaging materials nowadays reported as compostable (according to current legislation) require specific composting conditions (thermophilic: ~58 °C) to be fully and safely composted that are only feasible in industrial composting facilities [27,29,78,101]. Therefore, compostable bioplastic-based food packaging faces some obstacles to fulfill its purpose. A large volume of bioplastics does not enter composting systems due to the lack of industrial composting facilities or the poor sorting of these materials. Waste misdirection results from improper labeling, workers inexperience with biodegradable plastics treatment, a lack of appropriate sorting technologies in sorting facilities, or a lack of consumer education on waste separation or environmental protection. When biodegradable plastics wrongly end up in landfills, they can decompose anaerobically producing methane, a more harmful GHG than $CO_2$, hindering their sustainability [28,79]. By occasional littering or waste mistreatment, bioplastics can also enter water streams and marine environments in which they are not able to biodegrade or can degrade, posing a hazard to life in such ecosystems. Research has been conducted to prove biodegradable materials' ecotoxicity [29,101–104]. Zimmermann et al. reported that many bioplastics available on the market are just as toxic as conventional plastics, presenting clear bioluminescence inhibition of *Aliivibrio fischeri* [101]. Wang et al. thoroughly revised the available studies on plastics degradation in marine environments, highlighting that only a few works consider the toxicity of degradation products, including material fragments or debris and additives, which should not be toxic [29]. In this regard, the chemical safety of materials can be further achieved by bottom-up green chemistry design in the development of new biobased and biodegradable materials. On this matter, ecotoxicity assays of new biodegradable materials research is being conducted, especially after compost or soil degradation, usually showing promising results [103,105–107]. Furthermore, non-biodegradable and fossil-based plastics, biodegradable bioplastics can break into small particles, producing microplastic pollution and affecting different species of plant and animals adversely [29,102,108]. This points out the necessity of an in-depth understanding of the environmental degradability performance of available and upcoming biodegradable plastics.

In product-packaging design, it is also important to consider and understand the consumers perception and acceptance of the packaging. Some authors imply that consumers, tend to have a poor understanding of the benefits of packaging and mostly a negative image of its role to preserve food, e.g., removing the package prematurely to let the food "breathe" [64]. On the contrary, other research findings indicate that consumers recognize the importance of packaging plays in food safety and quality in relation to the information that it provides and show a pro-environmental consciousness [109]. A study carried out by Harpen et al. using an immersive 3D virtual supermarket environment revealed that customers tend to prefer unpacked fruits and vegetables [110]. Herbes et al. indicated that consumers mostly focus on the end-of-life stage of packaging and are mostly unaware of the negative impacts of packaging in the upstream and middle-stream supply chain [111]. Consumers' desire to pay for green or sustainable products grows as environmental awareness rises, yet the higher price of these products limits sales to a niche market [28,91]. Consequently, future research should contemplate consumers involved in packaging alternatives design as well as cost reduction for the increased accessibility of novel sustainable food packaging.

A systemic focus on understanding the relationship between packaging, products, supply chain networks, and physical processes is required in an era of multi-product physical supply chains that have a global footprint [95]. Padhi et al. identified sustainable design and development products, strategic commercialization, technology optimization, and sustainable product returns and recycling as the most important sustainable supply chain processes [96]. The conception of a new eco-design approach gives place to further innovation in packaging and products with a critical rethinking of the system and new perspectives towards sustainable production. From the literature review, some basic principles for sustainable design can be outlined [55,95,96,112]:

- A focus on real demand and problems and try to find solutions with social, environmental, and economic benefits altogether.
- A shift in the design dynamics from the application requirements up, thinking in terms of functions and services rather than the product itself.
- Consideration of realistic and updated life-cycle and process thinking, having an integrated view of the supply chain, thus taking into account the product manufacturing, distribution, consumption and end-of-life.
- Inclusion of users, stakeholders, and different experts in the design process as much as possible.
- Research and innovation need to be grounded on justifiable priorities within the available time frame and scope of the project.

A holistic understanding of the supply chain drivers, barriers, and opportunities is required in view of sustainable long-lasting packaging-product systems designed towards a circular economy.

## 3. Strategies to Improve Barrier and Hydrophobicity Properties for Food Packaging

With the main premise of preserving the food matrix quality, packaging industries and the scientific sector are striving to develop new materials to minimize deteriorative changes mainly due to physical and chemical modifications experienced by foods during their distribution and storage [113]. Food packaging is considered an integral part of the preservation system and, therefore, the package provides a barrier between the food matrix and the external environment, protecting it against physical, chemical, and biological damage [114]. To control the chemical and physical reactions in the matrix, packaging material must be able to act as a barrier to external factors that affect the food quality, such as the gaseous atmosphere, water activity, light, and temperature. In this sense, the barrier properties of a material are related to the protection of the food matrix inside the package by controlling mass transfer [115]. The controlled transference of diverse compounds, such as gasses, water vapor, and volatile molecules, is relevant to minimize the reactions that cause food degradation by creating a favorable atmosphere around the packed product [116]. Food stability is characterized by its chemically unstable nature and hence requires to be protected from spoilage, lipid deterioration, and microbial contamination. Therefore, polymeric materials must prevent the penetration of compounds from the surrounding environment to guarantee a "high barrier" [117]. A schematic representation of how packaging protects food from external agents, such as gaseous compounds, water vapor, UV radiation, and microorganisms, is shown in Figure 2. In this regard, some examples of alternatives to develop materials with enhanced barrier properties are also addressed.

As it is well known that the most common degradation processes of packaged foods are caused by oxygen, such as lipid oxidation, microorganism growth, enzymatic browning reactions, and vitamin loss, among others. According to Zabihzadeh-Khajavi et al. controlling the oxygen permeability of the packaging system can extend the shelf-life of the packed food [118]. Ethylene, on the other hand, is considered the aging hormone of plants since it is responsible for the growth and ripening of fruits [119]. A very high concentration of ethylene around the food can accelerate its decay. Therefore, the elimination of this gas from the packaging or the control of its permeability is relevant to increase the shelf-life of fresh products [120]. While carbon dioxide is one of the most employed components in the gas mixtures for modified atmosphere packaging (MAP) due to its antibacterial properties [121], this gas can prevent the microbial growth of fresh meat, cheese, or baked goods, and minimize the respiration rate of fruits and vegetables [120]. Moreover, moisture content alters the nutritional and organoleptic properties and safety of food products. The reaction rates of lipid oxidation, microbial growth, and browning are altered when the food moisture content changes. Effective packaging can play a major role in maintaining product moisture to extend the shelf-life of food.

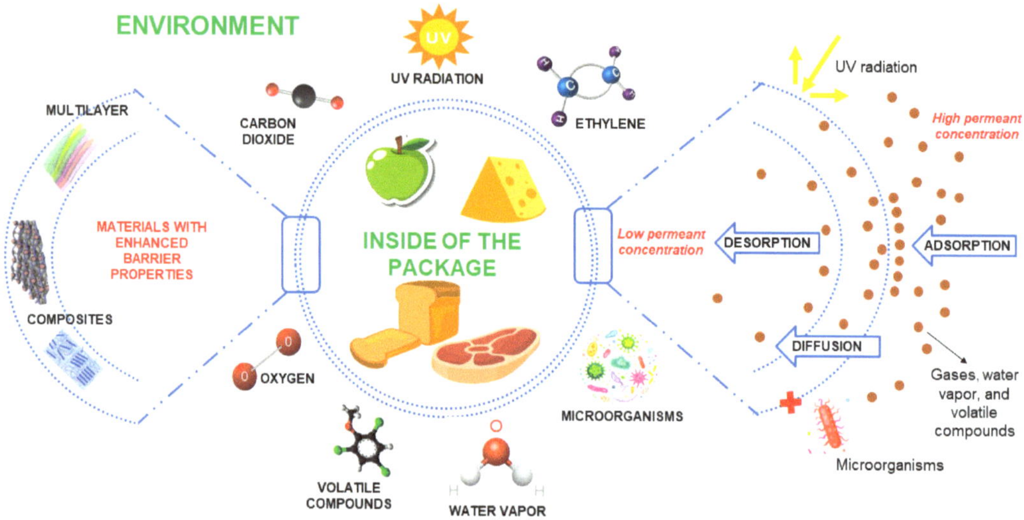


**Figure 2.** Barrier function of food packaging and hazardous agents for food safety present in the environment, along with some examples of enhanced barrier materials technologies.

Given the importance of quantifying the barrier properties, mainly to gasses and water vapor, it is relevant to use normalized or standardized methods that allow the comparison with other materials. As it was stressed by Baschetti and Minelli, numerous methodologies have been employed to study the permeability of different gasses and volatile compounds through polymeric matrices under diverse experimental conditions [120]. The most important international standards are American Standards and Table 1 summarizes the ASTM norms for the determination of gasses and water vapor permeability in polymers for food packaging [120,122]. Another important barrier property of food packaging is its capacity to block UV (200–400 nm) and visible (400–700 nm) radiation. The exposure of food products to UV light (natural or artificial) throughout their useful life (harvesting, elaboration, storage, distribution, sale, and preparation by consumers) can induce photo-oxidation and photo-degradation reactions that affect the quality of food [123]. Many photosensitive food components (proteins, vitamins, pigments, fats, and oils, etc.) are the substrates for these reactions in which free radicals and oxygen reactive compounds are produced, inducing food deterioration by the development of undesirable flavors and aromas, color loss, and a decrease of the nutritional value [124,125]. As an alternative to avoid photodegradation, different methodologies can be used, such as reflective layers, as well as the coating or incorporation of organic and inorganic absorbers to the polymeric matrices [115]. On the other hand, the study of the UV barrier capacity of food packages is relevant if the products, once packaged, will be submitted to microbial decontamination by light-based technologies, such as UV radiation. According to Cassar et al., this methodology can reduce the food microbial charge but also leads to product deterioration or the generation of undesirable residues [126]. Regarding visible radiation, transparent packaging allows consumers to visualize the packaged product and highlight some of the food organoleptic characteristics that condition the acceptability of the product. For this reason, opacity is a relevant optical property that conditions the quality of packaging films [127].

**Table 1.** ASTM standard methods for the determination of gasses and water vapor permeability in polymeric food packaging.

| Norm | Method | Permeant |
| --- | --- | --- |
| ASTM D 1434 | Manometric/volumetric | All gasses |
| ASTM D 7709 | Gravimetric | Water vapor |
| ASTM E 3985 | Dynamic with electrochemical sensor | $O_2$ |
| ASTM E 96 | Gravimetric | Water vapor |
| ASTM E 2945 | Static cells with analytic technique | All gasses |
| ASTM F 3136 | Accumulation method with optical sensor | $O_2$ |
| ASTM F 1249 | IR sensor | Water vapor |
| ASTM F 2622 | Dynamic with sensor | $O_2$ |
| ASTM F 3299 | Coulometric $P_2O_5$ sensor | Water vapor |
| ASTM F 2476 | Dynamic with IR sensor | $CO_2$ |

In accordance with other authors, it is relevant to find multifunctional high-performance barrier materials considering several issues, such as optical and mechanical properties, renewability, and bio-friendliness, to develop food packaging that assures food quality and fits all consumers' necessities [128–130].

### 3.1. Enhanced Gas and Water Vapor Barrier Properties

The permeation of low molecular weight gasses/vapors through films is measured by the ability of the polymer matrix to absorb and diffuse the penetrant. The main mechanisms that affect mass-transfer of substances across packaging materials are diffusivity, solubility, and permeability, which are closely related to the composition and structure of the polymer matrix [131]. Factors, such as the shape and size of the permeant, morphology, crystallinity, and chain orientation of the polymer, influence the diffusion and solubility coefficient. Other variables that modify the matter transfer coefficient are related to the processing methods, which can alter polymeric chain configuration or induce crystallization/orientation, including polymer blending, multi-layer coextrusion, casting, or applied nanotechnology [132].

The enhancement in the barrier properties in nanocomposite materials is attributed to the more tortuous path created by the presence of different nanofillers. This fact is explained considering that the nanofillers force the low molecular weight molecules to adopt particular tortuous pathways, producing a significant lag time. A higher aspect ratio of fillers compels the permeating gas molecule to follow a more tortuous path, leading to improved barrier properties [133]. Furthermore, gas diffusion through materials is also controlled by the crystal structure domains. It is generally assumed that ordered crystalline domains should act as an effective barrier to the diffusion of gasses and small molecules, making the amorphous phase the only pathway available for permeation. Moreover, penetrants cannot sorb in crystalline domains because their solubility coefficients are lower compared to those of their amorphous counterparts [117].

To achieve a better barrier performance, the assembly of nanoparticles to obtain nanocomposites constitute a strategy to improve barrier properties. Organic nanoparticles, such as carbon nanotubes, nanocrystals of cellulose or starch, and inorganic nanoparticles, such as nanoclays or montmorillonite (MMT), can physically and chemically interact with polymeric matrices to induce stronger and reinforced structures, enhancing both the barrier and mechanical properties [134]. MMT nano-clays can enhance the polymer barrier performance because of their exfoliated structures, small particle size, high aspect ratio, and exceptionally large surface areas that enlarge the tortuous path for small diffusing molecules [133]. Thus, the nanomaterials can be applied to improve the performance of conventional materials due to their particle size and their large surface area [135].

Other approaches have been tested to regulate the transport of water/gasses through the packaging, e.g., the use of different polymer mixtures, polymer crosslinking by chemical reactions to narrow the intermediate chains, assembly of micro and nanoparticles to obtain composites, nanostructured matrices formed by an electrodynamic process, and coatings or multi-layer films to design packaging materials with high water/gas barrier properties [115,136,137]. In this regard, multi-layer packaging constitutes an emerging technique that integrates the characteristics of different polymers or layers to create a package with enhanced performance in terms of properties [77]. These materials have improved properties as compared to typical single-layer films, exhibit high-barrier to water vapor and gasses, such as oxygen, carbon dioxide, and aromatics, as well as high mechanical strength and good sealing capacity [138].

Multi-layer films can be prepared by different methods, such as melt coextrusion, film-forming solution molded by casting, and compression molding of sandwich structures [132]. In the design of multi-layer film packaging different layers can be assembled. In this sense, to enhance the techno-functional properties of polymers, an inner barrier layer film usually consists of polymers with higher oxygen barrier properties, while polymers with higher water vapor barrier and resistance from a mechanical point of view act as an outer layer. The barrier layer is in direct contact with the external environment, acting as a barrier to substances that cause the degradation of packaged food, such as moisture, oxygen, and microorganisms [131].

According to Alias et al., multi-layered films based on biomass combined with synthetic biodegradable polymer films obtained from natural monomers showed better transparency, water solubility, and mechanical properties than single-layer films [138]. Meanwhile, multi-layer films based on biomass sources demonstrate better barrier properties in terms of water vapor permeability and offer advanced properties in terms of oxygen permeability, still exhibiting a significant improvement in terms of physical and mechanical properties [138].

In addition, to achieve high-performance biodegradable multi-layer films with tailor-made properties, knowledge of their microstructure and film processing steps is required. The challenge is to improve the protective barrier of the individual layers and therefore optimize the design of the matrices, so current research must be redirected in this way [132]. It is pertinent to emphasize that the most important factor is the adhesion between layers to combine different layers that improve the barrier protection of the individual layers and thus optimize the resulting system. In this sense, research considering the scale, interaction, and architecture of structural layers in multi-layer matrices and their influences on barrier capacity performance is hardly reported in the literature.

The layer-by-layer assembly generates a stratified structure with a significantly enhanced gas barrier and mechanical or even optical properties due to confinement and/or interfacial effects. Layer-by-layer assembly is based on different interactions, such as electrostatic attraction, hydrogen bonding, hydrophobic attraction, or entanglement between polymeric chains in neighboring layers. The interactions that are established depend on the characteristics of the polymers that make up the multi-layer system. Particularly, for some semi-crystalline polymers, the confinement generated by the stratified structure leads to a specific orientation of the crystals, nanoparticles, or compounds, inducing an increase in the tortuosity for diffusion [131,139].

### 3.2. *Enhanced Hydrophobicity*

Many biodegradable and renewable biopolymers are hydrophilic since they are conformed by polar molecules. This inherent hydrophilicity often results in moisture absorption that leads to the deterioration of the mechanical properties of the material and affects its dimensional stability, both undesirable effects for packaging applications [140].

To reduce the hydrophilicity of biodegradable and natural polymers, several methodologies can be carried out. One of these alternatives is to modify the chemical structure of the biopolymers to improve their hydrophobicity. According to this, Wang et al. stressed that chemical modifications of starch may be carried out to achieve adequate physicochemical characteristics by blocking or adding functional groups, improving the hydrophobicity of starch-based materials [52]. For instance, Petronilho et al. have worked on hydrophobic starch-based films by transesterification with sunflower oil in alkaline medium, showing increased water resistance (lower solubility and water vapor transmission, and higher contact angles) [137]. In the case of cellulose, for instance, chemical modification by silylation, esterification, amidation, and grafting are some of the reactions that can be used to obtain more hydrophobic cellulose [141].

Plasma treatment is another methodology usually used to increase the hydrophobicity of natural polymers. Pankaj et al. highlighted plasma treatment as a novel, more environmentally friendly technology to obtain more hydrophobic starch derivatives [142]. The effects of plasma occur on the surface of the material without altering the properties of the bulk involving the generation of reactive species, such as ions, radicals, electrons, photons, and other excited species. Generally, the plasma induces different reactions, such as surface cleansing, removal of organic contaminants, degradation (etching), cross-linking of polymer chains, and modification of the functional groups present on the surface. The different physical and chemical changes that the surface experiences depend on the gas used to generate the plasma [143].

Goiana et al. studied how a dielectric barrier discharge (DBD) plasma treatment affects the hydrophilicity, WVP, and mechanical properties of corn starch-based films [144]. The authors concluded that this treatment produces more hydrophobic starch materials with improved mechanical performance, equally interesting properties for food packaging. Following this idea, the use of cold plasma treatment increases the surface roughness of biopolymer films [145]. The magnitude of the effect on the film's roughness depends on some factors, such as power supply (voltage), exposure time, and uniformity of the exposed energy of the plasma species onto the film surface [146]. Among the effects that induce cold plasma treatment, the reduction of the contact angle between the film's surface and the water drops can be mentioned. Despite plasma treatments representing an effective method to reduce the hydrophilic character of natural polymers, they can reduce the biodegradability of materials [147].

A more innovative method to enhance the biopolymers hydrophobicity is based on the reduction of the contact angle between water droplets and the material surface considering its topographical features. Therefore, materials with a specific surface topography can include air compartments that hold water droplets, minimizing their contact with the material surface. In this regard, bioinspired hierarchical surfaces are very interesting mostly because of their wetting behavior properties [148]. For example, lotus leaves and rose petals present superhydrophobic surfaces with different wetting behaviors [67,149,150]. Two methodologies can be employed to obtain materials with the hierarchical structures of lotus leaves or rose petals: directly molding from hierarchical templates or growing nanowires or nanorods onto designed microstructures [148]. Following the first methodology, Luís et al. proposed a method to mimic the lotus leaf surface to fabricate zein-based films [151]. The authors demonstrated that films produced using the lotus negative template presented lotus-leaf-like rugosities, resulting in very hydrophobic surfaces. Likewise, de Oliveira Gama et al. "decorated" the surfaces of films based on thermoplastic starch (TPS) and blends of TPS/low-density polyethylene (LDPE) to improve their hydrophilic properties [152]. These authors employed micro- and nanocomponents to decorate the materials surfaces and they demonstrated the feasibility of creating topographical patterns to generate hydrophobic and even superhydrophobic features that can be useful in minimizing the water absorption.

Other techniques, such as electrospinning, have been used in the design of superhydrophobic nanostructure fibers and surfaces. Pardo-Figuerez et al. developed an innovative PLA and nanostructured silica ($SiO_2$) superhydrophobic multi-layer material with PET by electrospinning and electrospraying [68]. Moreover, Zhang et al. prepared a fully biobased thermo-resistant edible super-hydrophobic coating from coffee lignin and beeswax with a promising application for food packaging of food that require high-temperature sterilization [69]. Likewise, fully biobased materials from food industry waste with enhanced hydrophobic properties were developed and studied by Gonçalves et al. [153]. The authors used recovered starch, oil, and wax from potato chips manufacturing to produce starch-based films with enhanced flexibility and wettability with tailored roughness. For a further and thorough overview of the most recent studies on the application of superhydrophobic coatings and surfaces for sustainable food packaging, the authors recommend the work of Ruzi et al. [154].

### 3.3. Enhanced UV-Light Barrier Characteristics

A strategy to enhance UV-light barrier properties includes the assembly of light stabilizers into the film matrix to deactivate the reactive degradation elements, preventing their occurrence by consuming the products responsible for initiating the deterioration reactions and hindering free radicals spread. Aluminum foil represents the best material for UV-visible light blocking, but it poses some disadvantages related to its recyclability, high cost, and non-transparency. In this sense, the smart selection and application of UV absorbers to functionalize materials for specific functions constitute a promising strategy [123].

Inorganic materials are based mainly on the metal oxide particles, such as titanium dioxide ($TiO_2$), cerium oxide ($CeO_2$), iron oxide ($Fe_2O_3$), zirconium dioxide ($ZrO_2$), and zinc oxide (ZnO), that impart enhanced light barrier properties by scattering the incident light [123,155]. However, there are concerns regarding the diffusion of metal nanoparticles from film packaging to the food matrix and to the human body after ingestion. For this reason, the European Union legislation (Commission Regulation, 2011 (EU) N° 10/2011 on Plastic Materials and Articles Intended to Come into Contact with Food) established the type and quantity of additives in materials intended to remain in direct contact with food, though further advances in research are needed to find safe alternatives [156].

According to Song et al., inorganic nanoparticles can easily agglomerate, having a negative influence on the barrier UV performance of matrices [59]. This fact limited their large-scale industrial production and consequently the practical applications of UV-blocking films. In contrast, organic UV-blocking alternatives are identified as phenolic-type UV-absorbers which exhibit good photostability, because the compounds can absorb light energy and make it less harmful, and are typically involved with hydrogen bonds from either O–H–O bridges or O–H–N bridges [123]. Organic adsorbents are generally based on triazine, benzotriazole, benzophenone, anthranilates, dibenzoylmethane, and light amine stabilizers [157]. However, the dangers of organic absorbents for human health restrict their use by promoting the search for green alternatives. In this sense, molecules, such as flavonoids or natural compounds that act as green radiation absorbers (such as tannins, kombucha tea, lignin, grape syrup, pine extract, and others), are interesting substitutes to metal nanoparticles and organic absorbents, blocking radiation in a broad range of wavelengths and decreasing the risks associated with potential migration to the food matrices [59,157–160].

Another alternative to improve UV-light barrier characteristics is the design of a hybrid material, defined as a material that comprises different systems. In this sense, hybrid UV-blocking absorbers can result from the combination of organic UV absorbers and inorganic matrices or from the integration of inorganic UV-blocking absorbers with organic polymers [123].

## 4. Active and Intelligent Food Packaging

The increasing demand for fresh, healthy, and long-shelf-life foods require innovations in packaging design. In this sense, new, intelligent, and smart packaging able to sense and communicate information from the packaged food product has been developed. Before moving forward, it is necessary to clarify three terms that are sometimes used interchangeably: active, intelligent, and smart packaging (Figure 3).

Active materials are specifically designed to interact with the food or its surrounding environment, modifying their composition or characteristics to preserve the organoleptic or sensory characteristics of the product to ensure its quality for longer periods of time. Antimicrobials, antioxidants, flavor and gas scavengers, and light blockers are some examples of active substances usually used in food packaging [85]. Intelligent packaging materials, on the other hand, are aimed to sense changes within the food package and provide information about the status of the food inside [161]. Although limited to detecting and communicating, intelligent packaging can enhance security, safety, and convenience, providing real-time quantitative information on package integrity and food freshness, maturity, or contamination [48,85,162]. Finally, smart packaging results from the combination of both intelligent and active packaging technology, though it is sometimes used indistinctly to refer to either one or the other [163]. However, food packaging systems with a single function of freshness keeping or monitoring may not be able to meet all practical needs [164].

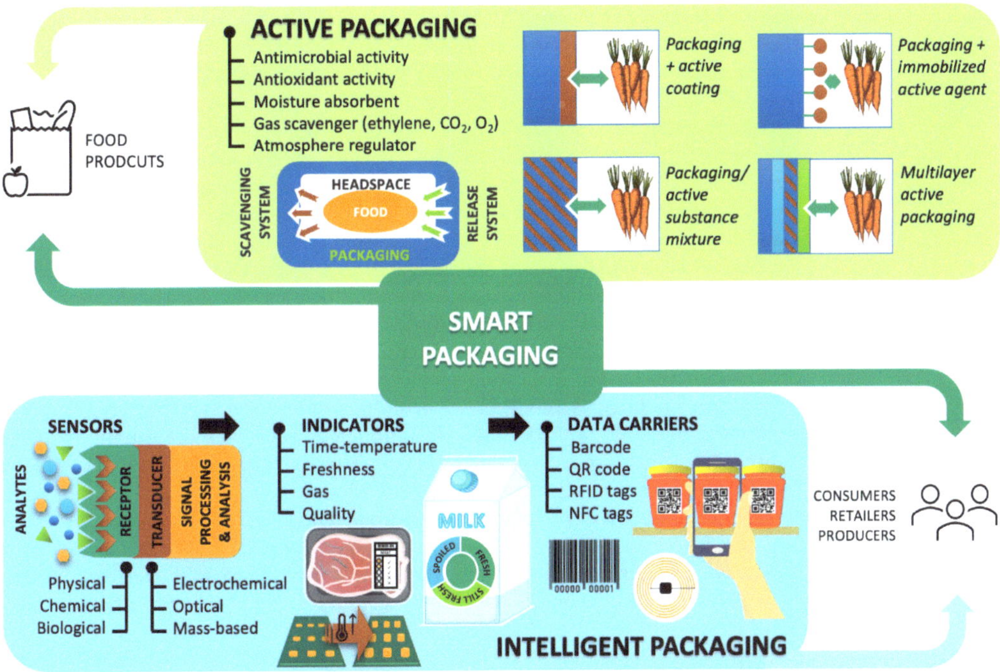


**Figure 3.** Classification of active, intelligent, and smart packaging, considering main properties, applications, interactions, and some examples.

The global market for active and intelligent packaging materials presents a growing tendency with a forecasted CARG of 5.5% from 2021 to 2030, estimated to reach $38.66 billion by 2030 [165], although the pandemic context has probably affected the development of this market.

### *4.1. Active Materials and Packaging Systems*

Active containers, commonly used in food packaging, are those that contain a substance capable of preserving the organoleptic or sensory characteristics of a product to ensure its quality. Of particular interest are active packaging containing natural antioxidants and antimicrobials that not only extend the shelf-life of packaged products by preventing rancidity reactions, but also prevent the growth of foodborne pathogens [87]. The EU Commission defined active compounds as any substance or device that can extend the shelf-life or maintain/improve the packaging environment [166]. According to European Regulation No 450/2009, the active agent can be an individual substance or a combination of substances [167]. These compounds are included in the packaging materials formulations and exert specific functions, such as releasing or absorbing $CO_2$, $O_2$, ethylene, odors, flavors, antioxidants, and antimicrobials. Recently, Atta et al., Amin et al., and

Pandey et al. extensively reviewed biodegradable active packaging materials for food applications [4,84,85].

Active materials can be categorized by active agent or function. There are numerous types of substances used in food packaging that can act as active agents. Thus, a classification based on the purpose of the active packaging proves clearer and more useful in terms of application. However, many active compounds can have more than one function, making their classification complex.

Active packaging materials containing food-approved synthetic active compounds were first developed. Among the synthetic antimicrobials, active films that include potassium sorbate, benzoate, and propionate in their formulation stand out [168–170]. Synthetic antioxidants, e.g., butylated hydroxytoluene (BHT) and butylated hydroxyanisole (BHA), have been also included in both polymeric and biodegradable film formulations and were used as active packaging to prevent lipid oxidation in foods [171].

Biobased materials have great potential to replace conventional synthetic ones for packaging applications [172–175]. The use of natural additives in food packaging formulations not only protects them from the environment, but also provides protection against microbial contamination and agents that stimulate oxidative rancidity reactions, among others, reducing the need of additives in the food matrix [4,55,84]. Some of the polymers and natural compounds can be obtained from industrial agri-food waste and their use would add value to this waste and keep it in circulation. Of particular interest is the use of chitosan or its derivatives due to intrinsic antimicrobial activity against a broad spectrum of microorganisms [176]. This biodegradable polymer derives from chitin that is mainly obtained from waste from the fishing industry, such as crustacean exoskeletons [177].

Likewise, by-products and waste from fruit and vegetable processing are an important source of bioactive compounds with high nutritional and functional value, such as vitamins, minerals, antioxidants, and antimicrobial compounds, although they are often discarded or derived for animal feed [178]. The essential oils (EOs) that can be obtained from these sources have been widely studied as additives for the development of active food packaging, mainly due to their antioxidant and antimicrobial capacity and their GRAS (generally recognized as safe) character [88,179]. Therefore, the use of active compounds derived from agricultural by-products not only contributes to the recovery of these compounds with specific activities, but also generates added value for them. For example, Bof et al. have developed and characterized active biodegradable films based on corn starch and chitosan with the addition of lemon essential oil (LEO) and grapefruit seed extracts (GSE) [180]. The inclusion of these active compounds, which are by-products or derivatives of citrus processing residues, did not affect the mechanical properties of the material and provided antimicrobial capacity on contact. Similarly, Kanmani and Rhim developed antimicrobial active films with GSE in a carrageenan matrix, with additional UV-barrier capacity, particularly important for UV-sensitive food packaging [181]. Moreover, Bof et al. have demonstrated that the biodegradable films based on corn starch and chitosan and the active film containing GSE reduced the post-harvest weight loss, without the incidence of rottenness, of packed blueberries during their refrigerated storage in comparison to clamshell PET containers [182]. Besides, their performance under real conditions of transport and commercialization was evaluated, also considering the costs and possible scaling of the process. Additionally, several EOs have been included in film formulations to impart antioxidant properties, such as rosemary extracts, tea polyphenols, eugenol, oregano, thyme extracts, red propolis, and green coffee oil extracts [183–190]. Thus, EOs exhibited combined and synergic actions since they can confer both antimicrobial and antioxidant capacities. This effect was demonstrated for gelatin films containing red propolis extract which exhibited both antioxidant capacity and inhibited the growth of *S. aureus*, *L. monocytogenes*, *S. enteriditis*, and *E. Coli* [186]. Similarly, chitosan films loaded with eugenol were also effective in the inhibition of *S. aureus* and *E. Coli* and presented antioxidant capacity [187].

Other natural compounds, e.g., α-tocopherol, ascorbic acid, lycopene, gallic acid and resveratrol, with proven antioxidant capacity have also been incorporated in polymer matrices to functionalize them [191–196]. Considering their action mechanism, antioxidant compounds can also be used as oxygen scavengers [197–200]. These compounds can be incorporated into the package as sachets, stickers, or embedded into the polymer matrix of the food packaging.

In addition, different strategies have been promoted to modulate the release kinetics of the active compound and thus ensure the effect of the active compound incorporated. Among them, encapsulation stands out, a technique of particular interest in the case of thermolabile compounds that will be incorporated into matrices that require processing at high temperatures. Likewise, novel methods have been developed for the design of micro and nano-systems that transport antioxidant and antimicrobial active ingredients. These structures allow the controlled release of the active components for different purposes [191]. The protection of active compounds by electrospinning or electrospraying are novel techniques with the potential to be scaled up [201]. Spray drying is a widely used technological tool to encapsulate active compounds due to its low cost and because it allows the continuous production of large amounts of product in short periods of time. However, it requires careful control of the operating conditions to avoid losses of the active compounds during the process [202].

In recent years, research reporting the formulation of active biomaterials, including metallic nanoparticles obtained by green synthesis techniques, has increased, highlighting Ag and Cu nanoparticles and ZnO nanorods [203–205]. In this regard, Ortega et al. have synthesized Ag nanoparticles by reduction with maltose in situ of the filmogenic suspension of starch-based biomaterials, characterizing the materials obtained and demonstrating their antimicrobial capacity [173,203]. They have also obtained nanocomposite materials based on corn starch by adding silver nanoparticles synthesized with reducing compounds present in lemon juice. These materials also exhibited the ability to inhibit the growth of fungi, yeasts, and bacteria [206]. Likewise, nanoparticles, e.g., those of $TiO_2$ and montmorillonite, as well as zeolites, are efficient as ethylene scavengers and were useful to reduce fruit decay and maintain quality of fruits during postharvest storage [171,207]. Zeolites have also been used as $CO_2$ scavengers in food packaging [208]. Nonetheless, even though progress has been made in nanoparticle migration studies and their cytotoxicity, it is still necessary to advance the framework legislation that regulates the use of nanocomposite active materials for their extensive use in contact with food [55,209].

Progress is still required in the design and large-scale manufacture of these active biomaterials for use in the food industry. Besides, the involved production costs should be considered since some biobased materials still require process optimization and the discovery of alternative raw materials for their production. Innovative research is being conducted on the preparation of composite materials with natural and synthetic additives using emerging technologies such as 3D printing, which could further lead to improved functionality of the obtained materials (See Section 5.2).

Future research in the field should be devoted to improving controlled release kinetics to obtain continuous release throughout the shelf-life of packaged food.

### *4.2. Intelligent and Smart Food Packaging*

Intelligent packaging can non-destructively and in situ monitor the food quality, detect storage conditions, expiration date, safety diagnostics, monitor microbial growth, and determine the freshness of food [55,164,210,211]. One aspect of this technology is based on the use of interactive indicators, dye-based compounds that allow the evaluation of the chemical and microbiological quality of foods. Temperature changes during storage, oxygen concentration, or microbial activity can be monitored in real-time. According to Moradi et al., there are three main types of commercially available, intelligent, and smart packaging: time-temperature, freshness, and gas-release indicators [212]. Those of freshness include acid-base indicators based on pigments whose colors are affected by the pH of the environment [213]. Recently, special attention has been paid to the use of natural colorimetric indicators sensitive to changes in pH, including anthocyanins from different vegetable sources [163,214,215].

Sensors are also used in the design of intelligent packaging, some of which can be obtained by 3D printing [164,216]. To this end, nanomaterial-based functional ink formulations with innovative electronic designs and architectures printed on packaging materials are used to achieve various functionalities, such as radio frequency communication, time-temperature indicators, gas sensors, and freshness indicators, among others. New concepts, such as integrated printed radio frequency antennas; printed temperature sensors; and printed supercapacitors, using electronic inks, are still under development [217].

Nowadays, several intelligent systems are commercially available [218,219]. These can be classified into three types: sensors, indicators, and data carriers [220]. A sensor is a device that responds to physical or chemical stimuli measuring it or simply detecting it by some sort of signal emission. Specific sensors named biosensors have been developed to detect hazardous substances, such as foodborne pathogens and allergens, residual antibiotics, or pesticides [221]. These sensors can detect, record, and convey information relevant to biological systems. Most of the commercial biosensors are a combination of antibody-based receptor and optical transducer [222]. For example, Food Sentinel System® developed by Sire Technologies Inc. (Crowthorne, UK) is an antibody-based biosensor where a membrane with immobilized antibodies is used as a part of the barcode that acts as the sensor. As the pathogens interact with antibodies, a localized dark bar is formed which renders the barcode unreadable. ToxinGuard® developed by Toxin Alert (Mississauga, ON, Canada) is another system where antibodies are printed on polyethylene packaging material. The interaction between pathogens and antibodies results in the production of a fluorescent signal that indicates a pathogenic attack. Besides, gas sensors are employed for the detection of gaseous analytes, e.g., oxygen, water vapor, carbon dioxide, ethylene, etc., inside the package. Papkovsky and Dimitriev described optical oxygen sensors based on the principle of quenching or luminescence upon gaseous analyte contact [223]. The use of pH-sensitive dyes, such as methyl red and curcumin, in starch-PVA films for the detection of basic volatile amine released from rotten meat and fish have been reported [224].

On the other hand, indicators can determine the presence or concentration of another substance, or the reaction between two or more substances, by giving characteristic changes in color. Among them, freshness indicators provide information about the product quality by determining the chemical changes resulting from the microbial growth within the product. For instance, COX Technologies (Louisville, KY, USA) markets FreshTag® which are colorimetric labels that indicate the production of volatile amines by fish and shellfish during storage [225]. Besides, the SensorQ® indicator produced from a polymer matrix containing a solution of the bromocresol green dye was used to monitor meat and poultry freshness [226].

Likewise, time-temperature indicators (TTIs) are labels that provide a visual indication of the temperature abuse of perishable products, mostly frozen foods, during distribution and storage from the point of production to that of consumption. The operating principle of these devices is mainly based on different reactions between two or more substances triggered or accelerated by time and temperature variations that result in (usually irreversible) color changes of the indicator [161]. TTIs are based on diffusion, enzymatic, microbial, or polymer-based systems, several of which are commercially available. Diffusion-based indicators rely on either phase-transition or rheology and diffusion rate dependence of colored substances with temperature. For example, Khairunnisa et al. studied different vegetable oils and their combination to tailor their viscosity and the activation energy of the TTIs [227]. The 3M Company (Maplewood, MN, USA) has commercialized diffusion-based TTIs by the trade name 3M Monitor Mark® and Freshness Check® for a wide range of exposure temperatures and times. Kim et al. used isopropyl palmitate on a microporous polymer structure over a colored layer, as the fatty acid ester fills the pores the reflection index of the material decreases letting the lower layer show. Indicating that the length of the

TTI device influences its application range. Enzymatic TTIs use the enzymatic hydrolysis reaction of the substrate, which is sensitive to temperature and or pH variations that can usually be identified by color change of the matrix [161,228]. Similarly, microbial TTIs use metabolites produced by microorganisms, namely yeast and lactic bacteria, under certain time and temperature conditions, generating pH changes, indicated by halochromic compounds [229–233]. These types of biosensors are being studied due to their sensitivity, safety and wide conditions range [228–231,233]. For instance, novel TTIs using laccase [233–236], lipase [237], and glucose oxidase [228] have been successfully developed. An example of commercial enzymatic TTI is VITSAB®, which is based on color change resulting from a pH drop due to controlled enzymatic hydrolysis of a lipid substrate [48]. Non-enzymatic browning reactions, such Maillard reactions, have been studied and modeled to tailor the reaction activation energy by varying the reactive concentrations or pH conditions of the TTI systems [238,239]. In addition, one of the most studied TTIs systems is based on polymerization reactions, typically poly-diacetylene (PDA) based. PDAs present thermochromism, among other chromisms, changing from blue to red upon stimulation with various sensing applications [240–243]. Polymer-based TTIs systems under the trademark Lifelines Freshness Monitor® are commercialized by Lifelines Technology Inc. (Leicester, MA, USA). Recently, Suppakul et al. have developed a diffusion based TTI temperature sensitive device by combining this simple technology with a thermochromic polydiacetylene/silica nanocomposite [244]. Likewise, FreshCode® (Varcode Ltd., Chicago, IL, USA) and Tempix® (Tempix AB, Gävle, Sweden) are based on barcodes printed with fading inks that disappear due to temperature abuse [245]. These inks are colored compounds that produce transparent or lightly colored products by reaction, whose kinetics are regulated by time and temperature, directly indicating changes of food shelf life [161]. Besides, the integrity indicator Ageless Eye® developed by Mitsubishi Gas Chemical Company (Tokyo, Japan) is an oxygen indicator tablet that turns pink when the oxygen concentration is less than 0.01% and blue when it goes beyond 0.5% [246]. Research and development in this area are constant, even more so than those related to commercially available alternatives.

The use of active and Intelligent packaging introduced new concepts and agents that required a clear updated regulatory framework on food contact materials. In fact, the lack of such a legal frame for many years led to reluctance by food packaging manufacturers to take on novel smart packaging systems which were not fully covered by the current legislation on food contact materials [220]. Regulations vary among countries, though they are mainly based on substance migration limits and toxicological properties [247]. The European approach is based on the theory that all materials should be explicitly cleared accordingly to listed hazardous substances, while in the United States substances that are not likely to cause any health problem are directly cleared (or deemed not to require regulation) on the basis of chemical or toxicological data extrapolation beneath minimal dietary exposure [248]. In addition to complying with current regulations, packaging manufacturers must also consider potential human health effects from the possible migration of contaminants, especially for intelligent devices that are placed in direct contact with food. Intelligent devices must also be properly labeled to increase consumer confidence in the safety of packaged foods as well as in the use of these technologies.

Finally, radio-frequency identification (RFID) tags provide the ability to identify, control and manage goods through the supply chain and have been successfully applied for this purpose. These are more advanced, reliable, and efficient than the conventional barcode tags for food traceability [48]. RFID tags for monitoring temperature, relative humidity, pressure, pH, and light exposure of the products are already available in the market which aid in enhancing food quality and safety. For example, Dogbone™, which detects and measures temperature and moisture levels in the environment, and Circus™ Tamper Loop, an NFC RFID tag that informs about authentication reordering, expiration, care, and label integration, were developed by Avery Dennison Smartrac (USA). Besides, this company developed the sensor AD-327 FCC®, which informs about supply chain management, inventory, and logistics [249]. According to a 2019 report by IDTechEx, the RFID market was \$11.6 billion and will grow to \$13 billion by 2022, which reflects the potential of this technology [249]. Although, the design of intelligent systems based on RFID technology still faces many challenges to overcome. Among them, the aspects related to the food safety, cost, reading range, multi-tag collision, multi-parameter sensors, recycling problems, security, and privacy of the RFID system should be solved [48]. Cost is an important factor to consider in intelligent and smart packaging development. If the cost is high in comparison to the product, a price rise could imply fewer sales and a consequent increase in product waste. Likewise, investments in these technologies may be too high from both economic and environmental perspectives [64]. Therefore, new, low-cost materials and methods for sensors, indicators, and data carriers are sought [250,251]. In this regard, 3D and 4D printing of biobased composite materials plays a key role, which may constitute the object of future research and development.

Intelligent packaging and collaboration in data sharing and information transparency may facilitate more synchronized supply chains with greater visibility and traceability, which can help reduce inventory and prevent food from going bad in stock by more precise demand forecast and sales data [64].

It is also imperative to study and control any possible negative effects of intelligent packaging on recycling systems. Bibi et al. remarked upon the need for recycling programs for RFID sensor tags [48]. Hence, materials used should be one of the most important considerations with the aim for low environmental impact elements. In this regard, bio-based smart packaging is a potential option that combines sustainable production and real-time monitoring of food quality to guarantee health safety, prevent food loss, and provide both economic and environmental benefits [252].

Table 2 summarizes recent novel biobased active materials and intelligent systems innovations as well as their applications in food packaging.

**Table 2.** Active and intelligent biodegradable films and their food applications.

| Active Biodegradable Films | | | |
|---|---|---|---|
| **Biopolymer** | **Active Compound** | **Food Application** | **References** |
| Soy protein isolate | Montmorillonite (0.5%wt) + clove essential oil (0.5%v) | Bluefin tuna filets | [253] |
| Chitosan and Corn starch | Lemon essential oil (1–3%wt) and grapefruit seed extract (1–3%wt) | Blueberries conservation simulating transport and commercial conditions | [182] |
| Corn starch | Green synthesized AgNPs in situ (143 ppm) | Cheese preservation | [173] |
| Tapioca starch | Grape pomace extracts (8% $v/v$) and cellulose nanocrystals (10% $v/v$) | Ready to eat chicken meats | [254] |
| PLA | Commercial nanoparticles: $TiO_2$ (3%wt); (2%wt) nano-$TiO_2$ + (1%wt) nano-Ag | Cottage cheese preservation | [255] |
| Curdlan + PVA | Thyme essential oil (1–2%wt) | Chilled pork meat preservation | [256] |
| Whey protein | Oregano and garlic essential oils (2%wt) | Kasar cheese | [257] |
| Chitosan-Cassava TPS bilayer films | Oregano and or cinnamon leaf essential oils (0.25%wt) | Sliced pork meat | [258] |
| Gelatin/Gellan gum | Red radish anthocyanins 5, 10, 15, and 20 mg/100 mL | Milk and fish quality | [259] |
| Zein | Laurel or rosemary leaves extracts (1–10%) | Cheese slices | [260] |
| Chitosan | Propionic acid | Pastry dough | [168] |

**Table 2.** *Cont.*

| Intelligent Biodegradable Films | | | |
|---|---|---|---|
| **Function** | **Intelligent System and Innovative Characteristic** | **Food Application** | **References** |
| Temperature sensor | A passive RFID tag modified with a copper-doped ionic liquid | Fresh products, for the identification of cold chain failures | [245] |
| Temperature abuse indicator | Au nanoparticles included in alginate hydrogel. Biobased, food safe, cost-effective, time sensible | Fresh products | [261] |
| Thermal insulation | Commercial pale-yellow carnauba wax. Biobased and biodegradable insulator | Beverages | [262] |
| pH-based freshness indicator | Biodegradable films containing anthocyanins from different sources. Real-time monitoring of food freshness. | Fresh products: cheese, yogurt, fish, pork, shrimp, and beef | [250,263–267] |
| pH sensitive | Natural compounds showing color changes with pH in biodegradable films | Fish and seafood products | [85,212,268] |
| $CO_2$ detector | Labels containing natural (anthocyanins) or commercial (bromothymol blue and tetrabutyl-ammonium) dyes. | Fermented products such as kimchi | [269] |
| Oxygen indicator | UV-light activated oxygen sensitive biobased film with methyl blue indicator | Suggested for food products packed in modified atmosphere | [270] |
| Hydrogen sulfide indicator | Biobased films containing silver nanoparticles (detect up to 0.81 μmole $H_2S$) or ferrous sulfate (detect 100 ppm $H_2S$) and had a fast response (3 min). | Meat and meat products. Chicken breast and silver carp | [85,271] |
| Humidity indicator | Colorimetric-based sensor on photonic cellulose nanocrystals | Suggested for pharmaceutical products, cereals, and grain seeds storage | [272] |

## 5. Biobased Packaging

One of the applications with the greatest impact on the growing demand for bioplastics is for containers and packaging, particularly biodegradable materials in food packaging. Before the COVID-19 pandemic, many multinationals, such as Danone, Coca-Cola, and Nestle, had committed to adopting bioplastics in their packaging [273]. In this sense, there are numerous investigations and developments of biodegradable and biobased materials due to their potential lower environmental impact. Numerous works have been reported and an extensive literature revision has been conducted by various researchers in the field of material sciences and the packaging industry [4,15,17,51,54,62,87,115,175,274,275]. A comparison of different biomass sources for bioplastics and biopolymers along with various processing technologies discussion can be found among available literature [55,61,113,114,276–279]. The latest works focus on new biomass sources to reduce bioplastics cost and climate impact due to extensive land use [14,18,19,22–24,46,47,55,56,280–290]. In view of more sustainable packaging systems, fully biobased packages are sought, considering all parts. Therefore, herein, a thorough overview of biobased adhesives, inks, and dyes for food packaging is presented.

### 5.1. Bioadhesives for Food Packaging

The term adhesive describes a formulation that can hold two or more specimens together. In other words, an adhesive can join materials by surface bonding (adhesion), with a bond possessing adequate internal strength (cohesion) [291,292] (Figure 4a). The materials which are to be joined are called adherents or substrates and the process of attaching one adherent to another is called bonding. The adhesives include a wide range of materials with very different specifications that produce adhesion through different mechanisms depending on the characteristics of the substrate. The effectiveness of the adhesive bonding capacity depends on several factors, which are intermolecular forces in the adhesive, wettability of the adhesive in the adherent, types of chemical bonds, functional groups involved, type of interface between the adhesive and the adherent, type of substrate, and surface tension generated [293]. Likewise, Dohr and Hirn reported that the strength of the adhesive bond depends on various physical properties of both the substrate and the adhesive [294].

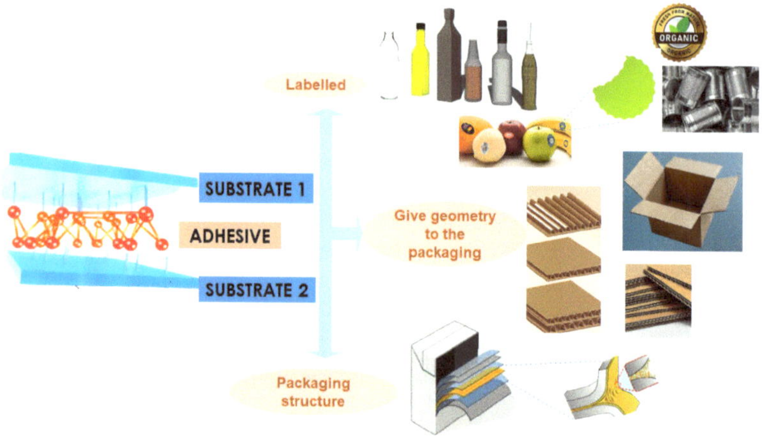


(a)

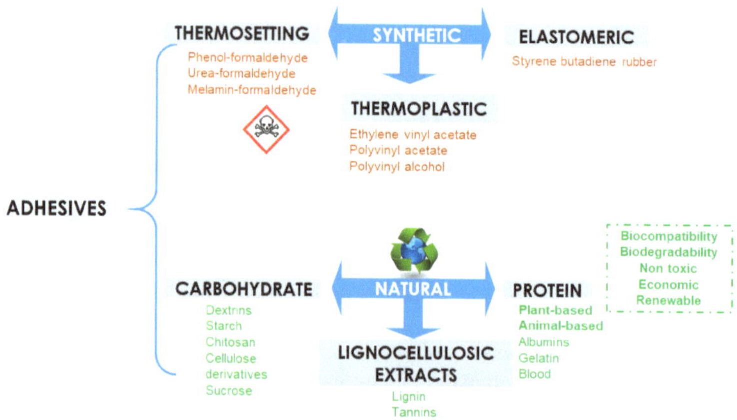


(b)

**Figure 4.** (**a**) Schematic representation of the function of adhesives in food packaging on different substrates; (**b**) Classification of adhesives according to the source of the polymers.

Both the way the adhesive is applied and its chemical composition determine the possible applications. Other factors, such as the dispersion of the adhesive over the substrate and the contact area, which in turn have a great impact on the adhesive forces that can develop, are also relevant [294].

Although in former times adhesives were prepared from natural sources, synthetic formulations of adhesive are nowadays based on thermoset resins: phenol-formaldehyde, an alkaline catalytic salt (PF), urea-formaldehyde, an acid catalytic salt (UF), melamine-formaldehyde (MF), and polymeric diphenylmethane diisocyanate resin (pMDI) [295,296]. These synthetic adhesives cause health problems during the manufacture process. For instance, formaldehyde was declared a carcinogen by the World Health Organization (WHO) in 2004. Similarly, the International Agency for Research on Cancer (IARC) reclassified formaldehyde as carcinogen category 1 in 2004 [297]. In addition, a comparative increase in price and the foreseen lack of availability for petroleum-based synthetic adhesives have prompted the development of green derivatives from economic and renewable resources.

Biobased adhesives are formulations based on natural raw materials, which are not derived from mineral or fossil sources. This term includes adhesives formulated with biopolymers obtained from plants, animals, and natural gums [298]. In the search to develop bio-based adhesives with properties similar to traditional formulations, proteins, tannins, lignin, and polysaccharides are eco-compatible biopolymers that could fit these requirements [299,300]. Their use is mainly limited to paper, cardboard, aluminum foil, and wood for construction applications. These adhesives develop stickiness quickly but exhibit low-strength properties. Most are soluble in water and use water as a solvent agent. They are supplied as liquids or dry powders to mix with water, though some are dispersions [300].

Figure 4b shows a classification of adhesives according to the source of the polymers from which they are prepared, as it was reported by Ebnesajjad and Landrock [301].

The global market for adhesives and sealants is growing rapidly and is expected to be worth $85.8 billion by 2026, with packaging being one of the most relevant applications [302]. Adhesives in packaging industries are critical to the structure of most paper and paperboard packaging. In this regard, paper-based materials have been broadly applied as packaging material for food products.

From a production point of view, adhesive selection can significantly affect process line efficiency and production performance [303]. Adhesives are commercialized in many shapes and types, and the choice will be determined by the substrates on which the adhesive for bonding will be applied, the machinery used in the process, and other factors, such as potential requirements for food-safe materials [303].

The types of adhesives used in the paper industry are water-based adhesives, both synthetic and biopolymer-based (starch, cellulose derivatives, proteins), solvent-based adhesives (polyurethane and acrylic base), and 100% solid adhesives, such as heat-sealing adhesive and hot melts. Gadhave and Gadhave reported the development of a water-based heat-expandable adhesive that has thermally insulating properties and has been used in protective food packaging [303].

Protein-based adhesives (animal or vegetable) were the first polymers to be used in ancient times as adhesives in contact with the skin due to their biocompatibility and biodegradability properties. Likewise, protein-based adhesive formulations have been developed as substitutes for formaldehyde resins, particularly urea-formaldehyde resins in applications for construction material adhesives and paper and coating manufacturing [298].

There is numerous research on adhesives to bond wood made from tannins [304], lignin [305], cellulose derivatives [306] and modified starches [302,307,308], and vegetable proteins, such as casein [309], gluten [310], and soy proteins [311]. Adhesives based on animal protein can be formulated from collagen, gelatin, or casein. Collagen is prepared from different parts of animal bodies, such as skins or bones, and gelatin is derived from the denaturation of collagen. The reversible gel-sol conversion by water absorption is an important property for the application of these adhesives. Casein is obtained from cow's milk by acid precipitation using different acid media, which can result in diverse properties depending on the resulting molecular structures. The primary structure of this protein can be chemically modified to achieve different rheological behaviors [292,309].

Research related to the development of biobased composite adhesives for applications in food packaging is still incipient. Fatty acids derived from vegetable oils were used as the base for monomers that once polymerized form pressure-sensitive adhesives [312].

Heinrich, Wang et al., Wu et al., Mahieu et al., and Ding et al. have reported the use of blends based on starches and proteins from different sources for green bioadhesives preparation [58,300,313–315]. Due to the high number of polar groups in both proteins and polysaccharides, the hydrophilicity of composite systems is often the biggest obstacle to overcome in adhesive formulations [300]. For proteins, this includes making functional groups available for crosslinking, for instance, by tertiary and quaternary structure modification. Chemical modifications can be made in the structure of the polysaccharides and proteins (starch, cellulose, gelatin, soy proteins, among others) to improve the crosslinking of the matrix using crosslinking agents. Kumar et al. have reported that the use of crosslinked starch with citric acid results in materials with good mechanical performance [316]. These properties are explained and revealed from a structural point of view by the interactions established between the carboxyl groups of citric acid and the hydroxyl groups in starch. On the other hand, Olomo informed that adhesives chemically treated using HCl as a cassava starch gelatinization modifier were of higher quality than those modified in presence of NaOH [317].

While native starch contains only hydroxyl groups and is limited in scope, chemically modified starch shows superior water resistance properties for adhesive applications [302]. Other modifications to prepare biobased adhesives with improved water resistance may involve esterification, transesterification, alkylation, acetylation, succinylation, or enzymatic reactions. The use of starch as a raw material in the manufacture of adhesives has the advantages of its renewability, biodegradability, and availability, in addition to its low cost and non-toxicity [317]. Cassava (*Manihot esculenta*) is produced mainly in the tropical and subtropical regions of Africa, South America, and Asia. The high starch content of cassava and its higher proportion of amylopectin, compared to other starch sources, make it an important source of biopolymers to be used in the development of biobased adhesives. It can be employed as native starch, but it can be modified by different means to improve its properties of consistency (viscosity) [315]. Cassava starch as a base to produce adhesives has many notable characteristics, including high paste viscosity, high paste clarity, and high freeze-thaw stability [318]. Besides, Li et al. worked on improving the adhesion-to-fibers and film properties of corn starch by starch sulfo-itaconation for a better application in warp sizing [299].

Biobased adhesives based on colloidal solutions in general dry slowly and therefore require a very long setting time. Bioadhesives containing starch, dextrin, and/or casein are primarily used in labeling applications, while starch-based formulations are also widely used in the obtention of corrugated boards [319]. Starch-based adhesives are mainly composed of water, starch, sodium hydroxide, borax or boric acid, and other additives, such as preservatives, adhesion enhancers and defoamers, among others [320]. Most starch-derived adhesives are used in the paper and textile industries as bonding agents and gluing materials. Corrugated cardboard is produced by the adhesion of a grooved layer of paper to another flat layer. To join them, a two-phase adhesive is usually employed: a liquid and a solid phase containing a mixture of starch and sodium hydroxide or native starch and borax, respectively. Developing biobased adhesive formulations for bonding different substrates, such as corrugated cardboard paper, glass, among others, to propose applications in the formation of film packaging and in the labeling process, constitute a great challenge.

Currently, most containers that are in contact with food have adhesive in their structure. Adhesive formulations can be found in three different forms (Figure 4a). It can form the structure of food packaging by combination with different materials (usually polymers, paper, cardboard, or glass), commonly known as laminate or multi-layer packaging. Moreover, the adhesive can help to provide the geometry of the container (box sealing) or can be used for labeling. The most common way to find adhesives on food packaging is of the first type; where the adhesive is applied on the total surface of the materials or substrates, joining different materials and forming multi-layer materials. Examples of practical applications are flexible film lamination, paper-film combinations, cardboard-film, aluminum-film, cardboard-aluminum, cardboard-aluminum-plastic, rigid multi-layer systems based on plastic, sacks, bags, among others. For such applications, the adhesive industry uses a wide variety of raw materials and formulations, combining different compounds to form special types of biobased adhesives.

There is currently a growing push in the packaging and adhesive industry to improve the sustainability of processes and products. Therefore, the general trend in the adhesive industry for use in food packaging is a reduction in the use of solvents as well as molecular weight components of the adhesive that could more easily migrate from the package to the food.

The use of biobased adhesives in the labeling of glass containers is an interesting alternative, since for this purpose there are few studies that contemplate the use of bioadhesives. The removal of the label-adhesive using only water is a desirable characteristic not only from an operational point of view, but also from an economic and environmental level. Easily removable labels would facilitate the sorting of recyclable waste packaging and therefore increase the recyclability of the whole package and reduce water and energy use. In addition, compared to the common industry practice for label removal using diluted NaOH solutions, water is a low-cost natural resource that, most importantly, could be reused, ultimately requiring simple effluent treatments before discharge.

Regarding the legal background surrounding adhesives intended to be used for food contact applications, normally, the adhesives are used to stick together packaging materials and not intended for direct food contact. However, the adhesives as components of the packaging material might contribute to the migration of constituents into the food matrix. Adhesives, as well as food contact materials, are regulated according to EU Framework Regulation (EC) No 1935/2004. Plastic materials and articles are additionally regulated by a specific measure, Regulation (EU) No 10/20115 on plastics and therefore harmonized at EU level. This regulation establishes, among other requirements, a list of authorized compounds. It is pertinent to emphasize that adhesives do not yet have such specific harmonized legislation. Alternatively, reference is made to the opinions of the European Authority focusing on aspects of Food Safety, Council of Europe resolutions, national legislation, and even non-European legislation for risk assessments [319]. The Commission Regulation (EC) No. 2023/2006, recommends procedures to assure the safety of adhesives for food contact applications, particularly on Good Manufacturing Practices for adhesives. Yet, there is currently no legal obligation for adhesive manufacturers to provide a declaration of conformity with Regulation (EC) No 1935/2004. However, if the adhesive falls under the Plastics Regulation, the adhesive manufacturer shall provide the specific information to enable the adhesive user to ensure compliance of substances with migration potential [319].

According to the specifications reported by Romero Zaliz et al., adhesives for returnable glass containers must be formulated with raw materials that are included in the positive lists of the FDA and the Argentine Food Code (Mercosur) for use in Food Containers and Equipment in Contact with Food (Argentine Food Code Chapter IV, FDA 21 CFR 175.105) [321]. For this reason, there is a high interest in the development of new strategies to produce new renewable materials that totally or partially replace petroleum-derived reagents, resulting in innovative products with special functionality, less toxicity, high biocompatibility, and/or biodegradability [322].

Using natural resources or bio-based materials as adhesive raw materials could help future societies become less dependent on hazardous chemicals, volatile organic compounds, and petroleum-derived chemicals; in addition to promoting safer working conditions. Consumer trends toward green products are prompting plastics industries to investigate more benign alternatives to petroleum-based polymers. Moreover, the recent classification of formaldehyde as a harmful substance has accelerated the investigation of more ecological and renewable alternatives, such as protein-based adhesives, to avoid harmful emissions both during production and during the lifetime. Furthermore, the use of bio-renewable or waste raw materials helps to reduce the carbon footprint, aligned with the current circular economy framework. As an added benefit, the inherent biodegradability of renewable materials, e.g., starch, polyhydroxyalkanoates, or cellulose, is often higher than that of synthetic materials, e.g., polypropylene and polyethylene.

The shift towards more ecofriendly alternatives has manifested itself in the adhesives industry first through the gradual change from solvent adhesives to water-based or high-solid content adhesives, and now by the renewed interest in the design of biobased adhesives. Nonetheless, it is convenient to modify the formulations based on natural polymers, such as polysaccharides and proteins, to improve their rheological properties, their adhesion capacity, and their mechanical resistance when applied for the bonding of different substrates to be used in food packaging. Consequently, composite adhesive formulations obtained from chemically modified biopolymers constitute an innovative proposal to overcome such difficulties.

Further research on the selection and compatibility of biobased adhesive formulations is essential since the adhesive characteristics presented will depend as well on the type of substrates applied to. Additionally, progress is still required in relation to the legislation that frames the adhesives that can be used in the manufacture of containers in contact with food.

### 5.2. Biobased Inks and Dyes in the Food Industry

The use of dyes and pigments dates back to 3500 BC, when the civilizations used natural extracts for coloring. Then, in 1856, the first synthetic dye was created by Perkin and a significant number of dyes were quickly discovered thereafter and adopted by industries [323,324]. Nowadays, the most used colorants (dyes, inks, and pigments) by industries are organic molecules derived from petrochemicals and other chemicals that are causing important environmental damage. According to The Synthetic Dye and Pigment Global Market Report 2022, the global synthetic dye and pigment market grew from $54.54 billion in 2021 to $59.82 billion in 2022 and is expected to grow to $79.45 billion in 2026 [325]. People are more inclined to be nature-friendly and health-conscious, which has created a revolution in research and development in eco-friendly and non-toxic colorants, pushing dyes and pigments manufacturers to shift back to natural dyes. Several countries have imposed a ban on the import of synthetic dyes. For example, the use of azo dyes is banned in India owing to their environmental and health impacts. In general, environmental considerations are becoming vital factors during the selection of consumer products.

Natural dyes are applied in several areas (Figure 5) because they have some special properties, e.g., soothing color, are biodegradable, non-hazardous, non-carcinogenic, and present antimicrobial resistance, among others [324,326,327]. In relation to the food industry, colorants are widely used since, in many products, they are a highly valued attribute by consumers. They are currently being extended to food packaging technologies by applying colorimetric indicators or sensors to exhibit color changes with variations in pH, temperature, and gas for control of the food's quality. In the last five years, especially in food packaging, the research of natural dyes and inks has increased according to the Scopus database (November 2022), although it is not a field that has been extensively studied.

Natural colorants can be classified according to their hue (red, yellow, brown, blue, purple, black, green, and orange), their origin (vegetal, animal, bacterial, fungal, etc.) or their chemical structure as follows: flavonoid derivatives (anthocyanins), isoprenoid derivatives (carotenoids), nitrogen heterocyclic derivatives (betalains) (Table 3). According to Iqbal and Ansari, these natural products present the advantages of being eco-friendly, biodegradable, renewable, not health hazardous or non-toxic; can be obtained from waste biomass, and in some cases present antibacterial and/or UV protective properties [324]. In contrast, natural dyes are expensive because of limited available sources, they are difficult to produce, and the reproducibility of shades is hard to control, among others.

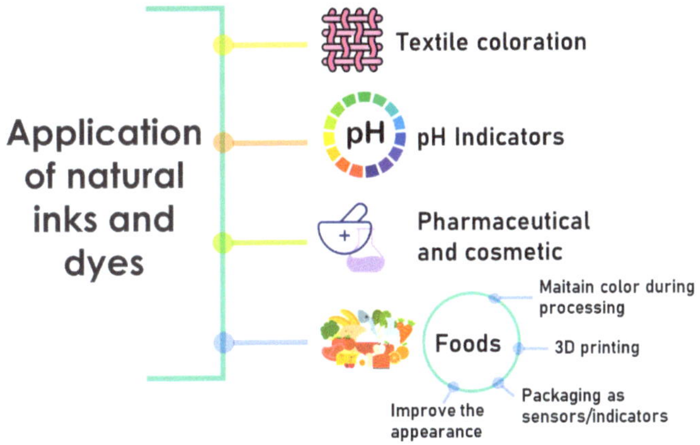


**Figure 5.** Areas of application of natural inks and dyes.

Colorants are widely used in the food industry as an essential ingredient since in some products the color is a relevant characteristic. Viera et al. established that color is responsible for 62–90% of the consumers' acceptability [328]. There is a growing worldwide concern for food quality and safety in the modern era, and given the availability of natural colorants, their use in the food industry is increasing. In this sense, there has been extensive research on the use of natural colorants in food products [326,329–331].

Another use of natural colors is associated with 3D printing technology (additive manufacturing), since this technology can increase the acceptability of certain foods because of the change in the form and presentation, yielding a more attractive product for consumers. Inks for 3D food printing are classified based on their ease of use, nutrition-related components (protein, carbohydrate, fat, fiber), and functional compounds, such as vitamins and antioxidants, that can be incorporated through pigments or natural inks derived from fruits and vegetables extracts. Besides, the additives applied to edible inks play an important role in improving the flow behavior, sedimentation, and lubrication properties of the material to be printed [332]. In this sense, there are some publications regarding the use of fruits and vegetables extracts to obtain more attractive presentations for kids. Lee et al. used spinach powder and xanthan gum to print spinach dispersions and Derossi et al. printed fruit snacks using banana, beans, mushrooms, and lemon juice, while Qiu et al. investigated the 3D printing of apple and edible rose blends as a dysphagia food [333–335].

Escalante-Aburto et al. suggest that the production of 3D food printing must be considered a zero-waste technique to reduce the environmental impact [332]. Thus, the development of bioinks should be focused on using low-carbon and low-water footprint food ingredients, leading to the introduction of a new market for novel and edible composites. Bioinks should be sold as a packaged ingredient that is inserted into the printer to obtain 3D-printed foods. The packaging material must be recyclable or biodegradable and innocuous to fit in with the sustainable technology and food safety concepts.

On the other hand, in 2013, Skylar Tibbits put forward the term 4D printing in his TED talk based on 3D printing [336]. Such 4D printing is based on smart materials and includes a fourth dimension: time, in addition to 3D spatial coordinates. Thus, 4D printing enables the shape, properties, or function of the 3D-printed products to change over time under environmental stimuli, such as temperature, concentration differences, water, pH, or light [336].

Color changes represent one of the most common applications of 4D printing food. He et al. studied the spontaneous color change induced by pH in ready-to-eat 4D foods with anthocyanin-rich purple sweet potatoes, while Ghazal et al. aimed to investigate changes in colors and flavors of 3D-printed healthy food products in response to an external or internal pH stimulus [336,337]. For the formulation, a combination of red cabbage juice, vanillin powder, potato starch, and different fruit juices was used. The changes in color, texture, flavor, and taste induced by the stimulus were determined, revealing that the color of the 3D-printed product changed from blue (control sample) to red, purple, violet, blue-green, and green-yellow colors when sprayed with pH solutions of different pH (2–10). In addition, dried 4D samples exhibited color and anthocyanin stability when stored at room temperature for three weeks. Likewise, Chen et al. discussed the possibility of using microwaves to stimulate the color change of 3D-printed curcumin lotus root gel [338]. Based on traditional 3D printing technology, this work uses microwaves as the stimulus to obtain the color change of printed products and provides a new method for producing colorful and attractive food through 4D printing.

Another use for 4D printed materials could be as quality sensors or indicators for food packaging.

The obtention and use of natural inks or dyes provide an advantage for food packaging. Because of their low molecular weights, coloring agents, photo-initiators, solvents, and oils may migrate from printing inks, favoring natural dyes over synthetic dyes. Studies to obtain natural dyes from beet, red beet, potato, red onion, quince, black carrot, and hibiscus to use them in areas where contact with the human body is frequent, such as textile, tissue engineering, and food packaging, are being carried out [339]. In addition, it is known that these plant extracts used as inks present antimicrobial activity, and for this reason, they are preferred for active and intelligent packaging development.

There are many studies concerning the use of natural dyes or inks in the development of intelligent packaging such as films or 4D printing sensors [212,214,337,340,341]. In this sense, Alizadeh-Sani et al. have critically reviewed pH-sensitive smart packaging films based on natural colorants for the monitoring of food quality and safety [342]. Meanwhile, Tracey et al. released an advanced 3D printing approach to intelligent food packaging [164]. Among the smart packages, those containing TTIs are the most studied. As mentioned in Section 4.2, these indicators adhere to the surface of the packaging and produce an irreversible color change in response to environmental conditions. Rachmelia and Imawan developed the TTI label using black corn extract and chitosan matrix. In this study, anthocyanin was obtained from black corn extract, while chitosan was used as a matrix [343]. The use of non-toxic ingredients was aimed at making the TTI label safe to apply in packaging products. The label's color was observed over time at temperatures of 10, 25, and 40 °C. The label changed color from purple, to blue, to yellow being the fastest color changes at higher temperatures (40 °C) and slowest at low temperatures (10 °C). Furthermore, Mataragas et al. developed a microbial TTI based on the violacein (a microbial violet pigment produced by *Janthinobacterium* sp.) formation for monitoring the shelf-life of minced beef-vacuum-packed cooked meat products. When the temperature varied from 0 to 15 °C, the color changed from purple to violet [231]. Other materials with a similar operation are pH and gas indicators, which due to these factors play an important role being indicative of food quality, shelf-life, microbial growth rate, and food deterioration [156,342,344]. Erna et al. obtained curcumin/rice starch films for sensitive detection of hypoxanthine in chicken and fish meat, since the deprotonation of curcumin occurs when the indicator is exposed to a pH level of 9 or above, where the indicator's color changes from yellow to a reddish-brown or wine-red color [345]; while Boccalon et al. developed potato starch composite films containing red onion skin extract as intelligent pH indicators for food packaging [346]. The films were tested by monitoring their color changes when applied to meat and milk storage.

According to Tracey et al., highly sensitive, self-indicating, multifunctional smart components using biocompatible, nontoxic materials via lower cost intelligent packaging systems and devices can be developed by 3D printing in comparison to conventional fabrication methods [164]. Zhou et al. used coaxial 3D printing followed by ionic crosslinking to create fruit freshness keeping and visual monitoring labels with high pH sensitivity and effective shelf-life extension capability [347]. Cellulose nanofibers (CNF)-based ink with blueberry anthocyanin was used to create the shell of fibers, exhibiting high formability and print fidelity as well as sensitive visual pH responsiveness for freshness monitoring. Chitosan containing 1-methylcyclopropene (1-MCP), an ethylene receptor inhibitor with good freshness-keeping performance and no toxicity, was loaded into the hollow microchannels of the fibers. The 1-MCP was trapped by the electrostatic effect of chitosan and CNF exhibited a sustained release behavior [347]. Finally, the 3D-printed labels prolonged the shelf-life of litchis by approximately six days. On the other hand, Li et al. developed an interlayer with chitosan, mulberry anthocyanin as a natural dye, and lemongrass essential oils as an antibacterial agent and antioxidant using a 3D printer, and cassava starch as a protective layer to form indicator films [341]. These were used to monitor the quality, freshness, and preservation of cold meats by observing the color changes of the indicators. When chilled pork spoiled, the color of the indicator films changed from red to gray-blue (RGB), and the RGB tone values could be analyzed by a smartphone application to determine pork freshness [341].

The use of smart packaging based on natural colorants and biodegradable films is intended as an attractive alternative for food packaging due to their low or nontoxicity, eco-friendliness, easy preparation, biodegradability, availability, renewability, and pollution-free properties. The main challenges of natural inks and biodegradable food packaging are related to the improvement of their functionality at reasonable costs. Likewise, progress in the assessment of their safety is needed. According to Bautista et al., this agrees with the global challenges of the packaging industry, some of the most important being: (1) increase sustainability of manufacturing processes, (2) improve the recyclability of materials, and (3) improve performance and functionality. In addition, the printing of labels with bioinks or the use of smart inks, including natural dyes, in the design of freshness indicator devices is a field that is still in its infancy with great potential [217].

**Table 3.** Classification of natural dyes according to their chemical structure and source.

<table>
<tr><th>Color</th><th>Classification</th><th>Source</th><th>Example</th><th>Reference</th></tr>
<tr><td rowspan="3">Yellow/Orange/Red</td><td>Curcumin</td><td>Plant/Vegetable</td><td>Turmeric (Curcuma longa L.)</td><td>[348]</td></tr>
<tr><td>Carotenoids</td><td>Plant/Vegetable</td><td>Carrot (Daucus carota L.), Annatto (Bixa orellana), Tomato (Solanum lycopersicum), Paprika (Capsicum annuum L.), petals of marigold (Tagetes erecta L.)</td><td>[349,350]</td></tr>
<tr><td>Aryl carotenoids</td><td>Microorganisms</td><td>Brevibacterium linens, Streptomyces mediolani, Mycobacterium aurum</td><td>[351]</td></tr>
<tr><td rowspan="3">Red/Pink</td><td>Betalains</td><td>Plant/Vegetable</td><td>Beetroot (Beta vulgaris L.), Opuntia lasiacantha</td><td>[339,349]</td></tr>
<tr><td>Carminic acid</td><td>Microorganisms</td><td>Cochineal (Dactylopius coccus)</td><td>[350]</td></tr>
<tr><td>Anthocyanins</td><td>Plant/Vegetable</td><td>Hibiscus rosa sinensis flowers</td><td>[352]</td></tr>
<tr><td rowspan="3">Blue/Purple</td><td>Anthocyanins</td><td>Plant/Vegetable</td><td>Grapes (Vitus labruscana L.), red cabbage (Brassica oleracea var. capitata f. rubra), cherry (Prunus cerasus), blueberry (Vaccinium sect. Cyanococcus), red onion skin (Allium cepa L.), Beetroot (Beta vulgaris L.)</td><td>[323,326,350]</td></tr>
<tr><td>Tyrian purple (6,6′-dibromoíndigo)</td><td>Animal</td><td>Mollusks Bolinus brandaris</td><td>[349]</td></tr>
<tr><td>Ultramarine Blue</td><td>Mineral</td><td>Lapis lazuli</td><td>[349]</td></tr>
<tr><td rowspan="3">Green</td><td>Chlorophylls</td><td>Plant/Vegetable</td><td>Spinach (Spinacia oleracea), kiwi pomace (Actinidiaceae), green beans (Phaseolus vulgaris), grass, alfalfa (Medicago sativa)</td><td>[57,350]</td></tr>
<tr><td>Terre-Verte (Green Earth)</td><td rowspan="2">Mineral</td><td>Mixture of hydrosilicates of Fe, Mg, Al, and K (gluconite and celadenite) but other minerals are likely to be present</td><td rowspan="2">[349]</td></tr>
<tr><td>Malachite</td><td>Copper carbonate hydroxide</td></tr>
</table>

## 6. Conclusions

Changes in lifestyle have directly influenced the type of food consumed, as well as consumption habits, which in turn has generated the need for the food industry to develop new containers and packaging. Today, practically any food product is marketed packaged, not only to contain the food, but also to protect it throughout the entire production chain, until it reaches the point of sale or consumption. Hence, the way we produce and consume food has substantial environmental, social, and economic impacts, requiring sustainable solutions for proper and efficient land use, better food preservation technologies during processing and packaging, and novel transport, distribution, and marketing systems to guarantee that these costs are well exceeded by benefits.

Food packaging fulfills very important functions for food preservation, protecting it from external agents, preventing physical, chemical, and/or microbiological contamination, as well as its possible adulteration. In this way, they fulfill the function of a barrier against the environment that surrounds them, protecting food from humidity, oxidation, UV radiation, and microorganisms. New packaging technologies aim to further extend the shelf-life of food products by active systems that can slow the natural oxidation process or avoid microbial growth. Intelligent and smart packaging can further monitor food quality and spoilage to guarantee food safety for the consumers and prevent food losses due to improper conservation or transportation, or inefficient logistics and marketing. Investments in packaging have the potential to reduce overall environmental impacts associated with food production, distribution, and consumption, aiming to minimize food waste.

However, systemic LCA approaches to determine the optimum product/package combination are needed. Effective assessment requires updated qualitative and quantitative information. Specialized sustainability agents within the industry, academia, and government are needed to develop a sustainability culture. Further efforts should also be made to raise awareness and educate food and packaging supply chain stakeholders and consumers on the role of packaging in extending food shelf-life and the opportunities to use it more effectively. Such opportunities entail: enhanced protection and larger shelf-life for fresh food products with potential tailored solutions; recovery of surplus and unsalable produce from farms to rescue supply chains; revalorization and use of agri-food industry byproducts and waste for packaging materials production; proper size (or portioning), function (i.e., easy-to-open, easy-to-empty, fit-for-purpose), reusable and/or recyclable food packaging design; new packaging materials and technologies to extend shelf-life (i.e., active, intelligent and smart packaging); unified labeling regulation for better use-by or best-before date indications for manufacturers, retailers, and consumers to avoid confusion regarding productions dates and prevent the unnecessary disposal of food; and intelligent packaging and data sharing to create more synchronized supply chains to reduce excess or out-of-date stock.

A gap exists between product/packaging system design, materials supply, manufacturing and commercialization, and the return flow from recyclable materials that enter the waste management stream, which hinders circularity. To close the cycle, a holistic understanding of the supply chain components, their opportunities, and limitations is required for transitioning sustainable production systems towards a circular economy.

**Author Contributions:** F.V., F.O., Y.M., S.R., O.V.L. and M.A.G. contributed to the study conceptualization, visualization, and design, as well as literature search and data analysis. A first draft of the manuscript was jointly written and commented on previous versions of the manuscript by F.V., F.O., Y.M., S.R., O.V.L. and M.A.G. All authors have read and agreed to the published version of the manuscript.

**Funding:** Projects PICT 2019-03380: PICT 2019-2827, and PICT 2018-3285 (ANPCyT, Argentina); PGI 24/ZQ18 (UNS and CONICET, Argentina), and PIP 2022-00493CO (CONICET, Argentina), and Drs. Florencia Ortega and Yuliana Monroy Post-Doctoral Fellowship CONICET (CONICET, Argentina).

**Data Availability Statement:** Data sharing is not applicable—no new data generated.

**Acknowledgments:** The authors would like to thank the National Council of Research in Argentina (CONICET), the National Agency for the Promotion of Science and Technology of Argentina (ANPCyT), the National University of the South (UNS) in Bahía Blanca and the National University of La Plata (UNLP).

**Conflicts of Interest:** The authors declare no conflict of interest. Commercial products and manufacturers' names listed in this review are not the only ones that have been commercialized and developed in the world. The authors are not related to any of the commercial products/manufacturers that are referenced in this review.

## References

1. UN. *World Population Prospects 2022—Summary of Results*; UN: New York, NY, USA, 2022.
2. FAO. *The Future of Food and Agriculture Trends and Challenges*; FAO: Roma, Italy, 2017.
3. Chakori, S.; Aziz, A.A.; Smith, C.; Dargusch, P. Untangling the Underlying Drivers of the Use of Single-Use Food Packaging. *Ecol. Econ.* **2021**, *185*, 107063. [CrossRef]
4. Atta, O.M.; Manan, S.; Shahzad, A.; Ul-Islam, M.; Ullah, M.W.; Yang, G. Biobased Materials for Active Food Packaging: A Review. *Food Hydrocoll* **2022**, *125*, 107419. [CrossRef]
5. Wohner, B.; Pauer, E.; Heinrich, V.; Tacker, M. Packaging-Related Food Losses and Waste: An Overview of Drivers and Issues. *Sustainability* **2019**, *11*, 264. [CrossRef]
6. Sasaki, Y.; Orikasa, T.; Nakamura, N.; Hayashi, K.; Yasaka, Y.; Makino, N.; Shobatake, K.; Koide, S.; Shiina, T. Determination of the Most Environmentally Friendly Packaging for Peach during Transportation by Modeling the Relationship between Food Loss Reduction and Environmental Impact. *J. Food Eng.* **2022**, *331*, 111120. [CrossRef]
7. Blakeney, M. Chapter 1: Food Loss and Waste and Food Security. In *Food Loss and Food Waste*; Edward Elgar Publishing: Cheltenham, UK, 2019; pp. 1–26. ISBN 9781788975391.
8. Bishop, G.; Styles, D.; Lens, P.N.L. Environmental Performance of Bioplastic Packaging on Fresh Food Produce: A Consequential Life Cycle Assessment. *J. Clean. Prod.* **2021**, *317*, 128377. [CrossRef]
9. Kan, M.; Miller, S.A. Environmental Impacts of Plastic Packaging of Food Products. *Resour. Conserv. Recycl.* **2022**, *180*, 106156. [CrossRef]
10. WEF. *The New Plastics Economy–Rethinking the Future of Plastics. Industry Agenda*; WEF: Geneva, Switzerland, 2016.
11. Brivio, E.; Petsa, I.; McPhie, T. Single-Use Plastics: New EU Rules to Reduce Marine Litter. Available online: https://ec.europa.eu/commission/presscorner/detail/en/IP_18_3927 (accessed on 23 November 2022).
12. Plastics Europe. *Plastics-the Facts 2022*; Plastics Europe: Brussels, Belgium, 2022.
13. Bandara, R.; Indunil, G.M. Food Packaging from Recycled Papers: Chemical, Physical, Optical Properties and Heavy Metal Migration. *Heliyon* **2022**, *8*, e10959. [CrossRef]
14. Hosen, M.D.; Hossain, M.S.; Islam, M.A.; Haque, A.N.M.A.; Naebe, M. Utilisation of Natural Wastes: Water-Resistant Semi-Transparent Paper for Food Packaging. *J. Clean. Prod.* **2022**, *364*, 132665. [CrossRef]
15. Markevičiūtė, Z.; Varžinskas, V. Plant-Origin Feedstock Applications in Fully Green Food Packaging: The Potential for Tree-Free Paper and Plant-Origin Bio-Plastics in the Baltic Sea Region. *Sustainability* **2022**, *14*, 7393. [CrossRef]
16. Garrido-Romero, M.; Aguado, R.; Moral, A.; Brindley, C.; Ballesteros, M. From Traditional Paper to Nanocomposite Films: Analysis of Global Research into Cellulose for Food Packaging. *Food Packag. Shelf Life* **2022**, *31*, 100788. [CrossRef]
17. Gonçalves de Moura, I.; Vasconcelos de Sá, A.; Lemos Machado Abreu, A.S.; Alves Machado, A.V. Bioplastics from Agro-Wastes for Food Packaging Applications. In *Food Packaging*; Elsevier: Amsterdam, The Netherlands, 2017; pp. 223–263.
18. Mostafa, N.A.; Farag, A.A.; Abo-dief, H.M.; Tayeb, A.M. Production of Biodegradable Plastic from Agricultural Wastes. *Arab. J. Chem.* **2018**, *11*, 546–553. [CrossRef]
19. Zhang, J.; Hori, N.; Takemura, A. Optimization of Agricultural Wastes Liquefaction Process and Preparing Bio-Based Polyurethane Foams by the Obtained Polyols. *Ind. Crops Prod.* **2019**, *138*, 111455. [CrossRef]
20. Yun, D.; Liu, J. Recent Advances on the Development of Food Packaging Films Based on Citrus Processing Wastes: A Review. *J. Agric. Food Res.* **2022**, *9*, 100316. [CrossRef]
21. Rajinipriya, M.; Nagalakshmaiah, M.; Robert, M.; Elkoun, S. Importance of Agricultural and Industrial Waste in the Field of Nanocellulose and Recent Industrial Developments of Wood Based Nanocellulose: A Review. *ACS Sustain. Chem. Eng.* **2018**, *6*, 2807–2828. [CrossRef]
22. Otles, S.; Kartal, C. Food Waste Valorization. In *Sustainable Food Systems from Agriculture to Industry*; Galanakis, C.M., Ed.; Elsevier: Amsterdam, The Netherlands, 2018; pp. 371–399.
23. Six, L.; Velghe, F.; Verstichel, S.; de Meester, S. Sustainability Considerations on the Valorization of Organic Waste. In *Biotransformation of Agricultural Waste and By-Products*; Poltronieri, P., D'Urso, O.F., Eds.; Elsevier: Amsterdam, The Netherlands, 2016; pp. 287–307. ISBN 9780128036488.
24. Jõgi, K.; Bhat, R. Valorization of Food Processing Wastes and By-Products for Bioplastic Production. *Sustain. Chem. Pharm.* **2020**, *18*, 100326. [CrossRef]

25. Vilaplana, F.; Strömberg, E.; Karlsson, S. Environmental and Resource Aspects of Sustainable Biocomposites. *Polym. Degrad. Stab.* **2010**, *95*, 2147–2161. [CrossRef]
26. Springle, N.; Li, B.; Soma, T.; Shulman, T. The Complex Role of Single-Use Compostable Bioplastic Food Packaging and Foodservice Ware in a Circular Economy: Findings from a Social Innovation Lab. *Sustain. Prod. Consum.* **2022**, *33*, 664–673. [CrossRef]
27. Gioia, C.; Giacobazzi, G.; Vannini, M.; Totaro, G.; Sisti, L.; Colonna, M.; Marchese, P.; Celli, A. End of Life of Biodegradable Plastics: Composting versus Re/Upcycling. *ChemSusChem* **2021**, *14*, 4167–4175. [CrossRef]
28. Moshood, T.D.; Nawanir, G.; Mahmud, F.; Mohamad, F.; Ahmad, M.H.; AbdulGhani, A. Sustainability of Biodegradable Plastics: New Problem or Solution to Solve the Global Plastic Pollution? *Curr. Res. Green Sustain. Chem.* **2022**, *5*, 100273. [CrossRef]
29. Wang, G.X.; Huang, D.; Ji, J.H.; Völker, C.; Wurm, F.R. Seawater-Degradable Polymers—Fighting the Marine Plastic Pollution. *Adv. Sci.* **2021**, *8*, 2001121. [CrossRef]
30. Suwanmanee, U.; Varabuntoonvit, V.; Chaiwutthinan, P.; Tajan, M.; Mungcharoen, T.; Leejarkpai, T. Life Cycle Assessment of Single Use Thermoform Boxes Made from Polystyrene (PS), Polylactic Acid, (PLA), and PLA/Starch: Cradle to Consumer Gate. *Int. J. Life Cycle Assess.* **2013**, *18*, 401–417. [CrossRef]
31. Saibuatrong, W.; Cheroennet, N.; Suwanmanee, U. Life Cycle Assessment Focusing on the Waste Management of Conventional and Bio-Based Garbage Bags. *J. Clean. Prod.* **2017**, *158*, 319–334. [CrossRef]
32. Cheroennet, N.; Pongpinyopap, S.; Leejarkpai, T.; Suwanmanee, U. A Trade-off between Carbon and Water Impacts in Bio-Based Box Production Chains in Thailand: A Case Study of PS, PLAS, PLAS/Starch, and PBS. *J. Clean. Prod.* **2017**, *167*, 987–1001. [CrossRef]
33. Banar, M.; Cokaygil, Z. A Comparative Life Cycle Analysis of Two Different Juice Packages. *Environ. Eng. Sci.* **2008**, *25*, 549–555. [CrossRef]
34. Tsiropoulos, I.; Faaij, A.P.C.; Lundquist, L.; Schenker, U.; Briois, J.F.; Patel, M.K. Life Cycle Impact Assessment of Bio-Based Plastics from Sugarcane Ethanol. *J. Clean. Prod.* **2015**, *90*, 114–127. [CrossRef]
35. Civancik-Uslu, D.; Puig, R.; Hauschild, M.; Fullana-i-Palmer, P. Life Cycle Assessment of Carrier Bags and Development of a Littering Indicator. *Sci. Total Environ.* **2019**, *685*, 621–630. [CrossRef]
36. Zabaniotou, A.; Kassidi, E. Life Cycle Assessment Applied to Egg Packaging Made from Polystyrene and Recycled Paper. *J. Clean. Prod.* **2003**, *11*, 549–559. [CrossRef]
37. Chen, L.; Pelton, R.E.O.; Smith, T.M. Comparative Life Cycle Assessment of Fossil and Bio-Based Polyethylene Terephthalate (PET) Bottles. *J. Clean. Prod.* **2016**, *137*, 667–676. [CrossRef]
38. Abejón, R.; Bala, A.; Vázquez-Rowe, I.; Aldaco, R.; Fullana-i-Palmer, P. When Plastic Packaging Should Be Preferred: Life Cycle Analysis of Packages for Fruit and Vegetable Distribution in the Spanish Peninsular Market. *Resour. Conserv. Recycl.* **2020**, *155*, 104666. [CrossRef]
39. Ferrara, C.; Migliaro, V.; Ventura, F.; de Feo, G. An Economic and Environmental Analysis of Wine Packaging Systems in Italy: A Life Cycle (LC) Approach. *Sci. Total Environ.* **2023**, *857*, 159323. [CrossRef]
40. Casson, A.; Giovenzana, V.; Frigerio, V.; Zambelli, M.; Beghi, R.; Pampuri, A.; Tugnolo, A.; Merlini, A.; Colombo, L.; Limbo, S.; et al. Beyond the Eco-Design of Case-Ready Beef Packaging: The Relationship between Food Waste and Shelf-Life as a Key Element in Life Cycle Assessment. *Food Packag. Shelf Life* **2022**, *34*, 100943. [CrossRef]
41. Razza, F.; Innocenti, F.D.; Dobon, A.; Aliaga, C.; Sanchez, C.; Hortal, M. Environmental Profile of a Bio-Based and Biodegradable Foamed Packaging Prototype in Comparison with the Current Benchmark. *J. Clean. Prod.* **2015**, *102*, 493–500. [CrossRef]
42. Bishop, G.; Styles, D.; Lens, P.N.L. Environmental Performance Comparison of Bioplastics and Petrochemical Plastics: A Review of Life Cycle Assessment (LCA) Methodological Decisions. *Resour. Conserv. Recycl.* **2021**, *168*, 105451. [CrossRef]
43. Accorsi, R.; Battarra, I.; Guidani, B.; Manzini, R.; Ronzoni, M.; Volpe, L. Augmented Spatial LCA for Comparing Reusable and Recyclable Food Packaging Containers Networks. *J. Clean. Prod.* **2022**, *375*, 134027. [CrossRef]
44. Briassoulis, D.; Pikasi, A.; Hiskakis, M. Recirculation Potential of Post-Consumer /Industrial Bio-Based Plastics through Mechanical Recycling—Techno-Economic Sustainability Criteria and Indicators. *Polym. Degrad. Stab.* **2021**, *183*, 109217. [CrossRef]
45. Briassoulis, D.; Pikasi, A.; Hiskakis, M. End-of-Waste Life: Inventory of Alternative End-of-Use Recirculation Routes of Bio-Based Plastics in the European Union Context. *Crit. Rev. Environ. Sci. Technol.* **2019**, *49*, 1835–1892. [CrossRef]
46. Ribeiro, T.B.; Oliveira, A.L.; Costa, C.; Nunes, J.; Vicente, A.A.; Pintado, M. Total and Sustainable Valorisation of Olive Pomace Using a Fractionation Approach. *Appl. Sci.* **2020**, *10*, 6785. [CrossRef]
47. Urbina, L.; Eceiza, A.; Gabilondo, N.; Corcuera, M.Á.; Retegi, A. Valorization of Apple Waste for Active Packaging: Multicomponent Polyhydroxyalkanoate Coated Nanopapers with Improved Hydrophobicity and Antioxidant Capacity. *Food Packag. Shelf Life* **2019**, *21*, 100356. [CrossRef]
48. Bibi, F.; Guillaume, C.; Gontard, N.; Sorli, B. A Review: RFID Technology Having Sensing Aptitudes for Food Industry and Their Contribution to Tracking and Monitoring of Food Products. *Trends Food Sci. Technol.* **2017**, *62*, 91–103. [CrossRef]
49. Gustavo, J.U.; Pereira, G.M.; Bond, A.J.; Viegas, C.V.; Borchardt, M. Drivers, Opportunities and Barriers for a Retailer in the Pursuit of More Sustainable Packaging Redesign. *J. Clean. Prod.* **2018**, *187*, 18–28. [CrossRef]
50. Muller, J.; González-Martínez, C.; Chiralt, A. Combination of Poly(Lactic) Acid and Starch for Biodegradable Food Packaging. *Materials* **2017**, *10*, 952. [CrossRef] [PubMed]

51. Shlush, E.; Davidovich-Pinhas, M. Bioplastics for Food Packaging. *Trends Food Sci. Technol.* **2022**, *125*, 66–80. [CrossRef]
52. Wang, X.; Huang, L.; Zhang, C.; Deng, Y.; Xie, P.; Liu, L.; Cheng, J. Research Advances in Chemical Modifications of Starch for Hydrophobicity and Its Applications: A Review. *Carbohydr. Polym.* **2020**, *240*, 116292. [CrossRef] [PubMed]
53. Ashter, S.A. Additives and Modifiers for Biopolymers. In *Introduction to Bioplastics Engineering*; Elsevier: Amsterdam, The Netherlands, 2016; pp. 153–178.
54. Jariyasakoolroj, P.; Leelaphiwat, P.; Harnkarnsujarit, N. Advances in Research and Development of Bioplastic for Food Packaging. *J. Sci. Food Agric.* **2020**, *100*, 5032–5045. [CrossRef] [PubMed]
55. Ortega, F.; Versino, F.; López, O.V.; García, M.A. Biobased Composites from Agro-Industrial Wastes and by-Products. *Emergent Mater.* **2022**, *5*, 873–921. [CrossRef]
56. Chakrapani, G.; Zare, M.; Ramakrishna, S. Biomaterials from the Value-Added Food Wastes. *Bioresour. Technol. Rep.* **2022**, *19*, 101181. [CrossRef]
57. Benucci, I.; Lombardelli, C.; Mazzocchi, C.; Esti, M. Natural Colorants from Vegetable Food Waste: Recovery, Regulatory Aspects, and Stability—A Review. *Compr. Rev. Food Sci. Food Saf.* **2022**, *21*, 2715–2737. [CrossRef]
58. Ding, X.; Dai, R.; Chen, H.; Shan, Z. Gelatin as Green Adhesive for the Preparation of a Multifunctional Biobased Cryogel Derived from Bamboo Industrial Waste. *Carbohydr. Polym* **2021**, *255*, 117340. [CrossRef]
59. Song, Y.; Chen, S.; Chen, Y.; Xu, Y.; Xu, F. Biodegradable and Transparent Films with Tunable UV-Blocking Property from Lignocellulosic Waste by a Top-down Approach. *Cellulose* **2021**, *28*, 8629–8640. [CrossRef]
60. Espinosa, E.; Rincón, E.; Morcillo-Martín, R.; Rabasco-Vílchez, L.; Rodríguez, A. Orange Peel Waste Biorefinery in Multi-Component Cascade Approach: Polyphenolic Compounds and Nanocellulose for Food Packaging. *Ind. Crops Prod.* **2022**, *187*, 115413. [CrossRef]
61. Castillo, L.A.; López, O.; Ninago, M.D.; Versino, F.; Barbosa, S.E.; García, M.A.; Villar, M.A. Composites and Nanocomposites Based on Starches. Effect of Mineral and Organic Fillers on Processing, Structure, and Final Properties of Starch. In *Starch-Based Materials in Food Packaging*; Villar, M.A., Barbosa, S.E., García, M.A., Castillo, L.A., López, O., Eds.; Academic Press: Cambridge, MA, USA, 2017; pp. 125–151. ISBN 9780128122570.
62. Versino, F.; López, O.V.; García, M.A. Green Biocomposites for Packaging Applications. In *Biocomposite Materials Design and Mechanical Properties Characterization*; Sultan, M.T.H., Majid, M.S.A., Azmi, M.R.M.J., Iskandar, A., Saba, N., Eds.; Springer Nature Singapore Pte Ltd.: Singapore, 2021; pp. 1–30. ISBN 9789813340916.
63. European Commission. Ecodesign Your Future: How Ecodesign Can Help the Environment by Making Products Smarter. Available online: https://op.europa.eu/en/publication-detail/-/publication/4d42d597-4f92-4498-8e1d-857cc157e6db (accessed on 13 December 2022).
64. Verghese, K.; Lewis, H.; Lockrey, S.; Williams, H. Packaging's Role in Minimizing Food Loss and Waste Across the Supply Chain. *Packag. Technol. Sci.* **2015**, *28*, 603–620. [CrossRef]
65. Wikström, F.; Verghese, K.; Auras, R.; Olsson, A.; Williams, H.; Wever, R.; Grönman, K.; Kvalvåg Pettersen, M.; Møller, H.; Soukka, R. Packaging Strategies That Save Food: A Research Agenda for 2030. *J. Ind. Ecol.* **2019**, *23*, 532–540. [CrossRef]
66. Heller, M.C.; Selke, S.E.M.; Keoleian, G.A. Mapping the Influence of Food Waste in Food Packaging Environmental Performance Assessments. *J. Ind. Ecol.* **2019**, *23*, 480–495. [CrossRef]
67. Cai, J.; Zhang, D.; Zhou, R.; Zhu, R.; Fei, P.; Zhu, Z.Z.; Cheng, S.Y.; Ding, W.P. Hydrophobic Interface Starch Nanofibrous Film for Food Packaging: From Bioinspired Design to Self-Cleaning Action. *J. Agric. Food Chem.* **2021**, *69*, 5067–5075. [CrossRef]
68. Pardo-Figuerez, M.; López-Córdoba, A.; Torres-Giner, S.; Lagaron, J.M. Superhydrophobic Bio-Coating Made by Co-Continuous Electrospinning and Electrospraying Polyethylene Terephthalate Films Proposed as Easy Emptying Transparent Food Packaging. *Coatings* **2018**, *8*, 364. [CrossRef]
69. Zhang, Y.; Bi, J.; Wang, S.; Cao, Q.; Li, Y.; Zhou, J.; Zhu, B.W. Functional Food Packaging for Reducing Residual Liquid Food: Thermo-Resistant Edible Super-Hydrophobic Coating from Coffee and Beeswax. *J. Colloid Interface Sci.* **2019**, *533*, 742–749. [CrossRef]
70. Lindh, H.; Williams, H.; Olsson, A.; Wikström, F. Elucidating the Indirect Contributions of Packaging to Sustainable Development: A Terminology of Packaging Functions and Features. *Packag. Technol. Sci.* **2016**, *29*, 225–246. [CrossRef]
71. Wikström, F.; Williams, H.; Venkatesh, G. The Influence of Packaging Attributes on Recycling and Food Waste Behaviour—An Environmental Comparison of Two Packaging Alternatives. *J. Clean. Prod.* **2016**, *137*, 895–902. [CrossRef]
72. Vanderroost, M.; Ragaert, P.; Devlieghere, F.; de Meulenaer, B. Intelligent Food Packaging: The next Generation. *Trends Food Sci. Technol.* **2014**, *39*, 47–62. [CrossRef]
73. Mendy, T.K.; Misran, A.; Mahmud, T.M.M.; Ismail, S.I. Application of Aloe Vera Coating Delays Ripening and Extend the Shelf Life of Papaya Fruit. *Sci. Hortic.* **2019**, *246*, 769–776. [CrossRef]
74. Agamuthu, P.; Visvanathan, C. Extended Producers' Responsibility Schemes for Used Beverage Carton Recycling. *Waste Manag. Res.* **2014**, *32*, 1–3. [CrossRef] [PubMed]
75. Geueke, B.; Groh, K.; Muncke, J. Food Packaging in the Circular Economy: Overview of Chemical Safety Aspects for Commonly Used Materials. *J. Clean. Prod.* **2018**, *193*, 491–505. [CrossRef]
76. Palombini, F.L.; Cidade, M.K.; de Jacques, J.J. How Sustainable Is Organic Packaging? A Design Method for Recyclability Assessment via a Social Perspective: A Case Study of Porto Alegre City (Brazil). *J. Clean. Prod.* **2017**, *142*, 2593–2605. [CrossRef]

77. Kaiser, K.; Schmid, M.; Schlummer, M. Recycling of Polymer-Based Multilayer Packaging: A Review. *Recycling* **2018**, *3*, 1. [CrossRef]
78. Körner, I.; Redemann, K.; Stegmann, R. Behaviour of Biodegradable Plastics in Composting Facilities. *Waste Manag.* **2005**, *25*, 409–415. [CrossRef]
79. Changwichan, K.; Silalertruksa, T.; Gheewala, S.H. Eco-Efficiency Assessment of Bioplastics Production Systems and End-of-Life Options. *Sustainability* **2018**, *10*, 952. [CrossRef]
80. López, O.V.; Versino, F.; Villar, M.A.; García, M.A. Agro-Industrial Residue from Starch Extraction of Pachyrhizus Ahipa as Filler of Thermoplastic Corn Starch Films. *Carbohydr. Polym.* **2015**, *134*, 324–332. [CrossRef]
81. Versino, F.; López, O.; García, M.A. Sunflower Oil Industry By-Product as Natural Filler of Biocomposite Foams for Packaging Applications. *J. Polym. Environ.* **2021**, *29*, 1869–1879. [CrossRef]
82. Bascón-Villegas, I.; Pereira, M.; Espinosa, E.; Sánchez-Gutiérrez, M.; Rodríguez, A.; Pérez-Rodríguez, F. A New Eco-Friendly Packaging System Incorporating Lignocellulose Nanofibres from Agri-Food Residues Applied to Fresh-Cut Lettuce. *J. Clean. Prod.* **2022**, *372*, 133597. [CrossRef]
83. dos Santos, J.W.S.; Garcia, V.A.d.S.; Venturini, A.C.; de Carvalho, R.A.; da Silva, C.F.; Yoshida, C.M.P. Sustainable Coating Paperboard Packaging Material Based on Chitosan, Palmitic Acid, and Activated Carbon: Water Vapor and Fat Barrier Performance. *Foods* **2022**, *11*, 4037. [CrossRef]
84. Pandey, S.; Sharma, K.; Gundabala, V. Antimicrobial Bio-Inspired Active Packaging Materials for Shelf Life and Safety Development: A Review. *Food Biosci.* **2022**, *48*, 101730. [CrossRef]
85. Amin, U.; Khan, M.K.I.; Maan, A.A.; Nazir, A.; Riaz, S.; Khan, M.U.; Sultan, M.; Munekata, P.E.S.; Lorenzo, J.M. Biodegradable Active, Intelligent, and Smart Packaging Materials for Food Applications. *Food Packag. Shelf Life* **2022**, *33*, 100903. [CrossRef]
86. Manfredi, M.; Fantin, V.; Vignali, G.; Gavara, R. Environmental Assessment of Antimicrobial Coatings for Packaged Fresh Milk. *J. Clean. Prod.* **2015**, *95*, 291–300. [CrossRef]
87. Jafarzadeh, S.; Jafari, S.M.; Salehabadi, A.; Nafchi, A.M.; Uthaya Kumar, U.S.; Khalil, H.P.S.A. Biodegradable Green Packaging with Antimicrobial Functions Based on the Bioactive Compounds from Tropical Plants and Their By-Products. *Trends Food Sci. Technol.* **2020**, *100*, 262–277. [CrossRef]
88. Sharma, S.; Barkauskaite, S.; Jaiswal, A.K.; Jaiswal, S. Essential Oils as Additives in Active Food Packaging. *Food Chem.* **2021**, *343*, 128403. [CrossRef]
89. Zindani, D.; Kumar, K.; Davim, J.P. Sustainability Manufacturing Systems Design. In *Encyclopedia of Renewable and Sustainable Materials*; Elsevier: Amsterdam, The Netherlands, 2020; pp. 512–518.
90. Karwasz, A.; Dostatni, E.; Diakun, J.; Grajewski, D.; Wichniarek, R.; Stachura, M. Estimating the Cost of Product Recycling with the Use of Ecodesign Support System. *Manag. Prod. Eng. Rev.* **2016**, *7*, 33–39. [CrossRef]
91. Döhler, N.; Wellenreuther, C.; Wolf, A. Market Dynamics of Biodegradable Bio-Based Plastics: Projections and Linkages to European Policies. *EFB Bioeconomy J.* **2022**, *2*, 100028. [CrossRef]
92. Dang, S.; Chu, L. Evaluation Framework and Verification for Sustainable Container Management as Reusable Packaging. *J. Bus Res.* **2016**, *69*, 1949–1955. [CrossRef]
93. Hellström, D.; Saghir, M. Packaging and Logistics Interactions in Retail Supply Chains. *Packag. Technol. Sci.* **2007**, *20*, 197–216. [CrossRef]
94. Bernstad Saraiva, A.; Pacheco, E.B.A.V.; Gomes, G.M.; Visconte, L.L.Y.; Bernardo, C.A.; Simões, C.L.; Soares, A.G. Comparative Lifecycle Assessment of Mango Packaging Made from a Polyethylene/Natural Fiber-Composite and from Cardboard Material. *J. Clean. Prod.* **2016**, *139*, 1168–1180. [CrossRef]
95. Meherishi, L.; Narayana, S.A.; Ranjani, K.S. Sustainable Packaging for Supply Chain Management in the Circular Economy: A Review. *J. Clean. Prod.* **2019**, *237*, 117582. [CrossRef]
96. Padhi, S.S.; Pati, R.K.; Rajeev, A. Framework for Selecting Sustainable Supply Chain Processes and Industries Using an Integrated Approach. *J. Clean. Prod.* **2018**, *184*, 969–984. [CrossRef]
97. Navaranjan, N.; Fletcher, G.C.; Summers, G.; Parr, R.; Anderson, R. Thermal Insulation Requirements and New Cardboard Packaging for Chilled Seafood Exports. *J. Food Eng.* **2013**, *119*, 395–403. [CrossRef]
98. Yin, D.; Mi, J.; Zhou, H.; Wang, X.; Tian, H. Fabrication of Branching Poly (Butylene Succinate)/Cellulose Nanocrystal Foams with Exceptional Thermal Insulation. *Carbohydr. Polym.* **2020**, *247*, 116708. [CrossRef]
99. Tyagi, P.; Salem, K.S.; Hubbe, M.A.; Pal, L. Advances in Barrier Coatings and Film Technologies for Achieving Sustainable Packaging of Food Products—A Review. *Trends Food Sci. Technol.* **2021**, *115*, 461–485. [CrossRef]
100. Majid, I.; Nayik, A.; Dar, M.; Nanda, V. Novel Food Packaging Technologies: Innovations and Future Prospective. *J. Saudi Soc. Agric. Sci.* **2018**, *17*, 454–462. [CrossRef]
101. Zimmermann, L.; Dombrowski, A.; Völker, C.; Wagner, M. Are Bioplastics and Plant-Based Materials Safer than Conventional Plastics? In Vitro Toxicity and Chemical Composition. *Environ. Int.* **2020**, *145*, 106066. [CrossRef]
102. Weinstein, J.E.; Dekle, J.L.; Leads, R.R.; Hunter, R.A. Degradation of Bio-Based and Biodegradable Plastics in a Salt Marsh Habitat: Another Potential Source of Microplastics in Coastal Waters. *Mar. Pollut. Bull.* **2020**, *160*, 111518. [CrossRef]
103. Emadian, S.M.; Onay, T.T.; Demirel, B. Biodegradation of Bioplastics in Natural Environments. *Waste Manag.* **2017**, *59*, 526–536. [CrossRef]

104. Magni, S.; Bonasoro, F.; della Torre, C.; Parenti, C.C.; Maggioni, D.; Binelli, A. Plastics and Biodegradable Plastics: Ecotoxicity Comparison between Polyvinylchloride and Mater-Bi® Micro-Debris in a Freshwater Biological Model. *Sci. Total Environ.* **2020**, *720*, 137602. [CrossRef]
105. Ortega, F.; Sobral, P.; Jios, J.L.; Arce, V.B.; García, M.A. Starch Nanocomposite Films: Migration Studies of Nanoparticles to Food Simulants and Bio-Disintegration in Soil. *Polymers* **2022**, *14*, 1636. [CrossRef]
106. Abe, M.M.; Branciforti, M.C.; Nallin Montagnolli, R.; Marin Morales, M.A.; Jacobus, A.P.; Brienzo, M. Production and Assessment of the Biodegradation and Ecotoxicity of Xylan- and Starch-Based Bioplastics. *Chemosphere* **2022**, *287*, 132290. [CrossRef]
107. Bandyopadhyay-Ghosh, S.; Ghosh, S.B.; Rodriguez, A.; Sain, M.M. Biosynthesis, Microstructural Characterisations and Investigation of In-Vitro Mutagenic and Eco-Toxicological Response of a Novel Microbial Exopolysaccharide Based Biopolymer. *J. Polym. Environ.* **2018**, *26*, 365–374. [CrossRef]
108. Shruti, V.C.; Kutralam-Muniasamy, G. Bioplastics: Missing Link in the Era of Microplastics. *Sci. Total Environ.* **2019**, *697*, 134139. [CrossRef]
109. Konstantoglou, A.; Folinas, D.; Fotiadis, T. Investigating Food Packaging Elements from a Consumer's Perspective. *Foods* **2020**, *9*, 1097. [CrossRef]
110. van Herpen, E.; van den Broek, E.; van Trijp, H.C.M.; Yu, T. Can a Virtual Supermarket Bring Realism into the Lab? Comparing Shopping Behavior Using Virtual and Pictorial Store Representations to Behavior in a Physical Store. *Appetite* **2016**, *107*, 196–207. [CrossRef]
111. Herbes, C.; Beuthner, C.; Ramme, I. Consumer Attitudes towards Biobased Packaging—A Cross-Cultural Comparative Study. *J. Clean. Prod.* **2018**, *194*, 203–218. [CrossRef]
112. Tischner, U.; Hora, M. Sustainable Electronic Product Design. In *Waste Electrical and Electronic Equipment (WEEE) Handbook*; Elsevier: Amsterdam, The Netherlands, 2019; pp. 443–482. ISBN 9780081021583.
113. Han, J.W.; Ruiz-Garcia, L.; Qian, J.P.; Yang, X.T. Food Packaging: A Comprehensive Review and Future Trends. *Compr. Rev. Food Sci. Food Saf.* **2018**, *17*, 860–877. [CrossRef]
114. Lamba, A.; Garg, V. Recent Innovations in Food Packaging: A Review. *Int. J. Food Sci. Nutr. Int.* **2019**, *4*, 123–129.
115. Mujtaba, M.; Lipponen, J.; Ojanen, M.; Puttonen, S.; Vaittinen, H. Trends and Challenges in the Development of Bio-Based Barrier Coating Materials for Paper/Cardboard Food Packaging; a Review. *Sci. Total Environ.* **2022**, *851*, 158328. [CrossRef]
116. Guillard, V.; Gaucel, S.; Fornaciari, C.; Angellier-Coussy, H.; Buche, P.; Gontard, N. The Next Generation of Sustainable Food Packaging to Preserve Our Environment in a Circular Economy Context. *Front. Nutr.* **2018**, *5*, 1–13. [CrossRef] [PubMed]
117. Marano, S.; Laudadio, E.; Minnelli, C.; Stipa, P. Tailoring the Barrier Properties of PLA: A State-of-the-Art Review for Food Packaging Applications. *Polymers* **2022**, *14*, 1626. [CrossRef] [PubMed]
118. Zabihzadeh Khajavi, M.; Ebrahimi, A.; Yousefi, M.; Ahmadi, S.; Farhoodi, M.; Mirza Alizadeh, A.; Taslikh, M. Strategies for Producing Improved Oxygen Barrier Materials Appropriate for the Food Packaging Sector. *Food Eng. Rev.* **2020**, *12*, 346–363. [CrossRef]
119. Blanke, M.M. Reducing Ethylene Levels along the Food Supply Chain: A Key to Reducing Food Waste? *J. Sci. Food Agric.* **2014**, *94*, 2357–2361. [CrossRef] [PubMed]
120. Giacinti Baschetti, M.; Minelli, M. Test Methods for the Characterization of Gas and Vapor Permeability in Polymers for Food Packaging Application: A Review. *Polym. Test* **2020**, *89*, 106606. [CrossRef]
121. Ameer, A.; Seleshe, S.; Kang, S.N. Effect of Modified Atmosphere Packaging Varying in $CO_2$ and $N_2$ Composition on Quality Characteristics of Dry Fermented Sausage during Refrigeration Storage. *Food Sci. Anim. Resour.* **2022**, *42*, 639–654. [CrossRef]
122. Turan, D. Water Vapor Transport Properties of Polyurethane Films for Packaging of Respiring Foods. *Food Eng. Rev.* **2021**, *13*, 54–65. [CrossRef]
123. Kwon, S.; Orsuwan, A.; Bumbudsanpharoke, N.; Yoon, C.; Choi, J.; Ko, S. A Short Review of Light Barrier Materials for Food and Beverage Packaging. *Korean J. Packag. Sci. Technol.* **2018**, *24*, 141–148. [CrossRef]
124. Csapó, J.; Prokisch, J.; Albert, C.; Sipos, P. Effect of UV Light on Food Quality and Safety. *Acta Univ. Sapientiae Aliment.* **2019**, *12*, 21–41. [CrossRef]
125. Duncan, S.E.; Hannah, S. Light-Protective Packaging Materials for Foods and Beverages. In *Emerging Food Packaging Technologies*; Elsevier: Amsterdam, The Netherlands, 2012; pp. 303–322.
126. Cassar, J.R.; Ouyang, B.; Krishnamurthy, K.; Demirci, A. Microbial Decontamination of Food by Light-Based Technologies: Ultraviolet (UV) Light, Pulsed UV Light (PUV), and UV Light-Emitting Diodes (UV-LED). In *Food Safety Engineering*; Elsevier: Amsterdam, The Netherlands, 2020; pp. 493–521.
127. Yadav, M.; Liu, Y.-K.; Chiu, F.-C. Fabrication of Cellulose Nanocrystal/Silver/Alginate Bionanocomposite Films with Enhanced Mechanical and Barrier Properties for Food Packaging Application. *Nanomaterials* **2019**, *9*, 1523. [CrossRef]
128. Nguyen, H.L.; Tran, T.H.; Hao, L.T.; Jeon, H.; Koo, J.M.; Shin, G.; Hwang, D.S.; Hwang, S.Y.; Park, J.; Oh, D.X. Biorenewable, Transparent, and Oxygen/Moisture Barrier Nanocellulose/Nanochitin-Based Coating on Polypropylene for Food Packaging Applications. *Carbohydr. Polym.* **2021**, *271*, 118421. [CrossRef]
129. Kim, T.; Tran, T.H.; Hwang, S.Y.; Park, J.; Oh, D.X.; Kim, B.S. Crab-on-a-Tree: All Biorenewable, Optical and Radio Frequency Transparent Barrier Nanocoating for Food Packaging. *ACS Nano* **2019**, *13*, 3796–3805. [CrossRef]
130. Kiryukhin, M.; Lau, H.H.; Goh, S.H.; Teh, C.; Korzh, V.; Sadovoy, A. A Membrane Film Sensor with Encapsulated Fluorescent Dyes towards Express Freshness Monitoring of Packaged Food. *Talanta* **2018**, *182*, 187–192. [CrossRef]

131. Wang, Q.; Chen, W.; Zhu, W.; McClements, D.J.; Liu, X.; Liu, F. A Review of Multilayer and Composite Films and Coatings for Active Biodegradable Packaging. *NPJ Sci. Food* **2022**, *6*, 18. [CrossRef]
132. Wu, F.; Misra, M.; Mohanty, A.K. Challenges and New Opportunities on Barrier Performance of Biodegradable Polymers for Sustainable Packaging. *Prog. Polym. Sci.* **2021**, *117*, 101395. [CrossRef]
133. Arrieta, M.P.; Peponi, L.; López, D.; López, J.; Kenny, J.M. *An Overview of Nanoparticles Role in the Improvement of Barrier Properties of Bioplastics for Food Packaging Applications*; Elsevier Inc.: Amsterdam, The Netherlands, 2017; ISBN 9780128043028.
134. Souza, V.G.L.; Mello, I.P.; Khalid, O.; Pires, J.R.A.; Rodrigues, C.; Alves, M.M.; Santos, C.; Fernando, A.L.; Coelhoso, I. Strategies to Improve the Barrier and Mechanical Properties of Pectin Films for Food Packaging: Comparing Nanocomposites with Bilayers. *Coatings* **2022**, *12*, 108. [CrossRef]
135. Bizymis, A.P.; Giannou, V.; Tzia, C. Improved Properties of Composite Edible Films Based on Chitosan by Using Cellulose Nanocrystals and Beta-Cyclodextrin. *Appl. Sci.* **2022**, *12*, 8729. [CrossRef]
136. Zhao, Y.; Sun, H.; Yang, B.; Fan, B.; Zhang, H.; Weng, Y. Enhancement of Mechanical and Barrier Property of Hemicellulose Film via Crosslinking with Sodium Trimetaphosphate. *Polymers* **2021**, *13*, 927. [CrossRef]
137. Petronilho, S.; Oliveira, A.; Domingues, M.R.; Nunes, F.M.; Coimbra, M.A.; Gonçalves, I. Hydrophobic Starch-Based Films Using Potato Washing Slurries and Spent Frying Oil. *Foods* **2021**, *10*, 2897. [CrossRef]
138. Alias, A.R.; Wan, M.K.; Sarbon, N.M. Emerging Materials and Technologies of Multi-Layer Film for Food Packaging Application: A Review. *Food Control* **2022**, *136*, 108875. [CrossRef]
139. Messin, T.; Marais, S.; Follain, N.; Guinault, A.; Gaucher, V.; Delpouve, N.; Sollogoub, C. Biodegradable PLA/PBS Multinanolayer Membrane with Enhanced Barrier Performances. *J. Membr. Sci.* **2020**, *598*, 117777. [CrossRef]
140. Hossain, K.M.Z.; Parsons, A.J.; Rudd, C.D.; Ahmed, I.; Thielemans, W. Mechanical, Crystallisation and Moisture Absorption Properties of Melt Drawn Polylactic Acid Fibres. *Eur. Polym. J.* **2014**, *53*, 270–281. [CrossRef]
141. Li, W.; Wang, S.; Wang, W.; Qin, C.; Wu, M. Facile Preparation of Reactive Hydrophobic Cellulose Nanofibril Film for Reducing Water Vapor Permeability (WVP) in Packaging Applications. *Cellulose* **2019**, *26*, 3271–3284. [CrossRef]
142. Pankaj, S.K.; Bueno-Ferrer, C.; Misra, N.N.; O'Neill, L.; Tiwari, B.K.; Bourke, P.; Cullen, P.J. Dielectric Barrier Discharge Atmospheric Air Plasma Treatment of High Amylose Corn Starch Films. *LWT* **2015**, *63*, 1076–1082. [CrossRef]
143. Luque-Agudo, V.; Hierro-Oliva, M.; Gallardo-Moreno, A.M.; González-Martín, M.L. Effect of Plasma Treatment on the Surface Properties of Polylactic Acid Films. *Polym. Test* **2021**, *96*, 107097. [CrossRef]
144. Goiana, M.L.; de Brito, E.S.; Alves Filho, E.G.; Miguel, E.d.C.; Fernandes, F.A.N.; de Azeredo, H.M.C.; Rosa, M.d.F. Corn Starch Based Films Treated by Dielectric Barrier Discharge Plasma. *Int. J. Biol. Macromol.* **2021**, *183*, 2009–2016. [CrossRef]
145. Hoque, M.; McDonagh, C.; Tiwari, B.K.; Kerry, J.P.; Pathania, S. Effect of Cold Plasma Treatment on the Packaging Properties of Biopolymer-Based Films: A Review. *Appl. Sci.* **2022**, *12*, 1346. [CrossRef]
146. Bahrami, R.; Zibaei, R.; Hashami, Z.; Hasanvand, S.; Garavand, F.; Rouhi, M.; Jafari, S.M.; Mohammadi, R. Modification and Improvement of Biodegradable Packaging Films by Cold Plasma; a Critical Review. *Crit. Rev. Food Sci. Nutr.* **2022**, *62*, 1936–1950. [CrossRef]
147. Cunha, A.G.; Gandini, A. Turning Polysaccharides into Hydrophobic Materials: A Critical Review. Part 1. *Cellulose. Cellulose* **2010**, *17*, 875–889. [CrossRef]
148. Wang, C.; Shao, R.; Wang, G.; Sun, S. Hierarchical Hydrophobic Surfaces with Controlled Dual Transition between Rose Petal Effect and Lotus Effect via Structure Tailoring or Chemical Modification. *Colloids Surf. A Physicochem. Eng. Asp.* **2021**, *622*, 126661. [CrossRef]
149. Wang, D.; Huang, J.; Guo, Z. Tomato-Lotus Inspired Edible Superhydrophobic Artificial Lotus Leaf. *Chem. Eng. J.* **2020**, *400*, 125883. [CrossRef]
150. Allred, T.P.; Weibel, J.A.; Garimella, S. The Petal Effect of Parahydrophobic Surfaces Offers Low Receding Contact Angles That Promote Effective Boiling. *Int. J. Heat Mass Transf.* **2019**, *135*, 403–412. [CrossRef]
151. Luís, Â.; Domingues, F.; Ramos, A. Production of Hydrophobic Zein-Based Films Bioinspired by the Lotus Leaf Surface: Characterization and Bioactive Properties. *Microorganisms* **2019**, *7*, 267. [CrossRef]
152. de Oliveira Gama, R.; Bretas, R.E.S.; Oréfice, R.L. Control of the Hydrophilic/Hydrophobic Behavior of Biodegradable Natural Polymers by Decorating Surfaces with Nano- and Micro-Components. *Adv. Polym. Technol.* **2018**, *37*, 654–661. [CrossRef]
153. Gonçalves, I.; Lopes, J.; Barra, A.; Hernández, D.; Nunes, C.; Kapusniak, K.; Kapusniak, J.; Evtyugin, D.; Lopes da Silva, J.A.; Ferreira, P.; et al. Tailoring the Surface Properties and Flexibility of Starch-Based Films Using Oil and Waxes Recovered from Potato Chips Byproducts. *Int. J. Biol. Macromol.* **2020**, *163*, 251–259. [CrossRef]
154. Ruzi, M.; Celik, N.; Onses, M.S. Superhydrophobic Coatings for Food Packaging Applications: A Review. *Food Packag. Shelf Life* **2022**, *32*, 100823. [CrossRef]
155. Guzman-Puyol, S.; Tedeschi, G.; Goldoni, L.; Benítez, J.J.; Ceseracciu, L.; Koschella, A.; Heinze, T.; Athanassiou, A.; Heredia-Guerrero, J.A. Greaseproof, Hydrophobic, and Biodegradable Food Packaging Bioplastics from C6-Fluorinated Cellulose Esters. *Food Hydrocoll.* **2022**, *128*, 107562. [CrossRef]
156. Mohammadian, E.; Alizadeh-Sani, M.; Jafari, S.M. Smart Monitoring of Gas/Temperature Changes within Food Packaging Based on Natural Colorants. *Compr. Rev. Food Sci. Food Saf.* **2020**, *19*, 2885–2931. [CrossRef]
157. Nikvarz, N.; Khayati, G.R.; Sharafi, S. Preparation of UV Absorbent Films Using Polylactic Acid and Grape Syrup for Food Packaging Application. *Mater. Lett.* **2020**, *276*, 128187. [CrossRef]

158. Guzman-Puyol, S.; Hierrezuelo, J.; Benítez, J.J.; Tedeschi, G.; Porras-Vázquez, J.M.; Heredia, A.; Athanassiou, A.; Romero, D.; Heredia-Guerrero, J.A. Transparent, UV-Blocking, and High Barrier Cellulose-Based Bioplastics with Naringin as Active Food Packaging Materials. *Int. J. Biol. Macromol.* **2022**, *209*, 1985–1994. [CrossRef]
159. Gonçalves, I.; Hernández, D.; Cruz, C.; Lopes, J.; Barra, A.; Nunes, C.; da Silva, J.A.L.; Ferreira, P.; Coimbra, M.A. Relevance of Genipin Networking on Rheological, Physical, and Mechanical Properties of Starch-Based Formulations. *Carbohydr. Polym.* **2021**, *254*, 117236. [CrossRef] [PubMed]
160. Oliveira, G.; Gonçalves, I.; Barra, A.; Nunes, C.; Ferreira, P.; Coimbra, M.A. Coffee Silverskin and Starch-Rich Potato Washing Slurries as Raw Materials for Elastic, Antioxidant, and UV-Protective Biobased Films. *Food Res. Int.* **2020**, *138*, 109733. [CrossRef] [PubMed]
161. Cheng, H.; Xu, H.; Julian McClements, D.; Chen, L.; Jiao, A.; Tian, Y.; Miao, M.; Jin, Z. Recent Advances in Intelligent Food Packaging Materials: Principles, Preparation and Applications. *Food Chem.* **2022**, *375*, 131738. [CrossRef] [PubMed]
162. Chen, S.; Wu, M.; Lu, P.; Gao, L.; Yan, S.; Wang, S. Development of PH Indicator and Antimicrobial Cellulose Nanofibre Packaging Film Based on Purple Sweet Potato Anthocyanin and Oregano Essential Oil. *Int. J. Biol. Macromol.* **2020**, *149*, 271–280. [CrossRef]
163. Jamróz, E.; Kulawik, P.; Krzyściak, P.; Talaga-Ćwiertnia, K.; Juszczak, L. Intelligent and Active Furcellaran-Gelatin Films Containing Green or Pu-Erh Tea Extracts: Characterization, Antioxidant and Antimicrobial Potential. *Int. J. Biol. Macromol.* **2019**, *122*, 745–757. [CrossRef]
164. Tracey, C.T.; Predeina, A.L.; Krivoshapkina, E.F.; Kumacheva, E. A 3D Printing Approach to Intelligent Food Packaging. *Trends Food Sci. Technol.* **2022**, *127*, 87–98. [CrossRef]
165. Digvijay, P.; Onkar, S. *Smart Packaging Market by Type: Global Opportunity Analysis and Industry Forecast, 2021–2030*; Allied Market Research: Pune, India, 2022.
166. European Commission. Regulation (EC) No 1935/2004 of the European Parliament and of the Council of 27 October 2004 on Materials and Articles Intended to Come into Contact with Food and Repealing Directives 80/590/EEC and 89/109/EEC. Available online: https://eur-lex.europa.eu/eli/reg/2004/1935/oj (accessed on 13 December 2022).
167. Azevedo, A.G.; Barros, C.; Miranda, S.; Machado, A.V.; Castro, O.; Silva, B.; Saraiva, M.; Silva, A.S.; Pastrana, L.; Carneiro, O.S.; et al. Active Flexible Films for Food Packaging: A Review. *Polymers* **2022**, *14*, 2442. [CrossRef]
168. Rivero, S.; Giannuzzi, L.; García, M.A.; Pinotti, A. Controlled Delivery of Propionic Acid from Chitosan Films for Pastry Dough Conservation. *J. Food Eng.* **2013**, *116*, 524–531. [CrossRef]
169. López, O.; Giannuzzi, L.; Zaritzky, N.E.; García, M.A. Potassium Sorbate Controlled Release from Corn Starch Films. *Mater. Sci. Eng. C* **2013**, *33*, 1583–1591. [CrossRef]
170. Villarruel, S.; Giannuzzi, L.; Rivero, S.; Pinotti, A. Changes Induced by UV Radiation in the Presence of Sodium Benzoate in Films Formulated with Polyvinyl Alcohol and Carboxymethyl Cellulose. *Mater. Sci. Eng. C* **2015**, *56*, 545–554. [CrossRef]
171. Dey, A.; Neogi, S. Oxygen Scavengers for Food Packaging Applications: A Review. *Trends Food Sci. Technol.* **2019**, *90*, 26–34. [CrossRef]
172. Liang, J.; Wang, J.; Li, S.; Xu, L.; Wang, R.; Chen, R.; Sun, Y. The Size-Controllable Preparation of Chitosan/Silver Nanoparticle Composite Microsphere and Its Antimicrobial Performance. *Carbohydr. Polym.* **2019**, *220*, 22–29. [CrossRef]
173. Ortega, F.; Giannuzzi, L.; Arce, V.B.; García, M.A. Active Composite Starch Films Containing Green Synthesized Silver Nanoparticles. *Food Hydrocoll.* **2017**, *70*, 152–162. [CrossRef]
174. Passaretti, M.G.; Ninago, M.D.; Paulo, C.I.; Petit, A.; Irassar, E.F.; Vega, D.A.; Armando, V.M.; Lopez, O.V. Biocomposites Based on Thermoplastic Starch and Granite Sand Quarry Waste. *J. Renew. Mater.* **2019**, *7*, 393–402. [CrossRef]
175. Youssef, A.M.; El-Sayed, S.M. Bionanocomposites Materials for Food Packaging Applications: Concepts and Future Outlook. *Carbohydr. Polym.* **2018**, *193*, 19–27. [CrossRef]
176. Castillo, L.A.; Farenzena, S.; Pintos, E.; Rodríguez, M.S.; Villar, M.A.; García, M.A.; López, O. Active Films Based on Thermoplastic Corn Starch and Chitosan Oligomer for Food Packaging Applications. *Food Packag. Shelf Life* **2017**, *14*, 128–136. [CrossRef]
177. Dima, J.B.; Sequeiros, C.; Zaritzky, N. Chitosan from Marine Crustaceans: Production, Characterization and Applications. In *Biological Activities and Application of Marine Polysaccharides*; InTech: London, UK, 2017; pp. 39–54.
178. Varghese, S.A.; Siengchin, S.; Parameswaranpillai, J. Essential Oils as Antimicrobial Agents in Biopolymer-Based Food Packaging—A Comprehensive Review. *Food Biosci.* **2020**, *38*, 100785. [CrossRef]
179. Atarés, L.; Chiralt, A. Essential Oils as Additives in Biodegradable Films and Coatings for Active Food Packaging. *Trends Food Sci. Technol.* **2016**, *48*, 51–62. [CrossRef]
180. Bof, M.J.; Jiménez, A.; Locaso, D.E.; García, M.A.; Chiralt, A. Grapefruit Seed Extract and Lemon Essential Oil as Active Agents in Corn Starch–Chitosan Blend Films. *Food Bioproc. Tech.* **2016**, *9*, 2033–2045. [CrossRef]
181. Kanmani, P.; Rhim, J.W. Development and Characterization of Carrageenan/Grapefruit Seed Extract Composite Films for Active Packaging. *Int. J. Biol. Macromol.* **2014**, *68*, 258–266. [CrossRef]
182. Bof, M.J.; Laurent, F.E.; Massolo, F.; Locaso, D.E.; Versino, F.; García, M.A. Bio-Packaging Material Impact on Blueberries Quality Attributes under Transport and Marketing Conditions. *Polymers* **2021**, *13*, 481. [CrossRef] [PubMed]
183. Zhang, W.; Jiang, W. Antioxidant and Antibacterial Chitosan Film with Tea Polyphenols-Mediated Green Synthesis Silver Nanoparticle via a Novel One-Pot Method. *Int. J. Biol. Macromol.* **2020**, *155*, 1252–1261. [CrossRef] [PubMed]
184. Abdollahi, M.; Rezaei, M.; Farzi, G. A Novel Active Bionanocomposite Film Incorporating Rosemary Essential Oil and Nanoclay into Chitosan. *J. Food Eng.* **2012**, *111*, 343–350. [CrossRef]

185. Romani, V.P.; Prentice-Hernández, C.; Martins, V.G. Active and Sustainable Materials from Rice Starch, Fish Protein and Oregano Essential Oil for Food Packaging. *Ind. Crops Prod.* **2017**, *97*, 268–274. [CrossRef]
186. Reyes, L.M.; Landgraf, M.; Sobral, P.J.A. Gelatin-Based Films Activated with Red Propolis Ethanolic Extract and Essential Oils. *Food Packag. Shelf Life* **2021**, *27*, 100607. [CrossRef]
187. Li, M.; Yu, H.; Xie, Y.; Guo, Y.; Cheng, Y.; Qian, H.; Yao, W. Fabrication of Eugenol Loaded Gelatin Nanofibers by Electrospinning Technique as Active Packaging Material. *LWT* **2021**, *139*, 110800. [CrossRef]
188. Darie-Niţă, R.N.; Vasile, C.; Stoleru, E.; Pamfil, D.; Zaharescu, T.; Tarţău, L.; Tudorachi, N.; Brebu, M.A.; Pricope, G.M.; Dumitriu, R.P.; et al. Evaluation of the Rosemary Extract Effect on the Properties of Polylactic Acid-Based Materials. *Materials* **2018**, *11*, 1825. [CrossRef]
189. Huang, X.; Ge, X.; Zhou, L.; Wang, Y. Eugenol Embedded Zein and Poly(Lactic Acid) Film as Active Food Packaging: Formation, Characterization, and Antimicrobial Effects. *Food Chem.* **2022**, *384*, 132482. [CrossRef]
190. Ren, J.; Li, Y.; Lin, Q.; Li, Z.; Zhang, G. Development of Biomaterials Based on Plasticized Polylactic Acid and Tea Polyphenols for Active-Packaging Application. *Int. J. Biol. Macromol.* **2022**, *217*, 814–823. [CrossRef]
191. Lamarra, J.; Giannuzzi, L.; Rivero, S.; Pinotti, A. Assembly of Chitosan Support Matrix with Gallic Acid-Functionalized Nanoparticles. *Mater. Sci. Eng. C* **2017**, *79*, 848–859. [CrossRef]
192. Mariño-Cortegoso, S.; Stanzione, M.; Andrade, M.A.; Restuccia, C.; Rodríguez-Bernaldo de Quirós, A.; Buonocore, G.G.; Barbosa, C.H.; Vilarinho, F.; Silva, A.S.; Ramos, F.; et al. Development of Active Films Utilizing Antioxidant Compounds Obtained from Tomato and Lemon By-Products for Use in Food Packaging. *Food Control* **2022**, *140*, 109128. [CrossRef]
193. Janani, N.; Zare, E.N.; Salimi, F.; Makvandi, P. Antibacterial Tragacanth Gum-Based Nanocomposite Films Carrying Ascorbic Acid Antioxidant for Bioactive Food Packaging. *Carbohydr. Polym.* **2020**, *247*, 116678. [CrossRef]
194. Wu, H.; Li, T.; Peng, L.; Wang, J.; Lei, Y.; Li, S.; Li, Q.; Yuan, X.; Zhou, M.; Zhang, Z. Development and Characterization of Antioxidant Composite Films Based on Starch and Gelatin Incorporating Resveratrol Fabricated by Extrusion Compression Moulding. *Food Hydrocoll.* **2023**, *139*, 108509. [CrossRef]
195. Talón, E.; Trifkovic, K.T.; Nedovic, V.A.; Bugarski, B.M.; Vargas, M.; Chiralt, A.; González-Martínez, C. Antioxidant Edible Films Based on Chitosan and Starch Containing Polyphenols from Thyme Extracts. *Carbohydr. Polym.* **2017**, *157*, 1153–1161. [CrossRef]
196. Lombo Vidal, O.; Barros Santos, M.C.; Batista, A.P.; Andrigo, F.F.; Baréa, B.; Lecomte, J.; Figueroa-Espinoza, M.C.; Gontard, N.; Guillard, V.; Rezende, C.M.; et al. Active packaging films containing antioxidant extracts from green coffee oil by-products to prevent lipid oxidation. *J. Food Eng.* **2021**, *136*, 110744.
197. Gaikwad, K.K.; Singh, S.; Lee, Y.S. Oxygen Scavenging Films in Food Packaging. *Environ. Chem. Lett.* **2018**, *16*, 523–538. [CrossRef]
198. Uruchaya, S.; Promsorn, J.; Bumbudsanpharoke, N.; Chonhenchob, V.; Harnkarnsujarit, N. Polyesters Incorporating Gallic Acid as Oxygen Scavenger in Biodegradable Packaging. *Encycl. Renew. Sustain. Mater.* **2022**, *14*, 5296. [CrossRef]
199. di Giuseppe, F.; Coffigniez, F.; Aouf, C.; Guillard, V.; Torrieri, E. Activated Gallic Acid as Radical and Oxygen Scavenger in Biodegradable Packaging Film. *Food Packag. Shelf Life* **2022**, *31*, 100811. [CrossRef]
200. Juan-Polo, A.; Maestre Pérez, S.E.; Monedero Prieto, M.; Sánchez Reig, C.; Tone, A.M.; Herranz Solana, N.; Beltrán Sanahuja, A. Oxygen Scavenger and Antioxidant LDPE/EVOH/PET-Based Films Containing β-Carotene Intended for Fried Peanuts (Arachis hypogaea L.) Packaging: Pilot Scale Processing and Validation Studies. *Polymers* **2022**, *14*, 3550. [CrossRef]
201. Lamarra, J.; Calienni, M.N.; Rivero, S.; Pinotti, A. Electrospun Nanofibers of Poly(Vinyl Alcohol) and Chitosan-Based Emulsions Functionalized with Cabreuva Essential Oil. *Int. J. Biol. Macromol.* **2020**, *160*, 307–318. [CrossRef]
202. Castro Coelho, S.; Nogueiro Estevinho, B.; Rocha, F. Encapsulation in Food Industry with Emerging Electrohydrodynamic Techniques: Electrospinning and Electrospraying—A Review. *Food Chem.* **2021**, *339*, 127850. [CrossRef] [PubMed]
203. Ortega, F.; García, M.A.; Arce, V.B. Nanocomposite Films with Silver Nanoparticles Synthesized in Situ: Effect of Corn Starch Content. *Food Hydrocoll.* **2019**, *97*, 105200. [CrossRef]
204. Nafchi, A.M.; Nassiri, R.; Sheibani, S.; Ariffin, F.; Karim, A.A. Preparation and Characterization of Bionanocomposite Films Filled with Nanorod-Rich Zinc Oxide. *Carbohydr. Polym.* **2013**, *96*, 233–239. [CrossRef] [PubMed]
205. Sarwar, N.; Humayoun, U.; Kumar, M.; Zaidi, S.F.A.; Yoo, J.H.; Ali, N.; Jeong, D.I.; Lee, J.H.; Yoon, D.H. Citric Acid Mediated Green Synthesis of Copper Nanoparticles Using Cinnamon Bark Extract and Its Multifaceted Applications. *J. Clean. Prod.* **2021**, *292*, 125974. [CrossRef]
206. Ortega, F.; Arce, V.B.; Garcia, M.A. Nanocomposite Starch-Based Films Containing Silver Nanoparticles Synthesized with Lemon Juice as Reducing and Stabilizing Agent. *Carbohydr. Polym.* **2021**, *252*, 117208. [CrossRef]
207. Hu, Q.; Fang, Y.; Yang, Y.; Ma, N.; Zhao, L. Effect of Nanocomposite-Based Packaging on Postharvest Quality of Ethylene-Treated Kiwifruit (Actinidia Deliciosa) during Cold Storage. *Food Res. Int.* **2011**, *44*, 1589–1596. [CrossRef]
208. Lee, D.S. Carbon Dioxide Absorbers for Food Packaging Applications. *Trends Food Sci. Technol.* **2016**, *57*, 146–155. [CrossRef]
209. Abreu, A.S.; Oliveira, M.; de Sá, A.; Rodrigues, R.M.; Cerqueira, M.A.; Vicente, A.A.; Machado, A.V. Antimicrobial Nanostructured Starch Based Films for Packaging. *Carbohydr. Polym.* **2015**, *129*, 127–134. [CrossRef]
210. Almasi, H.; Forghani, S.; Moradi, M. Recent Advances on Intelligent Food Freshness Indicators; an Update on Natural Colorants and Methods of Preparation. *Food Packag. Shelf Life* **2022**, *32*, 100839. [CrossRef]
211. Soltani Firouz, M.; Mohi-Alden, K.; Omid, M. A Critical Review on Intelligent and Active Packaging in the Food Industry: Research and Development. *Food Res. Int.* **2021**, *141*, 110113. [CrossRef]

212. Moradi, M.; Tajik, H.; Almasi, H.; Forough, M.; Ezati, P. A Novel PH-Sensing Indicator Based on Bacterial Cellulose Nanofibers and Black Carrot Anthocyanins for Monitoring Fish Freshness. *Carbohydr. Polym.* **2019**, *222*, 115030. [CrossRef]
213. Ezati, P.; Tajik, H.; Moradi, M.; Molaei, R. Intelligent PH-Sensitive Indicator Based on Starch-Cellulose and Alizarin Dye to Track Freshness of Rainbow Trout Fillet. *Int. J. Biol. Macromol.* **2019**, *132*, 157–165. [CrossRef]
214. Huang, S.; Xiong, Y.; Zou, Y.; Dong, Q.; Ding, F.; Liu, X.; Li, H. A Novel Colorimetric Indicator Based on Agar Incorporated with Arnebia Euchroma Root Extracts for Monitoring Fish Freshness. *Food Hydrocoll.* **2019**, *90*, 198–205. [CrossRef]
215. Wang, H.; Li, B.; Ding, F.; Ma, T. Improvement of Properties of Smart Ink via Chitin Nanofiber and Application as Freshness Indicator. *Prog. Org. Coat.* **2020**, *149*, 105921. [CrossRef]
216. Materon, E.M.; Wong, A.; Gomes, L.M.; Ibáñez-Redín, G.; Joshi, N.; Oliveira, O.N.; Faria, R.C. Combining 3D Printing and Screen-Printing in Miniaturized, Disposable Sensors with Carbon Paste Electrodes. *J. Mater. Chem. C Mater.* **2021**, *9*, 5633–5642. [CrossRef]
217. Bautista, L.; Molina, L.; Niembro, S.; García, J.M.; López, J.; Vílchez, A. Emerging Nanotechnologies in Food Science. In *Emerging Nanotechnologies in Food Science*; Elsevier Inc.: Amsterdam, The Netherlands, 2017; pp. 149–173.
218. Negi, Y.S.; Ruchir, P.; Kulshreshta, A. Intelligent Packaging Systems for Food Applications. Available online: https://packaging360.in/insights/intelligent-packaging-systems-for-food/ (accessed on 25 November 2022).
219. Azeredo, H.M.C.; Correa, D.S. Smart Choices: Mechanisms of Intelligent Food Packaging. *Curr. Res. Food Sci.* **2021**, *4*, 932–936. [CrossRef]
220. Ghaani, M.; Cozzolino, C.A.; Castelli, G.; Farris, S. An Overview of the Intelligent Packaging Technologies in the Food Sector. *Trends Food Sci. Technol.* **2016**, *51*, 1–11. [CrossRef]
221. Ye, Y.; Guo, H.; Sun, X. Recent Progress on Cell-Based Biosensors for Analysis of Food Safety and Quality Control. *Biosens. Bioelectron.* **2019**, *126*, 389–404. [CrossRef]
222. Fronczek, C.F.; Yoon, J.Y. Detection of Foodborne Pathogens Using Biosensors. In *Antimicrobial Food Packaging*; Elsevier Inc.: Amsterdam, The Netherlands, 2016; pp. 153–166. ISBN 9780128007235.
223. Papkovsky, D.B.; Dmitriev, R.I. Biological Detection by Optical Oxygen Sensing. *Chem. Soc. Rev.* **2013**, *42*, 8700. [CrossRef]
224. Chen, H.; Zhang, M.; Bhandari, B.; Yang, C. Novel PH-Sensitive Films Containing Curcumin and Anthocyanins to Monitor Fish Freshness. *Food Hydrocoll.* **2020**, *100*, 105438. [CrossRef]
225. Wu, D.; Zhang, M.; Chen, H.; Bhandari, B. Freshness Monitoring Technology of Fish Products in Intelligent Packaging. *Crit. Rev. Food Sci. Nutr.* **2021**, *61*, 1279–1292. [CrossRef] [PubMed]
226. Balbinot-Alfaro, E.; Craveiro, D.V.; Lima, K.O.; Costa, H.L.G.; Lopes, D.R.; Prentice, C. Intelligent Packaging with PH Indicator Potential. *Food Eng. Rev.* **2019**, *11*, 235–244. [CrossRef]
227. Khairunnisa, A.; Suyatma, N.E.; Adawiyah, D.R. Development of Smart TTI Label Based on Kinetics Diffusion of Vegetable Oils Blends for Cold Supply Chain Monitoring. In *IOP Conference Series: Earth and Environmental Science*; Institute of Physics Publishing: Bristol, UK, 2019; Volume 335.
228. Rahman, A.T.M.M.; Kim, D.H.; Jang, H.D.; Yang, J.H.; Lee, S.J. Preliminary Study on Biosensor-Type Time-Temperature Integrator for Intelligent Food Packaging. *Sensors* **2018**, *18*, 1949. [CrossRef] [PubMed]
229. Kim, Y.A.; Jung, S.W.; Park, H.R.; Chung, K.Y.; Lee, S.J. Application of a Prototype of Microbial Time Temperature Indicator (TTI) to the Prediction of Ground Beef Qualities during Storage. *Korean J. Food Sci. Anim. Resour.* **2012**, *32*, 448–457. [CrossRef]
230. Choi, D.Y.; Jung, S.W.; Kim, T.J.; Lee, S.J. A Prototype of Time Temperature Integrator (TTI) with Microbeads-Entrapped Microorganisms Maintained at a Constant Concentration. *J. Food Eng.* **2014**, *120*, 118–123. [CrossRef]
231. Mataragas, M.; Bikouli, V.C.; Korre, M.; Sterioti, A.; Skandamis, P.N. Development of a Microbial Time Temperature Indicator for Monitoring the Shelf Life of Meat. *Innov. Food Sci. Emerg. Technol.* **2019**, *52*, 89–99. [CrossRef]
232. Gao, T.; Tian, Y.; Zhu, Z.; Sun, D.W. Modelling, Responses and Applications of Time-Temperature Indicators (TTIs) in Monitoring Fresh Food Quality. *Trends Food Sci. Technol.* **2020**, *99*, 311–322. [CrossRef]
233. Kim, K.; Kim, E.; Lee, S.J. New Enzymatic Time-Temperature Integrator (TTI) That Uses Laccase. *J. Food Eng.* **2012**, *113*, 118–123. [CrossRef]
234. Jhuang, J.R.; Lin, S.; Chen, L.C.; Lou, S.N.; Chen, S.H.; Chen, H.H. Development of Immobilized Laccase-Based Time Temperature Indicator by Electrospinning Zein Fiber. *Food Packag. Shelf Life* **2020**, *23*, 100436. [CrossRef]
235. Jhuang, J.R.; Lou, S.N.; Lin, S.; Chen, S.H.; Chen, L.C.; Chen, H.H. Immobilizing Laccase on Electrospun Chitosan Fiber to Prepare Time-Temperature Indicator for Food Quality Monitoring. *Innov. Food Sci. Emerg. Technol.* **2020**, *63*, 102370. [CrossRef]
236. Lin, C.X.; Hsu, H.H.; Chang, Y.H.; Chen, S.H.; Lin, S.; Lou, S.N.; Chen, H.H. Expanding the Applicability of an Innovative Laccase TTI in Intelligent Packaging by Adding an Enzyme Inhibitor to Change Its Coloration Kinetics. *Polymers* **2021**, *13*, 3646. [CrossRef]
237. Wu, D.; Hou, S.; Chen, J.; Sun, Y.; Ye, X.; Liu, D.; Meng, R.; Wang, Y. Development and Characterization of an Enzymatic Time-Temperature Indicator (TTI) Based on Aspergillus Niger Lipase. *LWT* **2015**, *60*, 1100–1104. [CrossRef]
238. Chuang, W.-P.; Hsieh, B.-C. Development of a Gallic Acid Based Time Temperature Indicator with Adjustable Activation Energy. *Food Control* **2023**, *144*, 109396. [CrossRef]
239. Ye, B.; Chen, J.; Ye, H.; Zhang, Y.; Yang, Q.; Yu, H.; Fu, L.; Wang, Y. Development of a Time–Temperature Indicator Based on Maillard Reaction for Visually Monitoring the Freshness of Mackerel. *Food Chem.* **2022**, *373*, 131448. [CrossRef]
240. Jelinek, R.; Ritenberg, M. Polydiacetylenes-Recent Molecular Advances and Applications. *RSC Adv.* **2013**, *3*, 21192–21201. [CrossRef]

241. Lee, S.; Kim, J.Y.; Chen, X.; Yoon, J. Recent Progress in Stimuli-Induced Polydiacetylenes for Sensing Temperature, Chemical and Biological Targets. *Chem. Commun.* **2016**, *52*, 9178–9196. [CrossRef]
242. Dory, Y.L.; Caron, M.; Duguay, V.O.; Chicoine-Ouellet, L.; Fortin, D.; Baillargeon, P. Preparation and Single Crystal Structure Determination of the First Biobased Furan-Polydiacetylene Using Topochemical Polymerization. *Crystals* **2019**, *9*, 448. [CrossRef]
243. Qian, X.; Städler, B. Recent Developments in Polydiacetylene-Based Sensors. *Chem. Mater.* **2019**, *31*, 1196–1222. [CrossRef]
244. Suppakul, P.; Kim, D.Y.; Yang, J.H.; Lee, S.B.; Lee, S.J. Practical Design of a Diffusion-Type Time-Temperature Indicator with Intrinsic Low Temperature Dependency. *J. Food Eng.* **2018**, *223*, 22–31. [CrossRef]
245. Vivaldi, F.; Melai, B.; Bonini, A.; Poma, N.; Salvo, P.; Kirchhain, A.; Tintori, S.; Bigongiari, A.; Bertuccelli, F.; Isola, G.; et al. A Temperature-Sensitive RFID Tag for the Identification of Cold Chain Failures. *Sens. Actuators A Phys.* **2020**, *313*, 112182. [CrossRef]
246. Talegaonkar, S.; Sharma, H.; Pandey, S.; Mishra, P.K.; Wimmer, R. Bionanocomposites: Smart Biodegradable Packaging Material for Food Preservation. In *Food Packaging*; Elsevier: Amsterdam, The Netherlands, 2017; pp. 79–110.
247. Drago, E.; Campardelli, R.; Pettinato, M.; Perego, P. Innovations in Smart Packaging Concepts for Food: An Extensive Review. *Foods* **2020**, *9*, 1628. [CrossRef] [PubMed]
248. Restuccia, D.; Spizzirri, U.G.; Parisi, O.I.; Cirillo, G.; Curcio, M.; Iemma, F.; Puoci, F.; Vinci, G.; Picci, N. New EU Regulation Aspects and Global Market of Active and Intelligent Packaging for Food Industry Applications. *Food Control* **2010**, *21*, 1425–1435. [CrossRef]
249. Zuo, J.; Feng, J.; Gameiro, M.G.; Tian, Y.; Liang, J.; Wang, Y.; Ding, J.; He, Q. RFID-Based Sensing in Smart Packaging for Food Applications: A Review. *Future Foods* **2022**, *6*, 100198. [CrossRef] [PubMed]
250. Musso, Y.S.; Salgado, P.R.; Mauri, A.N. Smart Gelatin Films Prepared Using Red Cabbage (Brassica oleracea L.) Extracts as Solvent. *Food Hydrocoll.* **2019**, *89*, 674–681. [CrossRef]
251. Salgado, P.R.; di Giorgio, L.; Musso, Y.S.; Mauri, A.N. Recent Developments in Smart Food Packaging Focused on Biobased and Biodegradable Polymers. *Front. Sustain. Food Syst.* **2021**, *5*, 630393. [CrossRef]
252. Halonen, N.; Pálvölgyi, P.S.; Bassani, A.; Fiorentini, C.; Nair, R.; Spigno, G.; Kordas, K. Bio-Based Smart Materials for Food Packaging and Sensors—A Review. *Front. Mater.* **2020**, *7*, 82. [CrossRef]
253. Echeverría, I.; López-Caballero, M.E.; Gómez-Guillén, M.C.; Mauri, A.N.; Montero, M.P. Active Nanocomposite Films Based on Soy Proteins-Montmorillonite-Clove Essential Oil for the Preservation of Refrigerated Bluefin Tuna (Thunnus Thynnus) Fillets. *Int. J. Food Microbiol.* **2018**, *266*, 142–149. [CrossRef]
254. Xu, Y.; Rehmani, N.; Alsubaie, L.; Kim, C.; Sismour, E.; Scales, A. Tapioca Starch Active Nanocomposite Films and Their Antimicrobial Effectiveness on Ready-to-Eat Chicken Meat. *Food Packag. Shelf Life* **2018**, *16*, 86–91. [CrossRef]
255. Li, W.; Zhang, C.; Chi, H.; Li, L.; Lan, T.; Han, P.; Chen, H.; Qin, Y. Development of Antimicrobial Packaging Film Made from Poly(Lactic Acid) Incorporating Titanium Dioxide and Silver Nanoparticles. *Molecules* **2017**, *22*, 1170. [CrossRef]
256. Zhang, Y.; Zhou, L.; Zhang, C.; Show, P.L.; Du, A.; Fu, J.; Ashokkumar, V. Preparation and Characterization of Curdlan/Polyvinyl Alcohol/ Thyme Essential Oil Blending Film and Its Application to Chilled Meat Preservation. *Carbohydr. Polym.* **2020**, *247*, 116670. [CrossRef]
257. Seydim, A.C.; Sarikus-Tutal, G.; Sogut, E. Effect of Whey Protein Edible Films Containing Plant Essential Oils on Microbial Inactivation of Sliced Kasar Cheese. *Food Packag. Shelf Life* **2020**, *26*, 100567. [CrossRef]
258. Valencia-Sullca, C.; Vargas, M.; Atarés, L.; Chiralt, A. Thermoplastic Cassava Starch-Chitosan Bilayer Films Containing Essential Oils. *Food Hydrocoll.* **2018**, *75*, 107–115. [CrossRef]
259. Zhai, X.; Li, Z.; Zhang, J.; Shi, J.; Zou, X.; Huang, X.; Zhang, D.; Sun, Y.; Yang, Z.; Holmes, M.; et al. Natural Biomaterial-Based Edible and PH-Sensitive Films Combined with Electrochemical Writing for Intelligent Food Packaging. *J. Agric. Food Chem.* **2018**, *66*, 12836–12846. [CrossRef]
260. Göksen, G.; Fabra, M.J.; Ekiz, H.I.; López-Rubio, A. Phytochemical-Loaded Electrospun Nanofibres as Novel Active Edible Films: Characterization and Antibacterial Efficiency in Cheese Slices. *Food Control* **2020**, *112*, 107133. [CrossRef]
261. Wang, Y.C.; Lu, L.; Gunasekaran, S. Biopolymer/Gold Nanoparticles Composite Plasmonic Thermal History Indicator to Monitor Quality and Safety of Perishable Bioproducts. *Biosens. Bioelectron.* **2017**, *92*, 109–116. [CrossRef]
262. Fernandes, B.C.N.; de Mello Aquilino, M.; Prata, A.S. Biodegradable Thermoactive Packaging Using Phase Change Material Particles on Cellulosic Materials. *Cellulose* **2021**, *28*, 6427–6437. [CrossRef]
263. Zhao, L.; Liu, Y.; Zhao, L.; Wang, Y. Anthocyanin-Based PH-Sensitive Smart Packaging Films for Monitoring Food Freshness. *J. Agric. Food Res.* **2022**, *9*, 100340. [CrossRef]
264. Bi, F.; Zhang, X.; Bai, R.; Liu, Y.; Liu, J.; Liu, J. Preparation and Characterization of Antioxidant and Antimicrobial Packaging Films Based on Chitosan and Proanthocyanidins. *Int. J. Biol. Macromol.* **2019**, *134*, 11–19. [CrossRef]
265. Lee, K.; Park, H.; Baek, S.; Han, S.; Kim, D.; Chung, S.; Yoon, J.Y.; Seo, J. Colorimetric Array Freshness Indicator and Digital Color Processing for Monitoring the Freshness of Packaged Chicken Breast. *Food Packag. Shelf Life* **2019**, *22*, 100408. [CrossRef]
266. Hashim, S.B.H.; Tahir, H.E.; Lui, L.; Zhang, J.; Zhai, X.; Mahdi, A.A.; Ibrahim, N.A.; Mahunu, G.K.; Hassan, M.M.; Xiaobo, Z.; et al. Smart Films of Carbohydrate-Based/Sunflower Wax/Purple Chinese Cabbage Anthocyanins: A Biomarker of Chicken Freshness. *Food Chem.* **2023**, *399*, 133824. [CrossRef]
267. Pirsa, S.; Karimi Sani, I.; Pirouzifard, M.K.; Erfani, A. Smart Film Based on Chitosan/Melissa Officinalis Essences/ Pomegranate Peel Extract to Detect Cream Cheeses Spoilage. *Food Addit. Contam. Part A* **2020**, *37*, 634–648. [CrossRef] [PubMed]

268. Ezati, P.; Rhim, J.-W. PH-Responsive Chitosan-Based Film Incorporated with Alizarin for Intelligent Packaging Applications. *Food Hydrocoll.* **2020**, *102*, 105629. [CrossRef]
269. Lyu, J.S.; Choi, I.; Hwang, K.S.; Lee, J.Y.; Seo, J.; Kim, S.Y.; Han, J. Development of a BTB−/TBA+ Ion-Paired Dye-Based $CO_2$ Indicator and Its Application in a Multilayered Intelligent Packaging System. *Sens. Actuators B Chem.* **2019**, *282*, 359–365. [CrossRef]
270. Deshwal, G.K.; Panjagari, N.R.; Badola, R.; Singh, A.K.; Minz, P.S.; Ganguly, S.; Alam, T. Characterization of Biopolymer-Based UV-Activated Intelligent Oxygen Indicator for Food-Packaging Applications. *J. Packag. Technol. Res.* **2018**, *2*, 29–43. [CrossRef]
271. Zhai, X.; Li, Z.; Shi, J.; Huang, X.; Sun, Z.; Zhang, D.; Zou, X.; Sun, Y.; Zhang, J.; Holmes, M.; et al. A Colorimetric Hydrogen Sulfide Sensor Based on Gellan Gum-Silver Nanoparticles Bionanocomposite for Monitoring of Meat Spoilage in Intelligent Packaging. *Food Chem.* **2019**, *290*, 135–143. [CrossRef]
272. Bumbudsanpharoke, N.; Kwon, S.; Lee, W.; Ko, S. Optical Response of Photonic Cellulose Nanocrystal Film for a Novel Humidity Indicator. *Int. J. Biol. Macromol.* **2019**, *140*, 91–97. [CrossRef]
273. Choudhary, A.; Prasad, E. Bioplastic Market Outlook 2020–2027. In *Bioplastic Market by Biodegradable Type and Application: Opportunity Analysis and Industry Forecast, 2020–2027*; Allied Market Research: Pune, India, 2020.
274. Jabeen, N.; Majid, I.; Nayik, G.A.; Yildiz, F. Bioplastics and Food Packaging: A Review. *Cogent. Food Agric.* **2015**, *1*, 1117749. [CrossRef]
275. Mohammed, A.; Gaduan, A.; Chaitram, P.; Pooran, A.; Lee, K.Y.; Ward, K. Sargassum Inspired, Optimized Calcium Alginate Bioplastic Composites for Food Packaging. *Food Hydrocoll.* **2023**, *135*, 108192. [CrossRef]
276. Ashfaq, A.; Khursheed, N.; Fatima, S.; Anjum, Z.; Younis, K. Application of Nanotechnology in Food Packaging: Pros and Cons. *J. Agric. Food Res.* **2022**, *7*, 100270. [CrossRef]
277. Wang, J.; Euring, M.; Ostendorf, K.; Zhang, K. Biobased Materials for Food Packaging. *J. Bioresour. Bioprod.* **2022**, *7*, 1–13. [CrossRef]
278. Garrido, T.; Etxabide, A.; Leceta, I.; Cabezudo, S.; de La Caba, K.; Guerrero, P. Valorization of Soya By-Products for Sustainable Packaging. *J. Clean. Prod.* **2014**, *64*, 228–233. [CrossRef]
279. Kumari, S.V.G.; Pakshirajan, K.; Pugazhenthi, G. Recent Advances and Future Prospects of Cellulose, Starch, Chitosan, Polylactic Acid and Polyhydroxyalkanoates for Sustainable Food Packaging Applications. *Int. J. Biol. Macromol.* **2022**, *221*, 163–182. [CrossRef]
280. Nielsen, C.; Rahman, A.; Rehman, A.U.; Walsh, M.K.; Miller, C.D. Food Waste Conversion to Microbial Polyhydroxyalkanoates. *Microb. Biotechnol.* **2017**, *10*, 1338–1352. [CrossRef]
281. Amini, M.; Yousefi-Massumabad, H.; Younesi, H.; Abyar, H.; Bahramifar, N. Production of the Polyhydroxyalkanoate Biopolymer by Cupriavidusnecator Using Beer Brewery Wastewater Containing Maltose as a Primary Carbon Source. *J. Environ. Chem. Eng.* **2020**, *8*, 103588. [CrossRef]
282. Alsafadi, D.; Al-Mashaqbeh, O. A One-Stage Cultivation Process for the Production of Poly-3-(Hydroxybutyrate-Co-Hydroxyvalerate) from Olive Mill Wastewater by Haloferax Mediterranei. *N. Biotechnol.* **2017**, *34*, 47–53. [CrossRef]
283. Lammi, S.; le Moigne, N.; Djenane, D.; Gontard, N.; Angellier-Coussy, H. Dry Fractionation of Olive Pomace for the Development of Food Packaging Biocomposites. *Ind. Crops Prod.* **2018**, *120*, 250–261. [CrossRef]
284. Zhang, M.; Xie, L.; Yin, Z.; Khanal, S.K.; Zhou, Q. Biorefinery Approach for Cassava-Based Industrial Wastes: Current Status and Opportunities. *Bioresour. Technol.* **2016**, *215*, 50–62. [CrossRef]
285. Domingos, I.; Ferreira, J.; Cruz-Lopes, L.; Esteves, B. Polyurethane Foams from Liquefied Orange Peel Wastes. *Food Bioprod. Process.* **2019**, *115*, 223–229. [CrossRef]
286. Abol-Fotouh, D.; Hassan, M.A.; Shokry, H.; Roig, A.; Azab, M.S.; Kashyout, A.E.H.B. Bacterial Nanocellulose from Agro-Industrial Wastes: Low-Cost and Enhanced Production by Komagataeibacter saccharivorans MD1. *Sci. Rep.* **2020**, *10*, 3491. [CrossRef]
287. Valencia, G.A.; de Andrade, C.J.; Ienczak, J.L.; Monteiro, A.R.; Gutiérrez, T.J. *Bio-Valorization of Waste*; Shah, S., Venkatramanan, V., Prasad, R., Eds.; Environmental and Microbial Biotechnology; Springer: Singapore, 2021; ISBN 978-981-15-9695-7.
288. Tsang, Y.F.; Kumar, V.; Samadar, P.; Yang, Y.; Lee, J.; Ok, Y.S.; Song, H.; Kim, K.H.; Kwon, E.E.; Jeon, Y.J. Production of Bioplastic through Food Waste Valorization. *Environ. Int.* **2019**, *127*, 625–644. [CrossRef]
289. Liu, M.; Arshadi, M.; Javi, F.; Lawrence, P.; Davachi, S.M.; Abbaspourrad, A. Green and Facile Preparation of Hydrophobic Bioplastics from Tea Waste. *J. Clean. Prod.* **2020**, *276*, 123353. [CrossRef]
290. Chiarathanakrit, C.; Riyajan, S.A.; Kaewtatip, K. Transforming Fish Scale Waste into an Efficient Filler for Starch Foam. *Carbohydr. Polym.* **2018**, *188*, 48–53. [CrossRef] [PubMed]
291. Ocaña, R. Degradación Ambiental y En Condiciones Adversas de Adhesivos Estructurales: Análisis y Consideraciones Técnicas Para Su Aplicación Industrial. Doctoral Thesis, E.T.S.I. Diseño Industrial (UPM), Madrid, Spain, 2017.
292. Toenniessen, M. Packaging Materials: Adhesives for Food Packaging Applications. Available online: https://ilsi.eu/publication/packaging-materials-10-adhesives-for-food-packaging-applications/ (accessed on 20 December 2022).
293. Ortiz-Fernández, A.; Ríos-Soberanis, C.R.; Chim-Chi, Y.A.; Moo-Huchin, V.M.; Estrada-León, R.J.; Pérez-Pacheco, E. Optimization of Biodegradable Starch Adhesives Using Response Surface Methodology. *Polym. Bull.* **2021**, *78*, 3729–3749. [CrossRef]
294. Dohr, C.A.; Hirn, U. Influence of Paper Properties on Adhesive Strength of Starch Gluing. *Nord. Pulp Paper Res. J.* **2022**, *37*, 120–129. [CrossRef]

295. Norström, E.; Demircan, D.; Fogelström, L.; Khabbaz, F.; Malmström, E. Green Binders for Wood Adhesives. In *Applied Adhesive Bonding in Science and Technology*; InTech: London, UK, 2018.
296. Ferdosian, F.; Pan, Z.; Gao, G.; Zhao, B. Bio-Based Adhesives and Evaluation for Wood Composites Application. *Polymers* **2017**, *9*, 70. [CrossRef]
297. Solt, P.; Konnerth, J.; Gindl-Altmutter, W.; Kantner, W.; Moser, J.; Mitter, R.; van Herwijnen, H.W.G. Technological Performance of Formaldehyde-Free Adhesive Alternatives for Particleboard Industry. *Int. J. Adhes. Adhes.* **2019**, *94*, 99–131. [CrossRef]
298. Bacigalupe, A.; Escobar, M.M. Soy Protein Adhesives for Particleboard Production—A Review. *J. Polym. Environ.* **2021**, *29*, 2033–2045. [CrossRef]
299. Li, W.; Zhang, Z.; Wu, L.; Zhu, Z.; Xu, Z. Improving the Adhesion-to-Fibers and Film Properties of Corn Starch by Starch Sulfo-Itaconation for a Better Application in Warp Sizing. *Polym. Test* **2021**, *98*, 107194. [CrossRef]
300. Heinrich, L.A. Future Opportunities for Bio-Based Adhesives-Advantages beyond Renewability. *Green Chem.* **2019**, *21*, 1866–1888. [CrossRef]
301. Ebnesajjad, S.; Landrock, A.H. Material Surface Preparation Techniques. In *Adhesives Technology Handbook*; Elsevier: Amsterdam, The Netherlands, 2015; pp. 35–66.
302. Watcharakitti, J.; Win, E.E.; Nimnuan, J.; Smith, S.M. Modified Starch-Based Adhesives: A Review. *Polymers* **2022**, *14*, 2023. [CrossRef]
303. Gadhave, R.V.I.; Gadhave, C.R. Adhesives for the Paper Packaging Industry: An Overview. *Open J. Polym. Chem.* **2022**, *12*, 55–79. [CrossRef]
304. Zhang, J.; Liang, J.; Du, G.; Zhou, X.; Wang, H.; Lei, H. Development and Characterization of a Bayberry Tannin-Based Adhesive for Particleboard. *Bioresources* **2017**, *12*, 6082–6093. [CrossRef]
305. Ji, X.; Guo, M. Preparation and Properties of a Chitosan-Lignin Wood Adhesive. *Int. J. Adhes. Adhes.* **2018**, *82*, 8–13. [CrossRef]
306. Tenorio-Alfonso, A.; Sánchez, M.C.; Franco, J.M. Preparation, Characterization and Mechanical Properties of Bio-Based Polyurethane Adhesives from Isocyanate-Functionalized Cellulose Acetate and Castor Oil for Bonding Wood. *Polymers* **2017**, *9*, 132. [CrossRef]
307. Monroy, Y.; Rivero, S.; García, M.A. Sustainable Panels Design Based on Modified Cassava Starch Bioadhesives and Wood Processing Byproducts. *Ind. Crops Prod.* **2019**, *137*, 171–179. [CrossRef]
308. Monroy, Y.; Seré, P.; Rivero, S.; García, M.A. Sustainable Panels Based on Starch Bioadhesives: An Insight into Structural and Tribological Performance. *Int. J. Biol. Macromol.* **2020**, *148*, 898–907. [CrossRef]
309. Schwarzenbrunner, R.; Barbu, M.C.; Petutschnigg, A.; Tudor, E.M. Water-Resistant Casein-Based Adhesives for Veneer Bonding in Biodegradable Ski Cores. *Polymers* **2020**, *12*, 1745. [CrossRef]
310. Li, X.; Li, J.; Luo, J.; Li, K.; Gao, Q.; Li, J. A Novel Eco-Friendly Blood Meal-Based Bio-Adhesive: Preparation and Performance. *J. Polym. Environ.* **2018**, *26*, 607–615. [CrossRef]
311. Zeng, Y.; Xu, P.; Yang, W.; Chu, H.; Wang, W.; Dong, W.; Chen, M.; Bai, H.; Ma, P. Soy Protein-Based Adhesive with Superior Bonding Strength and Water Resistance by Designing Densely Crosslinking Networks. *Eur. Polym. J.* **2021**, *142*, 110128. [CrossRef]
312. Maaßen, W.; Oelmann, S.; Peter, D.; Oswald, W.; Willenbacher, N.; Meier, M.A.R. Novel Insights into Pressure-Sensitive Adhesives Based on Plant Oils. *Macromol. Chem. Phys.* **2015**, *216*, 1609–1618. [CrossRef]
313. Mahieu, A.; Vivet, A.; Poilane, C.; Leblanc, N. Performance of Particleboards Based on Annual Plant Byproducts Bound with Bio-Adhesives. *Int. J. Adhes. Adhes.* **2021**, *107*, 102847. [CrossRef]
314. Wang, S.; Xia, P.; Wang, S.; Liang, J.; Sun, Y.; Yue, P.; Gao, X. Packaging Films Formulated with Gelatin and Anthocyanins Nanocomplexes: Physical Properties, Antioxidant Activity and Its Application for Olive Oil Protection. *Food Hydrocoll.* **2019**, *96*, 617–624. [CrossRef]
315. Wu, H.; Lei, Y.; Lu, J.; Zhu, R.; Xiao, D.; Jiao, C.; Xia, R.; Zhang, Z.; Shen, G.; Liu, Y.; et al. Effect of Citric Acid Induced Crosslinking on the Structure and Properties of Potato Starch/Chitosan Composite Films. *Food Hydrocoll.* **2019**, *97*, 105208. [CrossRef]
316. Kumar, R.; Ghoshal, G.; Goyal, M. Biodegradable Composite Films/Coatings of Modified Corn Starch/Gelatin for Shelf Life Improvement of Cucumber. *J. Food Sci. Technol.* **2021**, *58*, 1227–1237. [CrossRef] [PubMed]
317. Olomo, V.D. Influence of Processing Variables on Some Physico-Chemical Properties and Quality of Manioc Starch-Based Adhesives. *Open J. Polym. Chem.* **2022**, *12*, 1–12. [CrossRef]
318. Monroy, Y.; Rivero, S.; García, M.A. Microstructural and Techno-Functional Properties of Cassava Starch Modified by Ultrasound. *Ultrason. Sonochem.* **2018**, *42*, 795–804. [CrossRef]
319. Toenniessen, M. *ILSI Europe Report Series: Adhesives for Food Packaging*; International Life Sciences Institute Europe (ILSI): Bruxelles, Belgium, 2018.
320. Altan, A.; Akıncı, E.; Arca, M. Comparing the Performances of Aqueous Solution of Boric Acid / Borax and Powder Boric Acid/Borax in Corrugated Board Production Process. *Biol. Chem. Res.* **2019**, *6*, 150–154.
321. Romero Zaliz, L.; Labombarda, J.; Wildner Fox, C. Etiquetado de Botellas Retornables de Vidrio. Available online: https://www.tecnicomadhesivos.com.ar/wp-content/uploads/2020/04/Tecnicom_Etiquetado-de-Botellas-Retornables-de-Vidrio.pdf (accessed on 20 December 2022).
322. Luque, G.C.; Stürtz, R.; Passeggi, M.C.G.; Gugliotta, L.M.; Gonzalez, V.D.G.; Minari, R.J. New Hybrid Acrylic/Collagen Nanocomposites and Their Potential Use as Bio-Adhesives. *Int. J. Adhes. Adhes.* **2020**, *100*, 102624. [CrossRef]

323. Phan, K.; Raes, K.; van Speybroeck, V.; Roosen, M.; de Clerck, K.; de Meester, S. Non-Food Applications of Natural Dyes Extracted from Agro-Food Residues: A Critical Review. *J. Clean. Prod.* **2021**, *301*, 126920. [CrossRef]
324. Iqbal, S.; Ansari, T.N. Extraction and Application of Natural Dyes. In *Sustainable Practices in the Textile Industry*; Rather, L.J., Shabbir, M., Haji, A., Eds.; Scrivener Publishing LLC: Beverly, MA, USA, 2021; Volume 4, pp. 218–233. ISBN 978-1-119-81891-5.
325. The Business Research Company. Synthetic Dye and Pigment Global Market Report. 2022. Available online: https://www.thebusinessresearchcompany.com/report/synthetic-dye-and-pigment-global-market-report (accessed on 27 December 2022).
326. Echegaray, N.; Guzel, N.; Kumar, M.; Guzel, M.; Hassoun, A.; Lorenzo, J.M. Recent Advancements in Natural Colorants and Their Application as Coloring in Food and in Intelligent Food Packaging. *Food Chem.* **2023**, *404*, 134453. [CrossRef]
327. Ezati, P.; Rhim, J.-W. Starch and Agar-Based Color-Indicator Films Integrated with Shikonin for Smart Packaging Application of Shrimp. *ACS Food Sci. Technol.* **2021**, *1*, 1963–1969. [CrossRef]
328. Viera, I.; Pérez-Gálvez, A.; Roca, M.; Barros, L.; Ferreira, I.C.F.R. Green Natural Colorants. *Molecules* **2019**, *24*, 154. [CrossRef]
329. Lin, W.S.; He, P.H.; Chau, C.F.; Liou, B.K.; Li, S.; Pan, M.H. The Feasibility Study of Natural Pigments as Food Colorants and Seasonings Pigments Safety on Dried Tofu Coloring. *Food Sci. Hum. Wellness* **2018**, *7*, 220–228. [CrossRef]
330. Jurić, S.; Jurić, M.; Król-Kilińska, Ż.; Vlahoviček-Kahlina, K.; Vinceković, M.; Dragović-Uzelac, V.; Donsì, F. Sources, Stability, Encapsulation and Application of Natural Pigments in Foods. *Food Rev. Int.* **2022**, *38*, 1735–1790. [CrossRef]
331. Teixeira, V.M.C.; da Silva, R.F.G.; Gonçalves, O.H.; Pereira, C.; Barros, L.; Ferreira, I.C.F.R.; Bona, E.; Leimann, F.V. Chemometric Approaches to Evaluate the Substitution of Synthetic Food Dyes by Natural Compounds: The Case of Nanoencapsulated Curcumin, Spirulina, and Hibiscus Extracts. *LWT* **2022**, *154*, 112786. [CrossRef]
332. Escalante-Aburto, A.; Trujillo-de Santiago, G.; Álvarez, M.M.; Chuck-Hernández, C. Advances and Prospective Applications of 3D Food Printing for Health Improvement and Personalized Nutrition. *Compr. Rev. Food Sci. Food Saf.* **2021**, *20*, 5722–5741. [CrossRef]
333. Lee, J.H.; Won, D.J.; Kim, H.W.; Park, H.J. Effect of Particle Size on 3D Printing Performance of the Food-Ink System with Cellular Food Materials. *J. Food Eng.* **2019**, *256*, 1–8. [CrossRef]
334. Qiu, L.; Zhang, M.; Bhandari, B.; Chitrakar, B.; Chang, L. Investigation of 3D Printing of Apple and Edible Rose Blends as a Dysphagia Food. *Food Hydrocoll.* **2023**, *135*, 108184. [CrossRef]
335. Derossi, A.; Caporizzi, R.; Azzollini, D.; Severini, C. Application of 3D Printing for Customized Food. A Case on the Development of a Fruit-Based Snack for Children. *J. Food Eng.* **2018**, *220*, 65–75. [CrossRef]
336. He, C.; Zhang, M.; Guo, C. 4D Printing of Mashed Potato/Purple Sweet Potato Puree with Spontaneous Color Change. *Innov. Food Sci. Emerg. Technol.* **2020**, *59*, 102250. [CrossRef]
337. Ghazal, A.F.; Zhang, M.; Bhandari, B.; Chen, H. Investigation on Spontaneous 4D Changes in Color and Flavor of Healthy 3D Printed Food Materials over Time in Response to External or Internal PH Stimulus. *Food Res. Int.* **2021**, *142*, 110215. [CrossRef] [PubMed]
338. Chen, C.; Zhang, M.; Guo, C.; Chen, H. 4D Printing of Lotus Root Powder Gel: Color Change Induced by Microwave. *Innov. Food Sci. Emerg. Technol.* **2021**, *68*, 102605. [CrossRef]
339. Ozcan, A.; Arman Kandirmaz, E. Natural Ink Production and Printability Studies for Smart Food Packaging. *Color Res. Appl.* **2020**, *45*, 495–502. [CrossRef]
340. Latos-Brozio, M.; Masek, A. The Application of Natural Food Colorants as Indicator Substances in Intelligent Biodegradable Packaging Materials. *Food Chem. Toxicol.* **2020**, *135*, 110975. [CrossRef] [PubMed]
341. Li, S.; Jiang, Y.; Zhou, Y.; Li, R.; Jiang, Y.; Alomgir Hossen, M.; Dai, J.; Qin, W.; Liu, Y. Facile Fabrication of Sandwich-like Anthocyanin/Chitosan/Lemongrass Essential Oil Films via 3D Printing for Intelligent Evaluation of Pork Freshness. *Food Chem.* **2022**, *370*, 131082. [CrossRef] [PubMed]
342. Alizadeh-Sani, M.; Moghaddas Kia, E.; Ghasempour, Z.; Ehsani, A. Preparation of Active Nanocomposite Film Consisting of Sodium Caseinate, ZnO Nanoparticles and Rosemary Essential Oil for Food Packaging Applications. *J. Polym. Environ.* **2021**, *29*, 588–598. [CrossRef]
343. Rachmelia, D.; Imawan, C. Time Temperature Indicator Label Using Black Corn Extract and Chitosan Matrix. *J. Phys. Conf. Ser.* **2018**, *1120*, 012041. [CrossRef]
344. Yousefi, H.; Su, H.-M.; Imani, S.M.; Alkhaldi, K.; Filipe, C.D.M.; Didar, T.F. Intelligent Food Packaging: A Review of Smart Sensing Technologies for Monitoring Food Quality. *ACS Sens.* **2019**, *4*, 808–821. [CrossRef]
345. Erna, K.H.; Felicia, W.X.L.; Rovina, K.; Vonnie, J.M.; Huda, N. Development of Curcumin/Rice Starch Films for Sensitive Detection of Hypoxanthine in Chicken and Fish Meat. *Carbohydr. Polym. Technol. Appl.* **2022**, *3*, 100189. [CrossRef]
346. Boccalon, E.; Viscusi, G.; Lamberti, E.; Fancello, F.; Zara, S.; Sassi, P.; Marinozzi, M.; Nocchetti, M.; Gorrasi, G. Composite Films Containing Red Onion Skin Extract as Intelligent PH Indicators for Food Packaging. *Appl. Surf. Sci.* **2022**, *593*, 153319. [CrossRef]
347. Zhou, W.; Wu, Z.; Xie, F.; Tang, S.; Fang, J.; Wang, X. 3D Printed Nanocellulose-Based Label for Fruit Freshness Keeping and Visual Monitoring. *Carbohydr. Polym.* **2021**, *273*, 118545. [CrossRef]
348. Cerreti, M.; Liburdi, K.; del Franco, F.; Esti, M. Heat and Light Stability of Natural Yellow Colourants in Model Beverage Systems. *Food Addit. Contam. Part A* **2020**, *37*, 905–915. [CrossRef]
349. Yusuf, M.; Shabbir, M.; Mohammad, F. Natural Colorants: Historical, Processing and Sustainable Prospects. *Nat. Prod. Bioprospect.* **2017**, *7*, 123–145. [CrossRef]
350. Coultate, T.; Blackburn, R.S. Food Colorants: Their Past, Present and Future. *Color. Technol.* **2018**, *134*, 165–186. [CrossRef]

351. Dufossé, L. Current and Potential Natural Pigments From Microorganisms (Bacteria, Yeasts, Fungi, Microalgae). In *Handbook on Natural Pigments in Food and Beverages: Industrial Applications for Improving Food Color*; Elsevier Inc.: Amsterdam, The Netherlands, 2016; pp. 337–354. ISBN 9780081003923.
352. Vankar, P.S.; Shukla, D. Natural Dyeing with Anthocyanins from Hibiscus Rosa Sinensis Flowers. *J. Appl. Polym. Sci.* **2011**, *122*, 3361–3368. [CrossRef]

MDPI AG
Grosspeteranlage 5
4052 Basel
Switzerland
Tel.: +41 61 683 77 34

*Foods* Editorial Office
E-mail: foods@mdpi.com
www.mdpi.com/journal/foods

www.ingramcontent.com/pod-product-compliance
Lightning Source LLC
Chambersburg PA
CBHW041413210926
51850CB00013B/3257
*9783725826391*